AMERICAN REGISTRY
OF PATHOLOGY
Publication & Education

ARP is pleased to inform you that when purchased from an approved seller, this fascicle comes with a digital version.

To access the online, searchable version:

- Go to www.arppress.org.
- Register your email.

If you do not have access after logging in, reach out to ARP based on your location:

- Customers based in the US, Canada, Latin America, and Asia should email admin@arppress.org.
- Customers based in Europe, Africa, the Middle East, and Oceania should email arp@eurospan.co.uk.

Non-Neoplastic Disorders of the Endocrine System

AFIP Atlases of Tumor and Non-Tumor Pathology

AMERICAN REGISTRY OF PATHOLOGY
Publication & Education

ARP PRESS

Arlington, Virginia

Director of Publications: Amy Goldenberg, PhD
Production Editor: Dian S. Thomas
Technical Editor: Elizabeth Tomlinson
Copyeditor: Audrey Kahn

American Registry of Pathology
Arlington, Virginia 22203
www.arppress.org
ISBN: 1-933477-25-3
978-1-933477-25-1

AFIP ATLASES OF TUMOR AND NON-TUMOR PATHOLOGY

Fifth Series
Fascicle 13

NON-NEOPLASTIC DISORDERS OF THE ENDOCRINE SYSTEM

by

Anthony J. Gill, AM, MD, FRCPA
University of Sydney
NSW Health Pathology Department of Anatomical Pathology,
Royal North Shore Hospital
Cancer Diagnosis and Pathology Group,
Kolling Institute of Medical Research
Sydney, New South Wales, Australia

Lori A. Erickson, MD
Department of Laboratory Medicine and Pathology
Mayo Clinic
Rochester, Minnesota

Talia L. Fuchs, MD, PhD, FRCPA
University of Sydney
Douglass Hanly Moir Pathology
Cancer Diagnosis and Pathology Group,
Kolling Institute of Medical Research
Sydney, New South Wales, Australia

Published by the
American Registry of Pathology
Arlington, Virginia
2022

AFIP ATLASES OF TUMOR AND NON-TUMOR PATHOLOGY

EDITOR
Jason L. Hornick, MD, PhD
Professor of Pathology, Harvard Medical School
Director of Surgical Pathology and Immunohistochemistry
Brigham and Women's Hospital
Boston, Massachusetts

EDITORIAL ADVISORY BOARD

Laura C. Collins, MBBS	Beth Israel Deaconess Medical Center Boston, Massachusetts
Toby C. Cornish, MD, PhD	University of Colorado School of Medicine Aurora, Colorado
Cristina Magi-Galluzzi, MD, PhD	University of Alabama Birmingham, Alabama
Robert P. Hasserjian, MD	Massachusetts General Hospital Boston, Massachusetts
Arie Perry, MD	University of California San Francisco, California
Lynette M. Sholl, MD	Brigham and Women's Hospital Boston, Massachusetts
Tomas Slavik, MD	Ampath Pathology Laboratories Pretoria, South Africa
Rhonda K. Yantiss, MD	Weill Cornell Medical College New York, New York
Matthew M. Yeh, MD, PhD	University of Washington Seattle, Washington

Manuscript reviewed by:
C. Christofer Juhlin, MD, PhD, BSc

EDITOR'S NOTE

The Atlases of Tumor Pathology have a long and distinguished history. They were first conceived at a cancer research meeting held in St. Louis in September 1947, as an attempt to standardize the nomenclature of neoplastic diseases. The first series was sponsored by the National Academy of Sciences-National Research Council. The organization of this formidable effort was entrusted to the Subcommittee on Oncology of the Committee on Pathology, and Dr. Arthur Purdy Stout was the first editor-in-chief. Many of the illustrations were provided by the Medical Illustration Service of the Armed Forces Institute of Pathology (AFIP), the type was set by the Government Printing Office, and the final printing was done at the Armed Forces Institute of Pathology. The American Registry of Pathology (ARP) purchased the Fascicles from the Government Printing Office and sold them virtually at cost. Over a period of 20 years, approximately 15,000 copies each of nearly 40 Fascicles were produced. The worldwide impact of these publications over the years has largely surpassed the original goal. They quickly became among the most influential publications on tumor pathology, primarily because of their overall high quality, but also because their low cost made them easily accessible the world over to pathologists and other students of oncology.

Upon completion of the first series, the National Academy of Sciences-National Research Council handed further pursuit of the project over to the newly created Universities Associated for Research and Education in Pathology (UAREP). The Second Series was started, generously supported by grants from the AFIP, the National Cancer Institute, and the American Cancer Society. Dr. Harlan I. Firminger became the editor-in-chief and was succeeded by Dr. William H. Hartmann. The Second Series Fascicles were produced as bound volumes instead of loose leaflets. They featured a more comprehensive coverage of the subjects, to the extent that the Fascicles could no longer be regarded as "atlases" but rather as monographs describing and illustrating in detail the tumors and tumor-like conditions of the various organs and systems.

Dr. Juan Rosai was appointed as editor-in-chief of the Third Series, and Dr. Leslie Sobin became associate editor. A distinguished Editorial Advisory Board was also convened, and these outstanding pathologists and educators played a major role in the success of this series, the first publication of which appeared in 1991 and the last (number 32) in 2003.

The same organizational framework applied to the Fourth Series, meticulously edited by Dr. Steven Silverberg with Dr. Ronald DeLellis as the associate editor. With UAREP and AFIP no longer functioning, ARP remained the responsible organization. The Fourth Series volumes were hardbound with illustrations almost exclusively in color. There was also an increased emphasis on the cytopathologic (intraoperative, exfoliative, or fine needle aspiration) and molecular features that are important in diagnosis and prognosis. At the time of the Fourth Series, ARP also produced Atlases of Non-Tumor Pathology; these volumes were numbered separately from the tumor volumes.

As in the prior series, the goal of the Fifth Series includes a continuous attempt to correlate, whenever possible, the nomenclature used in the Fascicles with that proposed by the World Health Organization Classification of Tumors, as well as to ensure a consistency of style. Including molecular diagnostics is more important than ever, as is the availability of an online component, a more nimble website (www.arppress.org), and a social media presence. Now in this series, the tumor and non-tumor volumes are combined and consecutively numbered as the Atlases of Tumor and Non-Tumor Pathology. Close cooperation between the various authors and their respective liaisons from the Editorial Board will continue to be emphasized in order to minimize unnecessary repetition and discrepancies in the text and illustrations.

Particular thanks are due to the members of the Editorial Advisory Board, our reviewers, the editorial and production staff, and the individual Fascicle authors for their ongoing efforts to ensure that this series is a worthy successor to the previous four.

Jason L. Hornick, MD, PhD

PREFACE AND ACKNOWLEDGEMENTS

It is indeed an honour to be asked to participate in a new edition of the timeless Fascicle series. We are indebted to many colleagues and friends, particularly at the Royal North Shore Hospital and the Mayo Clinic, who provided assistance with case selection, the provision of figures, and general advice throughout the preparation of this work.

In this second edition of the atlas of Non-Neoplastic Disorders of the Endocrine System, we have sought to build on the most excellent work produced by Doctors Lloyd, Douglas, and Young in the first edition published in 2002. We are particularly grateful to these three giants of their fields, wish to explicitly acknowledge how much of this latest edition is based on their work, and hope to be worthy of updating their legacy.

Too often as surgical pathologists, particularly in endocrine pathology, we find ourselves concentrating on the neoplastic at the expense of non-neoplastic disease processes. This is unfortunate as the non-neoplastic category of diseases is many times more common in the community at large. As in all fields of medicine, non-neoplastic endocrine diseases can closely mimic the neoplastic, and an awareness of all aspects of pathology underpins routine clinical practice. With this work, we hope that clinicians and pathologists will develop an increased understanding of nontumor endocrine pathology and that this informs patient care.

Anthony J. Gill, AM, MD, FRCPA
Lori A. Erickson, MD
Talia L. Fuchs, MD, PhD, FRCPA

CONTENTS

1 PITUITARY GLAND

NORMAL PITUITARY GLAND

Embryology

The human pituitary gland is in essence two separate but intimately related organs: the posterior pituitary (neurohypophysis) and the anterior pituitary (adenohypophysis). The gland is recognizable grossly by the third month of fetal development (1–3). The development of each component is embryologically distinct. The adenohypophysis derives from the Rathke pouch (adenohypophyseal primordium), a pocket arising from the ectoderm of the roof of the pharynx, just rostral to the oropharyngeal membrane (4). Rathke pouch is formed by day 29 of embryonic life and separates from the roof of the mouth by days 49 to 51 (4). Once detached from the buccal cavity, this portion of ectoderm grows upward and comes to lie in a depression of the sphenoid bone, the anlage of the sella turcica.

The neurohypophysis begins its development by days 33 to 41 as a thickening in the wall of the diencephalon (5). It is formed by the merging of the infundibular process of the primitive diencephalon (an extension of the hypothalamus) with the Rathke pouch, forming the pituitary infundibulum, the infundibular stem, and the posterior lobe proper (6).

By day 50 to 51, the epithelial and neural rudiments of the pituitary gland are located in positions corresponding to the lobes of the definitive pituitary gland (5). The adenohypophysis is composed of three anatomically distinct components: pars distalis, pars intermedia, and pars tuberalis. The pars distalis is the largest component and is formed by proliferation of cells in the anterior part of the Rathke pouch, which covers the neural rudiment anteriorly and is displaced sideways to form the lateral enlargements (5). The posterior wall of the pouch gives rise to the pars intermedia, which is an upward extension around the stalk of the neurohypophysis. The pars tuberalis is formed by rostral growth around the infundibular stem (6).

Although it was originally thought to be present in only a minority of individuals (7,8), the pharyngeal pituitary is now known to be present in most adults (9). It is formed by nests of adenohypophyseal cells trapped in the pharyngeal mucosa. The most common location of the pharyngeal pituitary is in the anterior pharyngeal roof, but other reported locations include the soft tissue just under the sella turcica or the posterior pharyngeal roof (9). Since its volume is approximately one-thousandth that of the sellar pituitary gland, the pharyngeal pituitary does not contribute meaningfully to hormone production (10). It is important, however, to recognize that ectopic pituitary neuroendocrine tumors (PitNETs) (formerly, pituitary adenoma) may arise from these rests of ectopic pituitary tissue (11).

The anterior lobe of the pituitary gland is responsible for the production of the pituitary hormones: growth hormone (GH), adrenocorticotropic hormone (ACTH), prolactin (PRL), thyroid-stimulating hormone (TSH), follicle-stimulating hormone (FSH), and luteinizing hormone (LH). The hormone-producing cells of the anterior pituitary are recognized early in development (12–15). Immunohistochemical and ultrastructural studies (12–14) have shown that ACTH-producing cells are seen as early as 6 weeks of gestation, GH-producing cells by 8 weeks, and cells producing most other hormones by 15 to 16 weeks. PRL-producing cells are variable in amount in the first and second trimester but are numerous near term.

The development of the anterior and posterior pituitary lobes and the generation of distinct hormone-producing endocrine cell lineages are regulated by a number of transcription factors expressed in the Rathke pouch as well as by

Table 1-1

TRANSCRIPTION FACTORS IMPORTANT FOR PITUITARY DEVELOPMENT

Factor	Target Cells
Pit-1/GHF1[a]	GH/PRL/TSH
Thyrotroph embryonic factor (TEF)	TSH
Corticotroph upstream transcription element-binding (CUTE) protein	ACTH
Steroidogenic factor Ad4BP/SF-1	LH/FSH
Estrogen receptor	PRL/Gonadotroph
Glucocorticoid receptor	ACTH

[a]Pit-1/GHF1 = pituitary-specific positive transcription factor 1/growth hormone factor 1; GH = growth hormone; PRL = prolactin; ACTH = adrenocorticotropic hormone; FSH/LH = follicle-stimulating hormone/luteinizing hormone; TSH = thyroid-stimulating hormone.

inductive signals from the ventral diencephalon/infundibulum (16). Secreted molecules, including fibroblast growth factor (FGF8), bone morphogenic protein (BMP4), and the Wnt protein (Wnt5A), are expressed in the ventral diencephalon and play important roles in controlling the growth of the Rathke pouch and the specification of various pituitary endocrine cell lineages (17–22). In addition, intricate cascades of transcriptional networks are involved in the formation of Rathke pouch and the generation of the various endocrine cell lineages in the anterior pituitary (22,23). Notably, the biocid-related transcription factors Pitx1 and Pitx2 are required for early cell proliferation, survival, and differentiation of the developing anterior pituitary lobe (24–26). Later in pituitary development, the transcription factor Pit1, which is regulated by the Notch signaling pathway and Prop1 (27), as well as the canonical Wnt/beta-catenin pathway (28), is required for terminal differentiation of specific endocrine cell lineages such as lactotrophs, somatotrophs, and thyrotrophs (29–31).

Other important transcription factors include steroidogenic factor 1 (SF-1), which gives rise to gonadotroph cells, and T-box transcription factor 19 (T-Pit), which gives rise to corticotroph cells (32). Transcription factors Lhx2 and Nkx2.1 recently have been found to play important roles in the development of the posterior pituitary (16). Some of the most important transcription factors for pituitary development are summarized in Table 1-1. Molecular and other defects in the expression of transcription factors can lead to specific endocrine disorders, although the recent addition of routine transcription factor immunohistochemistry has facilitated improved diagnosis and prognostication of pituitary tumors in clinical practice, particularly in cases with absent, low, or pluri-hormonal expression (32).

Anatomy

The pituitary gland is composed of an anterior epithelial lobe (anterior pituitary or adenohypophysis) and a posterior neural lobe (neurohypophysis). In humans, the wet weight of the gland is proportional to its volume (33). It increases from about 100 mg at birth to 500 to 600 mg during the second decade, depending upon ethnicity (34,35), and may even reach 800 mg in males and 900 mg in females (36). In females, the pituitary gland is generally about 20 percent larger than in males (34) and can double in weight during pregnancy, with a progressive increase in multiparity (36). These changes are restricted to the anterior lobe, while the size and weight of the posterior lobe remain constant throughout life and do not display sexual dimorphism (34,37,38). In general, the average adult pituitary gland measures 13x9x6 mm.

The gland is surrounded by a fibrous capsule that is organized into two layers. The first is an inner layer that gives rise to fibrous trabeculae that penetrate the adenohypophyseal parenchyma (39) and segments it into three main territories: a central mucoid wedge and two lateral wings (40). These fibrous trabeculae provide a passage for arterial branches that perfuse the adenohypophyseal parenchyma (41). The second is a thicker external layer that forms the lateral walls and roof of the sella turcica (42,43). This layer is an extension of the dural sheath and is presumed to be the true pituitary capsule (39).

The pituitary gland lies adjacent to many important structures including the internal carotid arteries, which lie lateral in the cavernous sinuses, the optic chiasm and optic tracts in front and above, the base of the diencephalon above, and the sphenoid air sinus in front and below. The sellar diaphragm is also present on the roof of the sella turcica and consists of a fold of dura mater extending transversely across the sella. Its center is perforated for the passage of the infundibulum.

The adenohypophysis, or anterior pituitary, constitutes about 70 to 80 percent of the gland. It is divided into three zones: the pars distalis, pars intermedia, and pars tuberalis. The pars distalis makes up the bulk of the anterior pituitary and contains the hormone-secreting cells of the adenohypophysis. The pars intermedia lies between the pars distalis and the posterior pituitary, representing the remnants of the posterior loop of Rathke pouch, which appear as a few colloid-filled cavities lined by epithelial cells. The pars tuberalis is composed of squamous cell nests and a few anterior pituitary cells, especially glycoprotein-producing hormone cells. It is subdivided into a superior part attached to the hypothalamic median eminence and an inferior part that surrounds the infundibular stem, but is separated from it by a fibrovascular lamina (septum tuberalis) that contains the portal vessels (41,44).

The neurohypophysis, or posterior pituitary, constitutes the remaining 20 to 30 percent of the gland. It is composed of: 1) the infundibulum or median eminence, which is attached to the hypothalamus and receives the hypothalamic peptidergic neurons with releasing and inhibiting hormones that regulate anterior pituitary cell function; 2) the pituitary stalk or infundibular stem, which is composed of unmyelinated nerve fiber tracts that originate in the hypothalamus and portal vessels. These fibers transport hypothalamic peptides to the anterior pituitary; and 3) the posterior lobe, composed of the infundibular process and pars nervosa, which stores vasopressin and oxytocin hormones.

The pituitary gland lies in the central part of the sella turcica (pituitary fossa), a saddle-shaped groove in the endocranial, superior surface of the sphenoid bone corresponding to the median part of the middle cranial fossa (33). It lies adjacent to the cavernous sinuses, the internal carotid arteries, and several cranial nerves. The pituitary gland is a highly vascular structure that receives both direct (arterial) and indirect (sinusoidal) vascularization. Direct vascularization originates from the supracavernous portion of the internal carotid arteries that give rise to the paired lateral and anterior superior hypophyseal arteries, and the intercavernous portion of the internal carotid artery that gives rise to the inferior hypophyseal arteries (45). The superior hypophyseal arteries give off numerous short branches that course along the infundibular stem and terminate in a dense capillary plexus. The superior hypophyseal artery also gives rise to branches that supply both the stalk and the anterior lobe (46).

Unlike the anterior pituitary, which has an indirect arterial blood supply, the posterior pituitary has a direct blood supply from branches of the internal carotid arteries. The inferior hypophyseal arteries divide into ascending and descending branches running in the groove between the anterior and posterior lobes, and anastomosing with each other, giving rise to an arterial circle that perfuses the posterior lobe (41,47). The middle hypophyseal arteries enter the anterior lobe and supply some of the adenohypophyseal cells at the periphery of the gland.

The most critical source of vascularization to the anterior pituitary is the pituitary portal system, originally termed the "portal veins" because they act as a link between the arterial blood of the hypophyseal arteries and the venous blood of the hypophyseal veins (48). These short portal vessels originate in the posterior lobe and distal portions of the stalk, and supply about 10 to 20 percent of the blood flow to the anterior lobe (45). The venous drainage from the pituitary gland is directed via the cavernous sinuses, inferior petrosal sinuses, and internal jugular veins.

The pars distalis has no direct nerve supply, except for a few sympathetic fibers that penetrate the anterior lobe alongside the capillaries. These pericapillary nerve fibers do not regulate anterior pituitary hormone secretion but may affect blood flow to the pituitary gland. The posterior pituitary is the lowest part of a continuous bundle of nerve fibers that extends from the hypothalamus, into the infundibular stem and terminates in the posterior pituitary (33). This bundle is composed of numerous unmyelinated and sparsely myelinated axons and axon terminals containing large neurosecretory vesicles, specialized glial cells called pituicytes, and a rich capillary network. The neurosecretory material is largely made up of vasopressin and oxytocin, together with their carrier proteins

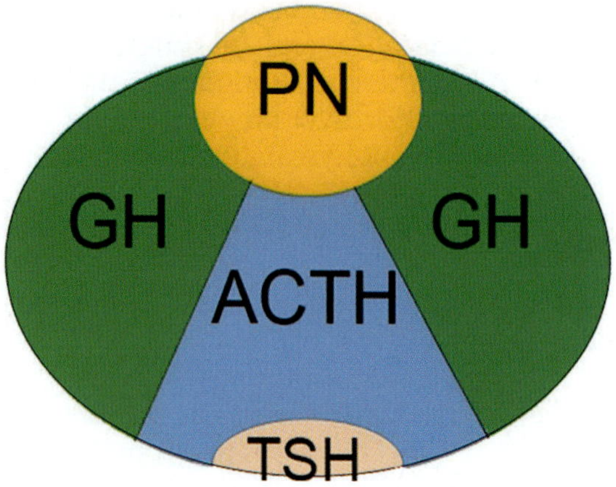

Figure 1-1

PITUITARY GLAND: NORMAL HISTOLOGY

The topographic distribution of various pituitary cell types in the adenohypophysis and the adjacent pars nervosa (PN) of the neurohypophysis. The growth hormone (GH)-producing acidophil cells are predominantly in the lateral wings, with the adrenocorticotropic hormone (ACTH)-producing basophil cells in the central mucoid wedge.

and several other substances that are co-contained in vasopressin and oxytocin terminals (33). The hypothalamic tracts from the supraoptic and paraventricular nuclei travel from the hypothalamus to the stalk of the posterior lobe as tracts of the supraopticohypophyseal and tuberohypophyseal fibers. Neural connections influence posterior pituitary secretion of oxytocin and vasopressin, as evidenced by the severe atrophy of the neurohypophysis after stalk sectioning or with injury to the axons originating in the supraoptic and paraventricular nuclei. In addition, neurons synthesizing dopamine and norepinephrine, originating from the hypothalamus and brainstem, respectively, also innervate the posterior pituitary (49).

Microscopic Anatomy and Physiology: Anterior Pituitary

As described above, the anterior pituitary is composed of three components: the pars distalis, the pars intermedia, and the pars tuberalis. The pars distalis is the main source of anterior pituitary hormone secretion. A horizontal section of the anterior pituitary reveals fibrous trabeculae in the mid-central portion, with lateral wings and a central mucoid wedge (fig. 1-1). The pars distalis is composed of nests or cords of large polygonal epithelial cells organized around venous sinusoids lined by fenestrated epithelium into which secretory products of the anterior pituitary are collected. The nesting pattern is apparent in reticulin-stained sections that show a delicate, continuous border around individual acini (fig. 1-2), differentiating normal anterior pituitary tissue from the disrupted reticulin framework of adenomatous tissue.

The cells of the pars distalis contain abundant cytoplasm that stains variably with a hematoxylin and eosin (H&E) stain, allowing distinction of three cell types: acidophils, basophils, and chromophobes. The acidophil cells stain red to pink and are present mainly in the lateral wings; the basophils stain blue to light purple and are in the mucoid wedge; and the chromophobe cells stain light blue and are widely dispersed (fig. 1-2). Several other histochemical stains have also been used previously, but immunohistochemistry is now used to differentiate the six major cell types of the anterior pituitary on the basis of their secretory products: somatotrophs (GH), lactotrophs (PRL), mammosomatotrophs (PRL and GH), corticotrophs (ACTH), thyrotrophs (TSH), and gonadotrophs (LH and FSH). Electron microscopic examination allows characterization of the unique features of different cell types but is now rarely used in clinical practice (33,50,51).

Each of these cell types is regulated by their respective hypothalamic releasing and inhibitory hormones after they are released into the portal vasculature for conveyance to pars distalis cells (Table 1-2) (33). This system is regulated by a negative feedback mechanism that operates between specific target organs, the pituitary gland, and the hypothalamus. For example, thyroid hormone exerts a negative feedback effect on the pituitary gland and hypothalamus to regulate the secretion of thyrotropin-releasing hormone (TRH) from hypothalamic neurons and TSH from the pituitary gland.

Somatotroph (GH) Cells

Somatotrophs comprise the largest cell population of the pars distalis, constituting 40 to 50 percent of secretory cells in the adult pituitary gland (33). The cells are round to oval, with round nuclei and acidophilic cytoplasm on H&E staining (fig. 1-2). Most somatotroph cells

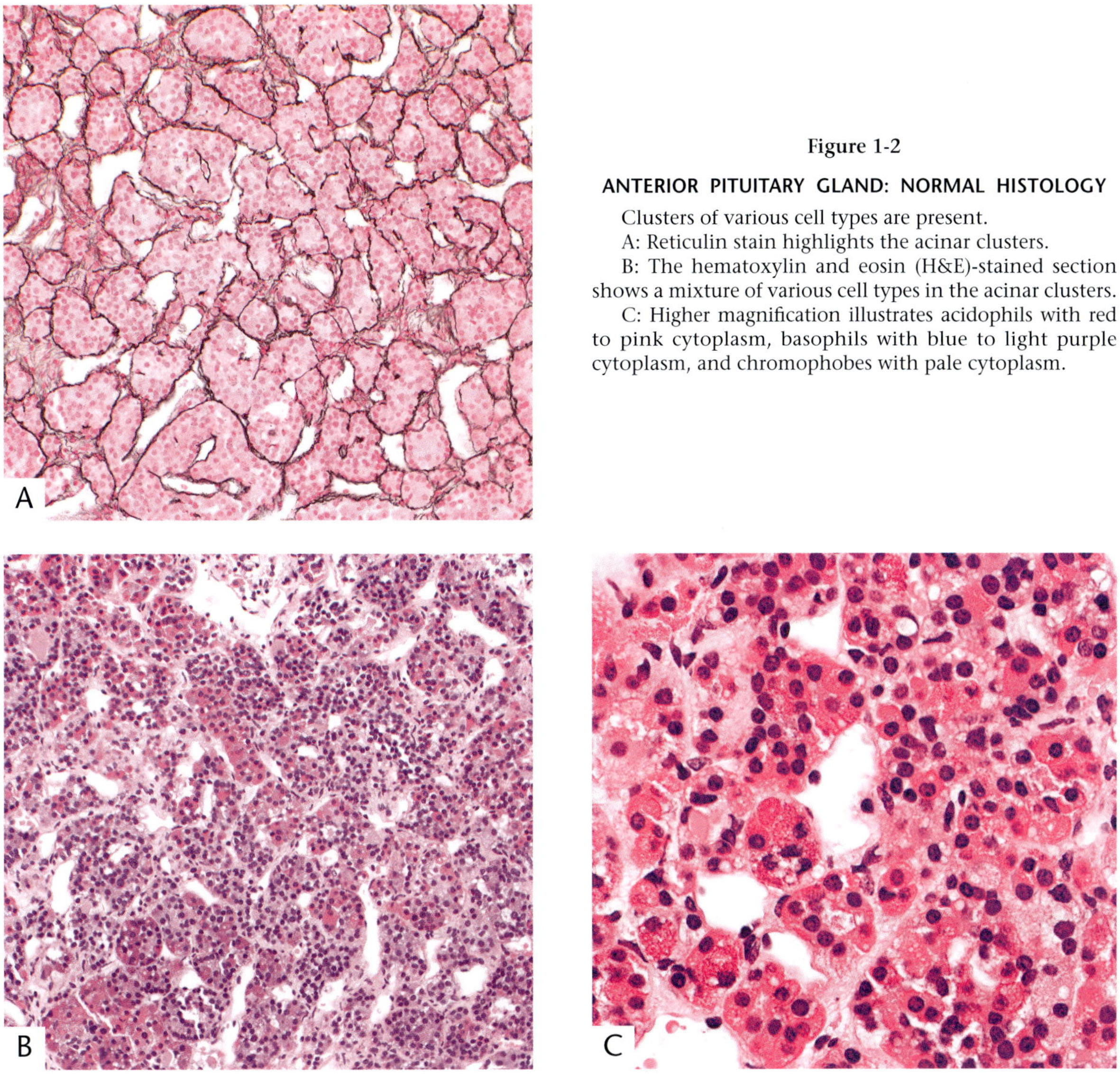

Figure 1-2

ANTERIOR PITUITARY GLAND: NORMAL HISTOLOGY

Clusters of various cell types are present.

A: Reticulin stain highlights the acinar clusters.

B: The hematoxylin and eosin (H&E)-stained section shows a mixture of various cell types in the acinar clusters.

C: Higher magnification illustrates acidophils with red to pink cytoplasm, basophils with blue to light purple cytoplasm, and chromophobes with pale cytoplasm.

are present in the lateral wings of the anterior pituitary (fig. 1-1). Immunostaining with GH antibodies shows particularly strong cytoplasmic immunopositivity and allows ready recognition of these cells. By electron microscopy, the entire cytoplasm is filled with dense secretory granules ranging from 150 to 800 nm in diameter, and there is prominent endoplasmic reticulum that changes with the activity of the cell (33,50). The *GH* gene, located on chromosome 17, is 2.5 kb long and directs the synthesis of a 191-amino acid single-chain peptide with two interchain disulfide bands. GH secretion is regulated by growth hormone-releasing hormone (GHRH) and somatostatin release-inhibiting factor (SRIF). The latter peptide inhibits growth hormone secretion. Insulin-like growth factor 1 (IGF-1), or somatonectin C, which is produced by the liver, mediates many of the effects of GH. IGF-1 exerts a negative feedback effect on GH secretion at the hypothalamic and pituitary levels.

Lactotroph (PRL) Cells

Lactotroph cells (mammotrophs or prolactin [PRL] cells) constitute 10 to 30 percent of the total cell population of the anterior pituitary gland (Table 1-3) (52). They are acidophilic or chromophobic, and are present in the highest density

Table 1-2

HYPOTHALAMIC HORMONES REGULATING PITUITARY SECRETION AND TARGET TISSUES

Peptide/Amine	Pituitary Hormone	Target Tissues
Stimulating		
Corticotropin-releasing hormone	ACTH[a]	Adrenal cortex
Gonadotropin-releasing hormone	FSH/LH	Ovaries, testes
Growth hormone–releasing hormone	GH	Many tissues
Thyrotropin-releasing hormone	TSH/PRL	Thyroid, breast, many other tissues
Vasoactive intestinal polypeptide	PRL	Breast, many other tissues
Inhibitory		
Somatostatin	GH	Many tissues
Dopamine	PRL	Breast, many other tissues
Others		
Vasopressin (ADH)	(Posterior pituitary)	Kidney
Oxytocin	(Posterior pituitary)	Uterus, breast

[a]GH = growth hormone; PRL = prolactin; ACTH = adrenocorticotropic hormone; FSH/LH = follicle-stimulating hormone/luteinizing hormone; TSH = thyroid-stimulating hormone; ADH = antidiuretic hormone.

Table 1-3

ANTERIOR PITUITARY HORMONES: CELL TYPE, DISTRIBUTION, AND FUNCTION

Cell Type	Approximate Cell Percentage in Pituitary	Function
GH[a]	40–50%	Stimulates linear growth
PRL	10–30%	Stimulates breast milk production
ACTH	10–20%	Stimulates cortisol synthesis
FSH/LH	10%	Stimulates estrogen and testosterone synthesis
TSH	5%	Stimulates thyroid hormone synthesis

[a]ACTH = adrenocorticotropic hormone; FSH/LH = follicle-stimulating hormone/luteinizing hormone; GH = growth hormone; TSH = thyroid-stimulating hormone; PRL = prolactin.

at the posterolateral and posteromedial edges of the adenohypophysis, but are located throughout the entire anterior lobe (fig. 1-3). The number of lactotrophs varies considerably, increasing significantly during pregnancy and lactation.

There are two types of lactotrophs. Most are small to medium in size and sparsely granulated. The secretory granules are spherical and evenly electron dense, measuring 150 to 350 nm in diameter. The other, less frequently occurring type, is densely granulated and the granules are larger (300 to 600 nm) (53). The cells usually have well-developed rough endoplasmic reticulum and Golgi complexes. During pregnancy, the cells undergo considerable hypertrophy: their Golgi complex enlarges and the endoplasmic reticulum becomes more extensive; multiple layers parallel to the cell membrane develop and the granules become larger (550 to 600 nm) and often irregular in outline (53).

The *PRL* gene, located on chromosome 6, is greater than 10 kb long, and the mRNA is about 1 kb long. The protein is a 198-amino acid peptide which has a common evolutionary origin with GH. One of the functions of PRL is to stimulate lactation, which normally begins when progesterone levels fall toward the end of pregnancy. Other actions include contributing to pulmonary surfactant synthesis in the fetal lungs at the end of gestation and maternal immune tolerance of the fetus during pregnancy. PRL also promotes neurogenesis in maternal and fetal brains (54,55). Receptors for PRL are widely distributed on a number of other cell types, although the function of the hormone in many tissues is uncertain.

In contrast to all other pituitary hormones, PRL secretion is tonically suppressed by the hypothalamus, meaning that it is only secreted when this hypothalamic inhibition is released. If the pituitary stalk is cut, PRL secretion increases, while secretion of all other pituitary hormones falls dramatically due to loss of hypothalamic-releasing hormones. Dopamine is the major inhibitor of PRL secretion. It is secreted into portal blood by hypothalamic neurons and binds to receptors on lactotrophs to inhibit both synthesis and secretion of PRL. PRL secretion is stimulated by several hormones, including TRH, gonadotropin-releasing hormone (GnRH), and vasoactive intestinal polypeptide (VIP).

Mammosomatotroph (PRL/GH) Cells

Mammosomatotrophs resemble somatotrophs but produce both GH and PRL, detected within the same cell, and even within the same secretory granule, by immunohistochemistry (33). These cells are uncommon in the normal pituitary gland and increase in number during pregnancy. Two types of mammosomatotrophs have been identified by electron microscopy: cells containing numerous small electron-dense granules (150 to 400 nm) and cells with large cigar-shaped granules measuring up to 2,000 nm with variable density (56). The regulation of this cell type is unknown, but tumor cells with features of mammosomatotrophs are common in large series of PitNETs examined ultrastructurally (50,51,57).

Corticotroph (ACTH) Cells

Corticotroph cells are basophilic and account for 10 to 20 percent of anterior pituitary cells. They are medium-sized cells with large, clear cytoplasmic vacuoles known as "enigmatic bodies" that are lysosomal complexes, readily identified on routine H&E sections. They are present mainly in the mucoid wedge, the area that on horizontal cross section is the central portion of the gland that abuts the pars nervosa (see fig. 1-1). Clusters of ACTH cells may be seen in the posterior pituitary, a phenomenon known as basophil invasion (fig. 1-4). The importance of recognizing this phenomenon is to distinguish it from corticotrophic PitNET, a particular pitfall on small biopsies (fig. 1-4) (51,58). ACTH cells stain for periodic acid–Schiff (PAS). In contrast, the GH and PRL cells are PAS negative. By electron microscopy, the cells contain 300- to 500-nm secretory granules along with large perinuclear phagolysosomes (corresponding to the enigmatic bodies seen on light microscopy), and bundles of 6- to 9-nm type 1 keratin filaments, which react with antibodies directed against keratins of 54 to 60 kd.

ACTH is produced in the pituitary gland in a process that also generates several other hormones. The process begins with a large precursor protein, pro-opiomelanocortin (POMC), which is synthesized and proteolytically cleaved to produce ACTH, lipotropin, beta-endorphin, met-enkephalin, and melanocyte-stimulating hormone. The *POMC* gene is located on chromosome 2 and consists of three exons. ACTH is a 39-amino acid single-chain peptide derived from the 31-kd POMC glycoprotein.

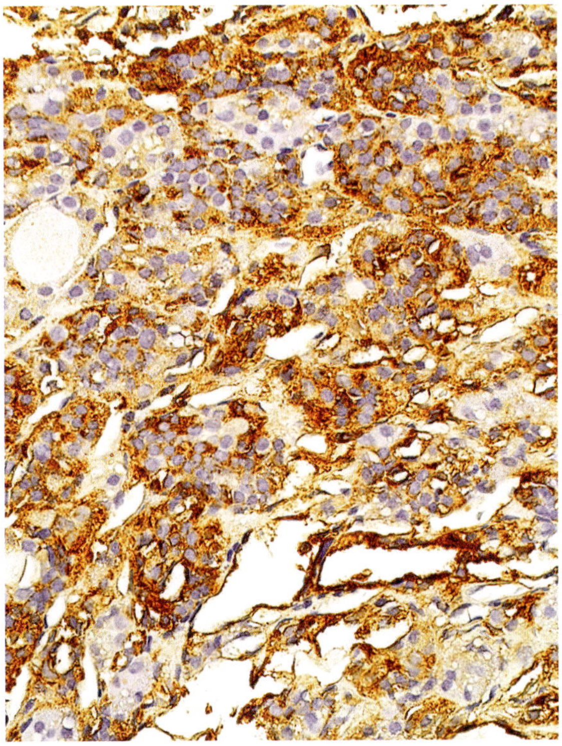

Figure 1-3

PROLACTIN (PRL) CELLS: IMMUNOHISTOCHEMISTRY

Immunohistochemical staining for PRL in the anterior pituitary. These cells constitute 10 to 30 percent of anterior pituitary cells.

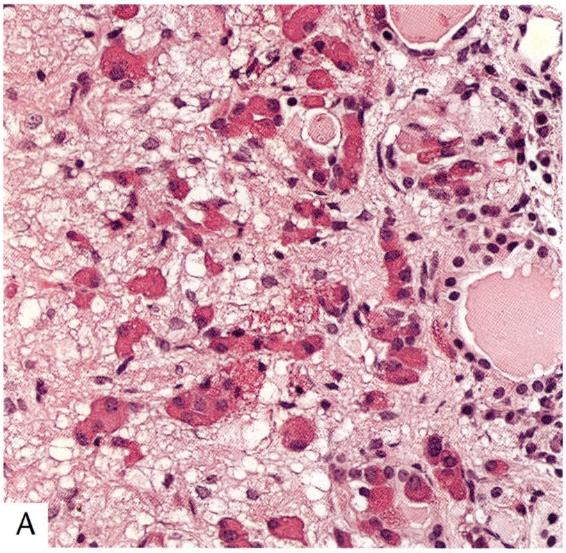

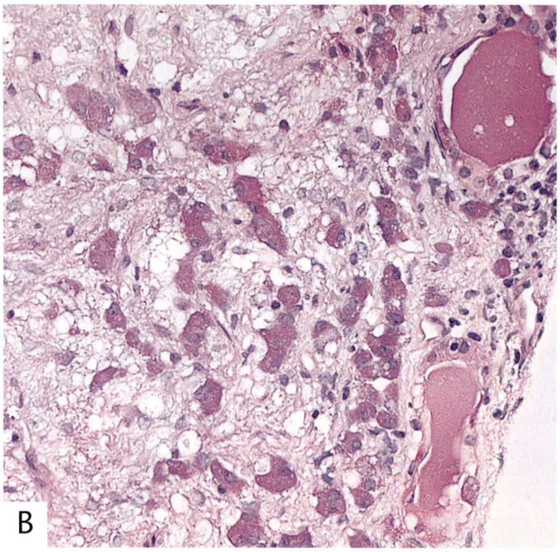

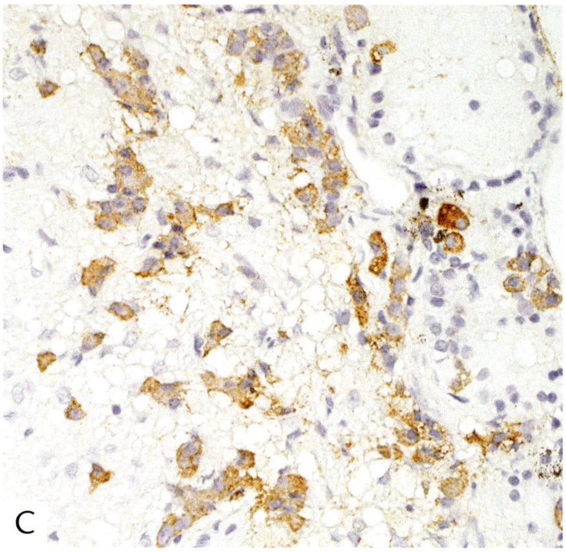

Figure 1-4

BASOPHIL INVASION

A: Basophil or corticotroph invasion into the pars nervosa. The "invading" cells have basophilic cytoplasm and are located among the processes of the pars nervosa.

B,C: The "invading" cells are positive for periodic acid–Schiff (PAS) (B) and ACTH immunohistochemistry (C). Some cells are arranged in small clusters, which may lead to misinterpretation as an ACTH-producing PitNET in a small biopsy.

ACTH is secreted from the anterior pituitary in response to corticotropin-releasing hormone (CRH), which is produced by the hypothalamus in response to many types of stress. ACTH exerts its effects by stimulating the adrenal cortex to secrete glucocorticoids, mineralocorticoids, and sex steroids. Glucocorticoids exert a direct negative feedback effect on both hypothalamic CRH and anterior pituitary ACTH secretion. In response to excess glucocorticoids, an accumulation of type 1 intermediate filaments in the cytoplasm displaces the secretory granules to the cell periphery and results in Crooke hyaline change, seen as pale pink, glassy, ring-like staining of the cytoplasm on H&E sections. These changes are readily identified by immunostaining for low molecular weight cytokeratins such as CAM5.2 and ring-like staining with ACTH, reflecting peripheral displacement of the secretory granules to the cell membrane (59).

Thyrotroph (TSH) Cells

Thyrotroph cells comprise less than 5 percent of anterior pituitary cells and are located principally in the anteromedial portion of the mucoid wedge (fig. 1-5). They are medium-sized, angulated cells that contain an eccentric nucleus and cytoplasmic PAS-positive droplets.

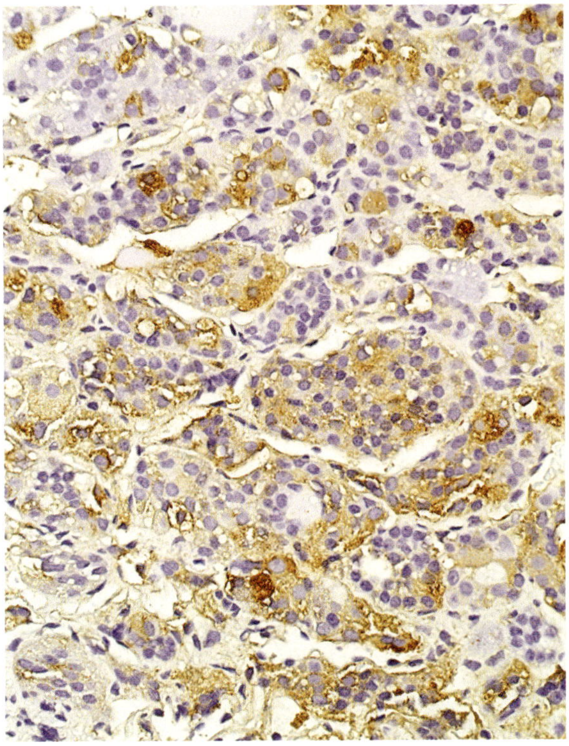

Figure 1-5

THYROID-STIMULATING HORMONE (TSH) CELLS: IMMUNOHISTOCHEMISTRY

TSH-producing cells in the anterior pituitary are highlighted by specific immunohistochemistry. These angular cells comprise less than 5 percent of anterior pituitary cells.

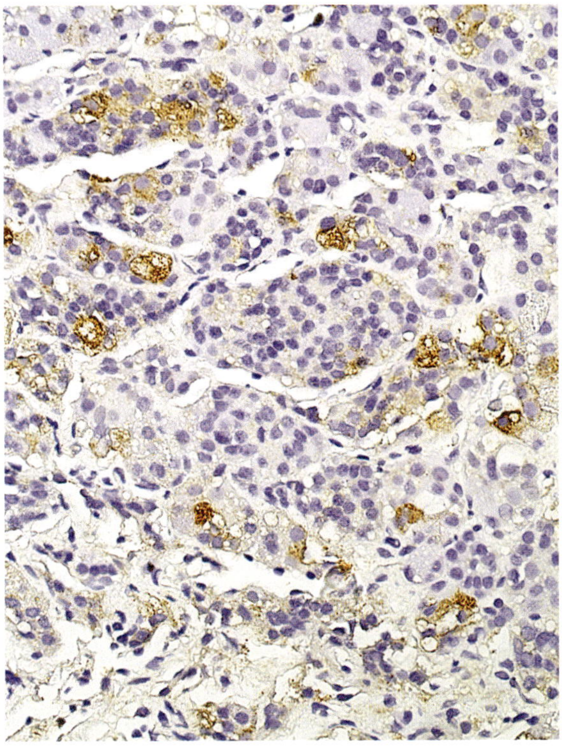

Figure 1-6

LUTEINIZING HORMONE (LH) CELLS: IMMUNOHISTOCHEMISTRY

Gonadotroph cells in the anterior pituitary immunostain for LH. These cells make up about 10 percent of anterior pituitary cells. Follicle-stimulating hormone (FSH) and LH are present in the same pituitary cells.

Ultrastructurally, the cells are filled with small secretory granules that measure 100 to 200 nm, short rough endoplasmic reticulum, and large Golgi complexes (33).

TSH is a glycoprotein hormone composed of two subunits that are noncovalently bound to one another. The alpha subunit gene is found on chromosome 6 and is 9.4 kb long with four exons. The alpha subunit protein consists of 92 amino acids. The beta subunit gene is located on chromosome 1 and consists of three exons. The TSH molecule is a 28-kd glycoprotein; the beta subunit that confers the specificity for TSH action ranges from 18 to 21 kd, depending on whether it contains one or two carbohydrate chains. The alpha subunit of TSH is also present in two other pituitary glycoprotein hormones, FSH and LH, each of which also has a unique beta subunit that provides receptor specificity.

TSH secretion is stimulated by TRH from the hypothalamus. Secretion of TRH, and hence, TSH, is inhibited by high serum levels of thyroid hormones in a classic negative feedback loop.

Gonadotroph (FSH/LH) Cells

Gonadotrophs can be bihormonal or contain only one of the gonadotropins. Most gonadotrophs produce both FSH and LH (fig. 1-6). The cells are basophilic, evenly distributed in the anterior pituitary, and comprise approximately 10 percent of the cells. They frequently show oncocytic change and squamous metaplasia in older individuals (33). Both FSH and LH are large glycoproteins composed of alpha and beta subunits. The alpha subunit is identical in FSH, LH, and TSH, while the beta subunit is unique and endows each hormone with its receptor specificity. The *FSH* beta gene is located on chromosome

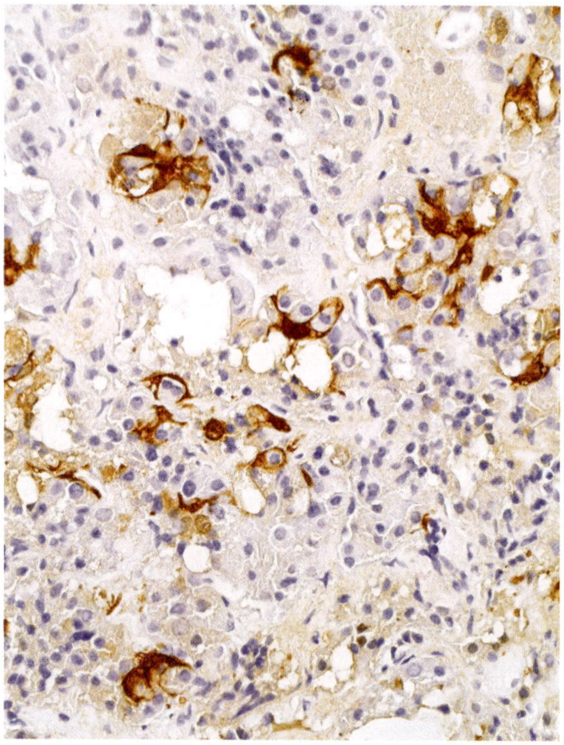

Figure 1-7

FOLLICULOSTELLATE CELLS

Folliculostellate cells are highlighted by immunohistochemistry for S-100 protein. These cells have cytoplasmic processes that extend between the hormone-producing cells.

11 and consists of three exons; the protein has 130 amino acids. The *LH* beta gene is located on chromosome 19 and consists of three exons; the LH beta protein consists of 145 amino acids.

As stated, most gonadotrophs are bihormonal, containing both FSH and LH proteins within small (200 nm) and larger (300 to 600 nm) secretory granules. The size of the secretory granules varies in males and females: smaller granules in males and larger granules in females (33). FSH stimulates maturation of ovarian follicles and spermatogenesis, while LH induces ovulation, luteinization of the ovarian follicles, and steroidogenesis. LH also stimulates Leydig cells to produce testosterone and the ovary to elaborate luteal phase hormones. The principle regulator of LH and FSH secretion is GnRH, which is synthesized and secreted by the hypothalamus. In a classic negative feedback loop, sex steroids inhibit secretion of GnRH and also appear to have direct negative effects on gonadotrophs (33). Removal of the gonadal organs results in hypertrophy and hyperplasia of the pituitary gonadotropic cells with formation of "gonadectomy" cells.

Folliculostellate Cells

These nonhormonal cells facilitate paracrine regulation and network functioning; they comprise approximately 5 to 10 percent of the pars distalis cell population (fig. 1-7) (33,60). They are small, chromophobic, stellate-shaped cells with long, delicate cytoplasmic processes creating a meshwork in which the glandular cells of the pars distalis reside (61). Because they are immunopositive for glial fibrillary acidic protein (GFAP), they are believed to be of glial origin (33). They are also strongly positive for S-100 protein (fig. 1-7).

Folliculostellate cells function as part of an intrapituitary regulatory center, modulating glandular cells of the pars distalis as a network capable of transmitting signals over long distances by electronic coupling to each other through gap junctions and intercellular zones (zona adherens) (62,63). They also make contact with hormonal cells in the pars distalis and, through the release of a variety of substances, including cytokines and growth factors, modify pituitary hormone secretion (62,64). For example, folliculostellate cells can release interleukin 6 (IL-6) in response to bacterial endotoxin, contributing to the increase in ACTH from corticotrophs in patients with sepsis (65).

Microscopic Anatomy and Physiology: Posterior Pituitary

The posterior pituitary is the lowest part of a continuous bundle of nerve fibers arising from hypothalamic neurons. This bundle is composed of numerous unmyelinated and sparsely myelinated axons and axon terminals containing large neurosecretory vesicles, specialized GFAP-positive glial cells called pituicytes, and a rich capillary network (fig. 1-8). The neurosecretory material is largely made up of the nonapeptides, vasopressin and oxytocin, and their carrier proteins, derived from magnocellular neurons in the hypothalamic paraventricular and supraoptic nuclei (33). The hormones are stored in secretory granules (Herring bodies) located in focal axonal dilatations.

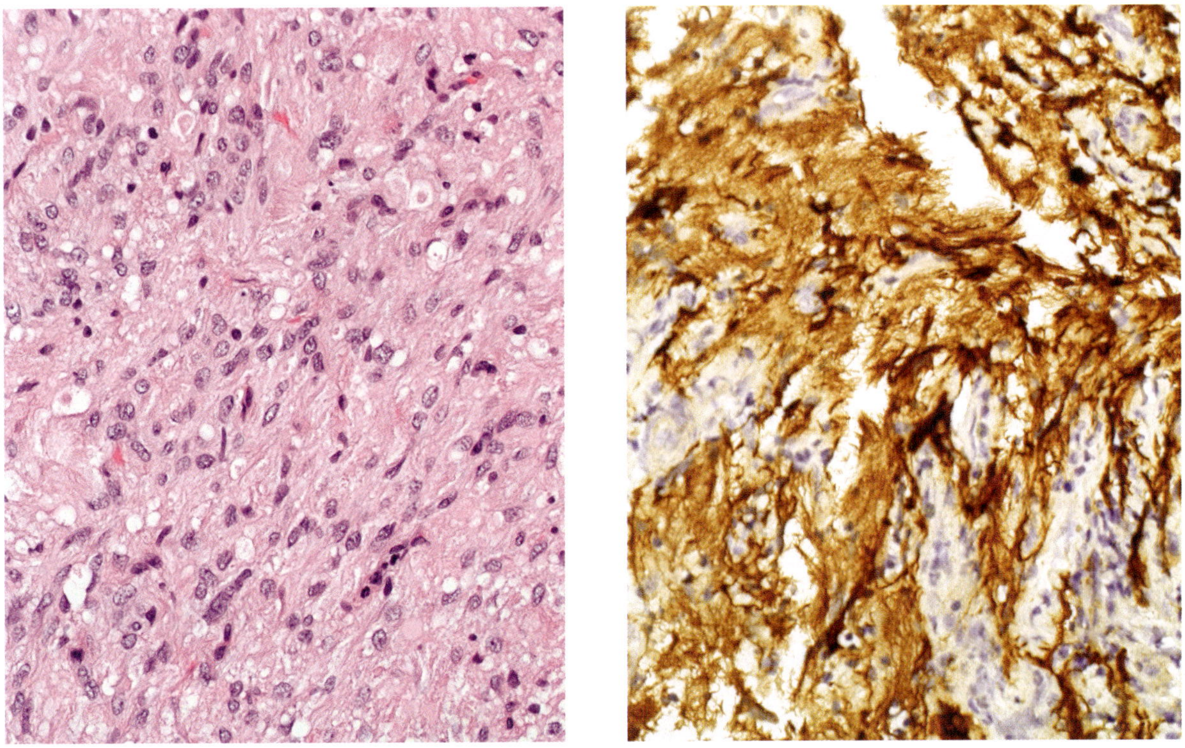

Figure 1-8

POSTERIOR PITUITARY

Left: The posterior pituitary is made up of pituicytes, glial cells, and nerve fibers.
Right: The glial cells and nerve fibers are positive for glial fibrillary acidic protein (GFAP).

REACTIVE CHANGES

Pregnancy

During pregnancy, pituitary size and function are modified mainly by placental hormone secretion. These changes play a crucial role both in mother and fetus during gestation, labor, and puerperium. Under the influence of PRL, the pituitary gland doubles in weight due to lactotroph hypertrophy and hyperplasia (66), reaching 12 mm in height immediately postpartum (67).

Stimulated by TRH (68), serum PRL levels gradually increase over the course of gestation, reaching up to 10 times nonpregnant upper limit values by the end of the third trimester (69,70). In addition, the placenta secretes a variant isoform of GH, which increases IGF-1 levels and subsequently inhibits pituitary GH secretion through a negative feedback loop (71–74).

A number of other changes occur as a result of the increased maternal hypothalamic-pituitary-adrenal axis function associated with normal pregnancy. Due to high concentrations of corticosteroid-binding globulin (CBG), induced by elevated estrogen levels, cortisol clearance is reduced, resulting in increased total plasma cortisol levels. In addition, placental CRH stimulates ACTH and thus, the production of cortisol by the adrenal gland increases significantly in the third trimester (75).

Because of the marked pituitary hyperplasia associated with pregnancy, the gland is more susceptible to infarcts, especially during delivery, which may be associated with hypotension (Sheehan syndrome). Pregnancy is also associated with an increased risk of lymphocytic hypophysitis (75). Although the pituitary gland decreases in weight after delivery, it remains larger in multiparous compared to nulliparous women.

Hormonal Syndromes

A number of reactive changes can occur in the pituitary gland in response to dysfunction of endocrine target organs. The most common reactive changes are pituitary cell hyperplasias, such as those seen in Addison disease and Schmidt syndrome (76,77). This is discussed later in the chapter. Most of the anterior pituitary cells producing trophic hormones directed at specific organs can be affected via a direct feedback effect due to decreased hormonal production. However, hyperfunction of specific endocrine target organs can also lead to reactive changes in the pituitary gland. Examples include hyperadrenocorticalism and hyperthyroidism.

Hyperadrenocorticalism. Glucocorticoid excess, resulting from either endogenous or exogenous sources, leads to specific morphologic changes in the anterior pituitary, known as *Crooke hyaline change* (78–82). In this process, corticotroph cells accumulate cytokeratin filaments in their cytoplasm in an annular or concentric fashion, resulting in displacement of the secretory granules to the periphery of the cell or around the nucleus (83). This imparts a glassy or hyalinized appearance to the cells on H&E-stained sections.

The presence of Crooke hyaline change is believed to reflect the degree and duration of hypercortisolism (84). Specific conditions leading to this change include: 1) adrenocortical hyperplasias, PitNETs, and carcinomas; 2) ectopic production of ACTH or CRH by tumors such as small cell lung carcinomas, bronchial carcinoids, thymomas, pheochromocytomas, and others; 3) treatment with glucocorticoids for autoimmune diseases, organ transplantation, and other conditions; and 4) production of excessive ACTH by a PitNET. The latter leads to Crooke changes predominantly in the non-neoplastic ACTH cells, although such changes also occur in the adenomatous cells themselves. Crooke changes are reversible once the stimulus is removed and glucocorticoid levels normalize.

Hyperthyroidism. Thyroid hormones directly block pituitary secretion of TSH and inhibit TRH production by the hypothalamus. Immunohistochemical studies have found decreased numbers of TSH cells in patients with thyrotoxicosis (85,86). However, it is believed that there are usually more TSH-positive cells with less immunoreactive hormone in this condition than previously realized, because the cells with small amounts of stored TSH do not stain with anti-TSH antibodies. The decreased immunostaining is probably related to the negative feedback effects of the increased triiodothyronine (T3) and thyroxine (T4) associated with hyperthyroidism. More recently, expression of TSH receptors has been demonstrated in the folliculostellate cells of the pituitary gland, where they are thought to play a role in paracrine feedback inhibition of TSH secretion (87,88).

Effects of Specific Drugs

Pituitary function may be influenced by several drugs, including antidepressants, opioids, glucocorticoids, chemotherapeutic agents, immunomodulators, and tyrosine kinase inhibitors. In most cases, treatment with these drugs negatively affects pituitary function, but in rare cases, activation of specific hypothalamic-pituitary axes is observed (89). High doses of estrogens are associated with PRL cell hyperplasia and hyperprolactinemia (82,90). Similarly, hyperprolactinemia due to antidopaminergic drugs is the most common pituitary side effect during treatment with antidepressant drugs (91). Antipsychotic agents, such as phenothiazines or tricyclic antidepressants, may result in a decreased response of TSH to TRH, with the opposite effect occurring with lithium treatment (92).

Somatostatin analogues, such as octreotide and lanreotide, are used in the treatment of many neuroendocrine tumors, including GH PitNET. In the normal pituitary gland, these drugs inhibit the secretion of GH, PRL, and TSH (93), and can induce variable degrees of anterior pituitary fibrosis (94). GnRH agonists are used in the treatment of prostatic and other cancers and isosexual precocious puberty in boys. These drugs initially stimulate secretion of FSH and LH by the pituitary gland. However, after a few days of continuous treatment, the pituitary becomes desensitized, loses its membrane receptors for GnRH, and stops releasing LH and FSH, leading to a decline in testosterone production and ending in levels similar to those achieved by orchidectomy (95).

Dopamine agonists such as bromocriptine and parlodel are often used in the treatment of prolactinomas (96). They act by inhibiting *PRL* gene transcription, as well as hormone synthesis and release; in addition, they cause reduction in the size of adenomatous cells, and suppress proliferation of both normal and neoplastic lactotrophs, leading to pituitary fibrosis (97–99).

Treatment with tyrosine kinase inhibitors (TKI) causes important derangements in the hypothalamic-pituitary axis, probably by directly inhibiting thyroid hormone feedback, thereby inducing an increase in TSH levels (89,100). In addition, chronic treatment with TKIs has been linked to the development of hypoglycemia (101) and GH deficiency (102), indicating that these drugs may have a direct impact on somatotroph cell function.

Specific pituitary side effects have also been described during treatment of cancer patients with immune checkpoint inhibitors (CTLA4 and PD-1/PD-L1 inhibitors), potent immune modulators that prevent the escape of neoplastic cells from the recognition by cytotoxic T lymphocytes (103,104). Drugs such as ipilimumab and tremelimumab cause autoimmune diseases in 61 to 77 percent of patients, and in some series, up to 17 percent of patients develop autoimmune hypophysitis (105,106). Pituitary inflammation with immune checkpoint inhibitor therapies appears to have a higher incidence in males and the elderly (107). The incidence is commonly noted prior to the third cycle of treatment and ranges from 5 to 36 weeks from the onset of therapy, with a mean duration of 9 weeks (108). The clinical presentation is similar to other forms of hypophysitis, with headache being the most frequently reported symptom (109). Visual disturbances and diabetes insipidus are rare (109–113). Endocrine deficiencies are noted in more than 70 percent of patients and include high rates of thyrotroph, gonadotroph and corticotroph insufficiency (114). Recovery of ACTH insufficiency is rare in CTLA4-induced hypophysitis (109,114).

HEREDITARY AND DEVELOPMENTAL DISORDERS

Agenesis of the Pituitary Gland

Pituitary agenesis (or *aplasia*) is the complete absence of the pituitary gland. Congenital absence of the pituitary gland is rare and most neonates with complete pituitary aplasia survive only a few hours (115). Pituitary aplasia may occur as an isolated anomaly or as part of a major brain malformation, such as septo-optic dysplasia, holoprosencephaly, or oral-facial-digital syndrome due to ciliary gene mutations (116). There is usually an associated hypoplasia of the adrenal glands, thyroid gland, and gonads because of the effects of the lack of trophic hormones. This developmental disorder arises because of the failure of Rathke pouch to fuse with the infundibular process (115,117–122). Several genes involved in pituitary development have been implicated in pituitary aplasia and hypoplasia, including *OTX2, LHX4,* and *HESX1* (123).

Pituitary aplasia may be difficult to diagnose, as it requires an examination of the pharyngeal side of the cranial base for adenohypophyseal tissue and the brain for neurohypophyseal tissue. In some cases, the sella turcica anatomy is abnormal and pituitary tissue is found in the osseous canal (124,125). In many cases of congenital absence of the anterior pituitary, posterior pituitary gland tissue may be present.

Pituitary Hypoplasia

Pituitary hypoplasia is the defective or incomplete development of the pituitary gland. It may occur as an isolated entity or in association with various (often overlapping) syndromes, such as septo-optic dysplasia, Kallmann syndrome, anencephaly, holoprosencephaly, Coffin-Siris syndrome, Worster-Drought syndrome, and Hall-Pallister syndrome (126–130). The pathogenesis of congenital pituitary hypoplasia remains largely unknown, with specific genetic abnormalities identified in fewer than 20 percent of cases (131). It has recently been demonstrated that germline loss of function mutations, and less commonly, gene deletions, of the *OTX2* gene may be at least partly responsible for this condition (132,133). Other genes involved in pituitary hypoplasia and related midline anomaly spectrum disorders include *PIT1, LHX3, SHH, HESX1, SOX2,* and *SOX3* (134–137).

Isolated pituitary hypoplasia most commonly affects the GH, thyroid, and prolactin axes. Most patients with pituitary hypoplasia present with short stature. Clinically, these syndromes are known as isolated *GH deficiency* or *combined pituitary hormone deficiency* (138–140). Hypoplasia of the pituitary gland and its related structures can be seen on magnetic resonance imaging (MRI) (141,142). The height of a hypoplastic pituitary gland, when present, is typically less than 3 mm; the normal height is greater than 5 mm (143). In patients with congenital GH deficiency, the triad of an ectopic posterior pituitary, pituitary aplasia/hypoplasia, and stalk defects correlates with the presence of other endocrine abnormalities (144). Pituitary hypoplasia can also be associated with agenesis of the internal carotid artery (145,146).

Hypoplasia of the pituitary gland also commonly occurs in association with anencephaly (147). Anencephaly is associated with a flattened sella turcica filled with spongy vascular tissue (148,149). In most cases, the number of anterior pituitary cells is within normal limits for age, but the number of corticotroph cells and gonadotroph cells is usually reduced (147). Anencephalic fetuses usually have hypoplastic adrenal cortices (150,151), and in males, hypoplastic external genitalia. In contrast, the numbers of other cell types and the weights of their target endocrine organs are within normal limits in anencephalic neonates (147).

Ataxia-Telangiectasia Syndrome

Ataxia-telangiectasia (AT), or *Louis-Bar syndrome,* is a complex autosomal recessive disorder with an incidence of 1 in approximately 30,000 to 100,000 live births (152–154). It is characterized by progressive cerebellar ataxia with onset in early childhood, oculocutaneous telangiectasias, immunodeficiency, susceptibility to infection, hypersensitivity to ionizing radiation and radiomimetic agents, and increased incidence of various neoplasms, primarily B-cell lymphomas and chronic lymphocytic leukemias of T-cell type (155). It is also associated with cytomegaly and nucleomegaly in multiple organ systems. The syndrome results from mutations in the ataxia-telangiectasia (*ATM*) gene (*ATM serine/threonine kinase* in full), which has been localized to the long arm of chromosome 11 (115,155). This gene plays a role in regulating cell cycle checkpoint signaling pathways, genomic stability, and cell response to DNA damage.

Microscopically, the pituitary gland of AT patients exhibits large cells with nucleomegaly and nuclear pleomorphism. These cells are most often immunoreactive with GH or ACTH (155). In the posterior lobe, scattered pituicytes are also enlarged and pleomorphic. DNA analysis shows the bizarre cells are aneuploid, with many nuclei having a DNA content greater than 8N. However, pituitary tumors do not usually develop in these glands, so the bizarre cells should not be considered preneoplastic.

Empty Sella Syndrome

Empty sella syndrome is characterized by a reduction in the volume of the sellar contents and may be of the primary or secondary type. The primary form results from downward herniation of the arachnoid membrane through an incompetent sellar diaphragm, resulting in chronic compression and flattening of the pituitary gland and enlargement of the hypophyseal fossa (156). Cerebrospinal fluid is present and replaces the volume of the compressed pituitary by almost filling the sella turcica. The pituitary gland is spread along the floor of the sella and the pituitary stalk reaches down to the compressed gland. The optic chiasm, which is anterior to the pituitary gland, forms a "V" shape as it descends with the hypothalamus into the sella. An association with intracranial hypertension is seen in approximately 75 percent of cases (157). Based on autopsy studies, nearly 50 percent of adults have a diaphragmatic defect measuring 5 mm or greater, with a fully developed empty sella present in approximately 5 percent (158).

The lesion is essentially an intrasellar arachnoidocele, and usually not associated with pituitary dysfunction. However, partial or panhypopituitarism may be present (159). Endocrine abnormalities are documented in around 19 percent of patients (160). GH deficiency is the most frequent pituitary deficit, both in the adult and pediatric population, occurring in 4 to 57 percent of patients with primary empty sella syndrome (156). Immunohistochemical findings usually correlate with the lack of endocrine abnormalities since all of the anterior pituitary cell types are readily identified. Autoantibodies to anterior pituitary cells have been detected in more than half of patients with primary empty sella syndrome (161).

Secondary empty sella syndrome results from several causes. These include spontaneous infarction of the gland, like that seen in postpartum pituitary infarction (Sheehan syndrome); surgical extirpation of a sellar tumor; hemorrhagic infarction of a PitNET (pituitary apoplexy) (162); and pituitary atrophy after irradiation of the sellar region.

Whatever the cause, empty sella syndrome may result in visual field defects, some of which are caused by chiasmal prolapse, or impaired pituitary function. Hyperprolactinemia may occur as a result of traction on the pituitary stalk (163). Diabetes insipidus is also encountered. With the advent of modern imaging techniques, patients infrequently come to surgery. When they do, the specimen often consists of little more than a fragment of arachnoid membrane (164).

Isolated Hormone Deficiency

Deficiencies of individual anterior pituitary hormones are uncommon. The most common isolated deficiency is *congenital isolated GH deficiency,* with a reported incidence of between 1 in 4,000 and 1 in 10,000 live births (165). Most cases are sporadic and of unknown cause, with a small proportion (3 to 30 percent) being familial (166,167). Short stature is the essential phenotypic feature in isolated GH deficiency. Heterozygous mutations in the *GH1* gene lead to the most common autosomal dominant form of GH deficiency (type II) (168). Other types are associated with mutations in the *GHRHR* (GHRH receptor) gene (165). Interestingly, morphologic and immunohistochemical studies in patients with isolated GH deficiency have shown abundant immunoreactive GH cells, suggesting another mechanism, such as defective functional and biologic activity of the secreted GH.

Congenital functional failure of a single lineage has been reported for all pituitary cell types, giving rise to isolated hormone deficiencies other than isolated GH deficiency. These include isolated TSH deficiency, isolated hypogonadotropic hypogonadism (LH and FSH deficiency), isolated ACTH deficiency, and rarely, isolated PRL deficiency (169). Isolated hypogonadotropic hypogonadism may result from hypothalamic defects in patients with Kallmann syndrome who have hypogonadism and anosmia. Although rare, isolated PRL deficiency clinically manifests only in women as puerperal alactogenesis (170). In isolated TSH deficiency, inadequate thyroid hormone biosynthesis occurs due to defective stimulation of the thyroid gland by TSH, resulting in central, or secondary, hypothyroidism (165). Isolated ACTH deficiency is a very rare and heterogenous condition, making diagnosis challenging due to the varied clinical presentation. It may be lethal due to the hypocortisolism and has also been associated with neonatal hypoglycemia, convulsions, hypercalcemia (171), and cholestasis, which are associated with a 20 percent mortality rate if unrecognized (172,173). Single cell type "knock outs" or loss of a specific cell type can also be found in patients with autoimmune hypophysitis.

Combined Pituitary Hormone Deficiency

Combined pituitary hormone deficiency (CPHD), involvement of more than one anterior pituitary hormone, is associated with severe morbidity and can be life-threatening. The clinical presentation depends on the age, number, and severity of hormone deficiencies. Many findings are nonspecific, especially in the newborn period, mandating a high index of suspicion, particularly in patients with midline defects (174).

Mutations in several genes encoding transcription factors have been implicated in CPHD. Among these are the homeobox expressed in embryonic stem cells 1 *(HESX1)*, LIM homeobox protein 3 and 4 (*LHX3/LHX4*), prophet of PIT1 (*PROP1*), POU domain class 1 transcription factor 1 (*POU1F1*), and zinc finger protein Gli2 *(GLI2)*.

HESX1 is a member of the class of homeobox genes and is one of the earliest markers of pituitary development. It is localized on chromosome 3p14.3 (175). Both heterozygous and homozygous mutations have been identified, with heterozygotes mostly demonstrating incomplete penetrance and milder phenotypes compared with homozygotes (176–179). The functional effects of mutations in *HESX1* vary depending on the specific genetic defect, ranging from isolated GH deficiency, to CPHD, to septo-optic dysplasia (SOD) (136,180).

LHX3 and *LHX4* are LIM homeodomain transcription factors important in early pituitary development and maintenance of mature anterior pituitary cells. The two genes are located on chromosomes 9q34.3 and 1q25.2, respectively (175). Missense and nonsense mutations, small deletions with frameshift, and partial and complete gene deletions of *LHX3* have been identified (181–183). Inheritance of

LHX3 deficiency syndrome is autosomal recessive. As such, all individuals with these mutations are homozygous and have CPHD, but in most cases, display normal ACTH levels. The pituitary gland is commonly hypoplastic. The inheritance pattern of *LHX4* is autosomal dominant (175). Similar to *LHX3* deficiency, patients have CPHD and/or IPHD, and affected patients show anterior pituitary hypoplasia.

PROP1 is a paired-like transcription factor restricted to the developing anterior pituitary. The gene is located on chromosome 5q35.3 and is required for the determination of somatotrophs, lactotrophs, and thyrotrophs, and differentiation of gonadotrophs (175). PROP1 expression precedes and is required for PIT1 (now called POU1F1) expression. In humans, *PROP1* mutations are the leading cause of CPHD, accounting for 30 to 50 percent of familial cases (175). Most patients have a hypoplastic or normal anterior pituitary. However, a hyperplastic pituitary with subsequent involution has also been reported (184,185). Abnormalities in *PROP1* typically result in GH, PRL, TSH, and LH/FSH deficiencies (186). Phenotypic variability, even among patients with the same mutation, has been described. The most consistent feature is short stature; however, normal growth has been reported (187,188).

POU1F1, also known as *PIT1*, encodes a POU domain protein essential to the terminal differentiation and expansion of somatotrophs, lactotrophs, and thyrotrophs. It functions as a transcription factor that regulates the transcription of itself and other pituitary hormones and their receptors, including GH, PRL, TSH beta subunit, TSH receptor, and GHRH receptor (175). *POU1F1* is located on chromosome 3p11 (189). Mutations of *POU1F1* have been shown to be responsible for GH, PRL, and TSH deficiencies. Both autosomal recessive and autosomal dominant inheritance patterns have been reported, and both normal and hypoplastic anterior pituitaries have been observed.

The GLI family of transcription factors is implicated as the mediators of Sonic hedgehog (Shh) signals in vertebrates. In humans, Shh signaling is associated with the forebrain defect holoprosencephaly. The *GLI2* gene is located on chromosome 2q14 (175). *GLI2* deficiency phenotype follows an autosomal dominant inheritance pattern and is associated with anterior pituitary malformations and panhypopituitarism (190).

CIRCULATORY DISORDERS

Pituitary Apoplexy

Pituitary apoplexy is acute hemorrhage and/or infarction of the pituitary gland. Although most often occurring in the setting of a preexisting PitNET, the condition also occurs in nonadenomatous or even normal pituitary glands, especially during pregnancy (191). In rare instances, pituitary apoplexy occurs in the context of therapeutic anticoagulation (192).

Pituitary apoplexy is a rare entity with an incidence ranging from 0.6 to 10.5 percent of patients with a pituitary tumor (193–205). In most cases, pituitary apoplexy is the first presentation of the preexisting pituitary tumor; the estimation of its true incidence is challenging due to diagnostic difficulties (206–209). In a review of 560 cases of pituitary tumors, Wakai et al. (209) estimated the incidence of pituitary apoplexy as 17 percent, symptomatic in 9 percent of the cases and asymptomatic in 8 percent of the cases. More recently, Charalampaki et al. (210) reported only 1 case of pituitary apoplexy in a series of 150 patients who underwent endoscopic pituitary surgery. In some autopsy series, infarctions involving 25 percent or more of the gland have been observed in up to 8 percent of patients (211–214).

Most patients with pituitary apoplexy are asymptomatic. In symptomatic patients, the clinical presentation is characterized by a sudden onset of headache, vomiting, decreased level of consciousness, mild or severe hormonal dysfunction, partial or complete ophthalmoplegia, and visual impairment from involvement of the optic nerve and/or chiasm (191). Because the pituitary gland has tremendous reserve, clinical hypopituitarism only occurs after 70 to 75 percent of the anterior pituitary is destroyed, and complete pituitary failure requires loss of more than 90 percent of pituitary tissue (191,211,212,215).

At presentation, most patients with apoplexy (about 80 percent) have a deficiency of one or more pituitary hormones (194,202,208). Changes in the serum levels of pituitary hormones can assist in the diagnosis. However, most patients with pituitary apoplexy harbor macroadenomas and many of them have preexisting pituitary insufficiency prior to the acute apoplexy episode.

Histologically, apoplexy is characterized by hemorrhagic and/or ischemic necrosis within the anterior pituitary gland. In a review of 324 PitNETs, Chacko et al. (216) found that 12 percent of patients had surgical or histopathologic evidence of hemorrhage, with or without necrosis.

Sheehan Syndrome

Sheehan syndrome refers to postpartum hypopituitarism due to pituitary necrosis occurring during severe hypotension or shock secondary to massive bleeding at or just after delivery (217). The frequency of Sheehan syndrome has gradually decreased, particularly in developed countries, because of improved obstetric care, including treatment of hemodynamic complications with rapid blood transfusion or fluid replacement (218).

The pathogenesis of Sheehan syndrome is not entirely certain, although there is no doubt that the basic process is infarction secondary to the arrest of blood flow to the anterior lobe of the pituitary gland. Whether this process results from vasospasm, thrombosis, or vascular compression is unclear (219). It has been suggested that the enlarged pituitary gland during pregnancy is more susceptible to ischemic necrosis after massive postpartum bleeding (220). However, in rare cases, Sheehan syndrome develops without any obvious postpartum bleeding (221,222).

The role of autoimmunity in the development of Sheehan syndrome is controversial. Several studies have demonstrated anti-pituitary autoantibodies in a subset of patients with Sheehan syndrome (223–225). In contrast, no anti-pituitary antibodies were detected in six patients with Sheehan syndrome in another study (226). Whether the antipituitary antibodies are the cause of postpartum pituitary necrosis or the result of pituitary antigen leakage from necrotic pituitary tissue is not clear. The role of autoimmunity in the pathogenesis of Sheehan syndrome remains to be established.

The resulting hypopituitarism can vary in severity, and the decreased endocrine function can be restricted to one hormone or affect all adenohypophyseal hormones (219,227). Clinical presentation varies and is dependent on the age of the patient, rapidity of onset, nature and causes of the pathologic process, and proportion of affected cells. Complete loss of adenohypophyseal function is life-threatening and requires immediate treatment. The posterior pituitary is usually not affected because of its rich arterial blood supply, which is independent of the portal vasculature.

Histologically, there may be focal necrosis of anterior pituitary cells, or with more extensive infarction, more than 90 percent of the gland may be lost. Normal adenohypophyseal cells may be replaced with fibrocollagenous scar tissue, ghost cells, necrotic debris, an inflammatory infiltrate, and hemosiderin (219).

Miscellaneous Conditions Causing Pituitary Infarction

Pituitary infarction resembling Sheehan syndrome *(Simmonds disease)* develops following severe trauma, massive cerebrovascular accidents, severe thrombocytopenia, disseminated intravascular coagulation, snakebites from the Burmese Russell viper (228), coronary artery bypass surgery (229), gastrointestinal hemorrhage (230), severe burns (231), and cocaine abuse (232).

Exogenous Injury

Hypopituitarism may be caused by various physical injuries, such as traumatic brain injury (TBI) or surgical trauma (233). Although initially thought to be a rare sequela of TBI, hypopituitarism is increasingly recognized in the literature and clinical practice, possibly affecting a third to half of all TBI survivors (233). There are several mechanisms that can cause pituitary dysfunction after TBI, including hemorrhage, raised intracranial pressure, edema, skull fracture, and as a direct insult to the gland (234).

A common finding in post-TBI autopsies is pituitary gland hemorrhage, infarction, and/or stalk laceration (235–237). It is conceivable that disruption of the pituitary stalk is associated with gland ischemia secondary to blood vessel tear or compromise of blood supply through a direct pressure effect or vessel thrombosis. The pituitary gland is extremely vulnerable due to its unique anatomic structure (238): the portal capillaries in the stalk and hypophyseal blood vessels are long and slender, and traverse the sellar diaphragm, making them particularly susceptible to the shearing injuries seen in TBI (234,236).

The incidence of acute post-traumatic hypopituitarism of any anterior pituitary hormonal axis is 50 to 76 percent (239–241), with the most commonly affected hormonal axes being the gonadotropin axis (41.6 percent), the GH axis (20.4 percent), the ACTH axis (9.8 percent), and the thyroid axis (5.8 percent) (241,242). The hormonal axis most often chronically affected in patients with traumatic hypopituitarism is the GH axis. The incidence of posterior pituitary dysfunction (diabetes insipidus) in the acute period following moderate or severe TBI is 21.6 percent; permanent diabetes insipidus develops in 6.9 percent of patients (243,244).

Hypothalamic pituitary (HP) *dysfunction* is frequently observed as a late effect in survivors of central nervous system (CNS) tumors and those with HP exposure to radiotherapy (245). The prevalence of HP dysfunction in a study of 748 adult childhood cancer survivors exposed to cranial radiotherapy and followed for a mean 27.3 years was 51.4 percent, with nearly 11 percent reporting multiple HP disorders (246). GH deficiency is the most frequent HP disorder in patients with a history of HP exposure to radiotherapy (246,247): the overall prevalence of GH deficiency is 12.5 percent in childhood CNS tumor survivors (245). GH deficiency is also reported in childhood cancer survivors who received total body irradiation for hematopoietic stem cell transplantation (248–254), and patients treated with radiotherapy for non-brain solid tumors of the head (255). Studies suggest that cranial irradiation is more likely to cause hypothalamic than pituitary dysfunction (256).

METABOLIC DISORDERS

Amyloid Deposits

Amyloids are highly organized, twisted, beta-pleated sheet fibrils that are fragments of larger protein precursors (257). They are associated with pathologic conditions including Alzheimer disease and type II diabetes (258).

Pituitary amyloid deposits occur in four settings. The gland may be affected in cases of systemic amyloidosis, in which it is one of many affected organs and tissues. Such deposits are present in the interstitium, vessel walls, and capsule of the pituitary gland and may derive from any of the known systemic precursor proteins that cause generalized amyloidosis, including serum amyloid A, transthyretin, beta-2-microglobulin, and kappa and lambda light chains (259–266).

There are also three local or organ-limited patterns of amyloid deposition in the pituitary gland, which occur in the absence of systemic or generalized amyloidosis (263). These patterns include interstitial amyloid deposits in the anterior lobe and amyloid deposits in PitNETs. Both have been interpreted as amyloid of endocrine origin, sharing characteristics with amyloid deposits of nonadenomatous endocrine tissues such as pancreatic islets, and endocrine tumors in other organs, such as pancreatic neuroendocrine tumors and medullary thyroid carcinoma (267). Local or organ-limited interstitial pituitary amyloid is usually considered a senile amyloid syndrome since it is an age-related disorder. Recent studies of senile interstitial amyloid deposits in the pituitary gland have noted glycosaminoglycans, basement membrane proteins, and apolipoprotein E, which are found in other amyloid syndromes such as AA-amyloidosis related to inflammatory conditions and Aβ-amyloidosis related to Alzheimer disease (266). The fourth pattern of pituitary amyloid occurs as intracellular deposits which are not associated with systemic amyloid deposition (265,268,269).

Several types of amyloid deposition have been described in pituitary tumors. The perivascular type is seen in all types of endocrine tumors. It is characterized by a fibrillary or crystalloid microstructure, and deposition around blood vessels (270). The spherical type is rare, occurring exclusively in PRL-secreting tumors and is characterized by coral-like spheroid accumulations with variable diameters (40 to 150 µm) (271,272). Studies have found that amyloid deposition in PitNETs is associated with local conditions, suggesting that the amyloid fibrils are produced by the tumor during degeneration (257,266,273). Bakhtiar et al. (274) suggest that deposition of amyloid may be related to the PRL synthesis process and may be enhanced by a naturally developed degenerative change in the PRL PitNET. Another possibility is that the amyloid is produced by histiocytes, and its accumulation is enhanced by dopamine agonist therapy (272,273,275). Studies have shown that treatment with the dopamine agonist bromocriptine is associated with extensive tumoral fibrosis and amyloid formation (274,276).

Iron Overload

Excessive accumulation of body iron stores can occur in hereditary hemochromatosis or as a result of multiple blood transfusions, prolonged intake of iron-containing supplements, in the setting of chronic liver disease, or from chronic ineffective erythropoiesis (e.g., thalassemia) (277). *Iron toxicity* occurs when iron overload causes nontransferrin-bound iron to accumulate in tissues as free iron, leading to organ dysfunction and damage (278,279). The liver, pancreas, heart, and pituitary gland are the most common sites of iron deposition. Endocrine dysfunction has been reported as the most common and earliest evidence of iron toxicity seen in patients with iron overload (280).

In the pituitary gland, iron deposits mostly in the gonadotroph (FSH/LH) cells of the anterior pituitary, leading to impaired cell function and abnormal hormone secretion (281). As a result, patients present with hypogonadism, with a reported prevalence ranging from 10 to 100 percent of patients with hemochromatosis in various studies (282–289). Clinically, the most common symptoms in men are reduced body hair, loss of libido and impotence, altered fat distribution, and rarely, gynecomastia and testicular atrophy (290). In contrast, panhypopituitarism and single hormone deficiencies besides hypogonadotropic hypogonadism are rare. Dysfunction of lactotroph (286,291) or somatotroph (285,287,291,292) cell secretory capacity has been demonstrated in a few studies in patients with hemochromatosis. In particular, a decreased PRL response to TRH stimulation (286,291) and an abnormal GH response to insulin-induced hypoglycemia (287) or a blunted GH response to GHRH stimulation (285,291,292) have been reported by some researchers, but have not been confirmed by others (282,289).

Miscellaneous Metabolic Conditions

The pituitary gland may be affected by various familial metabolic conditions such as lysosomal storage diseases, including Fabry disease, Hurler syndrome (or gargoylism), and Hunter syndrome, which are associated with abnormal storage of mucopolysaccharides in various tissues, including the anterior pituitary. Histologic examination shows membrane-bound vesicles with concentric osmophilic lamellae known as zebra bodies (293).

HYPOPHYSITIS

Hypophysitis is the collective term for conditions presenting with inflammation of the pituitary gland and infundibulum. Hypophysitis can occur as a primary entity or secondary to a predisposing local or systemic inflammatory condition. In clinical practice and in the literature, the term hypophysitis is used to describe a number of conditions that present with either primary or secondary inflammation of the pituitary gland. Numerous classification systems have evolved to categorize hypophysitis according to etiology, pattern of pituitary involvement, and/or histologic findings. In general, classification is based on etiology, with conditions described as either primary (isolated to the pituitary gland) or secondary (associated with an underlying local or systemic disorder) (294–296).

Primary hypophysitis is the most common form of pituitary inflammation. Three histopathologic conditions are now recognized within this category: lymphocytic hypophysitis, granulomatous hypophysitis, and xanthomatous hypophysitis (294). Secondary causes of hypophysitis include autoimmune conditions (e.g., systemic lupus erythematosus, autoimmune polyglandular syndrome), systemic inflammatory disorders (e.g., sarcoidosis, granulomatosis with polyangiitis, IgG4-related disease), infiltrative lesions (e.g., Langerhans cell histiocytosis, Erdheim-Chester disease), drug induced (e.g., immune checkpoint therapy, interferon alpha), and infection.

Regardless of the etiology, hypophysitis typically results in pituitary hormone deficiencies and enlargement of the pituitary gland, which may result in compression of the optic apparatus and consequent neuro-ophthalmic manifestations (294). Headache, with or without nausea, vomiting, and visual disturbances is the most common presenting symptom in large case studies (297,298). Cranial nerve palsies, cavernous sinus involvement, intracavernous carotid artery occlusion, and presentations mimicking meningitis and apoplexy have also been reported (298–302). The specific pattern of endocrine dysfunction may be influenced by etiology, and is discussed in more detail (later in this chapter).

INFECTIOUS DISEASES

Infections of the hypothalamic-pituitary region are rare, accounting for less than 1 percent of all pituitary lesions (303). They include bacterial infections (including pituitary abscess and tuberculosis), as well as fungal, viral, and parasitic infections. Infections may be confined to the pituitary gland, or occur in association with bacterial meningitis, otitis media, thrombophlebitis of the cavernous sinus, or secondary to a mass lesion (e.g., PiNET, Rathke cleft cyst, craniopharyngioma).

In *bacterial hypophysitis,* the pituitary gland is infiltrated by sheets of polymorphonuclear leukocytes, which may be walled off to form an abscess. Bacteria causing suppurative hypophysitis include *Pneumococcus,* group A *Streptococcus, Staphylococcus aureus, Enterococcus, Neisseria, Escherichia coli,* and others (304–306). The infection can be caused by hematogenous dissemination or by direct extension from surrounding structures. Pituitary surgery and immunosuppression are also risk factors for pituitary abscess (307,308).

Tuberculous meningitis has a tendency to affect basal parts of the brain, from where it can spread to the sellar region. In rare cases, CNS tuberculosis presents as *tuberculous hypophysitis* or a sella/suprasellar tuberculoma mimicking PitNET or pituitary apoplexy (309–311). It may occur in the absence of systemic tuberculosis, but most patients have a history of pulmonary tuberculosis or tuberculosis of other organs. Longstanding tuberculous hypophysitis with extensive scarring can lead to destruction of adenohypophyseal cells and hypopituitarism in some cases.

Syphilis is another well-recognized cause of hypopituitarism, manifesting as *granulomatous hypophysitis,* syphilitic gumma in the sellar region, or congenital syphilis causing hypothalamic-pituitary dysfunction. The infection is caused by the spirochete *Treponema pallidum* and is mostly seen in immunocompromised patients (e.g., human immunodeficiency virus [HIV]-infected patients) with syphilitic meningitis (312).

Fungal infections of the pituitary gland are extremely rare and usually occur in immunocompromised patients. There are only a few case reports of *Candida* and *Aspergillus* abscesses of the sella (313–318). Fungal sellar abscesses may occur in association with widespread CNS infections, such as meningitis, encephalitis, brain abscess, subdural abscess, and mycotic arteritis (319,320).

Hypothalamic-pituitary dysfunction may develop in the acute phase of viral infections of the CNS or in the late stage of these diseases (321–323). Infectious agents that cause CNS viral infections include herpes simplex virus, varicella zoster virus, enterovirus, cytomegalovirus, and neuroborreliosis. These infections mainly occur in immunocompromised patients with encephalitis (such as in the setting of HIV infection, Cushing syndrome, lymphoma, or immunosuppressive therapy) (324–328). Rare cases of *Toxoplasma hypophysitis* have also been reported in immunocompromised patients (329).

AUTOIMMUNE DISORDERS

Lymphocytic Hypophysitis

Definition. *Lymphocytic hypophysitis* is an autoimmune disease in which lymphocytes and plasma cells infiltrate the pituitary gland.

General Features. Despite being the most common chronic inflammatory condition of the pituitary gland, lymphocytic hypophysitis has an incidence of only 1 in 9 million persons per year (330). Although it may occur in both genders, most cases occur in women during the postpartum period (331–339). The condition is frequently associated with Hashimoto thyroiditis, adrenalitis, ovarian failure, atrophic gastritis, and pernicious anemia (340). It may also be associated with systemic lupus erythematosus (341). Antibody-mediated and cellular immune mechanisms appear to mediate the disorder, and antipituitary antibodies, although not specific to hypophysitis, have been demonstrated (342).

Clinical Findings. Though initially thought to occur only in adult women, lymphocytic hypophysitis has been described in both genders and at all ages (343–346). Nevertheless, the majority of cases occur in reproductive-aged women at the end of pregnancy or in the first few months after delivery (330,347). The condition is frequently associated with other autoimmune conditions including thyroiditis and oophoritis (338,340,348,349).

Patients present with symptoms of mass effect due to pituitary gland enlargement and pituitary dysfunction. Headache is the most common presenting symptom, occurring in about half of patients (350). Visual symptoms due to compression of the optic nerves and/or cranial nerves III, IV, and VI in the cavernous sinuses occur in some patients (330,347,351). Cavernous carotid artery occlusion is a rare complication (348,352–354). The onset of symptoms can be insidious, subacute, or acute, even mimicking apoplexy (351,355,356). Most patients have multiple anterior pituitary hormone deficiencies, and panhypopituitarism is common. The severity of hormone deficiencies may appear to be out of proportion to the radiographic findings (350). Serum PRL levels may be low, normal, or elevated (330,332,347,351). In contrast to clinically nonfunctioning PitNETs (357), there is not a clear hierarchy of anterior pituitary hormone deficiencies in patients with hypophysitis. Diabetes insipidus is also common and may occur in up to half of patients (330,332,347,351).

Radiologic findings include homogeneous enhancement of the pituitary, diffuse symmetric gland enlargement, midline stalk thickening, absence of a posterior pituitary bright spot, normal sella size, dural thickening, parasellar T2-weighted hypointensity, and parasellar mucosal thickening (350). However, radiologic findings are not specific for hypophysitis, especially compared to nonadenomatous sellar lesions (339,351).

In recent years, antibodies to pituitary tissue have been identified as a possible antigenic target in lymphocytic hypophysitis, even in patients not presenting with hormone deficiencies (358). Serum HLA DQ8 and DR53 may be useful diagnostic markers as they have been shown to be significantly elevated in patients with lymphocytic hypophysitis (359).

Gross Findings. The pituitary gland is invariably enlarged in this condition, but it usually does not have the soft consistency of a PitNET.

Microscopic Findings. The pituitary gland is diffusely infiltrated by inflammatory cells, predominantly lymphocytes that form lymphoid follicles, with some admixed plasma cells and histiocytes. There may be occasional neutrophils and eosinophils. Oncocytic metaplasia of the adenohypophyseal cells may be present. When the chronic inflammatory process results in focal or diffuse fibrosis, further destruction of anterior pituitary function can result.

Immunohistochemical Findings. Immunohistochemical staining shows a mixture of B- and T-lymphocytes. The B cells are invariably polyclonal. The macrophages may stain for CD68.

Differential Diagnosis. The differential diagnosis for lymphocytic hypophysitis is broad. Alternative diagnostic considerations include pituitary hyperplasia, solid and cystic sellar/suprasellar lesions (such as PitNET with or without apoplexy, Rathke cleft cyst, craniopharyngioma, pituicyte-derived tumors, hamartoma, dermoid or epidermoid cyst, gangliocytoma, lipoma), malignancies (germinoma, lymphoma, glioma, metastatic lesions, Langerhans cell histiocytosis), systemic inflammatory disorders (sarcoidosis, granulomatosis with polyangiitis, Crohn disease), and infections (tuberculosis, syphilis, mycoses).

The findings of hyperprolactinemia and a pituitary mass usually suggest a prolactinoma. In small biopsies, the presence of a mixture of anterior pituitary cells and preservation of the reticulin pattern in early lesions helps to exclude a PitNET. Immunohistochemical staining for pituitary hormones shows a heterogeneous mixed pattern of staining that helps exclude PitNET. Lymphomas of the adenohypophysis are extremely uncommon and are diagnostically excluded by the cytologic features of the lymphocytes and analysis of B and T cells by immunohistochemistry. Other inflammatory conditions, such as granulomatous disorders and sarcoidosis, can usually be excluded by the presence of multinucleated giant cells, which should not be present in lymphocytic hypophysitis.

A condition known as "secondary hypophysitis" in which the pituitary gland shows fibrosis, granulation tissue, and mixed B and T lymphocytes in equal amounts in the presence of a tumor in the sellar region, such as a craniopharyngioma or PitNET, has been reported and should be distinguished from lymphocytic hypophysitis (360). Detailed clinicopathologic studies and careful examination of all the biopsied tissues are needed to rule out this diagnosis.

Xanthomatous hypophysitis is the least common form of primary hypophysitis and may be confused with lymphocytic hypophysitis.

However, the presence of predominantly foamy histiocytes in addition to lymphocytes and some plasma cells allows histologic distinction.

Treatment and Prognosis. Treatment includes replacement of hormone deficiency (including antidiuretic hormone [ADH] if diabetes insipidus develops) and decision-making regarding conservative medical and surgical therapies. Observation and serial computerized imaging are reasonable if a pituitary mass is discovered in a clinical setting consistent with lymphocytic hypophysitis and if the symptoms of a mass (e.g., visual field defects) are absent. High-dose suppressive glucocorticoids remain the cornerstone of medical therapy, but a variety of immunosuppressive treatments have been used (361). Surgery is indicated for nonresponders, mass effect, headache, visual deficit, or when a tissue diagnosis is required. However, even when treatment is initially successful, disease recurrence is common (350). Radiotherapy may be useful when there is relapse of disease, and some patients require a multimodality approach (330,362).

Granulomatous Hypophysitis

Granulomatous hypophysitis is the second most common form of primary hypophysitis (351). Granulomatous inflammation can be seen as a primary entity (idiopathic primary granulomatous hypophysitis) or as a secondary manifestation of systemic conditions such as sarcoidosis, tuberculosis, or granulomatosis with polyangiitis (361). In its primary form, it is still unclear whether granulomatous hypophysitis forms a continuum of the inflammatory process that occurs in lymphocytic hypophysitis (363). The disease has a female preponderance (361) and the clinical presentation tends to be more severe, with a higher incidence of visual symptoms compared to lymphocytic hypophysitis (351,364). Systemic symptoms such as fever, nausea, and vomiting are reported to correlate with earlier presentation (364). In a review of 31 patients with granulomatous hypophysitis, Gutenberg et al. (363) reported that gonadotropin deficiency was present in 100 percent of cases.

Histologically, granulomatous hypophysitis is characterized by the presence of multinucleated giant cells, histiocytes, lymphocytes, and plasma cells in a background of fibrosis (330). Longstanding granulomatous inflammation with extensive scarring may lead to destruction of adenohypophyseal cells and hypopituitarism. The radiologic appearance does not help distinguish primary from secondary forms of granulomatous hypophysitis.

Sarcoidosis can involve the anterior pituitary, although hypothalamic and posterior pituitary involvement is more common. Sarcoidosis can cause destruction of the hypothalamic neurons that produce regulatory peptides. Histologically, sarcoidosis is usually distinguished from other granulomatous inflammatory conditions by its distinctive tight noncaseating granulomas (fig. 1-9) (365). However, special stains for fungi and acid-fast organisms should be performed because sarcoidosis is typically a diagnosis of exclusion.

Occasionally, the pituitary gland is involved by syphilis, which can lead to extensive destruction of the gland (366). Special stains and immunohistochemistry identify the spirochete organisms in these cases.

Xanthomatous Hypophysitis

Xanthomatous hypophysitis is the rarest type of primary hypophysitis, with only 18 cases reported since the entity was first described in 1998 (361,367,368). Similar to granulomatous hypophysitis, it is unclear whether xanthomatous hypophysitis constitutes a distinct entity or if it is an extension of the spectrum of autoimmune or lymphocytic hypophysitis. As for other types of xanthomatous inflammation, it is postulated that the infiltrate results from macrophage activation secondary to chronic inflammation (368). It is more commonly reported in females (369) and the clinical presentation is thought to be similar to lymphocytic hypophysitis, although visual symptoms are rare and the symptoms tend to be milder and present for a longer duration compared to other types of hypophysitis (368,369). Diabetes insipidus is rarely reported with xanthomatous hypophysitis (363).

Radiologically, xanthomatous lesions present as cystic sellar masses on MRI, with enhancement on postgadolinium contrast images. Macroscopically, the lesion appears as a cyst filled with thick orange fluid (368,369). Histologically, the pituitary gland is infiltrated by xanthoma cells or lipid-laden foamy histiocytes that can be highlighted with a CD68 immunostain.

Given the small number of cases, optimal treatment and prognosis of xanthomatous hypophysitis are yet to be established. It has been suggested that the condition may be less responsive to steroid therapy than lymphocytic hypophysitis, and most cases are treated surgically (368). Resolution of the symptoms of mass effect is variable, but endocrine deficiencies seldom recover (370,371).

Anti-Pit1 Antibody Syndrome

Anti-Pit1 antibody syndrome is a recently described entity that presents as an acquired deficiency of GH, TSH, and PRL, with detectable circulating anti-Pit1 antibodies (372,373). The condition is considered to be a cytotoxic T-cell–mediated autoimmune process, although further studies are required to establish the exact pathogenesis (374).

Autoimmune Diabetes Insipidus

Central diabetes insipidus is caused by a failure of secretion and/or release of ADH, or vasopressin, by supraoptic and paraventricular nuclei, in which it is synthesized, or from the posterior pituitary, in which it is stored (375). This ADH deficiency results in the clinical presentation of polyuria and polydipsia. Diabetes insipidus may have various etiologies, including genetic or developmental syndromes, surgical trauma, tumors, infections, inflammatory conditions, or autoimmune diseases, or it may be idiopathic. An autoimmune etiology has now been recognized in most cases previously thought to be idiopathic (376–378).

The diagnosis of autoimmune diabetes insipidus is suggested by the presence of isolated pituitary stalk thickening on MRI, which can disappear over time, after corticosteroid therapy, or spontaneously (379). Histologic examination reveals T-lymphocyte and plasma cell infiltration of the pituitary stalk and posterior pituitary (lymphocytic infundibulo-neurohypophysitis) (379).

The disease is mediated by antibodies to hypothalamic (supraoptic and paraventricular) vasopressin cells (AVPcAb) directed against cytoplasmic and cell surface antigens (380,381). Patients often also have a history of other autoimmune diseases (382–384) and/or organ-specific antibodies (384–386). As with autoimmune adenohypophysitis, patients may have mild hyperprolactinemia, probably secondary to disruption of dopaminergic signals to the anterior pituitary PRL cells.

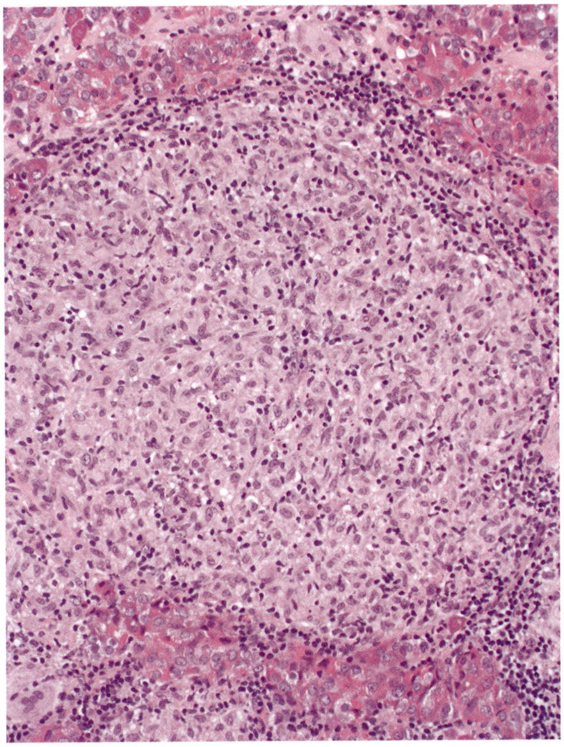

Figure 1-9

GRANULOMATOUS HYPOPHYSITIS

The well-formed compact granuloma is a clue to the diagnosis of sarcoid, despite the presence of background lymphocytic inflammation.

Autoimmune Polyglandular Syndromes

The *autoimmune polyglandular syndromes* (APS) are defined by the concurrence of at least two autoimmune-mediated endocrinopathies. There are two main types of APS. Type 1 (also called *juvenile APS*) consists of chronic candidiasis associated with chronic hypoparathyroidism and Addison disease, and is related to mutations in the autoimmune regulator (*AIRE*) gene (387). Type 2 (also called *adult APS*) includes any combination of at least two of the following: adrenal insufficiency, hyperthyroidism or hypothyroidism, type 1 diabetes mellitus, and primary hypogonadism. Several nonendocrine disorders also occur in patients with type 2 APS, such as myasthenia gravis, celiac disease, pernicious anemia, vitiligo, and alopecia. Type 2

APS is related to certain class II HLA haplotypes (388–391).

Pituitary function may be affected in patients with APS. Pituitary failure due to lymphocytic hypophysitis is reported in 5 to 58 percent of patients with type 1 APS (387). Similarly, lymphocytic hypophysitis and antipituitary antibodies are often reported in patients with autoimmune thyroid disease, which may occur in the context of type 2 APS (392). Moreover, hypofunction of the various endocrine organs may result in compensatory pituitary hyperplasia of various cell types, producing specific trophic hormones. These syndromes are discussed further in chapter 5.

IgG4-RELATED DISEASE

IgG4-related hypophysitis was first reported in 2004 (394). Although it is believed to be rare, retrospective histologic reviews have demonstrated that a significant proportion of cases previously diagnosed as lymphocytic hypophysitis may in fact be IgG4-related hypophysitis (395). While the exact etiology remains unclear, an association with autoimmune and atopic disorders has been reported (396,397).

IgG4-related disease occurs more frequently in men, with a peak incidence in the seventh decade of life (397–399). It presents clinically as mass-forming lesions in multiple organs (399–401), which show diffusely elevated uptake on fluorodeoxyglucose-positron emission tomography (FDG-PET) imaging (402). The most common coexistent IgG4-related pathology associated with pituitary disease is retroperitoneal fibrosis (403).

The histologic diagnosis of IgG4-related disease requires two out of the three following features: dense lymphoplasmacytic infiltrate, storiform fibrosis, and obliterative phlebitis (395,404). Specific diagnostic criteria have been proposed for IgG4-related hypophysitis. These include: 1) pituitary histopathology demonstrating mononuclear infiltration with greater than 10 IgG4-positive cells per high-power field; 2) MRI showing a sellar mass and/or stalk thickening plus biopsy-proven IgG4-related disease at another tissue site; or 3) sellar mass and/or stalk thickening plus a serum IgG4 level greater than 140 mg/dL and a radiologic and clinical response to treatment with glucocorticoids (401). However, recent studies have shown that neither serum IgG4 levels (405,406) nor IgG4-positive tissue staining (407–410) is necessarily sensitive nor specific for IgG4-related disease. As such, diagnosis is generally based upon the identification of two of the three aforementioned histologic features, with appropriate clinicopathologic correlation.

Steroids form the mainstay of treatment for IgG4-related hypophysitis. Nevertheless, despite a beneficial early response, recovery of endocrine function is thought to be uncommon (403). There is a potential role for the use of B-cell–depleting therapies (411,412).

LANGERHANS CELL HISTIOCYTOSIS

Definition. *Langerhans cell histiocytosis* (LCH) is a clonal proliferation of antigen-presenting dendritic cells.

General Features. LCH is a rare disease with an incidence of 1 to 2 cases per million adults (413, 414), although it is more common in the pediatric population. Historically, LCH was designated as histiocytosis X and was divided into three groups: Letterer-Siwe disease, Hand-Schüller-Christian disease, and eosinophilic granuloma (415). Although it was initially thought to be an inflammatory process, there is clear evidence that the neoplastic cells are clonal antigen-presenting dendritic cells and express human leukocyte antigen (HLA)-DR and CD1a (416).

The disease phenotype is heterogeneous, varying from single bone lesions to multisystem disease with organ failure and substantial mortality, often due to concurrent development of hemophagocytic lymphohistiocytosis (417,418). Liver, bone marrow, and/or splenic involvement impart a worse prognosis, as does a failure to respond to first-line therapy (419). There is a predilection for involvement of the hypothalamic-pituitary axis, with diabetes insipidus often presenting as one of the earliest disease manifestations (414). Anterior pituitary deficiencies are less common, but have been reported in approximately 20 percent of patients (418). GH deficiency is most common, followed by gonadotropin deficiency (420,421). PRL elevation is a possible result of stalk involvement (422). Patients with multisystem involvement are more likely to have diabetes insipidus associated with multiple endocrine insufficiencies.

Microscopic Findings. The Langerhans cells are characterized by abundant pale pink cytoplasm and large, folded nuclei with fine chromatin and indistinct nucleoli. The neoplastic cells are associated with giant cells, lymphocytes, plasma cells, variable numbers of eosinophils, and reactive astrocytes. The Langerhans cells are immunoreactive for CD1a, S-100 protein, and langerin (CD207). Ultrastructural studies often show cytoplasmic Birbeck granules (417).

Differential Diagnosis. LCH may mimic granulomatous lesions such as sarcoidosis and granulomatous hypophysitis, as well as other histiocytoses such as Rosai-Dorfman disease, Erdheim-Chester disease, and juvenile xanthogranulomatous disease (415,423). These conditions may also present with diabetes insipidus and loss of regulation of hypothalamic-releasing and -inhibitory hormones. The presence of eosinophils as well as histiocytes and lymphocytes, together with immunoreactivity for S-100 protein, CD1a, and langerin distinguish LCH from other granulomatous diseases and histiocytoses.

PITUITARY CYSTS

The sellar region is affected by a variety of cystic lesions that may be intrasellar or suprasellar in location. Many of these lesions produce no specific clinical symptoms and are discovered incidentally during autopsy or surgery for other conditions. Occasionally, cystic lesions are large enough to cause local symptoms. In these cases, the signs and symptoms are dominated by mass effects on the optic apparatus and pituitary gland, resulting in headaches, visual loss, and pituitary dysfunction. The most common endocrine abnormalities are gonadotropin deficiency, GH deficiency, and hyperprolactinemia (424,425). In most cases, however, the cysts do not produce any pituitary hormones and are not associated with endocrine symptoms or alterations in blood hormone levels.

Rathke Cleft Cyst

Definition. *Rathke cleft cyst* is an intrasellar or suprasellar cyst caused by failure of regression of the Rathke cleft during embryologic development.

General Features. Rathke cleft cysts are the most common cystic lesions of the pituitary gland (426–429). They are benign lesions that typically arise between the anterior and posterior lobes of the gland (430). They are remnants of the Rathke pouch, a structure of ectodermal origin formed during the fourth week of gestation (430). The Rathke pouch extends caudally to fuse with the infundibulum around the eighth week of gestation, forming the craniopharyngeal duct (431). During this time, the Rathke cleft is formed in the region of the pars intermedia. Failure of this cleft to regress during further development can lead to cystic dilatation and the formation of a Rathke cleft cyst (430,432). On rare occasions, an associated PiNET is identified, raising the possibility of a true transitional cell neoplasm from the hypophyseal duct. Despite their relatively high prevalence, Rathke cleft cysts only produce clinical symptoms in a minority of patients, comprising only 2 to 9 percent of all sellar lesions undergoing surgical resection (432–437).

Clinical Findings. Most patients are asymptomatic, and the cysts are incidental findings in 4 to 33 percent of autopsies (438–441). However, larger lesions, especially those greater than 1 cm in diameter, may cause mass effect on surrounding structures such as the pituitary gland and optic chiasm, leading to headache, pituitary dysfunction, or visual disturbance (432,433,442). Diabetes insipidus and hyperprolactinemia are common endocrine findings. Rarely, patients present with chemical meningitis, abscess, inflammatory hypophysitis, or intracystic hemorrhage and apoplexy (433,443).

Rathke cleft cysts show a slight female prepominance and typically present in the fifth and sixth decades of life (432,433,442). Uncommonly, Rathke cleft cysts present in children, potentially resulting in symptoms related to mass effect and gonadotropin deficiency (444).

On MRI, Rathke cleft cysts often appear as well-circumscribed, centrally located, spherical or ovoid lesions with a diameter ranging from 5 to 40 mm (mean diameter, 17 mm) (425,433,445). The normal pituitary gland may be displaced in any direction, including circumferentially if the cyst arises within and remains encased by the gland (446,447). In most cases, administration of gadolinium contrast demonstrates little or no enhancement of the cyst wall or contents on MRI (442,448,449). A thin peripheral rim of enhancement has been attributed to inflammation or squamous metaplasia of the cyst wall, or to a circumferential rim of displaced pituitary gland (433,446).

The cyst contents demonstrate high variability in signal intensity on T1 and T2 sequences, which has been reported to correlate with the nature of the cyst contents (450–452). Cysts lined by a simple epithelium which contain clear fluid may have a homogeneous density similar to cerebrospinal fluid (CSF). However, up to 40 percent of cysts are filled with mucus or more solid cellular debris, which typically fails to enhance following contrast administration and produces more heterogeneous signal characteristics (446,448,449).

Gross Findings. The cysts are thin-walled structures with marked variation in their contents, ranging from clear (CSF-like) to yellow, and mucoid to hemorrhagic. The cyst wall may be difficult to find in small biopsy specimens and any adherent pituitary tissue should be carefully examined for a cyst lining.

Microscopic Findings. The cysts are lined by simple columnar or cuboidal epithelium, often with ciliated and/or mucinous goblet cells (fig. 1-10). Pseudostratified columnar cells are sometimes observed. Occasional adenohypophyseal cells are also found in the epithelial lining or within the cyst wall. Squamous metaplasia and xanthomatous changes are seen, particularly when there is also evidence of old hemorrhage and inflammation. Squamous metaplasia is reported in 9 to 39 percent of cases and is associated with higher rates of cyst recurrence (432,433,453). On occasion, xanthomatous change may totally obscure the epithelial cyst lining (453).

Cytologic Findings. Cytologic preparations of the cyst fluid may show ciliated columnar cells and goblet cells, often in a myxoid or mucoid background.

Immunohistochemical Findings. The epithelial lining cells are positive for cytokeratins (including CK8 and CK20), epithelial membrane antigen (EMA), and carcinoembryonic antigen (CEA). Some of the cells are positive for glial fibrillary acidic protein (GFAP). Anterior pituitary cells may be present in the cyst wall and these stain for their hormonal products, most commonly ACTH.

Ultrastructural Findings. Epithelial cells with cilia and goblet cells with apical secretory granules and well-developed rough endoplasmic reticulum are present on ultrastructural examination (427). In areas with squamous metaplasia, the goblet cells and ciliated cells are replaced by flattened cells with abundant intracytoplasmic tonofilaments and well-developed desmosomes.

Differential Diagnosis. The differential diagnosis includes craniopharyngioma, cystic PitNET, arachnoid cyst, and epidermoid cyst. Squamous metaplasia may lead to an erroneous diagnosis of papillary craniopharyngioma, particularly in small biopsy specimens. Generous sampling may be required to identify the typical cuboidal or columnar epithelium of Rathke cleft cyst. Immunohistochemistry or molecular testing for the *BRAF* V600E mutation can also aid in this differential diagnosis, as most cases of papillary craniopharyngioma are positive for this mutation (454). The absence of calcification and lack of nuclear expression of beta-catenin in Rathke cleft cyst helps exclude an adamantinomatous craniopharyngioma. Expression of CK8 and CK20 in Rathke cleft cyst but not in craniopharyngioma also aids in differentiating these two entities (456).

Dermoid and epidermoid cysts also arise in the midline in the sellar and parasellar region. Their imaging characteristics are nonspecific and their differentiation from Rathke cleft cyst may be difficult. Dermoid cysts usually contain fat components and are heterogeneous on T1 images and hyperintense on T2 images (457–459). Epidermoid cysts contain keratin and are almost identical in appearance to CSF with no contrast enhancement (459). Similarly, arachnoid cysts are isointense to CSF on T1-weighted and T2-weighted sequences and are not enhanced by gadolinium (425).

Treatment and Prognosis. Most patients with a Rathke cleft cyst remain asymptomatic, do not require surgical management, and can be monitored with serial imaging studies. Endoscopic transsphenoidal operations are now widely used to fenestrate and drain Rathke cleft cysts. Complete resection of the cyst wall is not typically recommended, as it is associated with a higher incidence of postoperative diabetes insipidus (442). Previous studies have demonstrated high rates (over 90 percent) of complete resolution of the cyst following surgical drainage (432,442,460). Headache and visual symptoms frequently improve, but hypopituitarism

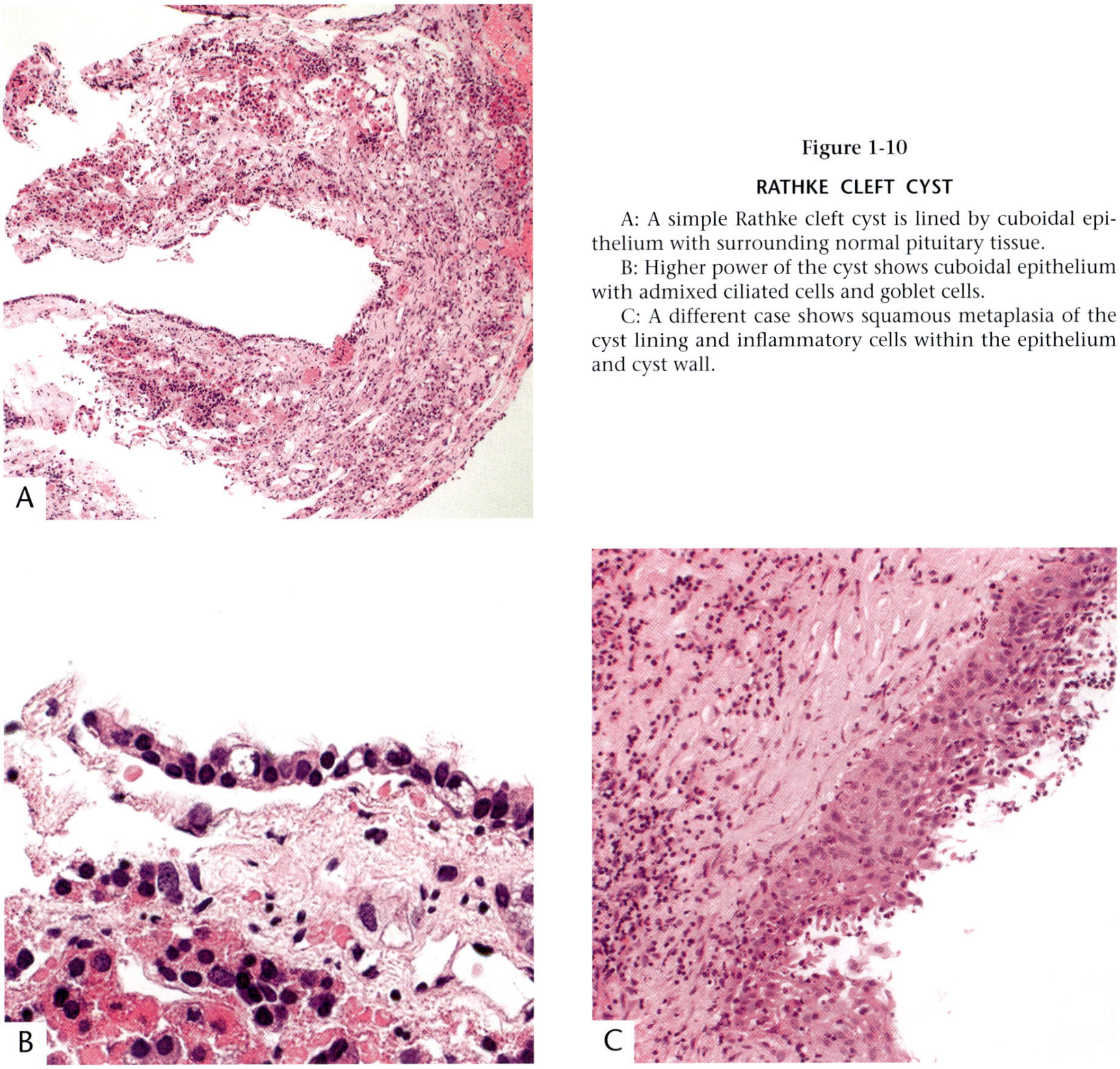

Figure 1-10

RATHKE CLEFT CYST

A: A simple Rathke cleft cyst is lined by cuboidal epithelium with surrounding normal pituitary tissue.

B: Higher power of the cyst shows cuboidal epithelium with admixed ciliated cells and goblet cells.

C: A different case shows squamous metaplasia of the cyst lining and inflammatory cells within the epithelium and cyst wall.

is less likely to improve following drainage (461). Abscess formation is a rare complication of Rathke cleft cyst (462). Long-term recurrence rates following fenestration or resection of Rathke cleft cysts have ranged from 3 to 33 percent, correlating with several factors such as radiologic enhancement pattern, presence of squamous metaplasia, and surgical technique (428,432,442,463,464). Serial imaging is recommended on a yearly basis following surgery.

Epidermoid Cyst

Within the cranial cavity, *epidermoid cysts* often arise in a paramedian location, most commonly in the region of the cerebellopontine angle (464,465). Occasionally, they present as suprasellar or parasellar lesions (466), where they typically occur in middle-aged patients with symptoms of mass effect, including headache, visual loss, and endocrine dysfunction (467–469). The cyst contents may be caustic to the surrounding tissue, and in some cases, cyst ruptures results in chemical meningitis, hemorrhage, seizure, neurologic deficits, and/or psychosis (470–475).

Histologically, epidermoid cysts typically exhibit a thin layer of squamous epithelium with a keratohyaline granular layer and flaky keratin formation, resting on a fibrocollagenous cyst wall. Rupture of the cyst may cause an exuberant inflammatory reaction in the surrounding parenchyma, resulting in secondary hypophysitis and meningitis.

Dermoid Cyst

Dermoid cysts typically arise in the midline (476). The most common intracranial location is in the posterior fossa, although they occasionally present as suprasellar or parasellar lesions (466). Their clinical presentation is essentially identical to that of epidermoid cysts. They are lined by keratinizing stratified squamous epithelium and contain skin adnexal structures such as hair follicles, sweat glands, and sebaceous glands. The cyst wall may contain lymphocytes, macrophages, and occasional foreign body type giant cells.

Arachnoid Cyst

Arachnoid cysts are developmental anomalies that can originate anywhere in the central nervous system. They are typically asymptomatic and rarely require surgical intervention. Among major series of transsphenoidal surgery cases, arachnoid cysts comprise 0.22 to 0.8 percent of lesions (434,477,478). The mean age of presentation is 47 years (477). The clinical presentation typically includes headache, amenorrhea or galactorrhea, anterior pituitary insufficiency, and/or visual disturbances (425,477,479,480). Hyperprolactinemia due to stalk effect is a common laboratory finding (477,480).

The histologic appearance of an arachnoid cyst wall is similar to that of normal arachnoid membrane. The cysts are lined by a single layer of flat to cuboidal cells resting on a delicate fibrovascular stroma. Unlike Rathke cleft cysts, the cells are devoid of cilia and other epithelial-like properties. The cells are immunoreactive for GFAP, S-100 protein, and CEA (481).

Colloid Cyst

Colloid cysts are benign lesions typically (99 percent) arising in the third ventricle near the foramen of Monro (482). In rare cases, they arise in the sellar or suprasellar region. They comprise less than 0.5 percent of surgically treated sellar lesions (434,483). They often arise in the region of the pars intermedia, located between the anterior and posterior lobes, and may be challenging to differentiate from Rathke cleft cysts (482). They can present clinically with signs of local mass effect, including headache, visual field defects, or hypopituitarism (484,485).

Colloid cysts are typically filled with a white or yellow thick colloid (mucopolysaccharide) substance (482). Areas of old hemorrhage and cholesterol accumulation may give the cyst contents a denser consistency. Calcification may be present.

Histologically, the cysts are lined by a single layer of cuboidal or columnar epithelium, with interspersed ciliated and goblet cells. They are often indistinguishable from Rathke cleft cysts and many consider them to be the same entity (486). However, Rathke cleft cysts often show squamous metaplasia, a finding never observed in colloid cysts (487). Moreover, colloid cysts may express EMA, but are consistently negative for cytokeratins and GFAP, which are expressed in Rathke cleft cysts (487,488).

PITUITARY HYPERPLASIA

Pituitary hyperplasia is defined as a non-neoplastic increase in one or more functionally distinct types of pituitary cells (489,490). The acini, although expanded, remain intact (489,491,492). This numerical increase in the number of cells may be physiologic or pathologic and may or may not be associated with an increase in hormone quantity (493). The hyperplastic cells, by definition, are polyclonal in nature (490) and the condition is often reversible if target hormone treatment is initiated. The hyperplasia may be: 1) idiopathic (*primary hyperplasia*); 2) caused by end-organ failure and lack of negative feedback stimulation (*secondary hyperplasia*); or 3) caused by direct stimulation by excess hypothalamic hormone production or by ectopic hormone production from neoplasms such as neuroendocrine tumors of the lungs, pancreas, and other sites (*tertiary hyperplasia*) (494,495). There are rare reports of pituitary hyperplasia caused by excess hormone production by gangliocytic hamartomas of the hypothalamus (496–500). These lesions most commonly produce GHRH and GnRH.

Surgical management is rarely indicated for pituitary hyperplasia and, in most cases, the diagnosis is made by clinical, hormonal, and radiologic findings. Classic features on MRI include diffuse and symmetrical pituitary enlargement, isointensity to gray matter, and homogeneous gadolinium uptake (495). Histologic hallmarks

Table 1-4

DISTINCTION OF PITUITARY HYPERPLASIA FROM PitNET

	Hyperplasia	PitNET
Reticulin stain	Preserved but expanded acini	Disruption of the acini
Microscopic appearance	Mixed diffuse and nodular pattern	Usually diffuse proliferation
Normal pituitary	Involved in hyperplastic processes	Compressed by expanding tumor
Immunohistochemistry	Generally weak staining for hormone	Variable but generally strong
Electron microscopy	Evidence of hyperactivity, well-developed RER and Golgi, decreased numbers of secretory granules	Variable patterns of RER and Golgi, cells usually well granulated

of pituitary hyperplasia include a polymorphous hypercellular population with enlarged acini and an intact reticulin network (Table 1-4, fig. 1-11) (493). The hyperplasia can be diffuse or nodular, unifocal or multifocal (489–492). Immunohistochemical staining can be used to identify the specific cell type undergoing hyperplasia.

ACTH Cell Hyperplasia

Primary Hyperplasia. Primary ACTH cell hyperplasia results from excessive secretion of CRH or is idiopathic (502–504). The hyperplasia is predominantly nodular but may be diffuse if CRH is stimulated. Ectopic ACTH and/or CRH secretion, causing Cushing syndrome, has been reported across a wide age range and from a wide variety of tumors from different organs (505–507). Tumors may be only a few millimeters in diameter.

The ectopic ACTH syndrome can be broadly classified into "overt," where the source of ACTH is clear on initial diagnostic workup (for example, small cell lung carcinoma); "covert," where the source is not apparent initially but following repeated investigation is finally disclosed; and "occult," where the source of the ACTH is not apparent (508). Neuroendocrine tumors causing the ectopic ACTH syndrome, most frequently bronchial carcinoid tumors, show a molecular phenotype close to that of corticotroph PitNETs, and may mimic Cushing disease in clinical and biochemical features. In pituitary hyperplasia due to ectopic hormone production, the adrenal glands become markedly enlarged and the combined weight may be up to 30 g.

Morphologically, ACTH cell hyperplasia is nearly always nodular (490,491). The hyperplastic cells are smaller than normal, amphophilic, PAS positive, and may display mild Crooke hyaline change (490,491).

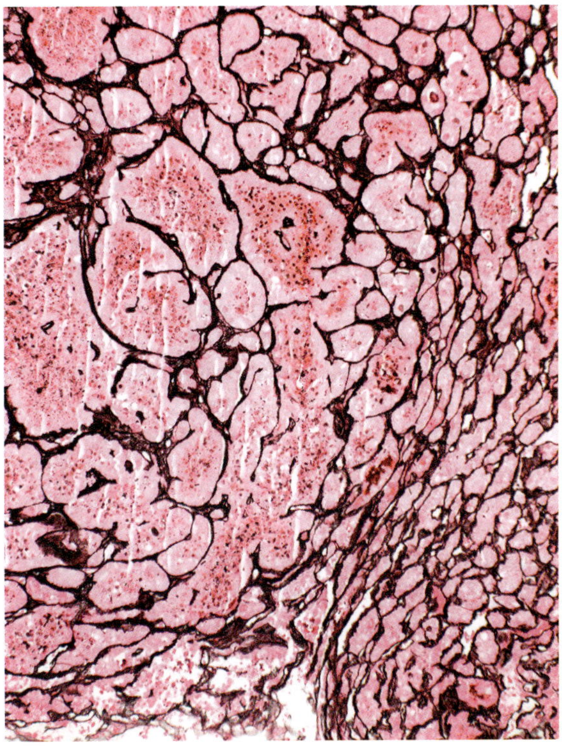

Figure 1-11

PITUITARY HYPERPLASIA

Hyperplasia is associated with expansion in the size of the acinar nests. Importantly, the reticulin framework is not disrupted.

ACTH cell hyperplasia may be difficult to diagnose in small biopsy specimens because of the variable pattern and distribution of ACTH cells in the normal pituitary gland. Because most ACTH cells are present in the mucoid wedge, if the biopsy is from this area, the pathologist should be cautious when making a diagnosis of hyperplasia. Biopsies from the pars nervosa may show basophil invasion with abundant ACTH-positive cells (490,491,509). ACTH cell hyperplasia and PitNETs should not be diagnosed in a biopsy showing only basophil invasion.

Pars intermedia-derived POMC cells also undergo hyperplasia, which typically occurs near the anterior rim, at the median wedge junction with the lateral wings, and in the posterior part of the median wedge (493). This is usually an incidental finding that is common in old age and may give rise to silent corticotroph PitNET, which must be distinguished from corticotroph hyperplasia (509).

Secondary Hyperplasia. Secondary ACTH hyperplasia occurs in longstanding Addison disease (493). In rats, postbilateral adrenalectomy there is an initial decline in the number of corticotrophs, which may be artifactual due to degranulation resulting from ACTH release. This is followed by regranulation and an increase in both cell size and number (492). The possible mechanisms are proliferation of the corticotrophs, and/or proliferation and differentiation of uncommitted cells (494). Transdifferentiation from other differentiated cells, although possible, is less likely since the upstream corticotroph transcription element is corticotroph specific (492). In a study of 18 patients with Addison disease, the degree of diffuse and nodular hyperplasia of the ACTH cells was found to be related to the duration of the disease (510). In rare instances, ACTH-producing PitNETs develop (511–513). An increased number of TSH cells have been noted in the pituitary glands of patients with Addison disease, which may have an immunologic etiology when the adrenal and thyroid glands become atrophic (Schmidt syndrome) (510).

GH Cell Hyperplasia

Primary Hyperplasia. Primary GH cell hyperplasia is associated with excess GHRH production from ectopic tumor-derived sources such as pheochromocytoma or various carcinoid tumors (514). Hypothalamic GHRH excess due to hypothalamic gangliocytoma or hypothalamic regulatory defects, as in cases of congenital gigantism or acromegaly, is extremely rare (491,492,515,516). Somatotroph hyperplasia is mostly of the diffuse morphologic type, with some nodularity.

The hyperplastic cells are strongly acidophilic and GH immunoreactive, although some are also PRL immunoreactive. Ultrastructural studies show densely granulated GH cells with larger secretory granules and a very prominent Golgi complex (502). Mammosomatotroph hyperplasia is rare and is associated with congenital gigantism and McCune-Albright syndrome (490,491). Although PitNET is not a frequent consequence of hyperplasia in most settings, there is a higher risk of neoplastic transformation of somatotroph hyperplasia in the settings of McCune-Albright syndrome and Carney complex (514).

Secondary Hyperplasia. Secondary GH cell hyperplasia usually does not occur since there is no unique target organ failure associated with GH secretion. This is supported by the finding that patients with Laron dwarfism, characterized by GH resistance resulting in low levels of IGF1, have no evidence of pituitary hyperplasia, suggesting that somatotrophs do not proliferate in the absence of negative feedback (517).

TSH Cell Hyperplasia

Primary TSH cell hyperplasia is exceedingly rare. The enlarged cells associated with thyroid deficiency contain abundant dilated rough endoplasmic reticulum and prominent Golgi complexes with many vesicles and are known as "thyroidectomy" cells. The secretory granules in primary and secondary TSH cell hyperplasia remain small, ranging from 150 to 250 nm in diameter (502). With longstanding hypothyroidism, there is weak immunoreactivity for TSH, associated with a decline in the numbers of cytoplasmic secretory granules.

Secondary TSH cell hyperplasia is a well-documented consequence of primary hypothyroidism, typically developing in women in the third or fourth decade of life (490,492,518–530). This response is usually reversible with the institution of appropriate thyroxine replacement therapy (490,491,518,519,522–524,530). TRH is thought to be largely responsible for initiating both thyrotroph proliferation and possibly transdifferentiation from lactotrophs and somatotrophs (491,531). In patients with secondary TSH cell hyperplasia, the pituitary gland often shows a nodular enlargement, corresponding to the numerical increase in TSH cells. The hyperplastic cells are large, ovoid, and pale, with vacuolated cytoplasm containing several lysosomes and variable TSH reactivity (490,491,519).

Morphologically, most cases of TSH cell hyperplasia are of the nodular type but some are diffuse or multifocal (490,491). In some cases, the hyperplasia is significant enough to mimic a tumor and compress adjacent structures (524,528,530,532). Concomitant hyperprolactinemia and lactotroph hyperplasia is common in cases of severe thyrotroph hyperplasia, frequently resulting in a misdiagnosis of prolactinoma (525,527). In cases of massive TSH cell hyperplasia there may be a decline in the number of somatotrophs (491). Both the transformation of thyrotroph hyperplasia into and its coexistence with thyrotroph PitNET have been reported (490).

Gonadotroph Hyperplasia

Secondary gonadotroph hyperplasia is a rare but well-documented response to primary hypogonadism of any etiology (506). It has been reported in response to castration in animal models and in humans with Klinefelter and Turner syndromes (490–492,534). It has also been reported in cases of secondary hypergonadotropic ovarian failure (535,536), in polycystic ovarian syndrome (537), and around the time of menopause (538).

Morphologically, gonadotroph hyperplasia tends to be of the diffuse type (490,491). The hyperplastic cells are large, ovoid, and vacuolated, with an eccentric nucleus, occasionally yielding a signet ring appearance (491,493). This cellular change is readily documented in gonadal dysgenesis (539), but it is uncommon following gonadectomy and rare in postmenopausal women (533). Ultrastructural studies show proliferation and dilatation of the rough endoplasmic reticulum, increased prominence of the Golgi complex, and gradual reduction in the numbers of secretory granules (502). The diagnosis of gonadotroph hyperplasia can be difficult to make due to the scattered distribution of cells in the pars distalis and the wide range of normal gonadotroph content and variation due to age and gender (490,491).

PRL Cell Hyperplasia

Physiologic hyperplasia of PRL cells during pregnancy and lactation has been discussed earlier.

Primary Hyperplasia. Idiopathic lactotroph (PRL) hyperplasia is rare but well documented; it may lead to hyperprolactinemia (490,491, 537,540). Nongestational lactotroph hyperplasia is morphologically either nodular, diffuse, or rarely, both in the same patient (490,491). The hyperplastic cells are large, with abundant chromophobic to slightly acidophilic cytoplasm that is PAS negative (490). Ultrastructural studies show sparsely granulated PRL cells with a well-developed rough endoplasmic reticulum and large Golgi complexes. Secretory granules are sparse, and there are few examples of reverse exocytosis or abnormal extrusion of secretory granules, a characteristic feature of PRL cells (502). The hyperplastic cells originate from lactotrophs and recruited (transdifferentiated) somatotrophs (493).

Secondary Hyperplasia. Lactotroph hyperplasia can occur with exogenous estrogen administration, pituitary space-occupying lesions due to stalk section effect resulting in loss of dopaminergic inhibition on the lactotrophs, in conjunction with corticotroph PitNET probably due to increased endorphin production, and in some patients with prolonged hypothyroidism probably as a consequence of elevated TRH (490–492,525). It has also been reported in patients treated with antipsychotics (541).

Differential Diagnosis of Pituitary Hyperplasia

The differential diagnosis of pituitary hyperplasia includes normal pituitary gland and PitNET (figs. 1-2, 1-12). The diffuse morphologic type is more likely to be confused with normal pituitary gland, while the nodular type can mimic PitNET. The hallmark of diffuse pituitary hyperplasia is the increased cell number of mostly one subtype of pituitary cell accompanied by variable enlargement of acini.

Pituitary hyperplasia is distinguished from PitNET using the features outlined in Table 1-4. Some key features of hyperplasia that are not seen in PitNET include: lack of disruption of the reticulin network, although this might occur focally in severe forms; hyperplastic mass of polymorphous cells; indistinctly demarcated area of hyperplasia; nonevident compression of adjacent structures; and formation of a pseudocapsule. In patients with idiopathic anterior pituitary hyperplasia not associated with excessive hormone production, multiple techniques, including imaging studies, H&E stains, reticulin stains, immunohistochemistry, and ultrastructural analyses, may be needed to establish the correct diagnosis.

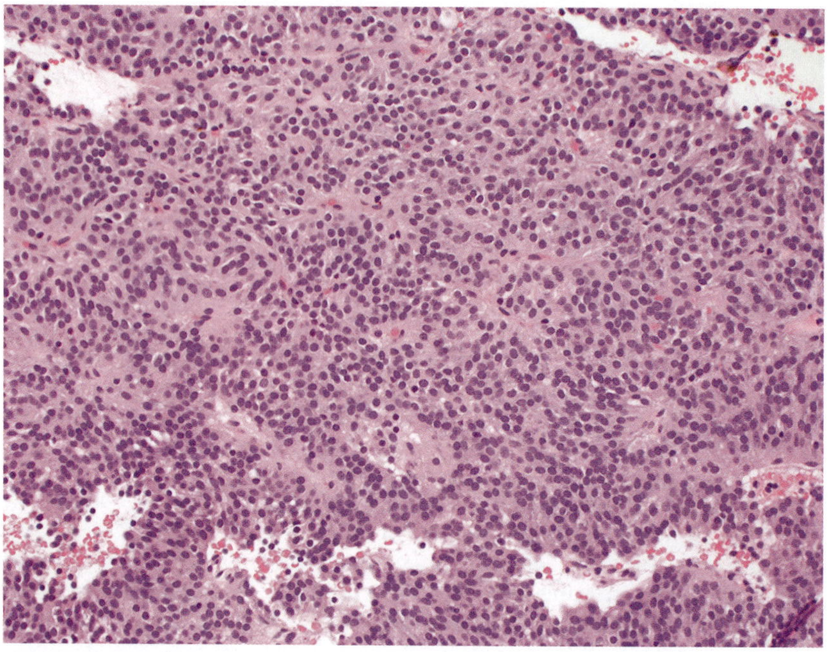

Figure 1-12

PITUITARY NEUROENDOCRINE TUMOR (PitNET)

PitNET is readily diagnosed on an H&E-stained section. The cells are all uniform. There is loss of the normal acinar architecture. There are no acidophilic or basophilic cells admixed with the sheets of chromophobe cells.

REFERENCES

Embryology, Anatomy, and Physiology

1. Asa SL. Tumors of the pituitary gland. AFIP Atlas of Tumor Pathology, 3rd Series, Fascicle 22. Washington DC: American Registry of Pathology; 1998.
2. Horvath E, Kovacs K. The adenohypophysis. In: Kovacs K, Asa SL, eds. Functional endocrine pathology. Boston: Blackwell Scientific; 1991:245-81.
3. Thorner MO, Vance ML, Horvath E, Kovacs K. The anterior pituitary. In: Wilson JD, Foster DW, eds. Williams textbook of endocrinology, 8th ed. Philadelphia: WB Saunders; 1992:221-310.
4. Schroeder JW, Vezina LG. Pediatric sellar and suprasellar lesions. Pediatr Radiol 2011;41:287-98.
5. Solov'ev GS, Bogdanov AV, Panteleev SM, Yanin VL. Embryonic morphogenesis of the human pituitary. Neurosci Behav Physiol 2008;38:829-33.
6. Donkelaar HJ, Lammens M, Hori A. Clinical neuroembryology: development and developmental disorders of the human central nervous system. New York: Springer Verlag; 2006.
7. Erdheim J. Über hypophysenganggeschwülste und hirncholesteatome. Kaiserl Akadem Wissensch Wien, mathemat wissensch klasse abt. III 1904;103:537-726. [German]
8. Romeis B. Innensekretorische Drüsen II Hypophyse. In: von Möllendorf W, ed. Handbuch der mikroskopischen Anatomie des Menschen, vol 6. Berlin: Springer; 1940:1-625. [German]
9. Hori A, Schmidt D, Rickels E. Pharyngeal pituitary: development, malformation, and tumorigenesis. Acta Neuropathol 1999;98:262-72.
10. McGrath P. Aspects of the human pharyngeal hypophysis sin normal and anencephalic fetuses and neonates and their possible significance in the mechanism of its control. J Anat 1978;127:65-81.
11. Lloyd RV, Chandler WF, Kovacs K, Ryan N. Ectopic pituitary adenomas with normal anterior pituitary glands. Am J Surg Pathol 1986;10:546-52.
12. Asa SL, Kovacs K, Horvath E, et al. Human fetal adenohypophysis. Electron microscopic and ultrastructural immunocytochemical analysis. Neuroendocrinology 1988;48:423-31.
13. Asa SL, Kovacs K, Laszlo FA, et al. Human fetal adenohypophysis. Histologic and immunocytochemical analysis. Neuroendocrinology 1986;43:308-16.
14. Asa SL, Kovacs K, Singer W. Human fetal adenohypophysis: morphologic and functional analysis in vitro. Neuroendocrinology 1991;53:562-72.

15. Ernst LM. Pituitary gland. In: Ernst L, Ruchelli ED, Huff DS, eds. Color atlas of fetal and neonatal histology. New York: Springer; 2011.
16. Zhao Y, Mailloux CM, Hermesz E, Palkovits M, Westphal H. A role of the LIM-homeobox gene Lhx2 in the regulation of pituitary development. Dev Biol 2010;337:313-23.
17. Cha KB, Douglas KR, Potok MA, Liang H, Jones SN, Camper SA. WNT5A signaling affects pituitary gland shape. Mech Dev 2004;121:183-94.
18. Ericson J, Norlin S, Jessell TM, Edlund T. Integrated FGF and BMP signaling controls the progression of progenitor cell differentiation and the emergence of pattern in the embryonic anterior pituitary. Development 1998;125:1005-15.
19. Norlin S, Nordstrom U, Edlund T. Fibroblast growth factor signaling is required for the proliferation and patterning of progenitor cells in the developing anterior pituitary. Mech Dev 2000;96:175-82.
20. Takuma N, Sheng HZ, Furuta Y, et al. Formation of Rathke's pouch requires dual induction from the diencephalon. Development 1998;125:4835-40.
21. Treier M, Gleiberman AS, O'Connell SM, et al. Multistep signaling requirements for pituitary organogenesis in vivo. Genes Dev 1998;12:1691-704.
22. Zhu X, Gleiberman AS, Rosenfeld MG. Molecular physiology of pituitary development: signaling and transcriptional networks. Physiol Rev 2007;87:933-63.
23. Scully KM, Rosenfeld MG. Pituitary development: regulatory codes in mammalian organogenesis. Science 2002;295:2231-5.
24. Charles MA, Suh H, Hjalt TA, Drouin J, Camper SA, Gage PJ. PITX genes are required for cell survival and Lhx3 activation. Mol Endocrinol 2005;19:1893-903.
25. Kioussi C, Briata P, Baek SH, et al. Identification of a Wnt/Dvl/betaCatenin–>Pitx2 pathway mediating cell-type-specific proliferation during development. Cell 2002;111:673-85.
26. Suh H, Gage PJ, Drouin J, Camper SA. Pitx2 is required at multiple stages of pituitary organogenesis: pituitary primordium formation and cell specification. Development 2002;129:329-37.
27. Zhu X, Zhang J, Tollkuhn J, et al. Sustained notch signaling in progenitors is required for sequential emergence of distinct cell lineages during organogenesis. Genes Dev 2006;20:2739-53.
28. Olson LE, Tollkuhn J, Scafoglio C, et al. Homeodomain-mediated beta-catenin-dependent switching events dictate celllineage determination. Cell 2006;125:593-605.
29. Camper SA, Saunders TL, Katz RW, Reeves RH. The Pit-1 transcription factor gene is a candidate for the murine Snell dwarf mutation. Genomics 1990;8:586-90.
30. Dasen JS, O'Connell SM, Flynn SE, et al. Reciprocal interactions of Pit1 and GATA2 mediate signaling gradient-induced determination of pituitary cell types. Cell 1999;97:587-98.
31. Li S, Crenshaw EB 3rd, Rawson EJ, Simmons DM, Swanson LW, Rosenfeld MG. Dwarf locus mutants lacking three pituitary cell types result from mutations in the POU-domain gene pit-1. Nature 1990;347:528-33.
32. Lenders NF, Wilkinson AC, Wong SJ, et al. Transcription factor immunohistochemistry in the diagnosis of pituitary tumours. Eur J Endocrinol 2021;184:891-901.
33. Lechan RM, Arkun K, Toni R. Pituitary anatomy and development. In: Tritos NA, Klibanski A, eds. Prolactin disorders: from basic science to clinical management. Cham: Humana Press; 2019:11-53.
34. Amar AP, Weiss MH. Pituitary anatomy and physiology. Neurosurg Clin N Am 2003;14:11-23.
35. Sahni D, Jit I, Harjeet N, Bhansali A. Weight and dimensions of the pituitary in northwestern Indians. Pituitary 2006;9:19-26.
36. Bergland RM, Ray BS, Torack RM. Anatomical variations in the pituitary gland and adjacent structures in 225 human autopsy cases. J Neurosurg 1968;28:93-9.
37. MacMaster FP, Keshavan M, Mirza Y, et al. Development and sexual dimorphism of the pituitary gland. Life Sci 2007;80:940-4.
38. Takano K, Utsunomiya H, Ono H, Ohfu M, Okazaki M. Normal development of the pituitary gland: assessment with three-dimensional MR volumetry. Am J Neuroradiol 1999;20:312-5.
39. Songtao Q, Yuntao L, Jun P, Chuanping H, Xiaofeng S. Membranous layers of the pituitary gland: histological anatomic study and related clinical issues. Neurosurgery 2009;64:1-9.
40. Horvath E, Kovacs K. The adenophyophysis. In: Kovacs K, Asa SL, eds. Fuctional endocrine pathology. Cambridge: Blackwell Scientific Publications; 1991:245-81. **[Duplicate of ref. 2]**
41. Stanfield JP. The blood supply of the human pituitary gland. J Anat 1960;94:257-73.
42. Ceylan S, Anik I, Koc K, et al. Microsurgical anatomy of membranous layers of the pituitary gland and the expression of extracellular matrix collagenous proteins. Acta Neurochir (Wein) 2011;153:2435-46.
43. Peker S, Kurtkaya-Yapicier O, Kilic T, Pamir MN. Microsurgical anatomy of the lateral walls of the pituitary fossa. Acta Neurochir (Wein) 2005;147:641-8.
44. Baker BL. Cellular composition of the human pituitary pars tuberalis as revealed by immunocytochemistry. Cell Tissue Res 1977;182:151-63.
45. Gorczyca W, Hardy J. Arterial supply of the human anterior pituitary gland. Neurosurgery 1987;20:369-78.

46. McConnell EM. The arterial blood supply of the human hypophysis cerebri. Anat Rec 1953; 115:175-203.
47. Xuereb GP, Prichard MM, Daniel PM. The arterial supply and venous drainage of the human hypophysis cerebri. Q J Exp Physiol Cogn Med Sci 1954;39:199-217.
48. Xuereb GP, Prichard ML, Daniel PM. The hypophysial portal system of vessels in man. Q J Exp Physiol Cogn Med Sci 1954;39:219-30.
49. Garten LL, Sofroniew MV, Dyball RE. A direct catecholaminergic projection from the brainstem to the neurohypophysis of the rat. Neuroscience 1989;33:149-55.
50. Horvath E, Kovacs K. Fine structural cytology of the adenohypophysis in rat and man. J Electron Microsc Technique 1988;8:401-32.
51. Kovacs K, Horvath E. Tumors of the pituitary gland. Atlas of Tumor Pathology, 2nd Series, Fascicle 21. Washington DC: Armed Forces Institute of Pathology; 1986.
52. Phelps C. The anterior pituitary and its hormones. In: Enna SJ, Bylund DB, eds. xPharm: the comprehensive pharmacology reference. Boston: Elsevier; 2007:1-12.
53. Halasz B. Pituitary gland anatomy and embryology. In: Martini L, ed. Encyclopedia of endocrine diseases; 2004:636-43.
54. Larsen CM, Grattan DR. Prolactin, neurogenesis, and maternal behaviors. Brain Behav Immun 2012;26:201-9.
55. Shingo T, Gregg C, Enwere E, et al. Pregnancy-stimulated neurogenesis in the adult female forebrain mediated by prolactin. Science 2003; 299:117-20.
56. Losinski NE, Horvath E, Kovacs K, Asa SL. Immunoelectron microscopic evidence of mammosomatotrophs in human adult and fetal adenohypophyses, rat adenohypophyses and human and rat pituitary adenomas. Anat Anz 1991;172:11-6.
57. Lloyd RV, Anagnostou D, Cano M, Barkan Al, Chandler WF. Analysis of mammosomatotropic cells in normal and neoplastic human pituitary tissues by the reverse hemolytic plaque assay and immunocytochemistry. J Clin Endocrinol Metabol 1988;66:1103-10.
58. Lloyd RV, D'Amato CJ, Thiny MT, Jin L, Hicks SP, Chandler WF. Corticotroph (basophil) invasion of the pars nervosa in the human pituitary: localization of pro-opiomelanocortin peptides, galanin and peptidylglycine-amidating monooxygenase-like immunoreactivities. Endocr Pathol 1993;4:86-94.
59. Mete O, Asa SL. Clinicopathological correlations in pituitary adenomas. Brain Pathol 2012;22:443-53.
60. Girod C, Trouillas J, Dubois MP. Immunocytochemical localization of S-100 protein in stellate cells (folliculostellate cells) of the anterior lobe of the normal human pituitary. Cell Tissue Res 1985;241:505-11.
61. Devnath S, Inoue K. An insight to pituitary folliculo-stellate cells. J Neuroendocrinol 2008;20:687-91.
62. Denef C. Paracrinicity: the story of 30 years of cellular pituitary crosstalk. J Neuroendocrinol 2008;20:1-70.
63. Vitale ML, Garcia CJ, Akpovi CD, Pelletier RM. Distinctive actions of connexin 46 and connexin 50 in anterior pituitary folliculostellate cells. PLoS One 2017;12:e0182495.
64. Le Tissier PR, Hodson DJ, Lafont C, Fontanaud P, Schaeffer M, Mollard P. Anterior pituitary cell networks. Front Neuroendocrinol 2012;33:252-66.
65. Arzt E, Pereda MP, Castro CP, Pagotto U, Renner U, Stalla GK. Pathophysiological role of the cytokine network in the anterior pituitary gland. Front Neuroendocrinol 1999;20:71-95.

Reactive Changes Including Pregnancy, Hormonal Symptoms, and Drug Effects

66. Goluboff LG, Ezrin C. Effect of pregnancy on the somatotropin and the PRL cell of the human adenohypophysis. J Clin Endocrinol Metab 1969;29:1533-8.
67. Elster AD, Sanders TG, Vines FS, Chen MY. Size and shape of the pituitary gland during pregnancy and postpartum: measurement with MR imaging. Radiology 1991;181:531-5.
68. Foyouzi N, Frisbaek Y, Norwitz ER. Pituitary gland and pregnancy. Obstet Gynecol Clin North Am 2004;31:873-92.
69. Ferriani RA, Silva de Sa MF, Lima Filho EC. A comparative study of longitudinal and cross-sectional changes in plasma levels of PRL and estriol during normal pregnancy. Braz J Med Biol Res 1986;19:183-8.
70. Rigg LA, Lein A, Yen SS. Pattern of increase in circulating PRL levels during human gestation. Am J Obstet Gynecol 1977;129:454-6.
71. Alsat E, Guibourdenche J, Luton D, Frankenne F, Evain-Brion D. Human placental growth hormone. Am J Obstet Gynecol 1997;177:1526-34.
72. Caufriez A, Frankenne F, Englert Y, et al. Placental growth hormone as a potential regulator of maternal IGF-1 during human pregnancy. Am J Physiol 1990;258(Pt 1):E1014-9.
73. Caufriez A, Frankenne F, Hennen G, Copinschi G. Regulation of maternal IGF-1 by placental GH in normal and abnormal human pregnancies. Am J Physiol 1993;265(Pt 1):E572-7.
74. Daughaday WH, Trivedi B, Winn HN, Yan H. Hypersomatotropism in pregnant women, as measured by a human liver radioreceptor assay. J Clin Endocrinol Metab 1990;70:215-21.
75. Glezer A, Bronstein MD. The pituitary gland in pregnancy. In: Melmed S, ed. The pituitary, 4th ed. Academic Press; 2016.

76. Carpenter CC, Solomon N, Silverberg SG, et al. Schmidt's syndrome (thyroid and adrenal insufficiency). A review of the literature and a report of fifteen new cases including ten instances of coexistent diabetes mellitus. Medicine 1964;43:153-80.
77. Crooke AC, Russell DS. The pituitary gland in Addison's disease. J Pathol Bacteriol 1935;40:255-83.
78. Crooke AC. A change in the basophil cells of the pituitary gland common to conditions which exhibit the syndrome attributed to basophil adenoma. J Pathol Bacteriol 1935;41:339-49.
79. Halmi NS, McCormick WF. Effects of hyperadrenocorticism on pituitary thyrotropic cells in man. Arch Pathol 1972;94:471-4.
80. Halmi NS, McCormick WF, Decker DA Jr. The natural history of hyalinization of ACTH-MSH cells in man. Arch Pathol 1971;91:318-26.
81. Saeger W. Surgical pathology of the pituitary in Cushing's disease. Pathol Res Pract 1991;187:613-6.
82. Scheithauer BW, Kovacs KT, Randall RV, Ryan N. Effects of estrogen on the human pituitary: a clinicopathologic study. Mayo Clin Proc 1989;64:1077-84.
83. Neumann PE, Horoupian DS, Goldman JE, Hess MA. Cytoplasmic filaments of Crooke's hyaline change belong to the cytokerain class. An immunocytochemical and ultrastructural study. Am J Pathol 1984;116:214-22.
84. Oldfield EH, Vance ML, Louis RG, Pledger CL, Jane JA, Lopes MS. Crooke's changes in Cushing's syndrome depends on degree of hypercortisolism and individual susceptibility. J Clin Endocrinol Metab 2015;100:3165-71.
85. Murray S, Ezrin C. Effect of Graves' disease on the "thyrotroph" of the adenohypophysis. J Clin Endocrinol 1966;26:287-93.
86. Scheithauer BW, Kovacs KT, Young WF Jr, Randall RV. The pituitary gland in hyperthyroidism. Mayo Clin Proc 1992;67:22-6.
87. Prummel MF, Brokken LJ, Meduri G, Misrahi M, Bakker O, Wiersinga WM. Expression of the thyroid-stimulating hormone receptor in the folliculo-stellate cells of human anterior pituitary. J Clin Endocrinol Metab 2000;85:4347-53.
88. Prummel MF, Brokken LJ, Wiersinga WM. Ultra short-loop feedback control of thyrotropin secretion. Thyroid 2004;14:825-9.
89. Zatelli MC, Ambrosio MR, Bondanelli M, Degli Uberti E. Pituitary side effects of old and new drugs. J Endocrinol Invest 2014;37:917-23.
90. White MC, Anapliotou M, Rosenstock J, et al. Heterogeneity of prolactin responses to oestradiol benzoate in women with prolactinomas. Lancet 1981;1:1394-6.
91. Tollin SR. Use of the dopamine agonists bromocriptine and cabergoline in the management of risperidone-induced hyperprolactinemia in patients with psychotic disorders. J Endocrinol Invest 2000;23:765-70.
92. Bou Khalil R, Richa S. Thyroid adverse effects of psychotropic drugs: a review. Clin Neuropharmacol 2011;34:248-55.
93. Gomes-Porras M, Cardenas-Salas J, Alvarez-Escola C. Somatostatin analogs in clinical practice: a review. Int J Mol Sci 2020;21:1682.
94. Barkan AL, Lloyd RV, Chandler WF, et al. Preoperative treatment of acromegaly with long-acting somatostatin analog SMS 201-995: shrinkage of invasive pituitary macroadenomas and improved surgical remission rate. J Clin Endocrinol Metab 1988;67:1040-8.
95. Avendano C, Menendez JC. Anticancer drugs that inhibit hormone action. In: Avendano C, Menendex JC. Medicinal chemistry of anticancer drugs, 1st ed. Amsterdam: Elsevier Science; 2008:53-91.
96. Stefaneanu L, Kovacs K, Scheithauer BW, et al. Effect of dopamine agonists on lactotroph adenomas of the human pituitary. Endocr Pathol 2000;11:341-52.
97. Lloyd HM, Meares JD, Jacobi J. Effects of oestrogen and bromocriptine on in vivo secretion and mitosis in prolactin cells. Nature 1975;255:497-8.
98. Tindall GT, Kovacs K, Horvath E, Thorner MO. Human prolactin-producing adenomas and bromocriptine: a histological, immunocytochemical, ultrastructural and morphometric study. J Clin Endocrinol Metab 1982;55:1178-83.
99. Wood DF, Johnston JM, Johnston DG. Dopamine, the dopamine D2 receptor and pituitary tumours. Clin Endocrinol 1991;35:455-66.
100. Braun D, Kim TD, le Coutre P, Kohrle J, Hershman JM, Schweizer U. Tyrosine kinase inhibitors noncompetitively inhibit MCT8-mediated iodothyronine transport. J Clin Endocrinol Metab 2012;97:E100-5.
101. Haap M, Gallwitz B, Thamer C, Mussig K, Haring HU, Kanz L, Hartmann JT. Symptomatic hypoglycemia during imatinib mesylate in a non-diabetic female patient with gastrointestinal stromal tumor. J Endocrinol Invest 2007;30:688-92.
102. Kebapcilar L, Bilgir O, Alacacioglu I, et al. Does imatinib mesylate therapy cause growth hormone deficiency? Med Princ Pract 2009;18:360-3.
103. Corsello SM, Barnabei A, Marchetti P, De Vecchis L, Salvatori R, Torino F. Endocrine side effects induced by immune checkpoint inhibitors. J Clin Endocrinol Metab 2013;98:1361-75.
104. Quandt Z, Young A, Perdigoto AL, Herold KC, Anderson MS. Autoimmune endocrinopathies: an emerging complication of immune checkpoint inhibition. Annu Rev Med 2021;72:313-30.

105. Hodi FS, O'Day SJ, McDermott DF, et al. Improved survival with ipilimumab in patients with metastatic melanoma. N Engl J Med 2010;363:711-23.
106. Robert C, Thomas L, Bondarenko I, et al. Ipilimumab plus dacarbazine for previously untreated metastatic melanoma. N Engl J Med 2011;364:2517-26.
107. Faje AT, Sullivan R, Lawrence D, et al. Ipilimumab-induced hypophysitis: a detailed longitudinal analysis in a large cohort of patients with metastatic melanoma. J Clin Endocrinol Metab 2014;99:4078-85.
108. Min L, Hodi FS, Giobbie-Hurder A, et al. Systemic high-dose corticosteroid treatment does not improve the outcome of ipilimumab-related hypophysitis: a retrospective cohort study. Clin Cancer Res 2015;21:749-55.
109. Caturegli P, Di Dalmazi G, Lombardi M, et al. Hypophysitis secondary to cytotoxic T-lymphocyte-associated protein 4 blockade: insights into pathogenesis from an autopsy series. Am J Pathol 2016;186:3225-35.
110. Barnard ZR, Walcott BP, Kahle KT, Nahed BV, Coumans JV. Hyponatremia associated with Ipilimumab-induced hypophysitis. Med Oncol 2012;29:374-7.
111. Dillard T, Yedinak CG, Alumkal J, Fleseriu M. Anti-CTLA-4 antibody therapy associated autoimmune hypophysitis: serious immune related adverse events across a spectrum of cancer subtypes. Pituitary 2010;13:29-38.
112. Joshi MN, Whitelaw BC, Palomar MT, Wu Y, Carroll PV. Immune checkpoint inhibitor-related hypophysitis and endocrine dysfunction: clinical review. Clin Endocrinol 2016;85:331-9.
113. Min L, Vaidya A, Becker C. Association of ipilimumab therapy for advanced melanoma with secondary adrenal insufficiency: a case series. Endocr Pract 2012;18:351-5.
114. Albarel F, Gaudy C, Castinetti F, et al. Long-term follow-up of ipilimumab-induced hypophysitis, a common adverse event of the anti-CTLA-4 antibody in melanoma. Eur J Endocrinol 2015;172:195-204.

Hereditary and Developmental Disorders

115. Moncrieff MW, Hill DS, Archer J, Arthur LJ. Congenital absence of pituitary gland and adrenal hypoplasia. Arch Dis Child 1972;47:136-7.
116. Adly N, Alhashem A, Ammari A, Alkuraya FS. Ciliary genes TBC1D32/C6orf170 and SCLT1 are mutated in patients with OFD Type IX. Hum Mutat 2014;35:36-40.
117. Bergeron C, Kovacs K, Bilbao JM. Primary empty sella: a histologic and immunocytologic study. Arch Intern Med 1979;139:248-9.
118. Brewer D. Congenital absence of the pituitary gland and its consequences. J Pathol Bacteriol 1957;73:59-67.
119. Kosaki K, Matsuo N, Tamai S, et al. Isolated aplasia of the anterior pituitary as a cause of congenital panhypopituitarism. Case report. Horm Res 1991;35:226-8.
120. Reid JD. Congenital absence of the pituitary gland. J Pediatr 1960;56:658-64.
121. Steiner MM, Boggs JD. Absence of pituitary gland, hypothyroidism, hypoadrenalism, and hypogonadism in a 17-year-old dwarf. J Clin Endocrinol Metab 1965;25:1591-8.
122. Tatsumi K, Miyai K, Notomi T, et al. Cretinism with combined hormone deficiency caused by a mutation in the PIT1 gene. Nature Genet 1992;1:56-8.
123. Davis SW, Ellsworth BS, Millan MI, et al. Pituitary gland development and disease: from stem cell to hormone production. Curr Top Dev Biol 2013;106:1-47.
124. Kjaer I, Fischer-Hansen B. The adenohypophysis and the cranial base in early human development. J Craniofac Genet Dev Biol 1995;15:157-61.
125. Kjaer I, Hansen BF. The prenatal pituitary gland—hidden and forgotten. Pediatr Neurol 2000;22:155-6.
126. Baban A, Moresco L, Divizia MT, et al. Pituitary hypoplasia and growth hormone deficiency in Coffin-Siris syndrome. Am J Med Genet A 2008; 146A:384-8.
127. Bas F, Darendeliler F, Yapici Z, et al. Worster-Drought syndrome (congenital bilateral perisylvian syndrome) with posterior pituitary ectopia, pituitary hypoplasia, empty sella and panhypopituitarism: a patient report. J Pediatr Endocrinol Metab 2006;19:535-40.
128. Cianfarani S, Vitale S, Stanhope R, Boscherini B. Imperforate anus, bilateral hydronephrosis, bilateral undescended testes and pituitary hypoplasia: a variant of Hall-Pallister syndrome or a new syndrome. Acta Paediatr 1995;84:1322-4.
129. Raivio T, Avbelj M, McCabe MJ, et al. Genetic overlap in Kallmann syndrome, combined pituitary hormone deficiency, and septo-optic dysplasia. J Clin Endocrinol Metab 2012;97:E694-9.
130. Tatsi C, Sertedaki A, Voutetakis A, et al. Pituitary stalk interruption syndrome and isolated pituitary hypoplasia may be caused by mutations in holoprosencephaly-related genes. J Clin Endocrinol Metab 2013;98:E779-84.
131. Fang Q, George AS, Brinkmeier ML, et al. Genetics of combined pituitary hormone deficiency: roadmap into the genome era. Endocr Rev 2016;37:636-75.

132. Matsumoto R, Suga H, Aoi T, et al. Congenital pituitary hypoplasia model demonstrates hypothalamic OTX2 regulation of pituitary progenitor cells. J Clin Invest 2020;130:641-54.
133. Nolen LD, Amor D, Haywood A, et al. Deletion at 14q22-23 indicates a contiguous gene syndrome comprising anophthalmia, pituitary hypoplasia, and ear anomalies. Am J Med Genet A 2006;140:1711-8.
134. Frisch H, Kim C, Häusler G, Pfäffle R. Combined pituitary hormone deficiency and pituitary hypoplasia due to a mutation of the Pit-1 gene. Clin Endocrinol (Oxf) 2000;52:661-5.
135. Pfäffle RW, DiMattia GE, Parks JS, et al. Mutation of the POU-specific domain of Pit-1 and hypopituitarism without pituitary hypoplasia. Science 1992;257:1118-21.
136. Thomas PQ, Dattani MT, Brickman JM, et al. Heterozygous HESX1 mutations associated with isolated congenital pituitary hypoplasia and septo-optic dysplasia. Hum Mol Genet 2001;10:39-45.
137. Zada G, Lopes MB, Mukundan S Jr, Laws ER Jr. Pituitary hypoplasia and other midline developmental anomalies. In: Zada G, Lopes MB, Mukundan S Jr, Laws ER Jr, eds. Atlas of sellar and parasellar lesions. Cham: Springer; 2016:493-6.
138. Costin G, Murphree AL. Hypothalamic-pituitary function in children with optic nerve hypoplasia. Am J Dis Child 1985;139:249-54.
139. Lettau M, Laible M. [Kallmann's syndrome with pituitary hypoplasia.] Rofo 2011;183:576-8. [German]
140. Margalith D, Tze WJ, Jan JE. Congenital optic nerve hypoplasia with hypothalamic-pituitary dysplasia. A review of 16 cases. Am J Dis Child 1985;139:361-6.
141. Acers TE. Optic nerve hypoplasia: septo-optic-pituitary dysplasia syndrome. Trans Am Ophthalmol Soc 1981;79:425-57.
142. Kaufman LM, Miller MT, Mafee MF. Magnetic resonance imaging of pituitary stalk hypoplasia. A discrete midline anomaly associated with endocrine abnormalities in septo-optic dysplasia. Arch Ophthalmol 1989;107:1485-9.
143. Murray RA, Maheshwari HG, Russell EJ, Baumann G. Pituitary hypoplasia in patients with a mutation in the growth hormone-releasing hormone receptor gene. Am J Neuroradiol 2000;21:685-9.
144. Deal C, Hasselmann C, Pfäffle RW, et al. Associations between pituitary imaging abnormalities and clinical and biochemical phenotypes in children with congenital growth hormone deficiency: data from an international observational study. Horm Res Paediatr 2013;79:283-92.
145. Inamo Y, Harada K. Agenesis of the internal carotid artery and congenital pituitary hypoplasia: proposal of a cause of congenital hypopituitarism. Eur J Pediatr 2003;162:610-2.
146. Shulman DI, Martinez CR. Association of ectopic posterior pituitary and anterior pituitary hypoplasia with absence of the left internal carotid. J Pediatr Endocrinol Metab 1996;9:539-42.
147. Pilavdzic D, Kovacs K, Asa SL. Pituitary morphology in anencephalic human fetuses. Neuroendocrinology 1997;65:164-72.
148. Osamura RY. Functional prenatal development of anencephalic and normal anterior pituitary glands. In human and experimental animals studied by peroxidase-labeled antibody method. Acta Pathol Jpn 1977;27:495-509.
149. Salazar H, MacAulay MA, Charles D, Pardo M. The human hypophysis in anencephaly. I. Ultrastructure of the pars distalis. Arch Pathol 1969;87:201-11.
150. Grumbach MM, Gluckman PD. The human fetal hypothalamus and pituitary gland: the maturation of neuroendocrine mechanisms controlling the secretion of fetal pituitary growth hormone, prolactin, gonadotropins, adrenocorticotropin-related peptides, and thyrotropin. In: Tulchinsky D, Little AB, eds. Maternal-fetal endocrinology, 2nd ed. Philadelphia: Saunders; 1994:193-261.
151. Kaplan SL, Grumbach MM, Aubert ML. The ontogenesis of pituitary hormones and hypothalamic factors in the human fetus: maturation of central nervous system regulation of anterior pituitary function. Recent Prog Horm Res 1976;32:161-243.
152. Gatti RA. Ataxia-telangiectasia. Dermatol Clin 1995;13:1-6.
153. Swift M, Morrell D, Cromartie E, Chamberlin AR, Skolnick MH, Bishop DT. The incidence and gene frequency of ataxia-telangiectasia in the United States. Am J Hum Genet 1986;39:573-83.
154. Swift A, Morrell D, Massey RB, Chase CL. Incidence of cancer in 161 families affected by ataxia-telangiectasia. N Engl J Med 1991;325:1831-6.
155. Kovacs K, Giannini C, Scheithauer BW, Stefaneanu L, Lloyd RV, Horvath E. Pituitary changes in ataxia-telangiectasia syndrome: an immunocytochemical, in situ hybridization, and DNA cytometric study of three cases. Endocr Pathol 1997;8:195-203.
156. Chiloiro S, Giampietro A, Bianchi A, et al. Primary empty sella: a comprehensive review. Eur J Endocrinol 2017;177:R275-85.
157. Maira G, Anile C, Mangiola A. Primary empty sella syndrome in a series of 142 patients. J Neurosurg 2005;103:831-6.

158. Bergland RM, Ray BS, Torack RM. Anatomical variations in the pituitary gland and adjacent structures in 225 human autopsy cases. J Neurosurg 1968;28:93-9.
159. Giustina A, Aimaretti G, Bondanelli M, et al. Primary empty sella: why and when to investigate hypothalamic-pituitary function. J Endocrinol Invest 2010;33:343-6.
160. De Marinis L, Bonadonna S, Bianchi A, Maira G, Giustina A. Primary empty sella. J Clin Endocrinol Metab 2005;90:5471-7.
161. Komatsu M, Kondo T, Yamauchi K, et al. Antipituitary antibodies in patients with primary empty sella syndrome. J Clin Endocrinol Metab 1988;67:633-8.
162. Weiss RE. Empty sella following spontaneous resolution of a pituitary macroadenoma. Horm Res 2003;60:49-52.
163. Durodoye OM, Mendlovic DB, Brenner RS, Morrow JS. Endocrine disturbances in empty sella syndrome: case reports and review of literature. Endocr Pract 2005;11:120-4.
164. Scheithauer BW, Lopes MB. Pituitary and sellar region. In: Mills SE, ed. Sternberg's diagnostic surgical pathology, 6th ed. Philadelphia: Wolters Kluwer Health; 2015.
165. Gregory LC, Dattani MT. Embryologic and genetic disorders of the pituitary gland. In: Kohn B, eds. Pituitary disorders of childhood: diagnosis and clinical management. Cham: Humana Press; 2019:3-27.
166. Alatzoglou KS, Kular D, Dattani MT. Autosomal dominant growth hormone deficiency (type II). Pediatr Endocrinol Rev 2015;12:347-55.
167. Alatzoglou KS, Webb EA, Le Tissier P, Dattani MT. Isolated growth hormone deficiency (GHD) in childhood and adolescence: recent advances. Endocr Rev 2014;35:376-432.
168. Lee MS, Wajnrajch MP, Kim SS, et al. Autosomal dominant growth hormone (GH) deficiency type II: the Del32-71-GH deletion mutant suppresses secretion of wild-type GH. Endocrinology 2000;141:883-90.
169. Douchi T, Nakae M, Yamamoto S, Iwamoto I, Oki T, Nagata Y. A woman with isolated prolactin deficiency. Acta Obstet Gynecol Scand 2001;80:368-70.
170. Kauppila A. Isolated prolactin deficiency. Curr Ther Endocrinol Metab 1997;6:31-3.
171. Ahn SW, Kim TY, Lee S, et al. Adrenal insufficiency presenting as hypercalcemia and acute kidney injury. Int Med Case Rep J 2016;9:223-6.
172. Alsaleem M, Saadeh L, Misra A, Madani S. Neonatal isolated ACTH deficiency (IAD): a potentially life-threatening but treatable cause of neonatal cholestasis. BMJ Case Rep 2016;bcr2016215032.
173. Bigos ST, Carnes TD. Isolated ACTH deficiency presenting as severe hypercalcemia. Am J Med Sci 1982;284:24-30.
174. Gangat M, Radovick S. Pituitary hypoplasia. Endocrinol Metab Clin North Am 2017;46:247-57.
175. Pang AL, Chan WY. Molecular basis of diseases of the endocrine system. In: Coleman WB, Tsongalis GJ, eds. Molecular pathology: the molecular basis of human disease, 2nd ed. Cambridge: Elsevier; 2018.
176. Coya R, Vela A, Perez de Nanclares G, et el. Panhypopituitarism: genetic versus acquired etiological factors. J Pediatr Endocrinol Metab 2007;20:27-36.
177. Kelberman D, Dattani MT. Hypothalamic and pituitary development: novel insights into the aetiology. Eur J Endocrinol 2007;157(Suppl 1):S3-14.
178. Reynaud R, Saveanu A, Barlier A, Enjalbert A, Brue T. Pituitary hormone deficiencies due to transcription factor gene alterations. Growth Horm IGF Res 2004;14:442-8.
179. Zhu X, Lin CR, Prefontaine GG, Tollkuhn J, Rosenfeld MG. Genetic control of pituitary development and hypopituitarism. Curr Opin Genet Dev 2005;15:332-40.
180. Dattani MT, Martinez-Barbera JP, Thomas PQ, et al. Mutations in the homeobox gene HESX1/Hesx1 associated with septo-optic dysplasia in human and mouse. Nat Genet 1998;19:125-33.
181. Bhangoo AP, Hunter CS, Savage CS, et al. Clinical case seminar: a novel LHX3 mutation presenting as combined pituitary hormonal deficiency. J Clin Endocrinol Metab 2006;91:747-53.
182. Pfäeffle RW, Savage JJ, Hunter CS, et al. Four novel mutations of the LHX3 gene cause combined pituitary hormone deficiencies with or without limited neck rotation. J Clin Endocrinol Metab 2007;92:1909-19.
183. Rajab A, Kelberman D, de Castro SC, et al. Novel mutations in LHX3 are associated with hypopituitarism and sensorineural hearing lsos. Hum Mol Genet 2008;17:2150-9.
184. Fofanova O, Takamura N, Kinoshita E, et al. MR imaging of the pituitary gland in children and young adults with congenital combined pituitary hormone deficiency associated with PROP1 mutations. AJR Am J Roentgenol 2000;174:555-9.
185. Mendonca BB, Osorio MG, Latronico AC, et al. Longitudinal hormonal and pituitary imaging changes in two females with combined pituitary hormone deficiency due to deletion of A301,G302 in the PROP1 gene. J Clin Endocrinol Metab 1999;84:942-5.

186. Rosenbloom AL, Almonte AS, Brown MR, Fisher DA, Baumbach L, Parks JS. Clinical and biochemical phenotype of familial anterior hypopituitarism from mutation of the PROP1 gene. J Clin Endocrinol Metab 1999;84:50-7.
187. Arroyo A, Pernasetti F, Vasilyev VV, Amato P, Yen SS, Mellon PL. A unique case of combined pituitary hormone deficiency caused by a PROP1 gene mutation (R120C) associated with normal height and absent puberty. Clin Endocrinol 2002;57:283-91.
188. Reynaud R, Barlier A, Vallette-Kasic S, et al. An uncommon phenotype with familial central hypogonadism caused by a novel PROP1 gene mutant truncated in the transactivation domain. J Clin Endocrinol Metab 2005;90:4880-7.
189. Ohta K, Nobukuni Y, Mitsubuchi H, et al. Characterization of the gene encoding human pituitary-specific transcription factor, Pit-1. Gene 1992;122:387-8.
190. Roessler E, Ermilov AN, Grange DK, et al. A previously unidentified amino-terminal domain regulates transcriptional activity of wild-type and disease-associated human GLI1. Hum Mol Genet 2005;14:2181-8.

Circulatory Disorders

191. Turgut M, Mahapatra AK, Powell M, Muthukumar N, eds. Pituitary apoplexy. Berlin: Springer; 2013.
192. Oo MM, Krishna AY, Bonavita GJ, Rutecki GW. Heparin therapy for myocardial infarction: an unusual trigger for pituitary apoplexy. Am J Med Sci 1997;314:351-3.
193. Arafah BM, Ybarra J, Tarr RW, Madhun ZT, Selman WR. Pituitary tumor apoplexy: pathophysiology, clinical manifestations and management. J Intensive Care Med 1997;12:123-34.
194. Ayuk J, McGregor EJ, Mitchell RD, Gittoes NJ. Acute management of pituitary apoplexy—surgery or conservative management? Clin Endocrinol (Oxf) 2004;61:747-52.
195. Bills DC, Meyer FB, Laws EE Jr, et al. A retrospective analysis of pituitary apoplexy. Neurosurgery 1993;33:602-9.
196. Cardosa ER, Peterson EW. Pituitary apoplexy: a review. Neurosurgery 1984;14:363-73.
197. Kaplan B, Day AL, Quisling R, Ballinger W. Hemorrhage into pituitary adenomas. Surg Neurol 1983;20:280-7.
198. McFadzean RM, Doyle D, Rampling R, Teasdale E, Teasdale G. Pituitary apoplexy and its effect on vision. Neurosurgery 1991;29:669-75.
199. Mohr G, Hardy J. Hemorrhage, necrosis, and apoplexy in pituitary adenomas. Surg Neurol 1982;18:181-9.
200. Nakahara K, Oka H, Utsuki S, et al. Pituitary apoplexy manifesting as diffuse subarachnoid hemorrhage. Neurol Med Chir (Tokyo) 2006;46:594-7.
201. Onesti ST, Wisniewski T, Post KD. Clinical versus subclinical pituitary apoplexy: presentation, surgical management, and outcome in 21 patients. Neurosurgery 1990;26:980-6.
202. Randeva HS, Schoebel J, Byrne J, Esiri M, Adams CB, Wass JA. Classical pituitary apoplexy: clinical features, management and outcome. Clin Endocrinol (Oxf) 1999;51:181-8.
203. Semple PL, Webb MK, de Villiers JC, Laws ER Jr. Pituitary apoplexy. Neurosurgery 2005;56:65-73.
204. Turgut M, Ozsunar Y, Basak S, Güney E, Kir E, Meteoglu I. Pituitary apoplexy: an overview of 186 cases published during the last century. Acta Neurochir (Wien) 2010;152:749-61.
205. Verrees M, Arafah BM, Selman WR. Pituitary tumor apoplexy: characteristics, treatment and outcomes. Neurosurg Focus 2004;16:E6.
206. Biousse V, Newman NJ, Oyesiku NM. Precipitating factors in pituitary apoplexy. J Neurol Neurosurg Psychiatry 2001;71:542-5.
207. Semple PL, Jane JA Jr, Laws ER Jr. Clinical relevance of precipitating factors in pituitary apoplexy. Neursurgery 2007;61:956-61.
208. Sibal L, Ball SG, Connoly V, et al. Pituitary apoplexy: a review of clinical presentation, management and outcome in 45 cases. Pituitary 2004;7:157-63.
209. Wakai S, Fukushima T, Teramoto A, Sano K. Pituitary apoplexy: its incidence and clinical significance. J Neurosurg 1981;55:187-93.
210. Charalampaki P, Ayyad A, Kockro RA, Perneczky A. Surgical complications after endoscopic transsphenoidal pituitary surgery. J Clin Neurosci 2009;16:786-9.
211. Kovacs K. Adenohypophysial necrosis in routine autopsies. Endokrinologie 1972;60:309-16.
212. Kovacs K. Necrosis of anterior pituitary in humans. I. Neuroendocrinology 1969;4:170-99.
213. McCormick WF, Halmi NS. The hypophysis in patients with coma dépassé ("respirator brain"). Am J Clin Pathol 1970;54:374-83.
214. Plaut A. Pituitary necrosis in routine necropsies. Am J Pathol 1952;28:883-99.
215. Horvath E, Kovacs K. Pathology of the pituitary gland. In: Ezrin C, Horvath E, Kaufman B, et al., eds. Pituitary diseases. Boca Raton: CRC Press; 1980:1-83.
216. Chacko AG, Chacko G, Seshadri MS, Chandy MJ. Hemorrhagic necrosis of pituitary adenomas. Neurol India 2002;50:490-3.
217. Sheehan HL. Postpartum necrosis of the anterior pituitary. J Pathol Bact 1937;45:189-214.

218. Kelestimur F. Sheehan's syndrome. Pituitary 2003;6:181-8.
219. Kovacs K. Sheehan syndrome. Lancet 2003;361: 520-2.
220. Daughaday WH. The anterior pituitary. In: Wilson JD, Foster DW, eds. Williams textbook of endocrinology. Philadelphia: W.B. Saunders; 1985:568-613.
221. Gupta D, Gaiha M, Mahajan R, Daga MK. Atypical presentation of Sheehan's syndrome without postpartum haemorrahage. J Assoc Physicians India 2001;49:386-7.
222. Roberts DM. Sheehan's syndrome. Am Fam Physician 1988;37:223-7.
223. Engelberth O, Jezkova Z, Bleha O, Málek J, Bendl J. [Autoantibodies in Sheehan' syndrome.] Lancet 1965;11:737-41. [Czech]
224. Goswami R, Kochupillai N, Crock PA, Jaleel A, Gupta N. Pituitary autoimmunity in patients with Sheehan's syndrome. J Clin Endocrinol Metab 2002;87:4137-41.
225. Nishiyama S, Takano T, Hidaka Y, et al. A case of postpartum hypopituitarism associated with empty sella: possible relation to postpartum autoimmune hypophysitis. Endocr J 1993;40:431-8.
226. Otsuka F, Kageyama J, Ogura T, Hattori T, Makino H. Sheehan's syndrome of more than 30 years' duration: an endocrine and MRI studies of 6 cases. Endocr J 1998;45:451-8.
227. Patel MC, Guneratne N, Haq N, West TE, Weetman AP, Clayton RN. Peripartum hypopituitarism and lymphocytic hypophysitis. QJM 1995;88:571-80.
228. Tun-Pe, Phillips RE, Warrell DA, et al. Acute and chronic pituitary failure resembling Sheehan's syndrome following bites by Russell's viper in Burma. Lancet 1987;2:763-7.
229. Zayour DH, Azar ST. Silent pituitary infarction after coronary artery bypass grafting procedure: case report and review of literature. Endocr Pract 2006;12:59-62.
230. Jones NS, Finer N. Pituitary infarction and development of the empty sella syndrome after gastrointestinal haemorrhage. Br Med J (Clin Res Ed) 1984;289:661-2.
231. Liu J, Zhang Q, Zheng JX, et al. Management of a patient with extensive burns and Sheehan's syndrome. Burns 2013;39:e17-9.
232. Insel JR, Dhanjal N. Pituitary infarction resulting from intranasal cocaine abuse. Endocr Pract 2004;10:478-82.
233. Zaben M, El Ghoul W, Belli A. Post-traumatic head injury pituitary dysfunction. Disabil Rehabil 2013;35:522-5.
234. Agha A, Phillips J, Thompson CJ. Hypopituitarism following traumatic brain injury (TBI). Br J Neurosurg 2007;21:210-6.
235. Della Corte F, Mancini A, Valle D, et al. Provocative hypothalamopituitary axis tests in severe head injury: correlations with severity and prognosis. Crit Care Med 1998;26:1419-26.
236. Kelly DF, Gonzalo IT, Cohan P, Berman N, Swerdloff R, Wang C. Hypopituitarism following traumatic brain injury and aneurysmal subarachnoid hemorrhage: a preliminary report. J Neurosurg 2000;93:743-52.
237. Woolf PD. Hormonal responses to trauma. Crit Care Med 1992;20:216-26.
238. Bondanelli M, Ambrosio MR, Zatelli MC, De Marinis L, degli Uberti EC. Hypopituitarism after traumatic brain injury. Eur J Endocrinol 2005;152:679-91.
239. Klose M, Juul A, Struck J, Morgenthaler NG, Kosteljanetz M, Feldt-Rasmussen U. Acute and long-term pituitary insufficiency in traumatic brain injury: a prospective single-centre study. Clin Endocrinol (Oxf) 2007;67:598-606.
240. Niederland T, Makovi H, Gál V, Andréka B, Abrahám CS, Kovács J. Abnormalities of pituitary function after traumatic brain injury in children. J Neurotrauma 2007;24:119-27.
241. Tanriverdi F, Senyurek H, Unluhizarci K, Selcuklu A, Casanueva FF, Kelestimur F. High risk of hypopituitarism after traumatic brain injury: a prospective investigation of anterior pituitary function in the acute phase and 12 months after trauma. J Clin Endocrinol Metab 2006;91:2105-11.
242. Tanriverdi F, De Bellis A, Ulutabanca H, et al. A five year prospective investigation of anterior pituitary function after traumatic brain injury: is hypopituitarism long-term after head trauma associated with autoimmunity? J Neurotrauma 2013;30:1426-33.
243. Agha A, Thornton E, O'Kelly P, Tormey W, Phillips J, Thompson CJ. Posterior pituitary dysfunction after traumatic brain injury. J Clin Endocrinol Metab 2004;89:5987-92.
244. Boughey JC, Yost MJ, Bynoe RP. Diabetes insipidus in the head-injured patient. Am Surg 2004;70:500-3.
245. Clement SC, Schouten-van Meeteren AY, Boot AM, et al. Prevalence and risk factors of early endocrine disorders in childhood brain tumor survivors: a nationwide, multicenter study. J Clin Oncol 2016;34:4362-70.
246. Chemaitilly W, Li Z, Huang S, et al. Anterior hypopituitarism in adult survivors of childhood cancers treated with cranial radiotherapy: a report from the St Jude Lifetime Cohort Study. J Clin Oncol 2015;33:492-500.

247. Laughton SJ, Merchant TE, Sklar CA, et al. Endocrine outcomes for children with embryonal brain tumors after risk-adapted craniospinal and conformal primary-site irradiation and high-dose chemotherapy with stem-cell rescue on the SJMB-96 trial. J Clin Oncol 2008;26:1112-8.
248. Brauner R, Adan L, Souberbielle JC, et al. Contribution of growth hormone deficiency to the growth failure that follows bone marrow transplantation. J Pediatr 1997;130:785-92.
249. Chemaitilly W, Boulad F, Heller G, et al. Final height in pediatric patients after hyperfractionated total body irradiation and stem cell transplantation. Bone Marrow Transplant 2007;40:29-35.
250. Chemaitilly W, Sklar CA. Endocrine complications of hematopoietic stem cell transplantation. Endocrinol Metab Clin North Am 2007;36:983-98.
251. Clement-De Boers A, Oostdijk W, Van Weel-Sipman MH, Van den Broeck J, Wit JM, Vossen JM. Final height and hormonal function after bone marrow transplantation in children. J Pediatr 1996;129:544-50.
252. Cohen A, Rovelli A, Bakker B, et al. Final height of patients who underwent bone marrow transplantation for hematological disorders during childhood: a study by the Working Party for Late Effects-EBMT. Blood 1999;93:4109-15.
253. Couto-Silva AC, Trivin C, Esperou H, et al. Final height and gonad function after total body irradiation during childhood. Bone Marrow Transplant 2006;38:427-32.
254. Papadimitriou A, Urena M, Hamill G, Stanhope R, Leiper AD. Growth hormone treatment of growth failure secondary to total body irradiation and bone marrow transplantation. Arch Dis Child 1991;66:689-92.
255. Clement SC, Schoot RA, Slater O, et al. Endocrine disorders among long-term survivors of childhood head and neck rhabdomyosarcoma. Eur J Cancer 2015;54:1-10.
256. Constine LS, Woolf PD, Cann D, et al. Hypothalamic pituitary dysfunction after radiation for brain tumors. N Engl J Med 1993;328:87-94.
Metabolic Disorders
257. Canda T, Sengiz S, Canda MS, Acar UD, Erbayraktar RS, Yilmaz HS. Histochemical and immunohistochemical features of a case showing association of meningioma and prolactinoma containing amyloid. Brain Tumor Pathol 2002;19:1-3.
258. Maji SK, Perrin MH, Sawaya MR, et al. Functional amyloids as natural storage of peptide hormones in pituitary secretory granules. Science 2009;325:328-32.
259. Bilbao JM, Kovacs K, Horvath E. Pituitary melanocorticotrophinoma with amyloid deposition. Can J Neurol Sci 1975;2:199-202.
260. Landolt AM, Kleihues P, Heitz PU. Amyloid deposits in pituitary adenomas. Differentiation of two types. Arch Pathol Lab Med 1987;111:453-8.
261. Rocken C, Paris D, Steusloff K, Saeger W. Investigation of the presence of apolipoprotein E, glycosaminoglycans, basement membrane proteins, and protease inhibitors in senile interstitial amyloid of the pituitary. Endocr Pathol 1997;8:205-14.
262. Rocken C, Saeger W. Amyloid deposits of the pituitary in old age: correlation with histopathological alterations. Endocr Pathol 1994;5:183-90.
263. Rocken C, Uhlig H, Saeger W, Linke RP, Fehr S. Amyloid deposits in pituitaries and pituitary adenomas: immunohistochemistry and in situ hybridization. Endocr Pathol 1995;6:135-43.
264. Tan SY, Pepys MB. Amyloidosis. Histopathology 1994;25:403-14.
265. Tashima T, Kitamoto T, Tateishi J, Ogomori K, Nakagaki H. Incidence and characterization of age related amyloid deposits in the human anterior pituitary gland. Virchows Arch A Pathol Anat Histopathol 1988;412:323-7.
266. Westermark P, Eriksson L, Engstrom U, Enestrom S, Sletten K. Prolactin-derived amyloid in the aging pituitary gland. Am J Pathol 1997;150:67-73.
267. Westermark P, Grimelius L, Polak JM, et al. Amyloid in polypeptide hormone-producing tumors. Lab Invest 1977;37:212-5.
268. Bohl J, Steinmetz H, Storkel S. Age-related accumulation of congophilic fibrillar inclusions in endocrine cells. Virchows Arch A Pathol Anat Histopathol 1991;419:51-8.
269. Eick B, Rocken C, Saeger W, Linke RP. Intrazellulires amyloid in der hypophyse und nebenniere: inzidenz, struktur und immunhistologie. Verh Dtsch Ges Pathol 1993;77:570.
270. Prabhu S, Prabhu S. Pituitary prolactinoma with amyloid deposits: surgery or dopamine agonists? Review of previous reports and new recommendations for managements. Asian J Neurosurg 2019;14:754-8.
271. Landolt AM, Heitz PU. Differentiation of two types of amyloid occurring in pituitary adenomas. Pathol Res Pract 1988;183:552-4.
272. Saitoh Y, Mori H, Matsumoto K, et al. Accumulation of amyloid in pituitary adenomas. Acta Neuropathol 1985;68:87-92.
273. Kubota T, Kuroda E, Yamashima T, Tachibana O, Kabuto M, Yamamoto S. Amyloid formation in prolactinoma. Arch Pathol Lab Med 1986;110:72-5.

274. Bakhtiar Y, Arita K, Hirano H, et al. Prolactin-producing pituitary adenoma with abundant spherical amyloid deposition masquerading as extensive calcification. Neurol Med Chir (Tokyo) 2010;50:1023-6.
275. Kuratsu J, Matsukado Y, Miura M. Prolactinoma of pituitary with associated amyloid-like substances. Case report. J Neurosurg 1983;59:1067-70.
276. Levine SN, Ishaq S, Nanda A, Wilson JD, Gonzalez-Toledo E. Occurrence of extensive spherical amyloid deposits in a prolactin-secreting pituitary macroadenoma: a radiologic-pathologic correlation. Ann Diagn Pathol 2013;17:361-6.
277. Shander A, Cappellini MD, Goodnough LT. Iron overload and toxicity: the hidden risk of multiple blood transfusions. Vox Sang 2009;97:185-97.
278. Andrews NC. Disorders of iron metabolism. N Engl J Med 1999;341:1986-95.
279. Porter J. Pathophysiology of iron overload. Hematol Oncol Clin 2005;19(Suppl 1):7-12.
280. Smiley D, Dagogo-Jack S, Umpierrez G. Therapy insight: metabolic and endocrine disorders in sickle cell disease. Nat Clin Pract Endocrinol Metab 2008;4:102-9.
281. Bergeron C, Kovacs K. Pituitary siderosis. A histologic, immunocytologic, and ultrastructural study. Am J Pathol 1978;93:295-309.
282. Charbonnel B, Chupin M, Le GA, Guillon J. Pituitary function in idiopathic haemochromatosis: hormonal study in 36 male patients. Acta Endocrinol (Copenh) 1981;98:178-83.
283. Duranteau L, Chanson P, Blumberg-Tick J, et al. Nonresponsiveness of serum gonadotropins and testosterone to pulsatile GnRH in hemochromatosis suggesting a pituitary defect. Acta Endocrinol (Copenh) 1993;128:351-4.
284. Kley HK, Stremmel W, Niederau C, et al. Androgen and estrogen response to adrenal and gonadal stimulation in idiopathic hemochromatosis: evidence for decreased estrogen formation. Hepatology 1985;5:251-6.
285. McNeil LW, McKee LC Jr, Lorber D, Rabin D. The endocrine manifestations of hemochromatosis. Am J Med Sci 1983;285:7-13.
286. Piperno A, Rivolta MR, D'Alba R, et al. Preclinical hypogonadism in genetic hemochromatosis in the early stage of the disease: evidence of hypothalamic dysfunction. J Endocrinol Invest 1992;15:423-8.
287. Stocks AE, Martin FI. Pituitary function in haemochromatosis. Am J Med 1968;45:839-45.
288. Tournaire J, Fevre M, Mazenod B, Ponsin G. Effects of clomiphene citrate and synthetic LHRH on serum luteinizing hormone (LH) in men with idiopathic hemochromatosis. J Clin Endocrinol Metab 1974;38:1122-4.
289. Walsh CH, Wright AD, Williams JW, Holder G. A study of pituitary function in patients with idiopathic hemochromatosis. J Clin Endocrinol Metab 1976;43:866-72.
290. Pelusi C, Gasparini DI, Bianchi N, Pasquali R. Endocrine dysfunction in hereditary hemochromatosis. J Endocrinol Invest 2016;39:837-47.
291. Lufkin EG, Baldus WP, Bergstralh EJ, Kao PC. Influence of phlebotomy treatment on abnormal hypothalamic-pituitary function in genetic hemochromatosis. Mayo Clin Proc 1987;62:473-9.
292. Hempenius LM, Van Dam PS, Marx JJ, Koppeschaar HP. Mineralocorticoid status and endocrine dysfunction in severe hemochromatosis. J Endocrinol Invest 1999;22:369-76.
293. Schochet SS Jr, McCormick WF, Halmi NS. Pituitary gland in patients with Hurler syndrome: light and electron microscopic study. Arch Pathol 1974;97:96-9.

Hypophysitis and Infectious Diseases

294. Joshi MN, Whitelaw BC, Carroll PV. Mechanisms in endocrinology: hypophysitis: diagnosis and treatment. Eur J Endocrinol 2018; 179:R151-63.
295. Lury KM. Inflammatory and infectious processes involving the pituitary gland. Top Magn Reason Imaging 2005;16:301-6.
296. Unluhizarci K, Bayram F, Colak R, et al. Distinct radiological and clinical appearance of lymphocytic hypophysitis. J Clin Endocrinol Metab 2001;86:1861-44.
297. Caturegli P, Lupi I, Landek-Salgado M, Kimura H, Rose NR. Pituitary autoimmunity: 30 years later. Autoimmun Rev 2008;7:631-7.
298. Honegger J, Schlaffer S, Menzel C, et al. Diagnosis of primary hypophysitis in Germany. J Clin Endocrinol Metab 2015;100:3841-9.
299. Fujiwara T, Ota K, Kakudo N, et al. Idiopathic giant cell granulomatous hypophysitis with hypopituitarism, right abducens nerve paresis and masked diabetes insipidus. Intern Med 2001;40:915-9.
300. Husain Q, Zouzias A, Kanumuri VV, Eloy JA, Liu JK. Idiopathic granulomatous hypophysitis presenting as pituitary apoplexy. J Clin Neurosci 2014;21:510-2.
301. Ikeda J, Kuratsu J, Miura M, Kai Y, Ushio Y. Lymphocytic adenohypophysitis accompanying occlusion of bilateral internal carotid arteries—case report. Neurol Med Chir (Tokyo) 1990;30:346-9.
302. Katoh N, Machida K, Satoh S, Yahikozawa H, Ikeda SI. [A clinically diagnosed lymphocytic hypophysitis presenting as recurrent meningitis.] Rinsho Shinkeigaku 2007;47:419-22. [Japanese]

303. Pekic S, Miljic D, Popovic V. Infections of the hypothalamic-pituitary region. In: Feingold KR, Anawalt B, Boyce A, et al., eds. Endotext [Internet]. South Dartmouth: MDText.com, Inc.; 2000-.
304. Domingue JN, Wilson CB. Pituitary abscesses. Report of seven cases and review of the literature. J Neurosurg 1977;46:601-8.
305. Doniach I. Histopathology of the pituitary. Clin Endocrinol Metab 1985;14:765-89.
306. Obenchain TG, Becker DP. Abscess formation in a Rathke's cleft cyst: case report. J Neurosurg 1972;36:359-62.
307. Flanagan DE, Ibrahim AE, Ellison DW, et al. Inflammatory hypophystis—the spectrum of disease. Acta Neurochir 2002;144:47-56.
308. Sautner D, Saeger W, Ludecke DK, Jansen V, Puchner MJ. Hypophysitis in surgical and autoptical specimens. Acta Neuropathol 1995;90:637-44.
309. Arunkumar MJ, Rajshekhar V. Intrasellar tuberculoma presenting as pituitary apoplexy. Neurol India 2001;49:407-10.
310. Husain N, Husain M, Rao P. Pituitary tuberculosis mimicking idiopathic granulomatous hypophysitis. Pituitary 2008;11:313-5.
311. Sharma MC, Arora R, Mahapatra AK, Sarat-Chandra P, Gaikwad SB, Sarkar C. Intrasellar tuberculoma—an enigmatic pituitary infection: a series of 18 cases. Clin Neurol Neurosurg 2000;102:72-7.
312. Spinner CD, Noe S, Schwerdtfeger C, et al. Acute hypophysitis and hypopituitarism in early syphilitic meningitis in a HIV-infected patient: a case report. BMC Infect Dis 2013;13:481-4.
313. Hong W, Liu Y, Chen M, Lin K, Liao Z, Huang S. Secondary headache due to aspergillus sellar abscesssimulating a pituitary neoplasm: case report and review of literature. Springerplus 2015;4:550.
314. Lim JC, Rogers TW, King J, Gaillard F. Successful treatment of pituitary sella Aspergillus abscess in a renal transplant recipient. J Clin Neurosci 2017;45:138-40.
315. Liu W, Chen H, Cai B, Li G, You C, Li H. Successful treatment of sellar aspergillus abscess. J Clin Neurosci 2010;17:1587-9.
316. Saffarian A, Derakhshan N, Taghipour M, Eghbal K, Roshanfarzad M, Dehghanian A. Sphenoid Aspergilloma with headache and acute vision loss. World Neurosurg 2018;115:159-61.
317. Strickland BA, Pham M, Bakhsheshian J, Carmichael J, Weiss M, Zada G. Endoscopic endonasal transsphenoidal drainage of a spontaneous candida glabrata pituitary abscess. World Neurosurg 2018;109:467-70.
318. Vijayvargiya P, Javed I, Moreno J, et al. Pituitary aspergillosis in a kidney transplant recipient and review of the literature. Transpl Infect Dis 2013;15:E196-200.
319. Hao L, Jing C, Bowen C, Min H, Chao Y. Aspergillus sellar abscess: case report and review of the literature. Neurol India 2008;56:186-8.
320. Iplikcioglu AC, Bek S, Bikmaz K, et al. Aspergillus pituitary abscess. Acta Neurochir (Wien) 2004;146:521-4.
321. Schaefer S, Boegershausen N, Meyer S, Ivan D, Schepelmann K, Kann PH. Hypothalamic-pituitary insufficiency following infectious diseases of the central nervous system. Eur J Endocrinol 2008;158:3-9.
322. Tanriverdi F, Alp E, Demiraslan H, et al. Investigation of pituitary functions in patients with acute meningitis: a pilot study. J Endocrinol Invest 2008;31:489-91.
323. Tsiakalos A, Xynos ID, Sipsas NV, Kaltsas G. Pituitary insufficiency after infectious meningitis: a prospective study. J Clin Endocrinol Metab 2010;957:3277-81.
324. Jones G, Muriello M, Patel A, Logan L. Enteroviral meningoencephalitis complicated by central diabetes insipidus in a neonate: a case report and review of the literature. J Pediatric Infect Dis Soc 2015;4:155-8.
325. Moses AM, Thomas DG, Canfield MC, Collins GH. Central diabetes insipidus due to cytomegalovirus infection of the hypothalamus in a patient with acquired immunodeficiency syndrome: a clinical, pathological, and immunohistochemical case study. J Clin Endocrinol Metab 2003;88:51-4.
326. Scheinpflug K, Schalk E, Reschke K, Franke A, Mogren M. Diabetes insipidus due to herpes encephalitis in a patient with diffuse large cell lymphoma. A case report. Exp Clin Endocrinol Diabetes 2006;114:31-4.
327. Torres AM, Kazee AM, Simon H, Knudson PE, Weinstock RS. Central diabetes insipidus due to herpes simplex in a patient immunosuppressed by Cushing's syndrome. Endocr Pract 2000;6:26-8.
328. Tsuji H, Yoshifuji H, Fujii T, et al. Visceral disseminated varicella zoster virus infection after rituximab treatment for granulomatosis with polyangiitis. Mod Rheumatol 2017;27:155-61.
329. Zhang X, Li Q, Hu P, Cheng H, Huang G. Two case reports of pituitary adenoma associated with toxoplasma gondii infection. J Clin Pathol 2002;55:965-6.

Autoimmune Disorders

330. Caturegli P, Newschaffer C, Olivi A, et al. Autoimmune hypophysitis. Endocr Rev 2005;26:599-614.
331. Asa SL, Bilbao JM, Kovacs K, Josse RG, Kreines K. Lymphocytic hypophysitis of pregnancy resulting in hypopituitarism: a distinct clinicopathologic entity. Ann Intern Med 1981;95:166-71.

332. Beressi N, Beressi JP, Cohen R, Modigliani E. Lymphocytic hypophysitis. A review of 145 cases. Ann Med Interne (Paris) 1999;150:327-41.
333. Hashimoto K, Takao T, Makino S. Lymphocytic adenohypophysitis and lymphocytic infundibuloneurohypophysitis. Endocr J 1997;44:1-10.
334. Jensen MD, Handwerger BS, Scheithauer BW, Carpenter PC, Mirakian R, Banks PM. Lymphocytic hypophysitis with isolated corticotropin deficiency. Ann Intern Med 1986;105:200-3.
335. Karlsson FA, Kämpe O, Winqvist O, Burman P. Autoimmune disease of the adrenal cortex, pituitary, parathyroid glands and gastric mucosa. J Int Med 1993;234:379-86.
336. Lee JH, Laws ER Jr, Guthrie BL, Dina TS, Nochomovitz LE. Lympocytic hypophysitis: occurrence in two men. Neurosurgery 1994;34:159-63.
337. Lury KM. Inflammatory and infectious processes involving the pituitary gland. Top Magn Reason Imaging 2005;16:301-6.
338. Meichner RH, Riggio S, Manz HJ, Earll JM. Lymphocytic adenohypophysitis causing pituitary mass. Neurology 1987;37:158-61.
339. Thodou E, Asa SL, Kontogeorgos G, Kovacs K, Horvath E, Ezzat S. Clinical case seminar: lymphocytic hypophysitis: clinicopathological findings. J Clin Endocrinol Metab 1995;80:2302-11.
340. Shimono T, Yamaoka T, Nishimura K, et al. Lymphocytic hypophysitis presenting with diabetes insipidus: MR findings. Eur Radiol 1999;9:1397-400.
341. Ji JD, Lee SY, Choi SJ, et al. Lymphocytic hypophysitis in a patient with systemic lupus erythematosus. Clin Exp Rheumatol 2000;18:78-80.
342. Takao T, Nanamiya W, Matsumoto R, Asaba K, Okabayashi T, Hashimoto K. Antipituitary antibodies in patients with lymphocytic hypophysitis. Horm Res 2001;55:288-92.
343. Gellner V, Kurschel S, Scarpatetti M, Mokry M. Lymphocytic hypophysitis in the pediatric population. Childs Nerv Syst 2008;24:785-92.
344. Guay AT, Agnello V, Tronic BC, Gresham DG, Freidberg SR. Lymphocytic hypophysitis in a man. J Clin Endocrinol Metab 1987;64:631-4.
345. Hoshimaru M, Hashimoto N, Kikuchi H. Central diabetes insipidus resulting from a nonneoplastic tiny mass lesion localized in the neurohypophyseal system. Surg Neurol 1992;38:1-6.
346. Levine SN, Benzel EC, Fowler MR, Shroyer 3rd JV, Mirfakhraee M. Lymphocytic adenohypophysitis: clinical, radiological, and magnetic resonance imaging characterization. Neurosurgery 1988;22:937-41.
347. Caturegli P, Lupi I, Landek-Salgado M, Kimura H, Rose NR. Pituitary autoimmunity: 30 years later. Autoimmun Rev 2008;7:631-7.
348. Leung GK, Lopes MB, Thorner MO, et al. Primary hypophysitis: a single-center experience in 16 cases. J Neurosurg 2004;101:262-71.
349. Tashiro T, Sano T, Xu B, et al. Spectrum of different types of hypophysitis: a clinicopathologic study of hypophysitis in 31 cases. Endocr Pathol 2002;13:183-95.
350. Faje A. Hypophysitis: evaluation and management. Clin Diabetes Endocrinol 2016;2:15.
351. Honegger J, Schlaffer S, Menzel C, et al. Diagnosis of primary hypophysitis in Germany. J Clin Endocrinol Metab 2015;100:3841-9.
352. Ikeda J, Kuratsu J, Miura M, Kai Y, Ushio Y. Lymphocytic adenohypophysitis accompanying occlusion of bilateral internal carotid arteries—case report. Neurol Med Chir (Tokyo) 1990;30:346-9.
353. Melgar MA, Mariwalla N, Gloss DS, Walsh JW. Recurrent lymphocytic hypophysitis and bilateral intracavernous carotid artery occlusion. An observation and review of the literature. Neurol Res 2006;28:177-83.
354. Peruzzotti-Jametti L, Strambo D, Sangalli F, De Bellis A, Comi G, Sessa M. Bilateral intracavernous carotid artery occlusion caused by invasive lymphocytic hypophysitis. J Stroke Cerebrovasc Dis 2012;21:918.
355. Husain Q, Zouzias A, Kanumuri VV, Eloy JA, Liu JK. Idiopathic granulomatous hypophysitis presenting as pituitary apoplexy. J Clin Neurosci 2014;21:510-2.
356. Minakshi B, Alok S, Hillol KP. Lymphocytic hypophysitis presenting as pituitary apoplexy in a male. Neurol India 2005;53:363-4.
357. Kravarusic J, Molitch ME. Lymphocytic hypophysitis and other inflammatory conditions of the pituitary. In: Wass JA, Stewart PM, eds. Oxford textbook of endocrinology and diabetes, 2nd ed. New York: Oxford University Press; 2011:259-66.
358. Smith CJ, Bensing S, Burns C, et al. Identification of TPIT and other novel autoantigens in lymphocytic hypophysitis: immunoscreening of a pituitary cDNA library and development of immunoprecipitation assays. Eur J Endocrinol 2012;166:391-8.
359. Heaney AP, Sumerel B, Rajalingam R, Bergsneider M, Yong WH, Liau LM. HLA markers DQ8 and DR53 are associated with lymphocytic hypophysitis and may aid in differential diagnosis. J Clin Endocrinol Metab 2015;100:4092-7.
360. Sautner D, Saeger W, Ludecke DK, Jansen V, Puchner MJ. Hypophysitis in surgical and autoptical specimens. Acta Neuropathol 1995;90:637-44.
361. Joshi MN, Whitelaw BC, Carroll PV. Mechanisms in endocrinology: hypophysitis: diagnosis and treatment. Eur J Endocrinol 2018;179:R151-63.

362. Bellastella G, Maiorino MI, Bizzarro A, et al. Revisitation of autoimmune hypophysitis: knowledge and uncertainties on pathophysiological and clinical aspects. Pituitary 2016;19:625-42.
363. Gutenberg A, Hans V, Puchner MJ, et al. Primary hypophysitis: clinical-pathological correlations. Eur J Endocrinol 2006;155:101-7.
364. Hunn BH, Martin WG, Simpson S Jr, McLean CA. Idiopathic granulomatous hypophysitis: a systematic review of 82 cases in the literature. Pituitary 2014;17:357-65.
365. Vesely DL, Maldonodo A, Levey GS. Partial hypopituitarism and possible hypothalamic involvement in sarcoidosis. Report of a case and review of the literature. Am J Med 1977;62:425-31.
366. Oelbaum MH. Hypopituitarism in male subjects due to syphilis, with a discussion of androgen treatment. Q J Med 1952;21:249-66.
367. Folkerth RD, Price DL Jr, Schwartz M, Black PM, De Girolami U. Xanthomatous hypophysitis. Am J Surg Pathol 1998;22:736-41.
368. Hanna B, Li YM, Beutler T, Goyal P, Hall WA. Xanthomatous hypophysitis. J Clin Neurosci 2015;22:1091-7.
369. Aste L, Bellinzona M, Meleddu V, Farci G, Manieli C, Godano U. Xanthomatous hypophysitis mimicking a pituitary adenoma: case report and review of the literature. J Oncol 2010:195323.
370. Burt MG, Morey AL, Turner JJ, Pell M, Sheehy JP, Ho KK. Xanthomatous pituitary lesions: a report of two cases and review of the literature. Pituitary 2003;6:161-8.
371. Gutenberg A, Buslei R, Fahlbusch R, Buchfelder M, Brück W. Immunopathology of primary hypophysitis: implications for pathogenesis. Am J Surg Pathol 2005;29:329-38.
372. Bando H, Iguchi G, Yamamoto M, Hidaka-Takeno R, Takahashi Y. Anti-PIT-1 antibody syndrome; a novel clinical entity leading to hypopituitarism. Pediatr Endocrinol Rev 2015;12:290-6.
373. Yamamoto M, Iguchi G, Takeno R, et al. Adult combined GH, prolactin, and TSH deficiency associated with circulating PIT-1 antibody in humans. J Clin Invest 2011;121:113-9.
374. Bando H, Iguchi G, Fukuoka H, et al. Involvement of PIT-1-reactive cytotoxic T lymphocytes in anti-PIT-1 antibody syndrome. J Clin Endocrinol Metab 2014;99:E1744-9.
375. De Bellis A, Bizzarro A, Bellastella A. Autoimmune central diabetes insipidus. In: Geenen V, ed. Immunoendocrinology in health and disease, 1st ed. Boca Raton: CRC Press; 2004.
376. Betterle C, Presotto F, Zanchetta R. Generalita` sulla malattie autoimmuni. In: Betterle C, ed. Le malattie autoimmuni. Padua: Piccin; 2001:9-36.
377. Doniach D, Bottazzo GF. Poliendocrine autoimmunity. In: Franklin EC, ed. Clinical immunology update. Amsterdam: Elsevier-North Holland; 1981:5-121.
378. Furmaniak J, Sanders J, Rees Smith B. Autoantigens in the autoimmune endocrinopathies. In: Volpe R, ed. Autoimmune endocrinopathies. Totowa: Humana Press; 1999:183-216.
379. Imura H, Nakao K, Shimatsu A, et al. Lymphocytic infundibuloneurohypophysitis as a cause of central diabetes insipidus. N Engl J Med 1993;329:683-9.
380. Scherbaum WA, Bottazzo JF. Autoantibodies to vasopressin cells in idiopathic diabetes insipidus: evidence for an autoimmune variant. Lancet 1983;1:897-901.
381. Scherbaum WA, Hauner H, Pfeiffer EF. Vasopressin cell surface antibodies in central diabetes insipidus detected on cultured human foetal hypothalamus. Horm Metab Res 1985;17:622.
382. Bhan GI, O'Brien TD. Autoimmune endocrinopathy associated with diabetes insipidus. Postgrad Med J 1982;58:165-6.
383. Nerup J. Addison's disease—clinical studies. A report of 108 cases. Acta Endocrinol 1974;76:127-41.
384. Scherbaum WA, Wass JA, Besser GM, Bottazzo JF, Doniach D. Autoimmune cranial diabetes insipidus: its association with other endocrine disorders and with histiocytosis X. Clin Endocrinol (Oxf) 1986;25:411-20.
385. De Bellis A, Bizzarro A, Di Martino S, et al. Association of arginin vasopressin-secreting cell, steroid-secreting cell, adrenal and islet cell antibodies in a patients presenting with central diabetes insipidus, empty sella, subclinical adrenocortical failure and impaired glucose tolerance. Horm Res 1995;44:142-6.
386. De Bellis A, Colao A, Bizzarro A, et al. Longitudinal study of vasopressin-cell antibodies and of hypothalamic-pituitary region on magnetic resonance imaging in patients with autoimmune and idiopathic complete central diabetes insipidus. J Clin Endocrinol Metab 2002;87:3825-9.
387. Betterle C, Sabbadin C, Scaroni C, Presotto F. Autoimmune polyendocrine syndromes (APS) or multiple autoimmune syndromes (MAS): an overview. In: Colao A, Jaffrain-Rea ML, Beckers A, eds. Polyendocrine disorders and endocrine neoplastic syndromes. Cham: Springer; 2019.
388. Eisenbarth GS, Gottlieb PA. Autoimmune polyendocrine syndromes. N Engl J Med 2004;350:2068-79.
389. Husebye ES, Anderson MS, Kampe O. Autoimmune polyendocrine syndromes. N Eng J Med 2018;378:1132-41.
390. Kahaly GJ. Polyglandular autoimmune syndromes. Eur J Endocrinol 2009;161:11-20.

391. Kahaly GJ, Frommer L. Polyglandular autoimmune syndromes. J Endocrinol Invest 2018;41:91-8.
392. Betterle C, Presotto F. Sindrome autoimmune multipla di tipo 3: una galassia in espansione. L'Endocrinologo 2009;10:132-42.
393. Beressi N, Cohen R, Beressi JP, et al. Pseudotumoral lymphocytic hypophysitis successfully treated by corticosteroid alone: first case report. Neurosurgery 1994;35:505-8.

IgG4-Related Disease

394. Vliet HJ, Perenboom RM. Multiple pseudotumors in IgG4- associated multifocal systemic fibrosis. Ann Intern Med 2004;141:896-7.
395. Stone JH, Chan JK, Deshpande V, Okazaki K, Umehara H, Zen Y. IgG4-related disease. Int J Rheumatol 2013;2013:532612.
396. Wallace ZS, Stone JH. An update on IgG4-related disease. Curr Opin Rheumatol 2015;27:83-90.
397. Shimatsu A, Oki Y, Fujisawa I, Sano T. Pituitary and stalk lesions (infundibulo-hypophysitis) associated with immunoglobulin G4-related systemic disease: an emerging clinical entity. Endocr J 2009;56:1033-41.
398. Stone JH, Zen Y, Deshpande V. IgG4-related disease. NEJM 2012;366:539-51.
399. Hori M, Makita N, Andoh T, et al. Long-term clinical course of IgG4-related systemic disease accompanied by hypophysitis. Endocr J 2010;57:485-92.
400. Leporati P, Landek-Salgado MA, Lupi I, Chiovato L, Caturegli P. IgG4-related hypophysitis: a new addition to the hypophysitis spectrum. J Clin Endocrinol Metab 2011;96:1971-80.
401. Lee J, Hyun SH, Kim S, et al. Utility of FDG PET/CT for differential diagnosis of patients clinically suspected of IgG4-related disease. Clin Nucl Med 2016;41:e237-43.
402. Bando H, Iguchi G, Fukuoka H, et al. The prevalence of IgG4-related hypophysitis in 170 consecutive patients with hypopituitarism and/or central diabetes insipidus and review of the literature. Eur J Endocrinol 2014;170:161-72.
403. Bernreuther C, Illies C, Flitsch J, et al. IgG4-related hypophysitis is highly prevalent among cases of histologically confirmed hypophysitis. Brain Pathol 2016;27:839-45.
404. Deshpande V, Zen Y, Chan JK, et al. Consensus statement on the pathology of IgG4-related disease. Mod Pathol 2012;25:1181-92.
405. Khosroshahi A, Cheryk LA, Carruthers MN, Edwards JA, Bloch DB, Stone JH. Brief report: spuriously low serum IgG4 concentrations caused by the prozone phenomenon in patients with IgG4- related disease. Arthritis Rheumatol 2014;66:213-7.
406. Wallace ZS, Deshpande V, Mattoo H, et al. IgG4-related disease: clinical and laboratory features in one hundred twenty-five patients. Arthritis Rheumatol 2015;67:2466-75.
407. Bando H, Iguchi G, Fukuoka H, et al. A diagnostic pitfall in IgG4-related hypophysitis: infiltration of IgG4-positive cells in the pituitary of granulomatosis with polyangiitis. Pituitary 2015;18:722-30.
408. Chang SY, Keogh KA, Lewis JE, et al. IgG4-positive plasma cells in granulomatosis with polyangiitis (Wegener's): a clinicopathologic and immunohistochemical study on 43 granulomatosis with polyangiitis and 20 control cases. Hum Pathol 2013;44:2432-7.
409. Nishioka H, Shibuya M, Haraoka J. Immunohistochemical study for IgG4-positive plasmacytes in pituitary inflammatory lesions. Endocr Pathol 2010;21:236-41.
410. Ohkubo Y, Sekido T, Takeshige K, et al. Occurrence of IgG4-related hypophysitis lacking IgG4-bearing plasma cell infiltration during steroid therapy. Intern Med 2014;53:753-7.
411. Carruthers MN, Topazian MD, Khosroshahi A, et al. Rituximab for IgG4-related disease: a prospective, open-label trial. Ann Rheum Dis 2015;74:1171-7.
412. Wallace ZS, Mattoo H, Mahajan VS, et al. Predictors of disease relapse in IgG4-related disease following rituximab. Rheumatology 2016;55:1000-8.

Langerhans Cell Histiocytosis

413. Baumgartner I, von Hochstetter A, Baumert B, Luetolf U, Follath F. Langerhans' cell histiocytosis in adults. Med Pediatr Oncol 1997;28:9-14.
414. Makras P, Alexandraki KI, Chrousos GP, Grossman AB, Kaltsas GA. Endocrine manifestations in Langerhans cell histiocytosis. Trends Endocrinol Metab 2007;18:252-7.
415. Favara BE. Langerhans' cell histiocytosis pathobiology and pathogenesis. Semin Oncol 1991;18:3-7.
416. Willman CL, Busque L, Griffith BB, et al. Langerhans' cell histiocytosis (histiocytosis X)—a clonal proliferative disease. N Engl J Med 1994;331:154-60.
417. Allen CE, Beverley PC, Collin M, et al. The coming of age of Langerhans cell histiocytosis. Nat Immunol 2020;21:1-7.
418. Girschikofsky M, Arico M, Castillo D, et al. Management of adult patients with Langerhans cell histiocytosis: recommendations from an expert panel on behalf of Euro-Histio-Net. Orphanet J Rare Dis 2013;8:72.
419. Gadner H, Minkov M, Grois N, et al. Therapy prolongation improves outcome in multisystem Langerhans cell histiocytosis. Blood 2013; 121:5006-14.

420. Donadieu J, Rolon MA, Thomas C, et al. Endocrine involvement in pediatric-onset Langerhans' cell histiocytosis: a population-based study. J Pediatr 2004;144:344-50.
421. Nanduri VR, Bareille P, Pritchard J, Stanhope R. Growth and endocrine disorders in multisystem Langerhans' cell histiocytosis. Clin Endocrinol 2000;53:509-15.
422. Su C. [Hypophysitis mimicking a pituitary adenoma (report of 3 cases).] Zhongguo Yi Xue Ke Xue Yuan Xue Bao 1991;13:376-9. [Chinese]
423. Gersey ZC, Zheng I, Bregy A, et al. Intracranial Langerhans cell histiocytosis: a review. Interdiscip Neurosurg 2020;21:100729.

Pituitary Cysts

424. Newman CB, New MI. Endocrine function in children with intrasellar and suprasellar neoplasms. Am J Dis Child 1981;33:99-105.
425. Shin JL, Asa SL, Woodhouse LJ, et al. Cystic lesions of the pituitary: clinicopathological features distinguishing craniopharyngioma, Rathke's cleft cyst, and arachnoid cyst. J Clin Endocrinol Metab 1999;84:3972-82.
426. Kucharczyk W, Peck WW, Kelly WM, Norman D, Newton TH. Rathke cleft cysts: CT, MR imaging, and pathologic features. Radiology 1987;165:491-5.
427. Matsushima T, Fukui M, Fujii K. Epithelial cells in symptomatic Rathke's cleft cysts. A light- and electron-microscopic study. Surg Neurol 1988;30:197-203.
428. Mukherjee JJ, Islam N, Kaltsas G, et al. Clinical, radiological and pathological features of patients with Rathke's cleft cysts and tumors that may recur. J Clin Endocrinol Metab 1997; 82:2357-62.
429. Steinberg GK, Koenig GH, Golden JB. Symptomatic Rathke's cleft cysts. Report of two cases. J Neurosurg 1982;56:290-5.
430. Prabhu VC, Brown HG. The pathogenesis of craniopharyngiomas. Childs Nerv Syst 2005;21: 622-7.
431. Potts MB, Jahangiri A, Lamborn KR, et al. Suprasellar Rathke cleft cysts: clinical presentation and treatment outcomes. Neurosurgery 2011;69:1058-69.
432. Aho CJ, Liu C, Zelman V, Couldwell WT, Weiss MH. Surgical outcomes in 118 patients with Rathke cleft cysts. J Neurosurg 2005;102:189-93.
433. Kim JE, Kim JH, Kim OL, et al. Surgical treatment of symptomatic Rathke cleft cysts: clinical features and results with special attention to recurrence. J Neurosurg 2004;100:33-40.
434. Saeger W, Ludecke DK, Buchfelder M, Fahlbusch R, Quabbe HJ, Petersenn S. Pathohistological classification of pituitary tumors: 10 years of experience with the German Pituitary Tumor Registry. Eur J Endocrinol 2007;156:203-16.
435. Voelker JL, Campbell RL, Muller J. Clinical, radiographic, and pathological features of symptomatic Rathke's cleft cysts. J Neurosurg 1991;74:535-44.
436. Zada G, Kelly DF, Cohan P, Wang C, Swerdloff R. Endonasal transsphenoidal approach for pituitary adenomas and other sellar lesions: an assessment of efficacy, safety, and patient impressions. J Neurosurg 2003;98:350-8.
437. Zada G, Liu CY, Fishback D, Singer PA, Weiss MH. Recognition and management of delayed hyponatremia following transsphenoidal pituitary surgery. J Neurosurg 2007;106:66-71.
438. Fager CA, Carter H. Intrasellar epithelial cysts. J Neurosurg 1966;24:77-81.
439. McGrath P. Cysts of sellar and pharyngeal hypophyses. Pathology 1971;3:123-31.
440. Shanklin WM. On the presence of cysts in the human pituitary. Anat Rec 1949;104:379-407.
441. Teramoto A, Hirakawa K, Sanno N, Osamura Y. Incidental pituitary lesions in 1,000 unselected autopsy specimens. Radiology 1994;193:161-4.
442. Benveniste RJ, King WA, Walsh J, Lee JS, Naidich TP, Post KD. Surgery for Rathke cleft cysts: technical considerations and outcomes. J Neurosurg 2004;101:577-84.
443. Shimoji T, Shinohara A, Shimizu A, Sato K, Ishii S. Rathke cleft cysts. Surg Neurol 1984;21:295-310.
444. Zada G, Ditty B, McNatt SA, McComb JG, Krieger MD. Surgical treatment of Rathke cleft cysts in children. Neurosurgery 2009;64:1132-7.
445. Nishioka H, Haraoka J, Izawa H, Ikeda Y. Magnetic resonance imaging, clinical manifestations, and management of Rathke's cleft cyst. Clin Endocrinol (Oxf) 2006;64:184-8.
446. Brassier G, Morandi X, Tayiar E, et al. Rathke's cleft cysts: surgical-MRI correlation in 16 symptomatic cases. J Neuroradiol 1999;26:162-71.
447. Ross DA, Norman D, Wilson CB. Radiologic characteristics and results of surgical management of Rathke's cysts in 43 patients. Neurosurgery 1992;30:173-8.
448. Binning MJ, Gottfried ON, Osborn AG, Couldwell WT. Rathke cleft cyst intracystic nodule: a characteristic magnetic resonance imaging finding. J Neurosurg 2005;103:837-40.
449. Byun WM, Kim OL, Kim D. MR imaging findings of Rathke's cleft cysts: significance of intracystic nodules. Am J Neuroradiol 2000;21:485-8.
450. Asari S, Ito T, Tsuchida S, Tsutsui T. MR appearance and cyst content of Rathke cleft cysts. J Comput Assist Tomogr 1990;14:532-5.
451. Hayashi Y, Tachibana O, Muramatsu N, et al. Rathke cleft cyst: MR and biomedical analysis of cyst content. J Comput Assist Tomogr 1999;23: 34-8.

452. Tominaga JY, Higano S, Takahashi S. Characteristics of Rathke's cleft cyst in MR imaging. Magn Reson Med Sci 2003;2:1-8.
453. Le BH, Towfighi J, Kapadia SB, Lopes MB. Comparative immunohistochemical assessment of craniopharyngioma and related lesions. Endocr Pathol 2007;18:23-30.
454. Brastianos PK, Taylor-Weiner A, Manley PE, et al. Exome sequencing identifies BRAF mutations in papillary craniopharyngiomas. Nat Genet 2014;46:161-5.
455. Zada G, Lin N, Ojerholm E, Ramkissoon S, Laws ER. Craniopharyngioma and other cystic epithelial lesions of the sellar region: a review of clinical, imaging, and histopathological relationships. Neurosurg Focus 2010;28:E4.
456. Xin W, Rubin MA, McKeever PE. Differential expression of cytokeratins 8 and 20 distinguishes craniopharyngioma from rathke cleft cyst. Arch Pathol Lab Med 2002;126:1174-8.
457. Freda PU, Post KD. Differential diagnosis of sellar masses. Endocrin Metab Clin North Am 1999;28:81-117.
458. Rennert J, Doerfler A. Imaging of sellar and parasellar lesions. Clin Neurol Neurosurg 2007; 109:111-24.
459. Spampinato MV, Castillo M. Congenital pathology of the pituitary gland and parasellar region. Top Magn Reson Imaging 2005;16:269-76.
460. El-Mahdy W, Powell M. Transsphenoidal management of 28 symptomatic Rathke's cleft cysts, with special reference to visual and hormonal recovery. Neurosurgery 1998;42:7-16.
461. Zada G. Rathke cleft cysts: a review of clinical and surgical management. Neurosurg Focus 2011;31:E1.
462. Israel ZH, Yacoub M, Gomori JM, et al. Rathke's cleft cyst abscess. Pediatr Neurosurg 2000;33:159-61.
463. Kasperbauer JL, Orvidas LJ, Atkinson JL, Abboud CF. Rathke cleft cyst: diagnostic and therapeutic considerations. Laryngoscope 2002;112:1836-9.
464. Sani S, Smith A, Leppla DC, Ilangovan S, Glick R. Epidermoid cyst of the sphenoid sinus with extension into the sella turcica presenting as pituitary apoplexy: case report. Surg Neurol 2005;63:394-7.
465. Zhou LF. Intracranial epidermoid tumours: thirty-seven years of diagnosis and treatment. Br J Neurosurg 1990;4:211-6.
466. Akdemir G, Daglioglu E, Ergungor MF. Dermoid lesion of the cavernous sinus: case report and review of the literature. Neurosurg Rev 2004;27:294-8.
467. Boggan JE, Davis RL, Zorman G, Wilson CB. Intrasellar epidermoid cyst. Case report. J Neurosurg 1983;58:411-5.
468. Lin CJ, Jou JR, Woung LC, Tu YK, Jeng YM, Chang HC. Suprasellar dermoid cyst presenting as acquired exotropia. J Pediatr Ophthalmol Strabismus 2003;40:47-50.
469. Oge K, Ozgen T. Transsphenoidal removal of an intra- and suprasellar epidermoid cyst. Neurochirurgia (Stuttg) 1991;34:94-6.
470. Ahmad I, Tominaga T, Ogawa A, Yoshimoto T. Ruptured suprasellar dermoid associated with middle cerebral artery aneurysm: case report. Surg Neurol 1992;38:341-6.
471. Cohen JE, Abdallah JA, Garrote M. Massive rupture of suprasellar dermoid cyst into ventricles. Case illustration. J Neurosurg 1997;87:963.
472. Mamata H, Matsumae M, Yanagimachi N, Matsuyama S, Takamiya Y, Tsugane R. Parasellar dermoid tumor with intra-tumoral hemorrhage. Eur Radiol 1998;8:1594-7.
473. Smith AS, Benson JE, Blaser SI, Mizushima A, Tarr RW, Bellon EM. Diagnosis of ruptured intracranial dermoid cyst: value MR over CT. Am J Neuroradiol 1991;12:175-80.
474. Takeuchi H, Kubota T, Kabuto M, Izaki K. Ruptured suprasellar dermoid cyst presenting olfactory delusion (Eigengeruchs erlebnis). Neurosurgery 1993;33:97-9.
475. Venkatesh SK, Phadke RV, Trivedi P, Bannerji D. Asymptomatic spontaneous rupture of suprasellar dermoid cyst: a case report. Neurol India 2002;50:480-3.
476. Caldarelli M, Massimi L, Kondageski C, Di Rocco C. Intracranial midline dermoid and epidermoid cysts in children. J Neurosurg 2004;100:473-80.
477. Dubuisson AS, Stevenaert A, Martin DH, Flandroy PP. Intrasellar arachnoid cysts. Neurosurgery 2007;61:505-13.
478. Fatemi N, Dusick JR, de Paiva Neto MA, Kelly DF. The endonasal microscopic approach for pituitary adenomas and other parasellar tumors: a 10-year experience. Neurosurgery 2008;63:244-56.
479. McLaughlin N, Vandergrift A, Ditzel Filho LF, et al. Endonasal management of sellar arachnoid cysts: simple cyst obliteration technique. J Neurosurg 2012;116:728-40.
480. Meyer FB, Carpenter SM, Laws ER Jr. Intrasellar arachnoid cysts. Surg Neurol 1987;28:105-10.
481. Inoue T, Matsushima T, Fukui M, Iwaki T, Takeshita I, Kuromatsu C. Immunohistochemical study of intracranial cysts. Neurosurgery 1988;23:576-81.
482. Bladowska J, Bednarek-Tupikowska G, Biel A, Sasiadek M. Colloid cyst of the pituitary gland. Case report and literature review. Pol J Radiol 2010;75:88-93.
483. Zada G, Lopes MB, Mukundan S Jr, Laws E Jr. Colloid cysts of the sellar region. In: Zada G, Lopes MB, Mukundan S Jr, Laws E Jr, eds. Atlas of sellar and parasellar lesions: clinical, radiologic, and pathologic correlations. New York: Springer; 2016:251-5.

484. Clarke PR, Rowbotham GF. Colloid cyst of the pituitary gland causing chiasmal compression. Br J Surg 1956;44:107-8.
485. Nomikos P, Buchfelder M, Fahlbusch R. Intra- and suprasellar colloid cysts. Pituitary 1999;2:123-6.
486. Lach B, Scheithauer BW, Gregor A, Wick MR. Colloid cyst of the third ventricle. A comparative immunohistochemical study of neuraxis cysts and choroid plexus epithelium. J Neurosurg 1993;78:101-11.
487. Gattuso P, Reddy VB, David O, Spitz DJ, Haber MH, eds. Central nervous system. Differential diagnosis in surgical pathology, 3rd ed. Philadelphia: Saunders; 2015:961-1023.
488. Guduk M, Sun HI, Sav MA, Berkman Z. Pituitary colloid cyst. J Craniofac Surg 2017;28:e166-8.

Hyperplasia

489. Horvath E. Pituitary hyperplasia. Pathol Res Pract 1988;183:623-5.
490. Scheithauer BW, Horvath E, Lloyd RV, Kovacs K. Pituitary hyperplasia. Pathol Case Rev 1998; 3:281-9.
491. Horvath E, Kovacs K, Scheithauer BW. Pituitary hyperplasia. Pituitary 1999;1:169-79.
492. McNicol AM, Carbajo-Perez E. Aspects of anterior pituitary growth, with special reference to corticotrophs. Pituitary 1999;1:257-68.
493. Al-Gahtany M, Horvath E, Kovacs K. Pituitary hyperplasia. Hormones (Athens) 2003;2:149-58.
494. De Sousa SM, Earls P, McCormack AI. Pituitary hyperplasia: a case series and literature review of an under-recognised and heterogeneous condition. Endocrinol Diabetes Metab J 2015:150017.
495. Shahani S, Nudelman RJ, Nalini R, Kim HS, Samson SL. Ectopic corticotropin-releasing hormone (CRH) syndrome from metastatic small cell carcinoma: a case report and review of the literature. Diagnostic Pathol 2010;5:56-60.
496. Asa SL, Kovacs K, Tindall GT, Barrow DL, Horvath E, Vecsei P. Cushing's disease associated with an intrasellar gangliocytoma producing corticotrophin-releasing factor. Ann Intern Med 1984;101:789-93.
497. Geddes JF, Jansen GH, Robinson SF, et al. 'Gangliocytomas' of the pituitary. A heterogeneous group of lesions with differing histogenesis. Am J Surg Pathol 2000;24:607-13.
498. Puchner MJ, Ludecke DK, Saeger W, Riedel M, Asa SL. Gangliocytomas of the sellar region—a review. Exp Clin Endocrinol Diabetes 1995; 103:129-49.
499. Scheithauer BW, Kovacs K, Randall RV, Horvath E, Okazaki H, Laws ER Jr. Hypothalamic neuronal hamartoma and adenohypophyseal neuronal choristoma: their association with growth hormone adenoma of the pituitary gland. J Neuropathol Exp Neurol 1983;42:648-63.
500. Towfighi J, Salam MM, McLendon RE, Powers S, Page RB. Ganglion cell-containing tumors of the pituitary gland. Arch Pathol Lab Med 1996;120:369-77.
501. Chanson P, Daujat F, Young J, et al. Normal pituitary hypertrophy as a frequent cause of pituitary incidentaloma: a follow-up study. J Clin Endocrinol Metab 2001;86:3009-15.
502. Horvath E, Kovacs K. Ultrastructural diagnosis of pituitary adenomas and hyperplasias. In: Lloyd RV, ed. Surgical pathology of the pituitary gland. Philadelphia: WB Saunders; 1993:52-84.
503. Lloyd RV, Chandler WF, McKeever PE, Schteingart DE. The spectrum of ACTH-producing pituitary lesions. Am J Surg Pathol 1986;10:618-26.
504. McKeever PE, Koppelman MC, Metcalf D, et al. Refractory Cushing's disease caused by multinodular, ACTH-cell hyperplasia. J Neuropathol Exp Neurol 1982;41:490-9.
505. Ilias I, Torpy DJ, Pacak K, Mullen N, Wesley RA, Nieman LK. Cushing's syndrome due to ectopic corticotropin secretion: twenty years' experience at the National Institutes of Health. J Clin Endocrinol Metab 2005;90:4955-62.
506. Isidori AM, Kaltsas GA, Pozza C, et al. The ectopic adrenocorticotropin syndrome: clinical features, diagnosis, management, and long-term follow-up. J Clin Endocrinol Metab 2006;91:371-7.
507. Salgado LR, Fragoso MC, Knoepfelmacher M, et al. Ectopic ACTH syndrome: our experience with 25 cases. Eur J Endocrinol 2006;155:725-33.
508. Newell-Price J. Etiologies of Cushing's syndrome. In: Bronstein M, ed. Cushing's syndrome. Totowa: Humana Press; 2010.
509. Horvath E, Kovacs K, Lloyd RV. Pars intermedia of the human pituitary revisited: morphologic aspects and frequency of hyperplasia of POMC-peptide immunoreactive cells. Endocr Pathol 1999;10:55-64.
510. Scheithauer BW, Kovacs K, Randall RV. The pituitary gland in untreated Addison's disease. A histologic and immunocytologic study of 18 adenohypophyses. Arch Pathol Lab Med 1983;107:484-7.
511. Krautli B, Muller J, Landolt AM, et al. ACTH-producing pituitary adenomas in Addison's disease: two cases treated by transsphenoidal microsurgery. Acta Endocrinol 1982;99:357-63.
512. Scheithauer BW, Kovacs K, Randall RV, Ryan N. Pituitary gland in hypothyroidism. Histologic and immunocytologic study. Arch Pathol Lab Med 1985;109:499-504.
513. Zhou J, Ruan L, Li H, et al. Addison's disease with pituitary hyperplasia: a case report and review of the literature. Endocr 2009;35:285-9.
514. Keil MF, Stratakis CA. Pituitary tumors in childhood: an update in their diagnosis, treatment and molecular genetics. Expert Rev Neurother 2008;8:563-74.

515. Asa SL, Scheithauer DW, Bilbao JM, et al. A case for hypothalamic acromegaly: a clinicopathological study of six patients with hypothalamic gangliocytomas producing growth hormone-releasing factor. J Clin Endocrinol Metab 1984;58:796-803.
516. Zimmerman D, Young WF Jr, Ebersold MJ, et al. Congenital gigantism due to growth hormone-releasing hormone excess and pituitary hyperplasia with adenomatous transformation. J Clin Endocrinol Metab 1993;76:216-22.
517. Kornreich L, Horev G, Schwarz M, Karmazyn B, Laron Z. Pituitary size in patients with Laron syndrome (primary GH insensitivity). Eur J Endocrinol 2003;148:339-41.
518. Adams C, Dean HJ, Israels SJ, et al. Primary hypothyroidism with intracranial hypertension and pituitary hyperplasia. Pediatr Neurol 1994;10:166-8.
519. Alkhani AM, Cusimano M, Kovacs K, et al. Cytology of pituitary thyrotroph hyperplasia in protracted primary hypothyroidism. Pituitary 1999;1:291-5.
520. Brinkmeier MR, Stahl JH, Gordon DF, et al. Thyroid hormone-responsive pituitary hyperplasia independent of somatostatin receptor 2. Mol Endocrinol 2001;15:2129-36.
521. Floyd JL, Dorwart RH, Nelson MJ, et al. Pituitary hyperplasia secondary to thyroid failure: CT appearance. Am J Neuroradiol 1984;5:469-71.
522. Hutchins WW, Crues JV III, Miya P, et al. MR demonstration of pituitary hyperplasia and regression after therapy for hypothyroidism. AJNR Am J Neuroradiol 1990;11:410.
523. Okuno T, Sudo M, Momoi T, et al. Pituitary hyperplasia due to hypothyroidism. J Comput Assist Tomogr 1980;4:600-2.
524. Papakonstantinou O, Bitsori M, Mamoulakis D, et al. MR imaging of pituitary hyperplasia in a child with growth arrest and primary hypothyroidism. Eur Radiol 2000;10:516-8.
525. Pioro EP, Scheithauer BW, Laws ER Jr, et al. Combined thyrotroph and lactotroph cell hyperplasia simulating prolactin-secreting pituitary adenoma in long-standing primary hypothyroidism. Surg Neurol 1988;29:218-26.
526. Shingyouchi H, Shindo M, Kobayashi M, et al. [Pituitary hyperplasia in primary hypothyroidism.] Rinsho Hoshasen 1990;35:529-32. [Japanese]
527. Williams RS, Williams JP, Davis MR, et al. Primary hypothyroidism with pituitary hyperplasia nd basal ganglia calcifications. Clin Imaging 1990;14:330-2.
528. Yamagishi S, Yokoyama-Ohta M. A rare case of pituitary hyperplasia with suprasellar extension due to primary myxoedema: case report. J Int Med Res 1999;27:49-52.
529. Yamamoto Y, Kunishio K, Sunami N, et al. [A case of pituitary hyperplasia associated with primary hypothyroidism.] No Shinkei Geka 1987;15:903-8. [Japanese]
530. Young M, Kattner K, Gupta K. Pituitary hyperplasia resulting from primary hypothyroidism mimicking macroadenomas. Br J Neurosurg 1999;13:138-42.
531. Vidal S, Horvath E, Kovacs K, et al. Transdifferentiation of somatotrophs to thyrotrophs in the pituitary of patients with protracted primary hypothyroidism. Virchows Arch 2000;436:43-51.
532. Kovacs K, Horvath E. The differential diagnosis of lesions involving the sella turcica. Endocr Pathol 2001;12:389-95.
533. Jentoft M, Scheithauer BW, Moshkin O, et al. Tumefactive postmenopausal gonadotroph cell hyperplasia. Endocr Pathol 2012;23:108-11.
534. Samaan NA, Stepanas AV, Danziger J, Trujillo J. Reactive pituitary abnormalities in patients with Klinefelter's and Turner's syndromes. Arch Intern Med 1979;139:198-201.
535. Okuda K, Yoshikawa M, Sugiyama S, et al. [Hypergonadotropic ovarian failure in three patients with pituitary hyperplasia.] Nippon Sanka Fujinka Gakkai Zasshi 1987;39:1579-84. [Japanese]
536. Okuda K, Yoshikawa M, Ushiroyama T, et al. Two patients with hypergonadotropic ovarian failure due to pituitary hyperplasia. Obstet Gynecol 1989;74:498-501.
537. Scheithauer BW, Moschopulos M, Kovacs K, Jhaveri BS, Percek T, Lloyd RV. The pituitary in Klinefelter syndrome. Endocr Pathol 2005;16:133-8.
538. Saeger W, Ludecke DK. Pituitary hyperplasia. Definition, light and electron microscopical structures and significance in surgical specimens. Virchows Archiv A Pathol Anat Histopathol 1983;399:277-87.
539. Tsunoda A, Okuda O, Sato K. MR height of the pituitary gland as a function of age and sex: especially physiological hypertrophy in adolescence and in climacterium. Am J Neuroradiol 1997;18:551-4.
540. Peillon F, Dupuy M, Li JY, et al. Pituitary enlargement with suprasellar extension in functional hyperprolactinemia due to lactotroph hyperplasia: a pseudotumoral disease. J Clin Endocrinol Metab 1991;73:1008-15.
541. Pariante CM, Dazzan P, Danese A, et al. Increased pituitary volume in antipsychotic-free and antipsychotic-treated patients of the AEsop first-onset psychosis study. Neuropsychopharmacology 2005;30:1923-31.

2 PARATHYROID GLAND

NORMAL PARATHYROID GLAND

Embryology and Anatomy

The parathyroid glands are derived from the third and fourth branchial pouches (1). The superior parathyroid glands are derived from proliferations along the lateral portion of the fourth branchial pouch. They are identified about 1 cm superior to the intersection of the recurrent laryngeal nerve and the medial thyroid artery, but can be seen in the thyroid gland and rarely in the retroesophageal or retropharyngeal spaces. The vasculature for the upper parathyroid glands in this location involves the superior thyroid artery and the superior or lateral thyroid vein.

The lower parathyroid glands are derived from the proliferation along the anterodorsal surface of the third branchial pouch, which separates from the pharynx at the lower pole of the thyroid gland. The vasculature of the lower parathyroid glands in this location involves the inferior thyroid artery and the lateral or inferior thyroid vein. The location of the lower parathyroid glands may be variable: they may be located adjacent to the lower thyroid gland or in the thymus, vagus nerve, mediastinum, or other ectopic locations.

Ectopic parathyroid glands are identified in 14 to 19 percent of cases of primary or secondary hyperparathyroidism (2,3). In addition to parathyroid autografts, ectopic or supranumerary parathyroid glands are a cause of persistent hyperparathyroidism in secondary hyperparathyroidism associated with renal disease (4–6). Ectopic parathyroid glands have been reported in unusual locations, including the vaginal wall, within a neck ganglion or paraganglion, at the carotid bifurcation, in the pyriform sinus, in the axillary region, in the sternohyoid muscle, or within the soft palate, trachea, or esophagus (7–17). Ectopic parathyroid glands can be involved by parathyroid disease, and may be associated with persistent or recurrent hyperparathyroidism if not identified initially (18).

Most individuals have four parathyroid glands, but 5 to 13 percent have five or more glands and 2 percent have less than four (19–21); individuals with more than 10 parathyroid glands have been reported. The most common location for supernumerary parathyroid is the thymus, but other locations include the retroesophageal groove, carotid sheath, mediastinum, and vagus nerve (19,22,23). In up to 30 percent of cases of secondary hyperparathyroidism, supernumerary parathyroid glands are identified at initial surgery, and may be responsible for persistent or recurrent disease (22,24).

Gross Findings

The parathyroid glands are usually ovoid but may be bilobed or multilobed, and are similar in size and shape to a kidney bean. They are usually dark reddish brown to light tan, depending on their fat content, and rarely are cystic (25). Normal parathyroid glands measure 2 to 4 mm in width, 3 to 6 mm in length, and 0.5 to 2.0 mm in thickness (25). They weigh 20 to 40 mg each; a parathyroid gland weighing more than 40 mg is considered abnormal (26). The lower parathyroid glands are often larger than the upper parathyroid glands (27). The combined weight of parathyroid glands in females is 142 mg (± 5.2 mg) and in males is 120 mg (± 3.5 mg) (28). Comparing parathyroid glands from 66 hospitalized patients with disease to 100 previously healthy individuals who died suddenly, those in hospitalized patients were heavier than those from individuals who were thought to be previously healthy (29). In this study, differences in parathyroid weights were reported for hospitalized patients of different races: among the previously healthy who died suddenly, black individuals had heavier parathyroid glands than white individuals. Also, gland weight

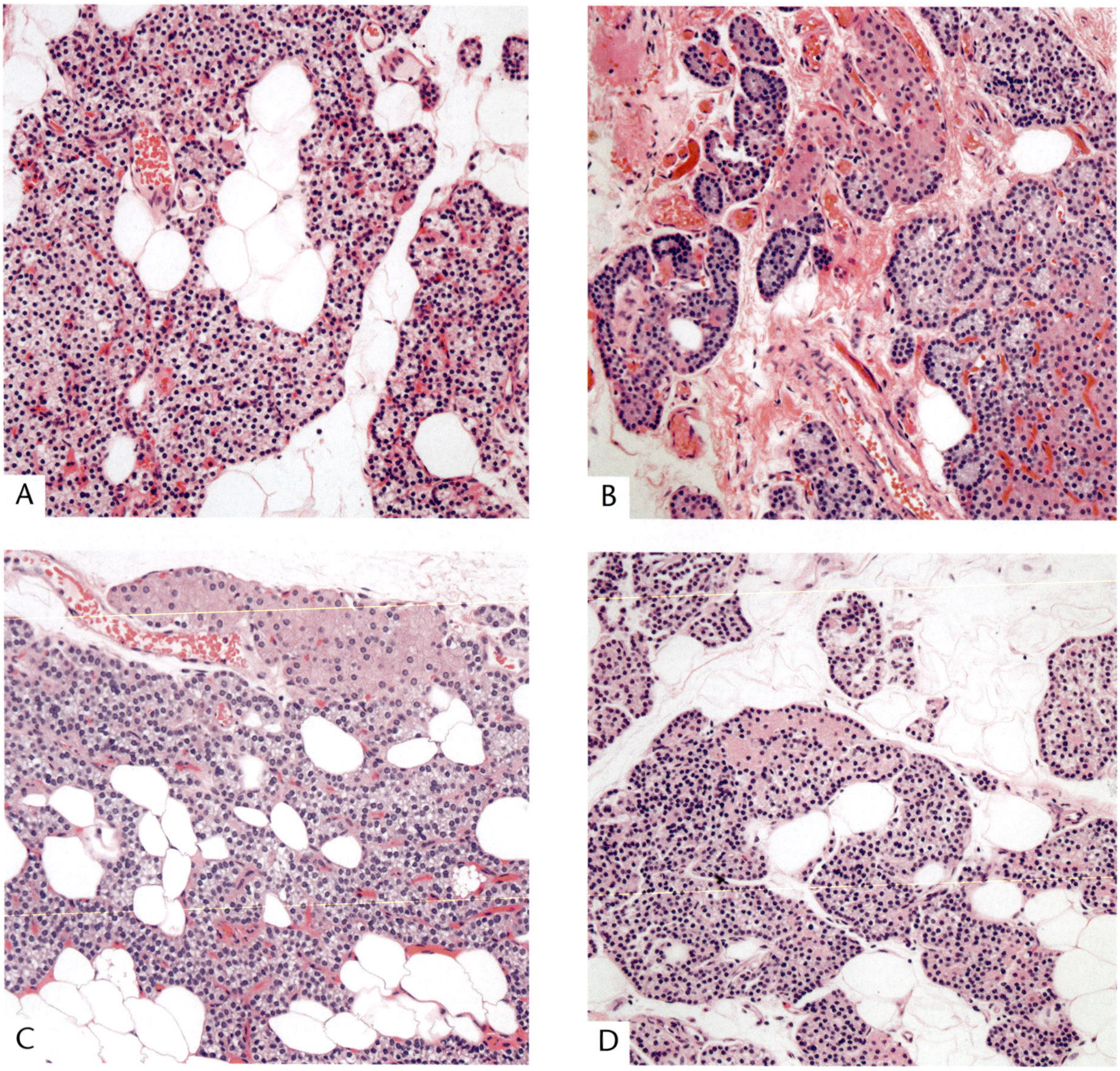

Figure 2-1

PARATHYROID GLAND

Parathyroid glands are composed of parenchymal cells and adipose tissue. Parenchymal cells include chief cells, oncocytic/oxyphilic cells, transitional cells, and clear (water clear) cells; mixtures of cell types can be seen (A–D). Parathyroid parenchymal cells have well-defined cytoplasmic membranes, a helpful feature in differentiating them from thyroid parenchymal cells.

varied somewhat with age, with a maximum weight found in the 41- to 60-year-old age group in males, but continued to increase until after age 70 in white women.

Histology

The parathyroid glands are composed of parenchymal cells and adipose tissue. The parenchymal cells are chief cells, oncocytic cells, transitional cells, and clear cells (fig. 2-1). Chief cells measure 8 to 10 µm and are polyhedral, with round central nuclei, eosinophilic to amphophilic cytoplasm, and well-defined cytoplasmic membranes.

Chief cells are the predominant cell type in infants and young children and may show some vacuolization (25). Chief cells in infants and young children are smaller (6 to 8 µm) than in adults and have less delineated cell membranes and less intracellular lipid. Fat granules (intracellular lipid) in parenchymal cells are generally not present at birth, but begin to appear after the third decade and increase with age (30,31). In a study of 250 autopsies, Kurokawa (31) examined 815 parathyroid glands from individuals ranging in age from 7 months to 80 years and noted that intracellular glycogen is present in parathyroid glands in young children, but intracellular lipid is usually not. In adults, chief cells often contain lipid droplets. Infants have less intracellular lipid in their chief cells than adults. Approximately 80 percent of normal chief cells in adults contain large cytoplasmic lipid droplets (32).

Well-defined cytoplasmic membranes are a characteristic feature of parathyroid parenchymal cells and can be useful in differentiating them from thyroid follicular cells. Cell membranes are less defined in infants than in adults.

Parathyroid oxyphil parenchymal cells are 12 to 20 µm in size and have oval nuclei and abundant granular eosinophilic cytoplasm reflecting prominent cytoplasmic mitochondria. Oxyphil cells are generally not present in infants and young children; they usually appear at puberty and increase with age (31). Oxyphil cells may form nodules and masses after 30 or 40 years of age (fig. 2-2) (31). Oxyphil cells do not contain fat or glycogen (25).

Nodules and masses of oxyphil cells can be mistaken for parathyroid disease. Oxyphil cells of the parathyroid gland also may be confused with Hurthle cells of the thyroid gland, especially in the diseased, and differentiating parathyroid oncocytic/oxyphilic adenoma from Hurthle cell lesion of the thyroid may be difficult.

Transitional parathyroid parenchymal cells are variants of oxyphil/oncocytic cells. They are smaller than chief and oxyphil cells, and have less eosinophilic cytoplasm.

Clear (water-clear, Wasserhelle) cells of the parathyroid gland are 10 to 15 µm and have clear cytoplasm (25). The nuclei are similar in size to those of chief cells, but are more hyperchromatic, often pyknotic, and eccentrically located (25). In 1935, Castleman and Mallory (25) did not identify clear cells in the parathyroid glands until puberty. Cytoplasmic clearing may be due to increased cytoplasmic vacuolization of chief or oxyphil cells and may be considered clear cell change. Clear cells are seen in clear cell parathyroid hyperplasia and occasionally in clear cell parathyroid adenoma.

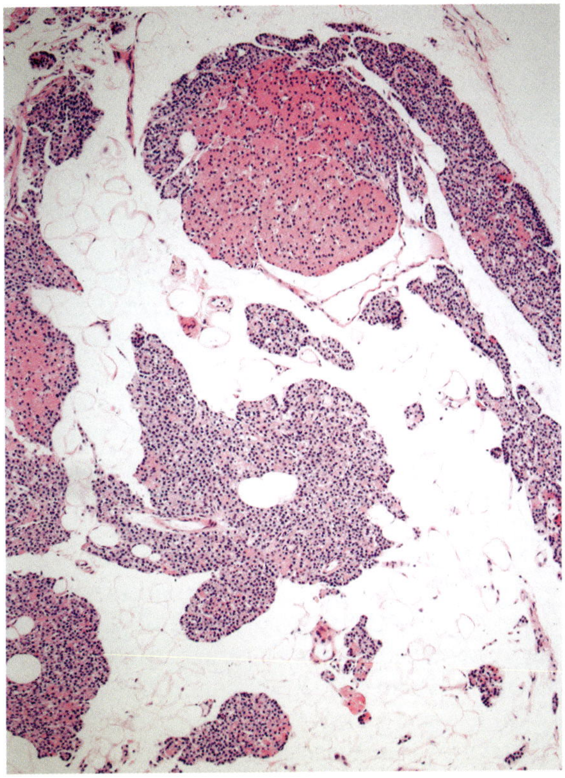

Figure 2-2

ONCOCYTIC/OXYPHILIC PARATHYROID PARENCHYMAL CELLS

Oncocytic/oxyphilic parathyroid parenchymal cells are not present in young children. These cells become apparent with increasing age. In older patients, oncocytic/oxyphil parathyroid parenchymal cells can form nodules and be mistaken for parathyroid disease.

Occasionally, parathyroid parenchymal cells form follicular- or glandular-like structures that mimic the follicular structures of thyroid parenchyma. Instead of colloid, as is present in the follicles of the thyroid gland, the eosinophilic material present in the parathyroid gland is congophilic and shows apple green birefringence (27). The material is also positive for parathyroid hormone (PTH). Unlike the colloid in thyroid parenchyma, the material is surrounded by the follicular- or acinar-type parathyroid cells, lacks oxalate crystals, and is negative for thyroglobulin. Mitotic figures are rare in thyroid parenchymal cells in adults.

Differentiating parathyroid from thyroid tissue can be difficult. Ectopic or supernumerary parathyroid glands are particularly difficult to identify clinically and histologically. Intrathyroidal parathyroid adenomas can be difficult to differentiate from thyroid lesions, and oxyphilic parathyroid lesions can be difficult to differentiate from Hurthle cell lesions of the thyroid gland. Additionally, differentiating parathyroid from thyroid tissue in small biopsies is a common problem for diagnostic pathologists. Parathyroid parenchymal cells are generally smaller than thyroid parenchymal cells. Parathyroid parenchymal cells have well-defined cytoplasmic membranes and may contain intracellular lipid (fat) droplets, while oxylate crystals may be identified in the colloid of thyroid parenchyma.

Normal parathyroid glands can become fibrotic with age. The interstitium of the parathyroid gland contains small amounts of collagen, which increases with age and may give the parathyroid gland a lobulated appearance.

Parathyroid glands are composed of a mixture of parenchymal cells and adipocytes, with adipocytes comprising 10 to 30 percent of the glandular volume (26,33,34). Historically, stromal fat was reported in up to 50 percent of parathyroid glands, but subsequent studies show that most parathyroid glands have less than 20 percent fat (33,34). However, the amount of stromal fat in the parathyroid glands is variable. The cellularity within and among parathyroid glands, even within a single individual, can vary. The poles of the parathyroid gland are generally less cellular and have more stromal fat than the central portion of the gland. Thus, judging cellularity of the parathyroid gland in small biopsies is extremely difficult, if not impossible, in many cases.

Fat cells are generally sparse in infants and children, appear just before puberty and increase until around the fourth decade (25). The amount of fat cells remains fairly constant from that time and does not appear to increase in the elderly (25). The amount of fat cells may even decrease somewhat in the eighth decade (25). In 1935, Castleman (25) noted that reductions in parathyroid gland size were attributable to a reduction in adipocyte content, with a relative preservation of parenchymal cell volume. That is, the size of a normally function parathyroid gland is mostly determined by the amount of intraparathyroidal fat not by the amount of adipose tissue (35).

The cellularity of parathyroid glands increases from birth until just before puberty, then decreases with increasing age. In addition to age, cellularity also varies by constitutional factors such as body fat, chronic illness, malignancy, genetic factors, race, and gender. Parathyroid glands in women have a higher percentage of fat than those in men. Capillaries, lymphatics, fibroblasts, mast cells, and rare lymphocytes are present in the interstitium.

Immunohistochemistry

Keratins. Parathyroid tissue is positive for cytokeratins (CK) 7, 8, 18, and 19 (26). CAM5.2 is also positive in parathyroid tissue, and is the most useful keratin in identifying epithelial origin in tumors with neuroendocrine differentiation. Some keratins, such as CK14, show differential expression in oxyphil parathyroid adenoma and some chief cells, but may not be expressed in oxyphilic parathyroid carcinomas (36). Keratins are positive in many other endocrine and neuroendocrine tissues and tumors, including normal thyroid tissue. Folliculogenic thyroid tissues and tumors are positive for keratins, as are thyroid C cells and medullary thyroid carcinomas. Thus, panels of immunostains are often required in the evaluation of endocrine and neuroendocrine tissues and tumors.

Neuroendocrine Markers. Parathyroid tissues express markers of neuroendocrine differentiation, with the most commonly used markers being chromogranin A and synaptophysin. Other markers may be positive in neuroendocrine tissues, such as neuron-specific enolase (NSE) and CD56, but the lack of specificity greatly limits their use.

Chromogranin A, the most specific neuroendocrine marker, may not stain parathyroid parenchymal cells uniformly. It stains chief cells more strongly than oxyphil cells. Additionally, non-neoplastic parathyroid tissue may stain more intensely for chromogranin A than parathyroid adenomas (37). Synaptophysin is more sensitive than chromogranin A in identifying neuroendocrine tissues and tumors, but it is less specific. Other neuroendocrine tissues and tumors in the neck may be positive for neuroendocrine markers, such as thyroid C cells, medullary thyroid carcinomas, and paragangliomas.

Parathyroid Hormone. Parathyroid hormone immunohistochemistry can be helpful in identifying parathyroid tissue and tumors, although its staining intensity can be variable. Parathyroid hormone may stain chief cells more intensely than oxyphil cells, and may stain normal parathyroid tissue more intensely than adenomatous or hyperfunctioning parathyroid tissue (37). A panel of stains, including parathyroid hormone, is useful in differentiating parathyroid tissue and tumors from other tissues and tumors in the neck.

Parathyroid hormone is positive in parathyroid tissues and generally negative in thyroid folliculogenic tissues and tumors, C cells, and medullary thyroid carcinomas. Folliculogenic thyroid tissues and tumors are positive for thyroglobulin, and thyroid C cells and medullary thyroid carcinomas are positive for calcitonin. Thyroglobulin and calcitonin are generally negative in parathyroid tissues and tumors; however, parathyroid tumors can occasionally express calcitonin and may even be associated with increased plasma calcitonin levels (38). Parathyroid hormone may be aberrantly expressed and secreted in a variety of other neoplasms, including: ovarian tumors, adenosquamous endometrial carcinoma, lymphoma, and prostate and gastric cancers (39–46).

Hypercalcemia can complicate many tumors, and secretion of parathyroid hormone-related peptide (PTHRP) is strongly associated with small cell carcinoma of the ovary, hypercalcemic type (SCCOHT) (47).

Transcription Factors. Glial cells missing 2 (GCM2) is a transcription factor involved in parathyroid development (48,49). Recent studies have identified *GCM2* mutations in both sporadic and familial parathyroid disease (48,50). Interestingly, immunohistochemistry for GCM2 is reportedly negative in thyroid and thymic tissues in addition to parathyroid cysts (49). Although not commonly used diagnostically, GCM2 expression is a sensitive and specific marker of parathyroid tissues and lesions, although the intensity of staining may vary (49).

Thyroid transcription factor-1 (TTF1) is one of the most commonly used transcription factors in diagnostic pathology. TTF1 is positive in thyroid tissues and tumors, both folliculogenic and C-cell–derived, and is helpful in differentiating thyroid from parathyroid tissue, which is negative for TTF1. TTF1 is also helpful in identifying lung adenocarcinomas and neuroendocrine carcinomas (51–55). High-grade neuroendocrine carcinomas from a variety of sites may be positive for TTF1. TTF1 also stains tumors of the ovary, breast, colon, kidney (nephroblastomas), and brain (gliomas), and thus it is not a highly specific marker (52,56–62). Differing clones of TTF1 vary in sensitivity and specificity: for example, clone SPT24 is more sensitive than clone 8G7G3, which is more specific (63,64).

PAX8 (paired-box gene 8) is a transcription factor involved in embryologic development of the thyroid gland, kidney, and mullerian tract, and is helpful in identifying tumors of these organs (65,66). While parathyroid tissue may show variably positive staining with polyclonal PAX8 antibodies, staining for monoclonal PAX8 is negative (65,66). Parathyroid tissue, however, is negative with monoclonal PAX8 antibody. Other neuroendocrine tumors, such as well-differentiated pancreatic neuroendocrine tumors (66) and tumors from the breast, lung (except squamous cell carcinoma), and head and neck are usually negative for PAX8 (65,66). It is important to use the monoclonal PAX8 antibodies directed against the C-terminus since the N-terminus polyclonal antibodies are nonspecific.

GATA3 is a transcription factor involved in transcription and activation of parathyroid hormone gene expression (67–69). In mice, GATA3 deficiency is associated with parathyroid abnormalities (69). In humans, the disorder of hypoparathyroidism, deafness, and renal dysplasia is associated with GATA3 haplo-insufficiency. GATA3 is expressed in most breast and urothelial carcinomas (70). Other neuroendocrine tumors do not express GATA3, although some folliculogenic thyroid and adrenal cortical tumors show infrequent GATA3 expression (70).

CDX2 is helpful in identifying neuroendocrine tumors from the gastrointestinal tract, such as midgut neuroendocrine tumors and a few other gastrointestinal neuroendocrine tumors, but parathyroid, pituitary, Merkel cell, and medullary thyroid tumors are usually CDX2 negative, as are paragangliomas and pheochromocytomas (71). A variety of other transcription factors are available (SF1, Pit-1), but are not used in the evaluation of parathyroid tissues or tumors.

HYPERPARATHYROIDISM

Hyperparathyroidism is an increase in parathyroid hormone secretion. In *primary hyperparathyroidism*, the increase in parathyroid hormone is due to overproduction by the parathyroid gland. *Secondary hyperparathyroidism* is caused by disease outside the parathyroid gland, such as chronic renal failure, which causes abnormally low serum calcium and vitamin D levels and results in excess secretion of parathyroid hormone by the parathyroid glands. In *tertiary hyperparathyroidism*, patients with secondary hyperparathyroidism develop an autonomously functioning parathyroid gland.

Primary hyperparathyroidism is the third most common endocrine disorder (19). With the advent of serum calcium screening in the early 1970s, the incidence of primary hyperparathyroidism increased, with a peak of 121.7 per 100,000 person-years (72,73). After the initial increase, the incidence decreased until 1998, with release of national (United States) osteoporosis guidelines and increased bone mineral density measurements, and a second peak (86.2 per 100,000 person-years) of hyperparathyroidism was identified (73). Historically, patients often presented with symptoms such as renal stones and bone pain. But with the advent of serum calcium screening, patients are often asymptomatic or have symptoms of lethargy or weakness.

Primary hyperparathyroidism is more common in women than in men, and the incidence increases with age (72,74,75), being particularly common in postmenopausal women. In a study of a racially mixed population, the incidence of primary hyperparathyroidism was highest among blacks, followed by whites; Asians, Hispanics, and other races had lower incidences of primary hyperparathyroidism than whites (75). The racial differences in primary hyperparathyroidism remained similar as the prevalence of primary hyperparathyroidism increased during the study period (75). Primary hyperparathyroidism is usually sporadic, but can also occur in the familial setting. The underlying genetic abnormalities in sporadic, hereditary, and syndrome-associated disease are being elucidated.

Secondary hyperparathyroidism is caused by low serum calcium due to vitamin D disorders (vitamin D deficiency, malabsorption, rickets), tissue resistance to vitamin D, disorders of phosphate metabolism (renal disease, malnutrition or malabsorption), hypomagnesemia, calcium deficiency, and pseudohypoparathyroidism. The most common cause of secondary hyperparathyroidism is chronic renal failure. Tertiary hyperparathyroidism occurs when an autonomously functioning parathyroid gland develops in the setting of secondary hyperparathyroidism.

FAMILIAL/HEREDITARY HYPERPARATHYROIDISM

Familial primary hyperparathyroidism accounts for 5 to 15 percent of primary hyperparathyroidism. The underlying pathogenesis and genetic bases of hereditary forms of hyperparathyroidism are better understood than their sporadic counterparts. A variety of genes are involved in hereditary hyperparathyroidism, and most cases are inherited in an autosomal dominant manner (Table 2-1). Hereditary primary hyperparathyroidism can occur in association with syndromes such as multiple endocrine neoplasia type 1 (MEN1), MEN2, MEN4, and hyperparathyroidism jaw-tumor syndrome, or in a nonsyndromic manner in isolated primary hyperparathyroidism as occurs in familial isolated hyperparathyroidism (FIHP) and familial hypocalciuric hypercalcemia (FHH) (76).

Germline *MEN1* (11q13) mutation causes MEN1 syndrome (77–79), which typically manifests as multiglandular parathyroid disease. However, not all patients with MEN1 syndrome have an identifiable mutation in the *MEN1* gene. A recently described syndrome, MEN4, caused by *CDKN1B* mutation, may account for some of the patients with a MEN1-like syndrome but without a *MEN1* mutation (80–82).

Germline mutation of the *RET* proto-oncogene (10q21) causes MEN2A. Parathyroid disease (hyperplasia or adenoma) occurs in 20 to 30 percent of MEN2A patients.

Heterozygous inactivating germline mutation in the calcium sensing receptor gene (*CASR*) is the cause of some cases of FHH. Homozygous inactivating mutation of *CASR* causes neonatal severe hyperparathyroidism, one of the few autosomal recessive types of hereditary hyperparathyroidism and one of the most life-threatening. The *CASR* gene is also involved in some types of familial hypoparathyroidism (Table 2-2).

Table 2-1

HEREDITARY SYNDROMIC AND NONSYNDROMIC (ISOLATED) HYPERPARATHYROIDISM

Disorder	Genetics	Other Manifestations
Multiple endocrine neoplasia type 1	*MEN1* germline mutation, chromosome 11q13 has 10 exons, and encodes menin, a 610-amino acid protein Autosomal dominant	Multiglandular parathyroid disease is most common manifestation of MEN1 (also adenomas and carcinomas) Other manifestations include PitNETs, pancreatic endocrine tumors, neuroendocrine (NE) tumors of duodenum, gastrinomas, NE tumors of thymus and lung, adrenal cortical adenomas, and hyperplasia
Multiple endocrine neoplasia type 2A	*RET* proto-oncogene (10q21) encodes a plasma membrane tyrosine kinase with extracellular domains and an intracellular domain involved in cell growth and differentiation Approximately 95% of MEN2A families have a *RET* mutation in exon 10 or 11 Autosomal dominant	Multiglandular parathyroid disease and/or adenomas in 20-30%; medullary thyroid carcinoma, pheochromocytomas
Multiple endocrine neoplasia type 4(X)		
Hyperparathyroidism-jaw-tumor syndrome	*CDC73/HRPT2* 1q21-31 Autosomal dominant	
Familial isolated hyperparathyroidism	*MEN1, CDC73, CASR,* but in most cases the etiology is unknown	Familial hyperparathyroidism without other syndromic features (parathyroid only disease)
Familial hypocalciuric hypercalcemia (FHH)	FHH1 (majority): heterozygous loss of function mutation in calcium sensing receptor *(CASR)* FHH2 (<5%): *GNA11* mutation causes loss of function in G-protein subunit alpha11 (Ga11) which decreases sensitivity of CASR receptor on cells FHH3 (>5%): *AP2S1* loss of function mutation Autosomal dominant	
Neonatal severe hyperparathyroidism	Calcium sensing receptor (*CASR*) mutation Homozygous inactivating mutation Autosomal recessive	Parathyroid glands become hyperplastic and infants symptomatically hypercalcemic

The *CDC73* gene (1q21-31), previously known as *HRPT2,* is strongly associated with familial parathyroid disease. The best known *CDC73*-related disorder is hyperparathyroidism jaw-tumor syndrome (HPT-JT). Patients with HPT-JT can have multiglandular parathyroid disease, adenoma, or carcinoma.

Familial hyperparathyroidism may occur in a nonsyndromic manner (FIHP), with hyperparathyroidism as the sole manifestation. Patients may have familial hyperparathyroidism and germline mutations in *MEN1, CDC73,* or *CASR* and lack other syndrome-associated features, but the underlying mechanism in most families with FIHP is unknown. The entities included as part of FHH are associated with mutations of *CASR, GNA11* (19p13), and *AP2S1* (3q21.1). Understanding the pathogenesis of hereditary syndromic and nonsyndromic (isolated) hyperparathyroidism provides a basis for clinical genetic testing that may lead to optimized management of patients and families with these disorders.

Multiple Endocrine Neoplasia Type 1

Multiple endocrine neoplasia type 1 (MEN1) is caused by a mutation in the *MEN1* tumor

Table 2-2

CALCIUM SENSING RECEPTOR (CASR) ABNORMALITIES IN NON-NEOPLASTIC PARATHYROID DISEASE

Parathyroid Disease	Genetic	Clinical
Familial hypocalciuric hypercalcemia	Inactivating (heterozygous loss of function) *CASR* mutation (3q13.3-q21)	Hyperparathyroidism
Neonatal severe hyperparathyroidism	Inactivating (homozygous loss of function) *CASR* mutation (3q13.3-q21)	Hyperparathyroidism
Familial isolated hypoparathyroidism	Various genes have reportedly been associated including *CASR* (3q13.3-q21)	Hypoparathyroidism
Sporadic idiopathic hypoparathyroidism	Unknown cause, but antibodies to calcium sensing receptor in some cases	Hypoparathyroidism
Autoimmune polyglandular syndrome 1	Mutations in the *APECED* (autoimmune polyendocrinolpathy-candidiasis-ectodermal-dystrophy) or *AIRE* (autoimmune regulator) gene (21q22.3) Autoantigens to calcium sensing receptor	Hypoparathyroidism

suppressor gene (11q13) (77–79), leading to the development of multifocal neoplastic endocrine lesions. MEN1 has also been historically referred to as multiple endocrine adenomatosis type 1, Wermer syndrome, and familial Zollinger-Ellison syndrome. MEN1 occurs in the familial setting as an autosomal dominant disorder with very high penetrance (over 95 percent by 40 to 50 years of age) (83). The incidence of MEN1 is 1 in 20,000 to 40,000, and females and males are similarly affected. Sporadic cases of *MEN1* result from a new mutation or somatic mosaicism. More than 400 mutations located over the entire coding and intronic sequence of the *MEN1* gene have been described.

MEN1 is defined clinically as the occurrence of two or more types of MEN1-related tumors (parathyroid disease, pituitary neuroendocrine tumor (PitNET), and entero-pancreatic neuroendocrine tumors) in an individual or the presence of one MEN-associated tumor in a patient with family members with known MEN1. The most common manifestation of MEN1 is multiglandular parathyroid disease (fig. 2-3). Single gland parathyroid disease also occurs but is less common. Although historically referred to as parathyroid hyperplasia, in MEN1 multiglandular parathyroid disease is currently referred to as multiglandular parathyroid disease or adenomatosis because the parathyroid glands are composed of multiple monoclonal cell proliferations.

Over 90 percent of patients with MEN1 have parathyroid disease. The onset of clinical hyperparathyroidism in MEN1 is often in the third decade, which is much younger than patients with sporadic hyperparathyroidism (84,85). Patients known to have familial MEN1 may have their hyperparathyroidism detected even earlier. In MEN1, males and females are affected equally, unlike sporadic disease in which females are more often affected than males. Of patients presumed to have sporadic primary parathyroid hyperplasia, at least 20 percent were found to have MEN1 historically. But the percentage of cases with an underlying genetic abnormality may be much greater as more genetic testing is being performed. Alternatively, the absence of a family history of hyperparathyroidism may be due to the disease not being investigated in a parent or a parent dying before showing symptoms (76). Thus, genetic testing of patients presumed to have sporadic primary parathyroid hyperplasia may be helpful in identifying those with a new *MEN1* mutation.

Asymmetric enlargement of parathyroid glands is often seen but is not specific for multiglandular involvement in MEN1. Histologically, the glands have increased parenchymal cells, usually chief cells, and a nodular or diffuse pattern of growth. Parathyroid adenomas in MEN1 are histologically indistinguishable from sporadic adenomas. In addition to parathyroid disease, MEN1 may also be associated with PitNETs; pancreatic, gastric, and small intestine neuroendocrine tumors; neuroendocrine tumors of the lung and thymus; adrenocortical adenomas; ependymomas; meningiomas; lipomas; uterine and esophageal leiomyomas; and mucocutaneous

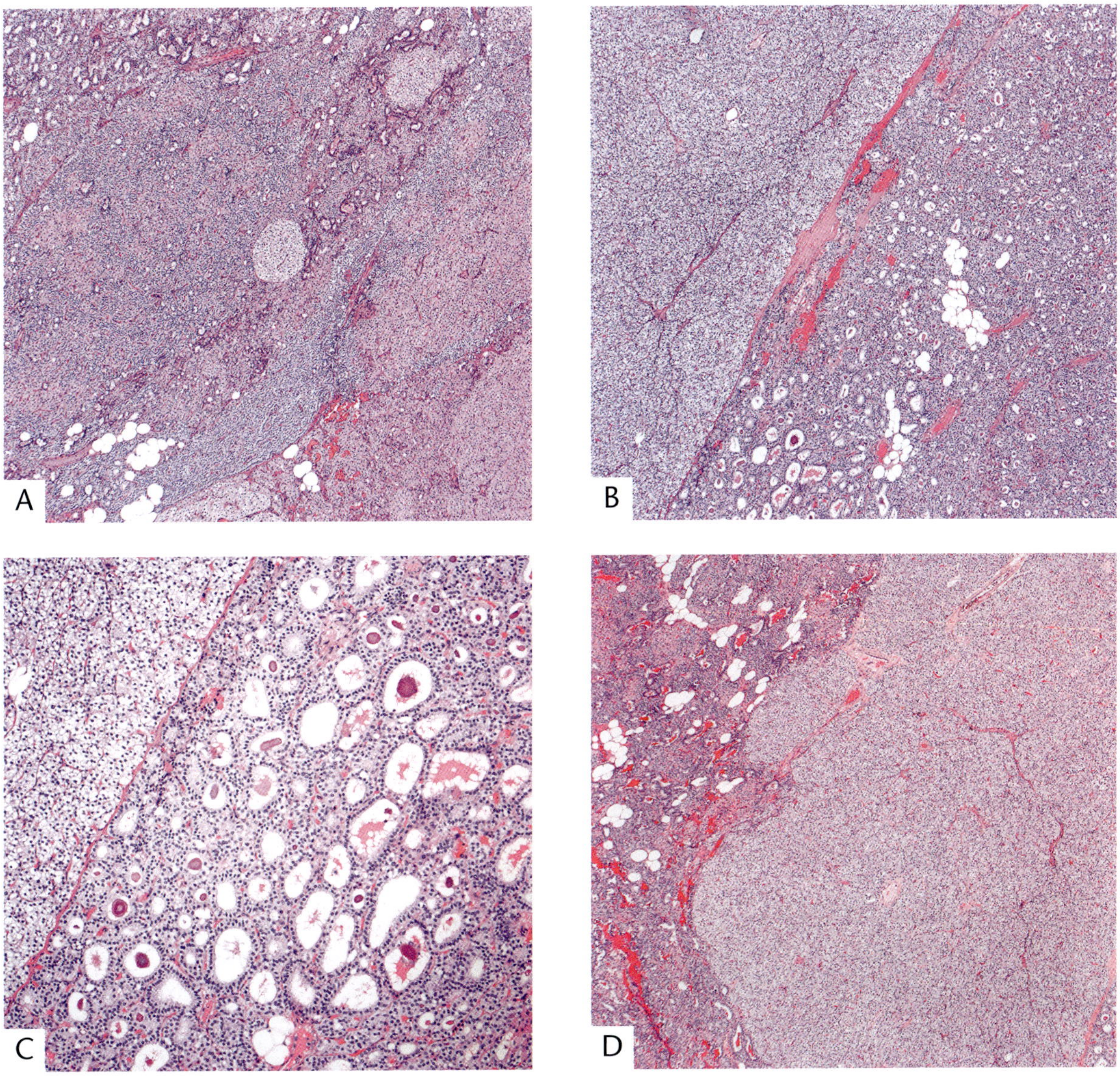

Figure 2-3

MULTIGLANDULAR PARATHYROID DISEASE IN MULTIPLE ENDOCRINE NEOPLASIA TYPE 1 (MEN1)

Hyperparathyroidism is the most common manifestation of MEN1. The parathyroid glands can have a diffuse or nodular growth pattern (A–D). Histologically, the parathyroid glands involved in hyperparathyroidism in MEN1 cannot be distinguished from those of sporadic hyperparathyroidism.

lesions (collagenomas, angiofibromas, café au lait macules, and gingival papules) (81,86,87).

Hyperparathyroidism is usually the first clinical manifestation of MEN1, and the incidence of hyperparathyroidism increases with age in affected individuals. Gross enlargement of all four parathyroid glands is usually seen. Histologically, the parathyroid glands are hypercellular, and consist of nests, cords, or a glandular pattern, and foci of solid sheets of parenchymal cells. The parenchymal hyperplasia is predominated by chief cells, but can also include transitional cells, oncocytic cells, or clear cells. The parathyroid glands are composed of multiple monoclonal cell proliferations (88).

Pituitary tumors are the first clinical manifestation of MEN1 in 10 to 25 percent of patients. In a multicenter study of 324 individuals with

MEN1, PitNETs were identified in 42 percent (89). Individuals with parathyroid disease or pituitary disease with MEN1 tend to be younger than their sporadic counterparts. Multiple pituitary adenomas (PitNETs) in MEN1 are identified in 5 percent of cases. The pituitary tumors are most often prolactinomas; the remainder are PitNETs of multihormonal, nonsecreting, growth hormone, adrenocorticotropin, and thyrotroph types.

Duodenal and pancreatic neuroendocrine tumors in MEN1 occur approximately 10 years earlier than sporadic tumors, and clinical manifestations are identified in approximately 40 percent (90–92). Some studies have suggested that up to 40 percent of individuals with MEN1 present with a functional pancreatic neuroendocrine tumor (91,93,94). In this setting, the tumor is often first thought to be sporadic. With gastrinomas, peripancreatic and/or periduodenal lymph node metastases are frequent. Mortality in MEN1 is most often due to gastrinomas of the small bowel and pancreas, and bronchial and thymic neuroendocrine tumors (90).

Individuals with pancreatic or duodenal gastrinomas may have Zollinger-Ellison syndrome. Multicentric gastrin-producing neuroendocrine tumors are usually identified in the upper duodenum, are multiple, and are small. MEN1 accounts for 20 to 30 percent of cases of Zollinger-Ellison syndrome.

Nonfunctioning pancreatic neuroendocrine tumors occur in 20 to 40 percent of individuals with MEN1. Glucagonoma, VIPomas, and other pancreatic neuroendocrine tumors occur in less than 5 percent of individuals with MEN1. Diffuse microadenomatosis associated with one or multiple macrotumors (over 0.5 cm) is a characteristic pattern seen in MEN1 (91,93–95). Duodenal endocrine tumors are well-circumscribed mucosal or submucosal nodules with a trabecular or pseudoglandular pattern. The cells are predominantly immunoreactive with gastrin. There may be associated focal hyperplasia of gastric and somatostatin cells in duodenal crypts and Brunner glands. Thymic and bronchial neuroendocrine tumors occur in 5 to 10 percent of individuals with MEN1. Gastric ECLomas, often small and multiple, can also occur.

Adrenal cortical lesions are identified in 20 to 40 percent of individuals with MEN1 (96). Bilateral adrenal cortical hyperplasia may be identified, and adrenal cortical carcinomas also occur. Nonendocrine cutaneous and soft tissue proliferations are present in 40 to 80 percent of individuals with MEN1 (83,97–104). Collagenomas, café au lait macules, angiofibromas, hypopigmented macules, lipomas, and gingival papules may involve the skin or mucosa. Soft tissue tumors seen in MEN1 include esophageal leiomyomas, renal angiomyolipomas, malignant gastrointestinal stromal tumors, visceral and intrathoracic lipomas, and malignant peripheral nerve sheath tumors. Central nervous system meningiomas, ependymomas, and astrocytomas also occur (83,97–104).

The *MEN1* gene encodes menin, a 610-amino acid protein, which is truncated and inactivated with *MEN1* mutation or other genetic alterations (77–79,105). Individuals with MEN1 have an inactivating mutation in one *MEN1* allele in all somatic cells, predisposing them to the development of tumors. Tumors occur when there is loss of the remaining copy of the *MEN1* gene (83,106). Mice with germline homozygous *MEN1* deletion die in utero (107–111). Mice with germline heterozygous deletion survive but develop tumors with loss of heterozygosity (LOH) of the other *MEN1* allele in some tissues, similar to that seen in humans with MEN1 (107–111).

MEN1 mutations are identified in 80 to 94 percent of cases of familial MEN1 and 65 to 88 percent of patients with sporadic MEN1 (78,86,112). Somatic *MEN1* mutations occur in 15 to 35 percent of sporadic parathyroid adenomas (113–115), and in some sporadic parathyroid carcinomas (116,117). Sporadic parathyroid tumors may also show gain or loss of chromosome 11 and overexpression of cyclinD1/PRAD1 (118,119).

Menin is ubiquitously expressed, localized primarily in the nucleus but also in the cell membrane and cytoplasm (120–122). It has been ascribed numerous functions and is thought to be particularly important in transcriptional regulation and chromatin modification as well as in cell signaling, division, adhesion, motility, and DNA repair (76,83,123–127). It interacts with the AP1 transcription factor JunD, which may interfere with the RAS signaling pathway (128). Menin may also affect the WNT/B-catenin signaling pathway by interacting with beta-catenin and T-cell factor 3, an associated transcription factor (129).

Menin is involved in both suppressing and promoting tumorigenesis (130). It may enhance C-MYC-mediated transcription to promote cancer progression (130). It is involved in the formation of the MLL methyltransferase complex, a role thought to promote tumorigenesis, but may also enhance the activity of MYC, independent of H3K4me3 (130). Menin is an oncogenic factor in MLL fusion leukemia. It interacts with SMAD proteins which, along with the transcription factor Pit-1, are involved in prolactin expression (131). Menin and SMADs are regulators of pituitary processes (132). Menin also interacts with bone morphogenic protein-regulated SMADs and osteoblast regulator Runx2 to regulate osteoblasts (131,133). It is involved in transcriptional regulation of the *ESR1* gene, which is involved in breast and pituitary cancers (123,134).

The function of menin has been described as linking "gene-specific transcription factors to chromatin modification in a cell-specific context"(123). The loss of menin in MEN1-associated tumors is associated with "disruption of antiproliferative gene expression programs and the development of endocrine tumors"(123). Further studies on the role of menin will lead to a greater understanding of disease and possibly to new treatment targets or strategies.

Although there is no phenotype-genotype correlation within MEN1, and there are over 1,300 different mutations, the detection rate of *MEN1* mutation is greater in the presence of a positive family history, a greater number of MEN1-related tumors, and in patients with parathyroid and pancreatic involvement compared with those lacking pancreatic involvement (135–137). The prevalence of single gland parathyroid disease and pituitary tumors may be higher in patients fulfilling the diagnostic criteria of MEN1 but without a family history of MEN1-related manifestations compared to those with familial MEN1 (138). Patients suspected of having an *MEN1* mutation should undergo genetic counseling and mutational analysis. If MEN1 is confirmed, then family members should also undergo molecular testing. Generally, genetic counseling and germline *MEN1* genetic testing is offered to patients with MEN1 and their first-degree relatives, whether symptomatic or not, as well as individuals with an atypical MEN1 phenotype (e.g., multigland hyperparathyroidism) (139).

MEN1 should be distinguished from MEN4, which is also an autosomal dominant disorder, but is caused by *CDKN1B* mutations. Individuals with MEN4 also have involvement of the parathyroid glands, pituitary gland, and pancreas.

HPT-JT syndrome is an autosomal dominant disorder caused by mutation of the *CDC73* gene and is associated with hyperparathyroidism and a significant risk of parathyroid cancer. HPT-JT is also associated with fibro-osseous jaw tumors and renal cysts and hamartomas, but is not usually associated with pituitary or pancreatic/duodenal disease.

MEN2 is an autosomal dominant tumor syndrome that affects the parathyroid and adrenal glands, in addition to medullary thyroid carcinoma. MEN2 is caused by *RET* mutation.

Familial isolated pituitary adenoma is an autosomal dominant disorder involving the *AIP* gene in a subset of cases. Carney complex can involve multiple endocrine and nonendocrine tissues, but usually does not involve the parathyroid glands or pancreas/duodenum. Carney complex is caused by a *PRKAR1A* mutation.

Multiple Endocrine Neoplasia Type 2A

Multiple endocrine neoplasia type 2A (MEN2A) is an autosomal dominant disorder with high penetrance caused by activating mutation of the *RET* proto-oncogene (10q21). Medullary thyroid carcinoma is the most common and deadly manifestation of MEN2A, occurring in over 90 percent of cases, with pheochromocytoma occurring in 50 percent and parathyroid disease (adenoma or multiglandular disease) in 20 to 30 percent (76).

The hyperparathyroidism in MEN2A is usually mild. A single parathyroid gland or multiple parathyroid glands are involved. The hyperparathyroidism is usually diagnosed after the medullary thyroid carcinoma. The parathyroid disease in most individuals is asymptomatic at diagnosis, but hypercalciuria and renal caliculi can occur. Symptoms of hyperparathyroidism may be more pronounced in longstanding disease.

MEN2A is diagnosed clinically by the presence of at least two of these tumors in an individual. The parathyroid glands in MEN2A often show asymmetric enlargement, and an increase in chief cells in a nodular or diffuse pattern. Unlike MEN1, which may show somatic MEN1 abnormalities in sporadic parathyroid lesions, sporadic *RET* mutations are usually not seen in sporadic parathyroid adenomas or hyperplasia, but have been identified in up to 50 percent of sporadic medullary thyroid carcinomas and in a subset of sporadic pheochromocytomas (140–144).

MEN2A is the most common subtype of MEN2, which includes *MEN2A*, *MEN2B* (also known as *MEN3*), and *familial isolated medullary thyroid carcinoma*, all of which are associated with *RET* mutations. All three subtypes are associated with medullary thyroid carcinoma. Parathyroid disease is seen in 20 to 30 percent of MEN2A patients, but is generally not seen in MEN2B (145–150). MEN2B accounts for approximately 5 percent of MEN2 cases and is associated with medullary thyroid carcinoma, pheochromocytoma, Marfanoid habitus, mucocutaneous neuromas, and gastrointestinal ganglioneuromas (145–150). Familial isolated medullary thyroid carcinoma is not associated with other features. Although the clinical features of these syndromes are helpful in their classification, treatment decisions are based on the specific codon involved in an individual case.

Approximately 95 percent of MEN2A families have mutations in exon 10 or 11 of the *RET* gene (151–153). Genotype-phenotype correlations with *RET* mutations, the aggressiveness of medullary thyroid carcinoma, and the incidence of pheochromocytoma, hyperparathyroidism, and other features in MEN2 syndromes are used in clinical decision making (154). The American Thyroid Association has published risk categories and treatment guidelines for hereditary medullary thyroid carcinoma (and MEN2-associated diseases) based on the specific *RET* mutation (154). The highest risk category involves mutation of *RET* codon M918T, which occurs in MEN2B, with other high risk categories involving mutations of *RET* codon C634 and A883F which occur in MEN2A. Other *RET* codon mutations are moderate risk and are generally seen in hereditary isolated medullary thyroid carcinoma (154).

The genotype-phenotype correlations extend beyond predicting the aggressiveness of medullary thyroid carcinoma in MEN2. The incidence of hyperparathyroidism is highest with mutation of codon 634, while the incidence of pheochromocytoma is highest with mutation of codons 634, 631, 883, and 918 (142,150,154–156). Codon 634 mutation is also associated with high penetrance of disease (157). Patients with medullary thyroid carcinoma are offered genetic counseling and *RET* mutation testing. If the patient is found to have a *RET* mutation, then their relatives are also offered genetic counseling and *RET* mutation testing.

Multiple Endocrine Neoplasia Type 4

Multiple endocrine neoplasia type 4 (MEN4) is caused by a *CDKN1B* mutation (12p13) (80). *CDKN1B* is a putative tumor suppressor gene encoding p27 (*KIP1*), a cyclin-dependent kinase inhibitor involved in the progression from the G1 phase to the S phase of the cell cycle (80). MEN4 clinically resembles MEN1. Up to 10 percent of patients with familial or sporadic MEN1-like tumors clinically do not have an *MEN1* mutation (81,82,137) and MEN4 comprises 1.5 to 3.7 percent of individuals with a MEN1-like clinical phenotype.

MEN4 is characterized by a variety of tumors, but the most common neuroendocrine lesions involve the parathyroid glands, pituitary gland, and pancreas. Hyperparathyroidism occurs in 75 to 80 percent of patients with MEN4 (82), and may occur at an older age (56 years) than in MEN1 (20 to 25 years) (82). PitNETs occur in 37 percent of MEN4 patients and may be functioning or nonfunctioning (82). Pancreatic neuroendocrine tumors are also seen. A variety of other tumors in MEN4 involve the bronchus, stomach, duodenum, cervix, breast, and thyroid gland. Although the manifestations of MEN4 are more variable than MEN1, the lesions grossly and histologically appear similar.

A MEN-like syndrome was described in rats in 2002 without an underlying *MEN1* or *RET* mutation (158). The gene locus was mapped to the distal part of chromosome 4 (80,159). A homozygous frameshift mutation encoding *CDKN1B* was identified, and the rats had a marked decrease in p27 protein. This recessive MEN-like syndrome in rats was termed MENX (80). Pellegata et al. (80) identified a germline *CDKN1B* mutation in a human patient with pituitary and parathyroid tumors and expanded the pedigree analysis to identify multiple generations with MEN1-like tumors. They concluded that germline mutation in *CDKN1B* can predispose to MEN-like tumors in humans and in rats (80). Subsequent reports of patients with MEN-like tumors but without *MEN1* or *RET* mutations followed.

In a review published in 2017 of the 19 MEN4 cases to date, hyperparathyroidism was the most common manifestation, occurring in 15 of 19 cases (80 percent) (82). Hyperparathyroidism occurs at an older age in patients with MEN4 than MEN1 (82,160). PitNETs were identified in 7 of 19 (37 percent) patients with established MEN4 and may be associated with less aggressive disease in this setting. Gastrointestinal and pancreatic neuroendocrine tumors were identified in 7 of 19 (37 percent) individuals with MEN4, occurring in the pancreas, duodenum, and stomach. Both functional and nonfunctional neuroendocrine tumors occur. Additionally, a neuroendocrine carcinoma of the cervix of one patient and a bronchial carcinoid were described. Although adrenocortical disease is reported in rodents with MENX, only 1 of the 19 humans with MEN4 had adrenocortical disease (bilateral nonfunctional adrenal masses).

Individuals with MEN4 do not appear to have the skin manifestations seen in MEN1 (82). Additionally, somatic and germline *CDKN1B* mutations have been identified in individuals with sporadic hyperparathyroidism, breast cancer, lymphoma, and neuroendocrine tumors, suggesting that *CDKN1B* may function as a susceptibility gene for other tumors (82). Since some patients thought to have MEN1 clinically but do not have an *MEN1* mutation may have a *CDKN1B* mutation, patients with MEN-like tumors who are negative for *MEN1* mutation and *RET* mutation may benefit from genetic counseling and evaluation for a possible *CDKN1B* mutation. Unlike *MEN1* mutations, which are associated with a high penetrance, *CDKN1B* mutations may show incomplete penetrance, as suggested by Alrezk et al. (82), since relatives with *CDKN1B* mutations may not manifest features of MEN4. However, the small number of confirmed patients with MEN4 limits our understanding of this syndrome.

CDC73/HRPT2 Disorders

CDC73 (1q21-q31) is a putative tumor suppressor gene, also referred to as *HRPT2,* which is inactivated in CDC73-related disorders such as HPT-JT syndrome and some cases of familial isolated hyperparathyroidism (161–168). HPT-JT is an autosomal dominant disorder associated with hyperparathyroidism, fibro-osseous jaw tumors, renal cysts, hamartomas, and Wilms tumors. Hyperparathyroidism is the most common manifestation of HPT-JT, affecting 80 percent of individuals. The parathyroid disease often occurs by late adolescence. Patients may have parathyroid adenoma or carcinoma (169). The hyperparathyroidism is aggressive, with severe hypercalcemia and an increased incidence of parathyroid carcinoma; up to 15 percent of individuals with HPT-JT develop parathyroid carcinoma. The unaffected parathyroid glands are normal. A variety of features have been reported in the abnormal parathyroid glands, including eosinophilic cytoplasm, perinuclear cytoplasmic clearing, nuclear enlargement, and a sheet-like growth pattern (169).

Inactivating *CDC73* mutations are seen in benign and malignant parathyroid lesions in HPT-JT, and are common in sporadic parathyroid carcinomas (75 percent), but uncommon in sporadic parathyroid adenomas (170,171). Germline *CDC73* mutations are present in a subset of patients with parathyroid carcinomas originally thought to be sporadic (169). Additional genes may be involved in parathyroid malignancy.

CDC73 encodes parafibromin (161,168). Loss of nuclear immunoreactivity is seen in syndrome-associated and sporadic parathyroid carcinomas and in syndrome-associated parathyroid adenomas (172–179). Parafibromin (CDC73 protein) is a tumor suppressor protein that interacts with RNA polymerase II as part of the Paf1 transcription regulatory complex; it mediates H3K9 methylation, silencing the expression of cyclin D1 (180–182). Thus, parafibromin is involved in inhibiting cell proliferation by inducing cell cycle arrest (183). Parafibromin may have antiproliferative effects by inhibition of C-MYC (182). Cytoplasmic parafibromin destabilizes p53 messenger RNA and p53-mediated apoptosis (184). It also mediates interferon (IFN)-gamma (type II IFN)-stimulated signaling pathways that facilitate STAT1 tyrosine phosphorylation mediated by JAK1/2 (185). Dephosphorylated parafibromin is a transcriptional coactivator of the Wnt/Hedgehog/Notch pathways (186).

Familial Isolated Hyperparathyroidism

Individuals with *familial isolated hyperparathyroidism* have hyperparathyroidism, but do not have other features that are classically seen in syndromic familial hyperparathyroidism. Familial primary hyperparathyroidism accounts for 5 to 15 percent of primary hyperparathyroidism cases, and familial isolated hyperparathyroidism accounts for approximately 1 percent of primary hyperparathyroidism cases (50,187). Although individuals with familial isolated hyperparathyroidism usually have parathyroid hyperplasia or adenoma, these individuals are at an increased risk of parathyroid carcinoma (165,188,189).

Some cases of familial isolated hyperparathyroidism represent an incomplete manifestation of a syndromic form of familial hyperparathyroidism since germline mutations in *MEN1, CDC73*, and *CASR* have been reported in some affected families (76,103,138,165,166,187,188,190–194). *MEN1* mutations accounted for 13 percent and *CDC73* mutations accounted for 7 percent of familial isolated hyperparathyroidism cases in a cohort study of 15 familial isolated hyperparathyroid kindreds (195). In another study, 1 of 15 kindreds had a *CASR* mutation (138). A further ten kindreds of familial isolated hyperparathyroidism have been reported due to *CASR* mutation (196). Of 32 kindreds with familial isolated hyperparathyroidism without an *MEN1* or *CASR* mutation, only 1 had a *HRPT2* mutation (165). Familial isolated hyperparathyroidism associated with *MEN1* or *CASR* mutations is associated with younger age and multiglandular disease, while *CDC73*-associated kindreds have an increased risk of parathyroid carcinoma (167,196–199).

GCM2 encodes a transcription factor important in parathyroid development. *GCM2* activating mutations (6p24.2) are identified in some cases of familial isolated hyperparathyroidism (50,189). Individuals with *GCM2* mutation-associated familial isolated hyperparathyroidism have higher preoperative parathyroid hormone levels and a lower rate of biochemical cure than those with sporadic primary hyperparathyroidism (189). Also, the rates of multigland parathyroid disease and parathyroid carcinoma are higher in *GCM2* mutation-associated hyperparathyroidism than in sporadic primary hyperparathyroidism (189).

Familial Hypocalciuric Hypercalcemia

Familial hypocalciuric hypercalcemia (FHH) is a term used to encompass three autosomal dominant conditions characterized by lifelong moderately elevated serum calcium, elevated or inappropriately normal plasma parathyroid hormone, and low urinary calcium due to decreased renal calcium excretion (200,201). FHH types 1, 2, and 3 involve loss of function mutations of *CASR, GNA11,* and *APS1,* respectively.

Most patients with FHH type 1 and type 2 are asymptomatic, but those with type 3 may have symptoms of hypercalcemia, low bone mineral density, and childhood cognitive deficits (202,203). In approximately 20 percent of individuals, plasma parathyroid hormone levels are elevated and patients can be mistaken for having primary hyperparathyroidism. In difficult cases, genetic testing may be helpful, since parathyroidectomy does not result in normalization of hypercalcemia in these cases (202,203).

FHH type 1 is the most common type (65 percent of cases). It is caused by a heterozygous, inactivating *CASR* mutation (204-208). The *CASR* gene encodes a G-protein-coupled receptor that is present on parathyroid, kidney, intestinal, bone, and thyroid C cells. The receptor senses alterations in calcium levels and modulates parathyroid hormone secretion and urinary calcium excretion (209–212). The *CASR* mutation in FHH type 1 causes decreased ability of the calcium sensing receptor to sense calcium, and thus does not appropriately release parathyroid hormone (213). Autoimmune hypocalciuric hypercalcemia is caused by antibodies directed against calcium sensing receptors and can mimic FHH (214).

The *CASR* gene is involved in many parathyroid disorders, that can result in either hyperparathyroidism or hypoparathyroidism. FHH type 1 is caused by heterozygous inactivating *CASR* mutations, neonatal severe hyperparathyroidism is caused by homozygous inactivating or compound heterozygous loss of function *CASR* mutations, and autosomal dominant hypoparathyroidism and familial hypocalcemia are caused by activating *CASR* mutations (205,207,215). Additionally, antibodies to the calcium sensing receptor may be involved in a subset of sporadic idiopathic hypoparathyroidism and autoimmune polyglandular syndrome 1 cases.

FHH type 2 is caused by loss of function mutation of *GNA11* (19p13.3), which encodes a G-protein subunit alpha11 and results in decreased sensitivity of calcium sensing receptors to changes in extracellular calcium levels (208). Gain of function *GNA11* mutations cause autosomal dominant hypocalcemia type 2 (208).

FHH type 3 is caused by loss of function mutation of the *AP2S1* gene (19q13.3) that encodes the adaptor protein 2 (AP2) sigma subunit (203). Although FHH types 1 and 2 are usually asymptomatic, individuals with type 3 may have symptoms of hypercalcemia (more than 20 percent), low bone mineral density (more than 50 percent), and childhood cognitive deficits (more than 75 percent) (202,203).

Neonatal Severe Hyperparathyroidism

Neonatal severe hyperparathyroidism is an autosomal recessive (or autosomal dominant) disorder caused by a homozygous or compound heterozygous inactivating mutation of *CASR* (calcium sensor receptor gene; 3q13.3-q21) (205, 207,215). This is a life-threatening disorder of hypercalcemia in infants, who are often symptomatic from severe hypercalcemia and markedly elevated serum parathyroid hormone (216). The parathyroid glands are enlarged and markedly hypercellular (216). Homozygous inactivating *CASR* mutation-associated hyperparathyroidism is associated with more markedly hypercellular parathyroid glands than in conditions involving heterozygous mutations (216).

SPORADIC PRIMARY HYPERPARATHYROIDISM

Sporadic primary hyperparathyroidism is usually caused by a parathyroid adenoma (85 percent), with multiglandular parathyroid disease accounting for approximately 15 percent of cases and carcinoma for less than 1 percent (217). Sporadic primary hyperparathyroidism is more common in women than men. It usually occurs in the fifth decade, which is an older age than individuals with hereditary hyperparathyroidism.

Primary hyperparathyroidism is usually sporadic, and the underlying etiology remains unknown in most cases. Long-term lithium ingestion (218) and ionizing radiation to the head and neck in childhood (219) are associated with an increased prevalence of hyperparathyroidism. Sporadic mutations in some of the same genes involved in hereditary and syndrome-associated hyperparathyroidism have a role in sporadic hyperparathyroidism. For example, germline *MEN1* mutations occur in the syndrome of hereditary MEN1, but somatic *MEN1* mutations also occur in a subset of sporadic parathyroid adenomas (115).

PRIMARY CHIEF CELL HYPERPLASIA

Definition. *Primary chief cell parathyroid hyperplasia* is an increase in parathyroid parenchymal mass with an increase predominantly in chief cells in the absence of a stimulus for parathyroid hormone secretion. There may also be an increase in transitional cells, oncocytic cells, and clear cells and multiple parathyroid glands can be involved.

Clinical Features. Primary chief cell hyperplasia accounts for approximately 15 percent of cases of primary hyperparathyroidism. Most cases (80 to 85 percent) are caused by parathyroid adenomas, while parathyroid carcinoma causes less than 1 percent of cases.

The majority (75 percent) of cases of chief cell parathyroid hyperplasia are sporadic, with familial primary parathyroid hyperplasia accounting for about 25 percent of cases (19). As an underlying genetic abnormality is becoming increasingly evaluated for and identified in multiglandular parathyroid disease, the actual number of truly sporadic cases is likely less than historically reported. Heritable parathyroid hyperplasia can occur in the setting of MEN1, MEN4, MEN2A, HPT-JT syndrome, FHH, neonatal severe hyperparathyroidism, and familial isolated hyperparathyroidism. Parathyroid hyperplasia occurs in over 90 percent of individuals with MEN1 and 30 percent with MEN2A (76). Twenty percent of individuals with primary parathyroid hyperplasia have MEN.

The incidence of parathyroid hyperplasia increases with age. Patients with sporadic parathyroid hyperplasia often present in the fifth decade, while those with familial disease may present three decades earlier (84,85). Sporadic hyperparathyroidism is more common in women than in men, but familial hyperparathyroidism usually affects women and men similarly.

The clinical presentation of hyperparathyroidism has changed from individuals with symptoms, nephrocalcinosis, and osteopenia to now asymptomatic patients with elevated screening serum calcium and parathyroid hormone levels, or with weakness and lethargy. Patients are more likely to be asymptomatic early in the disease and may become more symptomatic with increasing serum calcium levels.

The underlying pathogenesis and genetic abnormalities in sporadic idiopathic primary parathyroid hyperplasia are not as well defined as in hereditary forms of parathyroid hyperplasia. Primary parathyroid hyperplasia is usually polyclonal, but nodular areas may show clonality. Additionally, syndrome-associated parathyroid hyperplasia, such as MEN1, have been shown to be monoclonal (88). The terms adenomatosis, multiglandular parathyroid tumors, and multiple adenomas have been suggested to reflect the monoclonal nature of many of these cases.

Gross Findings. In primary parathyroid hyperplasia, multiple parathyroid glands are enlarged, but may not be enlarged equally, as occurs in asymmetric hyperplasia. Three patterns of gross findings have been reported: classic type, in which all glands are uniformly enlarged; pseudoadenomatous type, in which there is variation in extent of involvement of parathyroid glands; and occult type, in which the parathyroid glands are only slightly enlarged and slightly hypercellular (220).

Pseudoadenomatous, or asymmetric hyperplasia, may be difficult, if not impossible, to differentiate from parathyroid adenoma. Asymmetric hyperplasia is very common, with studies reporting that up to two thirds of cases of parathyroid hyperplasia show enlargement of less than all four glands (221). MEN1-associated hyperplasia/multiglandular disease is often multicentric. Occult hyperplasia may show subtle parathyroid gland enlargement and hypercellularity and may be difficult to distinguish from normal parathyroid glands.

Microscopic Findings. The parathyroid glands in primary hyperplasia show an increase in parathyroid parenchymal cells, predominantly chief cells, but transitional cells, oncocytic cells, and clear cells may also be present. Sporadic and hereditary parathyroid hyperplasias are usually indistinguishable histologically. There is a relative decrease in adipose tissue. The cellularity can vary, however, and determinations of cellularity in small biopsies may be highly problematic. Thus, clinical and radiographic correlation is always important, as is intraoperative parathyroid hormone monitoring.

Although primary chief cell hyperplasia is often nodular (fig. 2-4), the growth pattern may also be diffuse (fig. 2-5) (19). The cells show sheet-like growth or patterns such as nests, glandular formations, cords, palisading of cells, papillary growth, microcysts, and cribriform growth (fig. 2-6). Degenerative and edematous features may be present (fig. 2-7). The nests and nodules are usually composed of a single cell type.

Diffuse hyperplasia is more commonly seen in younger patients with moderate hypercalcemia who have moderately enlarged glands that are more uniform in size and morphology (222) Nodular growth is more common in elderly patients with asymmetrically enlarged glands with variable cellularity (222). Interestingly, in clinically normal parathyroid glands, oxyphilic cells increase until around the fourth decade, and they may form small nodules that should not be mistaken for the hypercellularity seen in hyperplasia or adenoma. The growth pattern in MEN1-associated parathyroid hyperplasia may be nodular or diffuse, and the glands often show asymmetric hypercellularity.

The parenchymal cells in parathyroid hyperplasia usually show little variation in size and shape, but foci of nuclear pleomorphism, often referred to as "endocrine atypia," are seen and may be a degenerative finding. Mitotic figures are present in the majority of parathyroid glands in hyperplasia, but the mitotic rate is usually less than 1 per 10 high-power fields. Atypical mitotic figures are usually only seen in parathyroid carcinoma.

Cystic change is uncommon, but can be seen in hyperplasic parathyroid glands (fig. 2-8) (223). Cystic change is more common in larger parathyroid glands (224), and in HPT-JT-associated parathyroid disease. Fibrosis and hemosiderin deposition (fig. 2-9) are also seen in hyperplastic parathyroid glands, particularly in very large glands.

An unusual type of parathyroid hyperplasia, referred to as *lipohyperplasia,* occurs when multiple parathyroid glands show abundant mature adipose tissue, with only scattered nests of parenchyma involving multiple hyperfunctional parathyroid glands. Myxoid change can be prominent in cases with fewer lipomatous components (fig. 2-10).

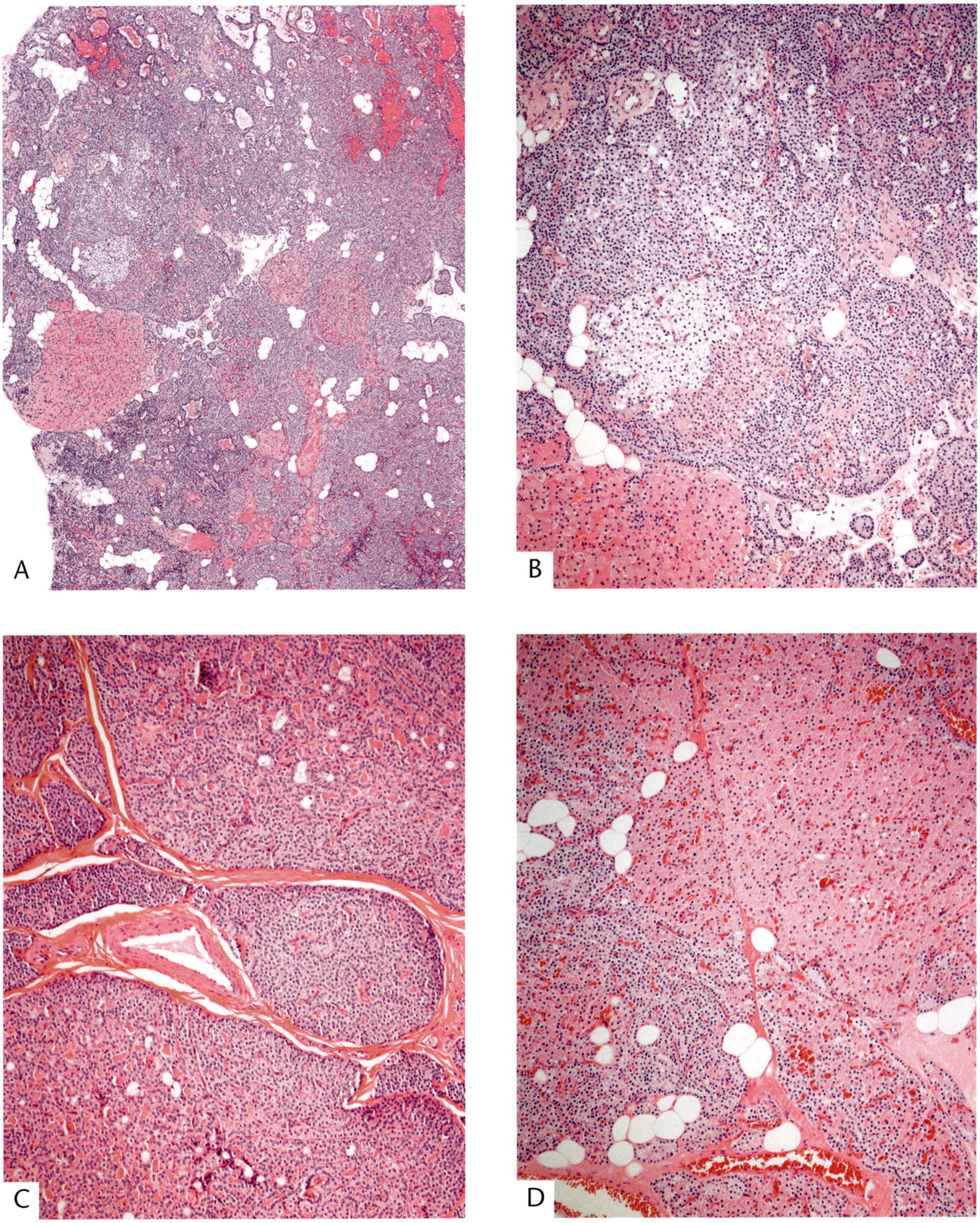

Figure 2-4

SPORADIC PRIMARY PARATHYROID HYPERPLASIA WITH NODULAR GROWTH

Nodular growth is more often seen in older patients with asymmetrically enlarged parathyroid glands (A–D). Primary parathyroid hyperplasia often shows predominantly chief cells, but may also include transitional cells, oncocytic cells, and clear cells.

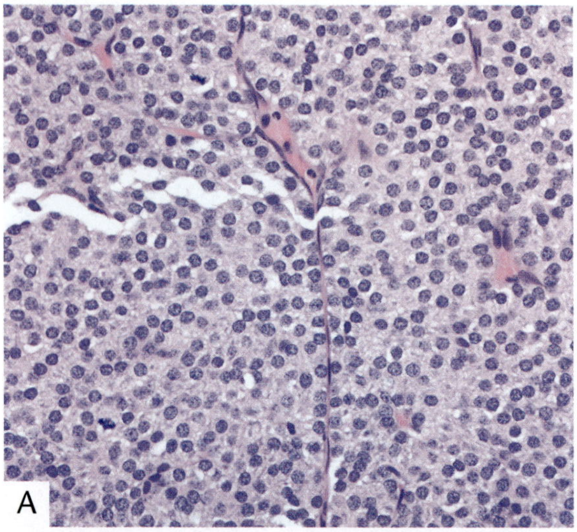

Figure 2-5

SPORADIC PRIMARY PARATHYROID HYPERPLASIA WITH DIFFUSE GROWTH

Primary parathyroid hyperplasia is an increase in parathyroid parenchymal mass, with an increase in predominantly chief cells, but may also include transitional cells, oncocytic cells, and clear cells. It involves multiple parathyroid glands in the absence of a stimulus for parathyroid hormone secretion. Diffuse hyperplasia is more common in younger patients with more uniformly enlarged parathyroid glands, and nodular growth is more common in older patients with asymmetrically enlarged parathyroid glands. These parathyroid glands show a diffuse growth pattern (A–C).

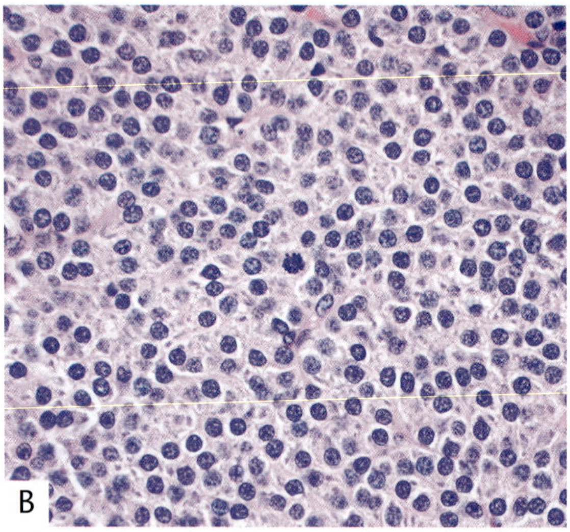

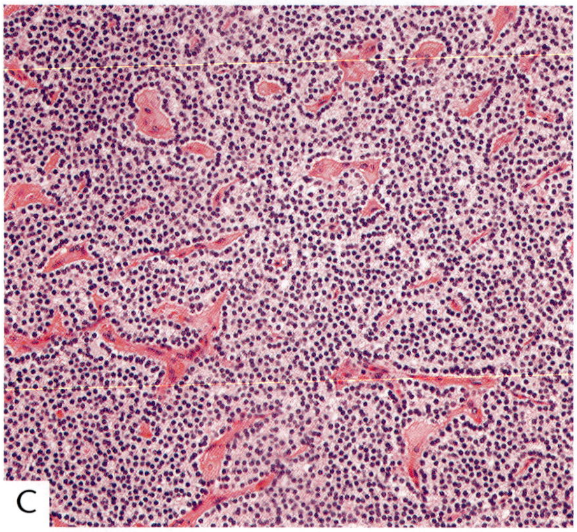

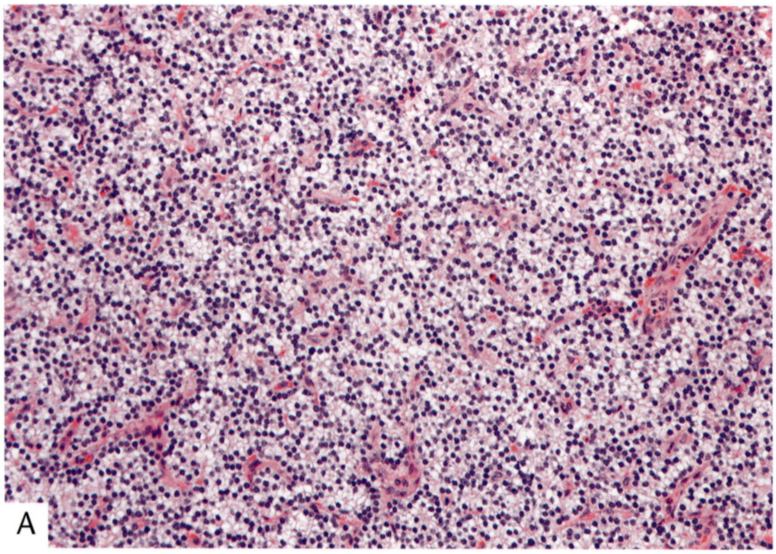

Figure 2-6

PARATHYROID HYPERPLASIA

There are many growth patterns in parathyroid hyperplasia including sheet-like growth, acinar growth, follicular growth, nests, glandular formations, cords, palisading of cells, papillary growth, microcysts, and cribriform growth.

A: These parathyroid glands show sheet-like growth.

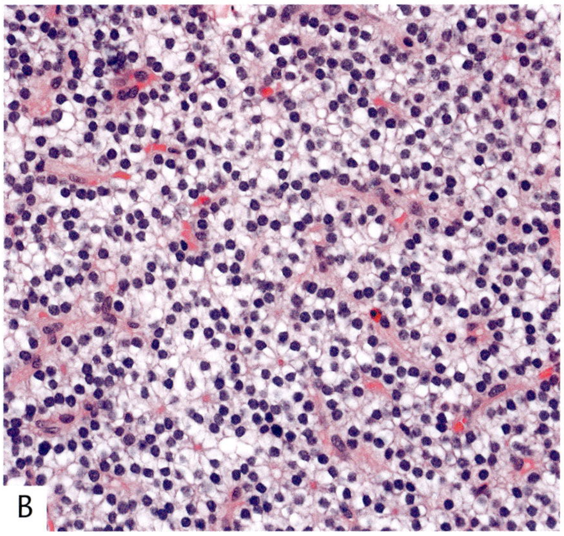

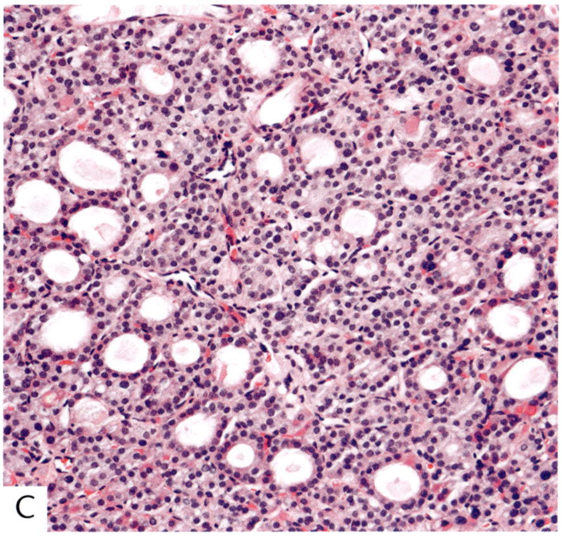

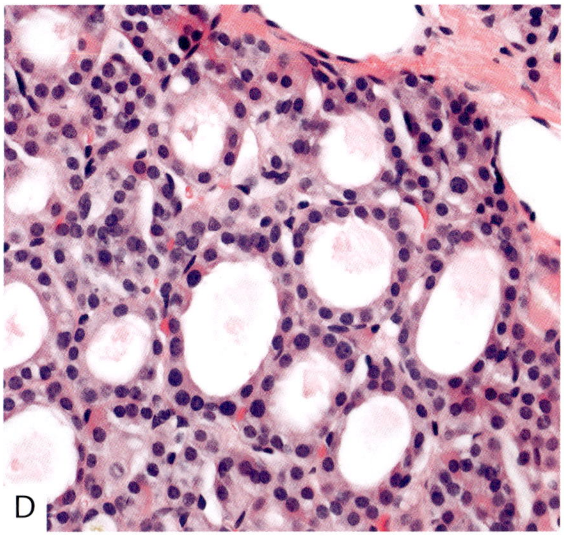

Figure 2-6, continued

B: Sheet-like growth is seen.
C,D: Acinar growth in parathyroid hyperplasia.
E,F: Parathyroid gland showing follicular growth.

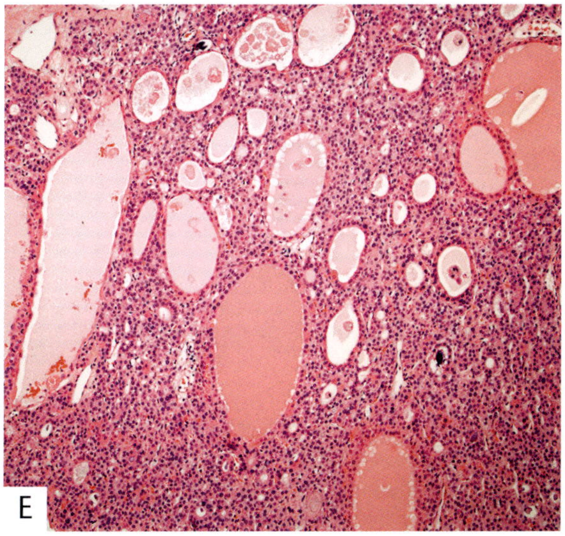

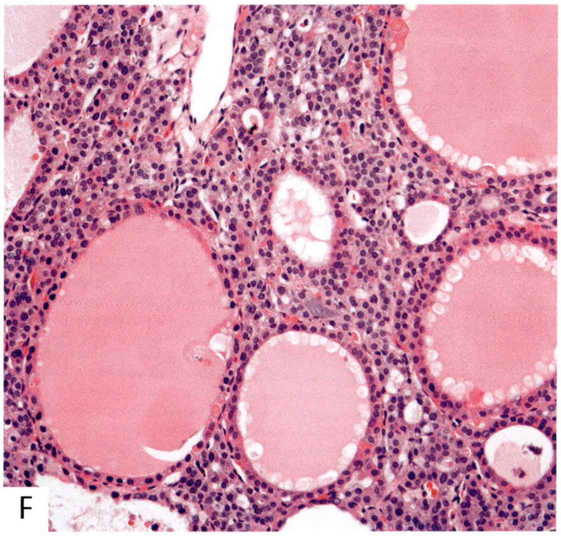

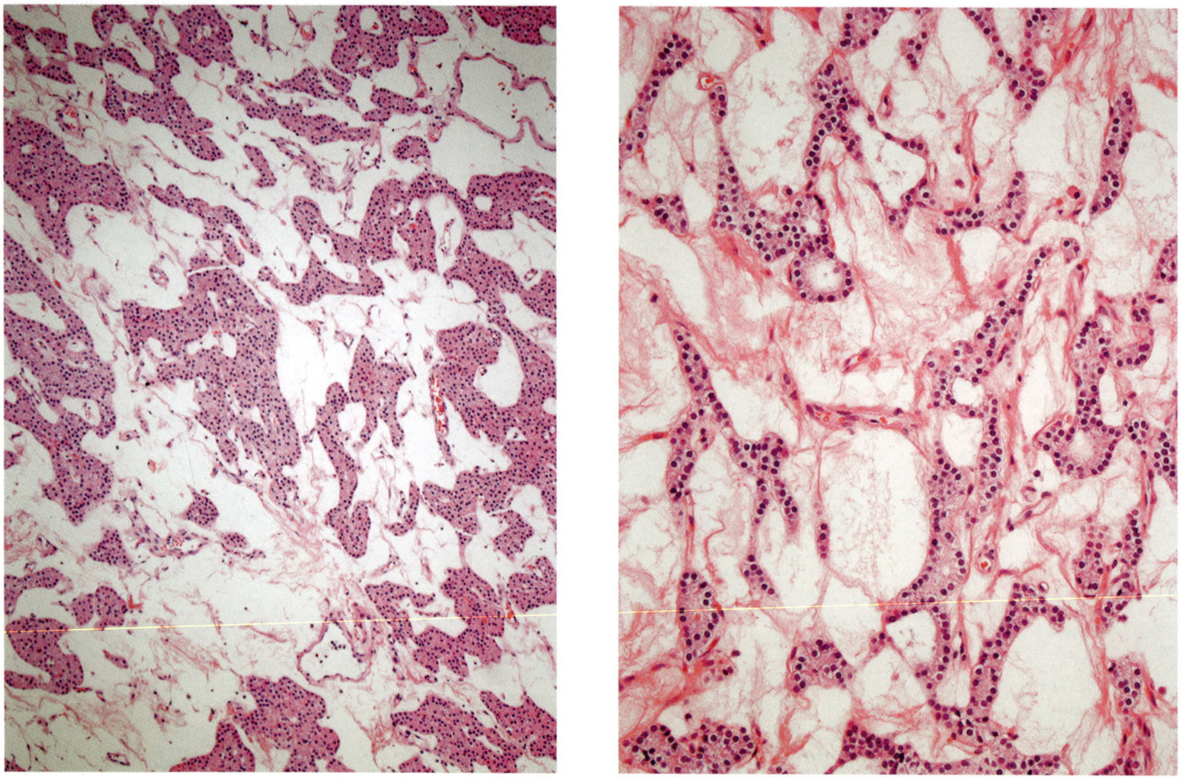

Figure 2-7

PARATHYROID HYPERPLASIA

These parathyroid glands show degenerative edematous changes (left, right).

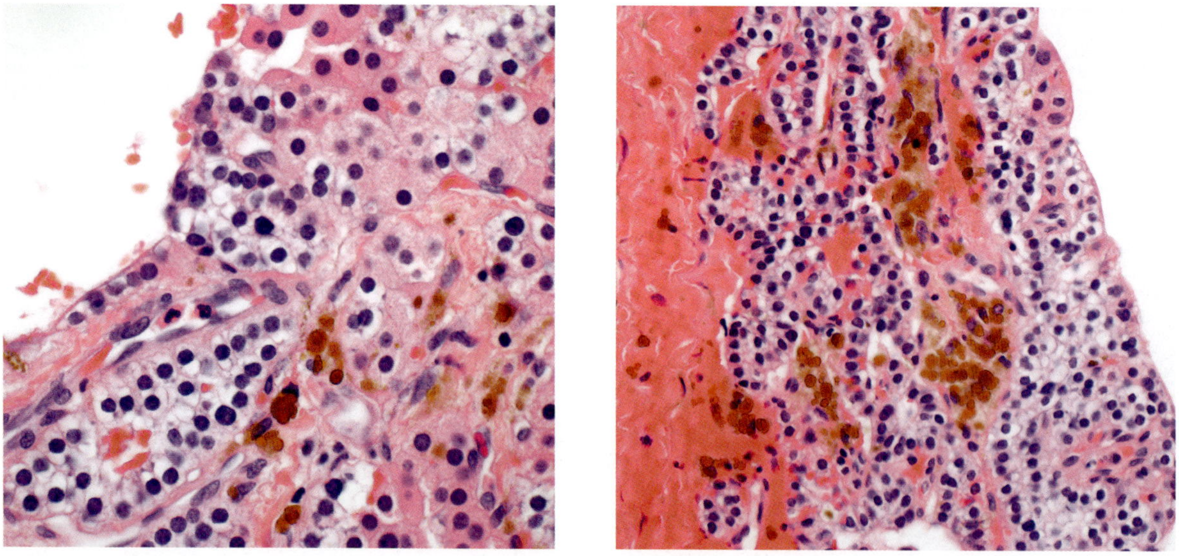

Figure 2-8

HYPERCELLULAR PARATHYROID GLAND: CYSTIC CHANGE

The parathyroid glands can show hemorrhage and cystic change, particularly in larger glands and in those with hyperparathyroidism-jaw tumor–associated disease more so than sporadic hyperparathyroidism (left, right).

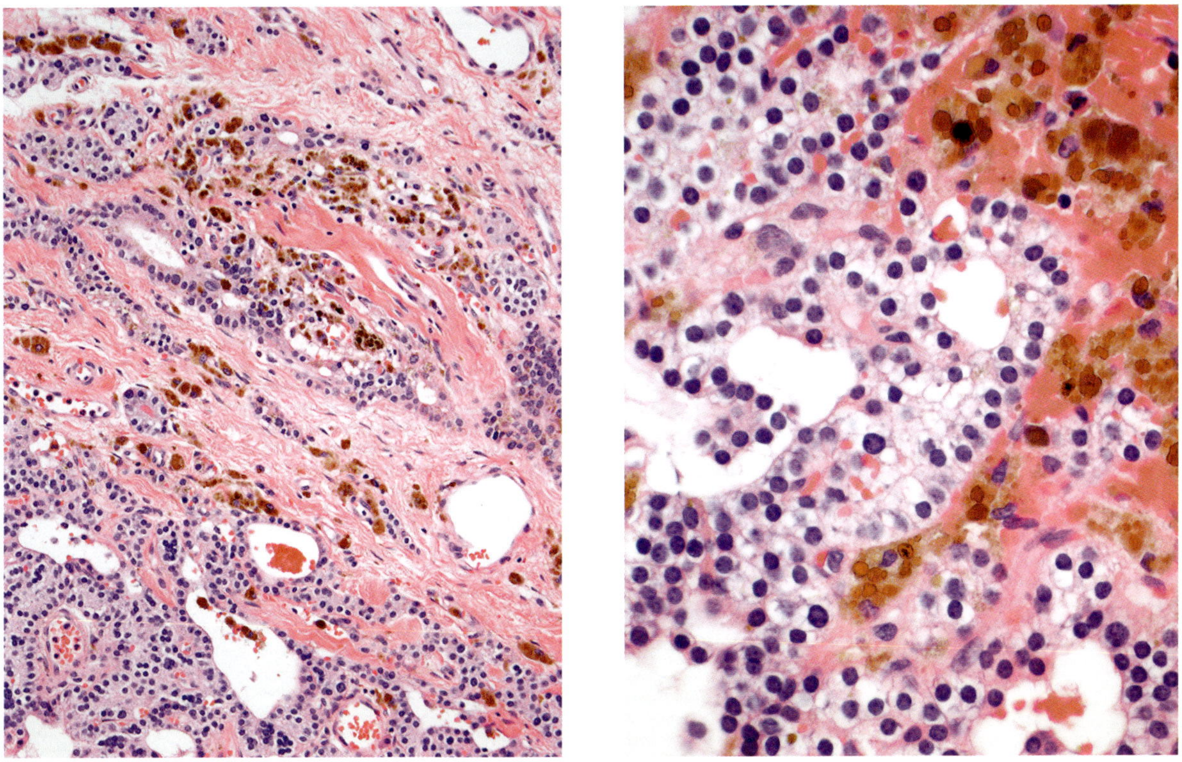

Figure 2-9

HYPERCELLULAR PARATHYROID TISSUE: HEMOSIDERIN

Parathyroid gland showing hemorrhage and hemosiderin deposition (left, right).

Immunohistochemical Findings. Parathyroid glands involved by primary hyperplasia are immunoreactive for neuroendocrine markers (chromogranin A, synaptophysin), low molecular weight keratins, and parathyroid hormone. Parathyroid tissues and tumors are negative for TTF1 and thyroglobulin and are usually negative for calcitonin. Parathyroid hyperplasias and adenomas have greater Ki-67 (evaluated with MIB1) proliferative activity than normal parathyroid tissue and less proliferative activity than parathyroid carcinoma (225). p27 expression is highest in normal parathyroid tissues, followed by hyperplasia and adenoma, with parathyroid carcinomas showing the lowest levels (225). Parafibromin shows loss of nuclear and/or nucleolar expression in parathyroid carcinoma and in syndrome-associated parathyroid disease, but it is generally intact in sporadic parathyroid hyperplasia. Fatty acid synthase is overexpressed in hyperplastic parathyroid glands (226). Endoglin (CD105), vascular endothelial growth factor (VEGF), and VEGF-R2 show differential staining in hyperplastic parathyroid glands compared to neoplastic glands (227).

Differential Diagnosis. Parathyroid hyperplasia must be distinguished from parathyroid adenoma. Parathyroid hyperplasia generally involves multiple parathyroid glands, but the involvement may be asymmetric and difficult to differentiate from adenoma in the evaluation of a single parathyroid gland. As size and cellularity may vary greatly among the parathyroid glands in an individual patient, an asymmetrically enlarged parathyroid gland can be misinterpreted as a parathyroid adenoma. Histologically, parathyroid adenomas often possess rims of normal-appearing parathyroid tissue, however, approximately 10 percent of hyperplasic parathyroid glands have rim-like areas of normal-appearing parathyroid tissue. Thus the presence of a possible "rim" of normal tissue should not be used to definitively distinguish parathyroid adenoma from hyperplasia. Clinical information, imaging studies,

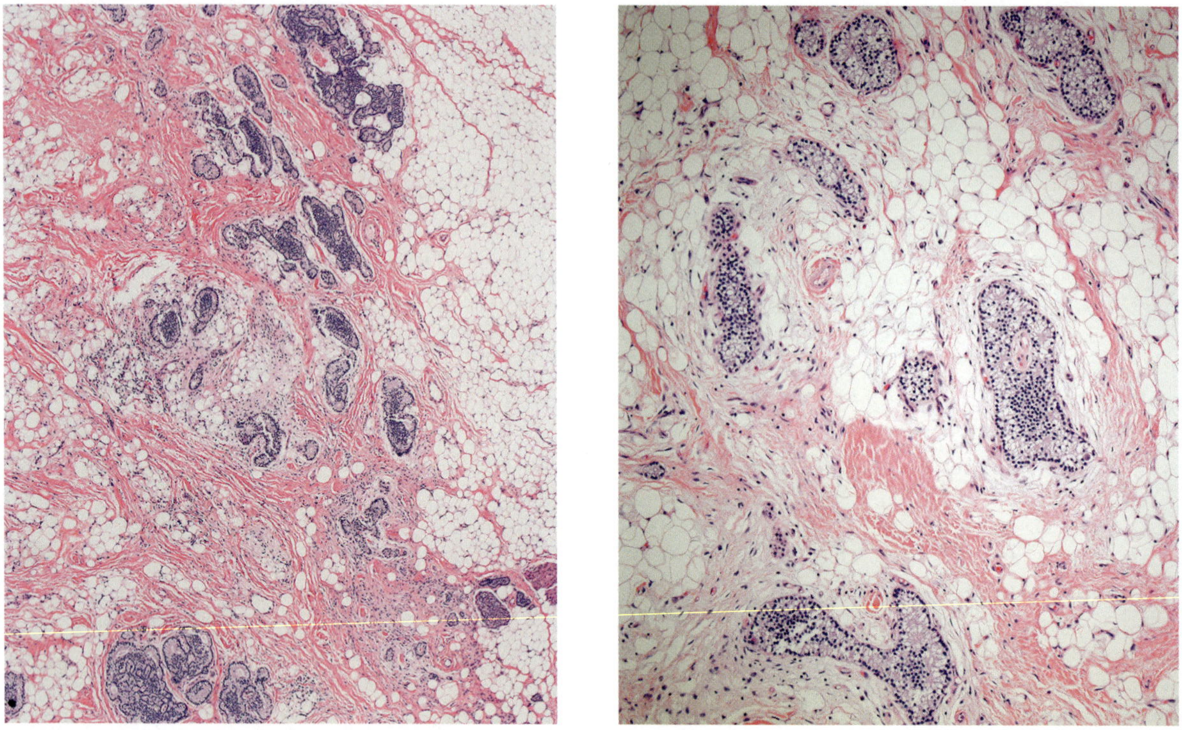

Figure 2-10

PARATHYROID LIPOHYPERPLASIA

Left: Parathyroid lipohyperplasia is an uncommon type of parathyroid hyperplasia in which multiple parathyroid glands show abundant mature adipose tissue with only scattered nests of parenchyma involving multiple hyperfunctional parathyroid glands.

Right: Parathyroid lipohyperplasia usually shows abundant mature adipose tissue (and may show myxoid change), with nests of parenchymal cells involving multiple hyperfunctional parathyroid glands.

intraoperative parathyroid hormone monitoring, and pathologic evaluation of parathyroid gland(s) are used in an integrated manner in the evaluation of an individual with hyperparathyroidism.

Primary parathyroid hyperplasia is differentiated from secondary parathyroid hyperplasia by the absence of a known stimulus of parathyroid hormone secretion. Although the glands in secondary parathyroid hyperplasia may be more diffusely hyperplastic, they can become nodular with time, and the parathyroid glands in primary hyperplasia can be nodular or diffuse.

Differentiating parathyroid tissue from thyroid tissue in small biopsies, including intraoperatively, can be difficult. Parathyroid parenchymal cells are generally smaller than thyroid parenchymal cells. Parathyroid cells have well-defined cell membranes and may contain intracytoplasmic lipid, while oxylate crystals may be seen in the colloid of thyroid parenchyma. In small biopsy specimens it may be difficult or impossible to differentiate hypercellular parathyroid tissue from normal parathyroid tissue because normal parathyroid tissue can be highly variable in cellularity. Cellularity can vary both within and among parathyroid glands in a single individual. Thus, correlation with the clinical and radiographic findings and the intraoperative parathyroid hormone levels is essential. Parathyroid oxyphil/oncocytic cells can become more prominent and nodular with age, and must be distinguished from hyperplasia.

Treatment. Primary parathyroid hyperplasia may be treated surgically by subtotal parathyroidectomy: removing three glands and leaving a vascularized remnant of the fourth parathyroid gland in place. Alternatively, a total parathyroidectomy is performed with autotransplantation of a portion of a parathyroid gland into the forearm or neck. The residual or autotransplanted parathyroid tissue may become hyperplastic and require further removal. Intraoperative parathyroid hormone monitoring has decreased the risk of missing additional glands and helps confirm removal of diseased parathyroid glands.

PRIMARY CLEAR CELL HYPERPLASIA

Definition. *Primary clear cell (water-clear, Wasserhelle) parathyroid hyperplasia* is an absolute increase in parathyroid parenchymal cells with abundant clear cytoplasm involving multiple parathyroid glands in the absence of a known stimulus for parathyroid hormone secretion.

Clinical Features. Primary clear cell parathyroid hyperplasia was described in 1934 by Albright et al. (228). Unlike primary chief cell hyperplasia (described in 1958 by Cope et al. [229]), primary clear cell hyperplasia is rare in modern times (229). Most cases were described in the historical literature (217,228,230): individuals often presented with symptoms of hyperparathyroidism such as nephrolithiasis and bone disease. Primary clear cell parathyroid hyperplasia has not been studied with molecular techniques because of its rarity. It is not known to be associated with familial hyperparathyroidism.

Gross Findings. Grossly, the parathyroid glands are enlarged, but they may vary in size within an individual. Usually all four glands are involved, but supernumerary parathyroid glands may also be involved (231). The total gland weights vary from less than 10 g to over 100 g. Of the 19 cases of clear cell parathyroid hyperplasia reported by Castleman et al. (217), none were under 1 g, 21 percent were 1 to 5 g, 32 percent were 5 to 10 g, and 47 percent were over 10 g. There may be a correlation between gland weights, serum calcium levels, and symptoms. The upper parathyroid glands are often larger than the lower glands. Pseudopods may extend from the glands. The parathyroid glands are red-brown or brown, and can show hemorrhage, fibrosis, and cystic change.

Microscopic Findings. Histologically, there is a diffuse increase in clear cells rather than nodular growth (fig. 2-11). The cells may grow in a sheet-like pattern or have a glandular or tubular architecture. Cystic structures may be noted. The cells are polyhedral, round to ovoid, and usually 15 to 20 μm in size but can range from 10 to 40 μm (232). The nuclei are located basally at the pole of the cell next to stroma and vessels. The cells have distinct plasma membranes and hyperchromatic nuclei, which may be multiple (232). The cytoplasm is clear, with small cytoplasmic vacuoles (0.8 μm) that may be Golgi vessels, as noted on electron microscopy studies.

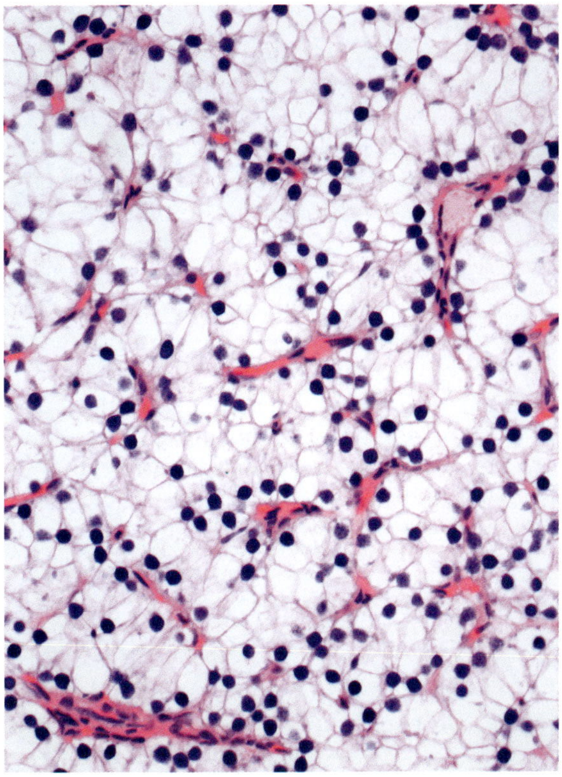

Figure 2-11

CLEAR CELL PARATHYROID HYPERPLASIA

Clear cell parathyroid hyperplasia is an absolute increase in parathyroid parenchymal cells with abundant clear cytoplasm involving multiple parathyroid glands in the absence of a known stimulus for parathyroid hormone secretion.

Immunohistochemical Findings. Parathyroid tissues are positive for neuroendocrine markers such as chromogranin A and synaptophysin, keratin, and parathyroid hormone, and are negative for TTF1 and thyroglobulin.

Differential Diagnosis. Primary clear cell hyperplasia can be mistaken for clear cell parathyroid adenoma based on histologic features alone. Thus, correlation with clinical, radiographic, and surgical findings as well as intraoperative parathyroid hormone monitoring are helpful. Clear cell parathyroid disease can also be mistaken for thyroid tumors that show clear cell change. Metastases, such as from clear cell renal cell carcinoma, can also be confused with a primary clear cell lesion. In difficult cases, immunohistochemical studies are helpful.

Treatment. Primary clear cell parathyroid hyperplasia is treated similarly to primary chief cell hyperplasia: surgically, with subtotal parathyroidectomy removing three glands and leaving a vascularized remnant of the fourth parathyroid gland in place or with a total parathyroidectomy with autotransplantation of a portion of a parathyroid gland into the forearm or neck. The residual or autotransplanted parathyroid tissue may become hyperplastic and require further removal. Intraoperative parathyroid hormone monitoring has decreased the risk of missing additional glands and helps confirm removal of diseased parathyroid glands.

SECONDARY HYPERPARATHYROIDISM

Definition. *Secondary hyperparathyroidism* is an increase in parathyroid hormone secretion and an adaptive increase in parathyroid parenchymal mass. It is caused by an increase in predominantly chief cells, but may also include transitional cells, oncocytic cells, and clear cells involving multiple parathyroid glands in the presence of a stimulus for parathyroid hormone secretion.

Clinical Features. The increase in parathyroid hormone occurring in secondary hyperparathyroidism results from low serum calcium levels caused by disorders of vitamin D (vitamin D deficiency, malabsorption, Ricketts), tissue resistance to vitamin D, disorders of phosphate metabolism (malnutrition, malabsorption, renal disease), pseudohypoparathyroidism, hypomagnesemia, and calcium insufficiency (233–239). The most common cause of secondary hyperparathyroidism is chronic renal failure.

Patients present with elevated serum calcium levels, bone pain, osteitis fibrosa cystica, osteomalacia, soft tissue and visceral calcifications, and periarticular calcification with arthritis. A rare but life-threatening complication is calciphylaxis, which can be associated with skin and soft tissue necrosis (240–242).

Gross Findings. In secondary parathyroid hyperplasia, multiple parathyroid glands are enlarged, but may vary in size and weight. They are usually yellow to gray. Early in the disease, the glands are more uniformly increased in size than glands in primary parathyroid hyperplasia. Over time, however, the glands may become more nodular. Also, the size of the parathyroid glands may become more variable later in disease. Hemorrhage, cystic change, and fibrosis may also become more prominent over time.

Microscopic Findings. The parathyroid glands in secondary hyperplasia are hypercellular. The cellularity is increased in a diffuse manner early in the disease (fig. 2-12). Chief cells predominate in early disease, and oxyphil cells increase over time and may become the predominant cell type. Later, nodules of chief, oncocytic, and transitional cells become more prominent, and the nodularity may increase (fig. 2-13) with increasing renal failure; this may be difficult to differentiate from primary or tertiary hyperparathyroidism. Scattered mitotic figures are seen, but atypical mitoses are generally not present.

Immunohistochemical Findings. Parathyroid tissues are positive for neuroendocrine markers such as chromogranin A and synaptophysin, keratin, and parathyroid hormone, and are negative for TTF1 and thyroglobulin. Parathyroid hyperplasias and adenomas have greater proliferative Ki-67 activity (evaluated with MIB1) than normal parathyroid tissue and less proliferative activity than parathyroid carcinoma (225). p27 expression is highest in normal parathyroid tissues, followed by hyperplasia and adenoma, with parathyroid carcinomas showing the lowest levels of p27 expression (225). Parafibromin shows loss of nuclear and/or nucleolar expression in parathyroid carcinoma and in syndrome-associated parathyroid disease, but it is generally intact in parathyroid hyperplasia. However, metastases of parathyroid carcinoma arising in the setting of secondary parathyroid hyperplasia have shown loss of parafibromin.

Differential Diagnosis. Secondary parathyroid hyperplasia is differentiated from primary parathyroid hyperplasia by the presence of a known stimulus for parathyroid hormone secretion. Differentiation from adenoma is discussed with the differential diagnosis of primary chief cell hyperplasia.

Treatment. Surgical approaches to the treatment of secondary hyperparathyroidism include subtotal parathyroidectomy with removal

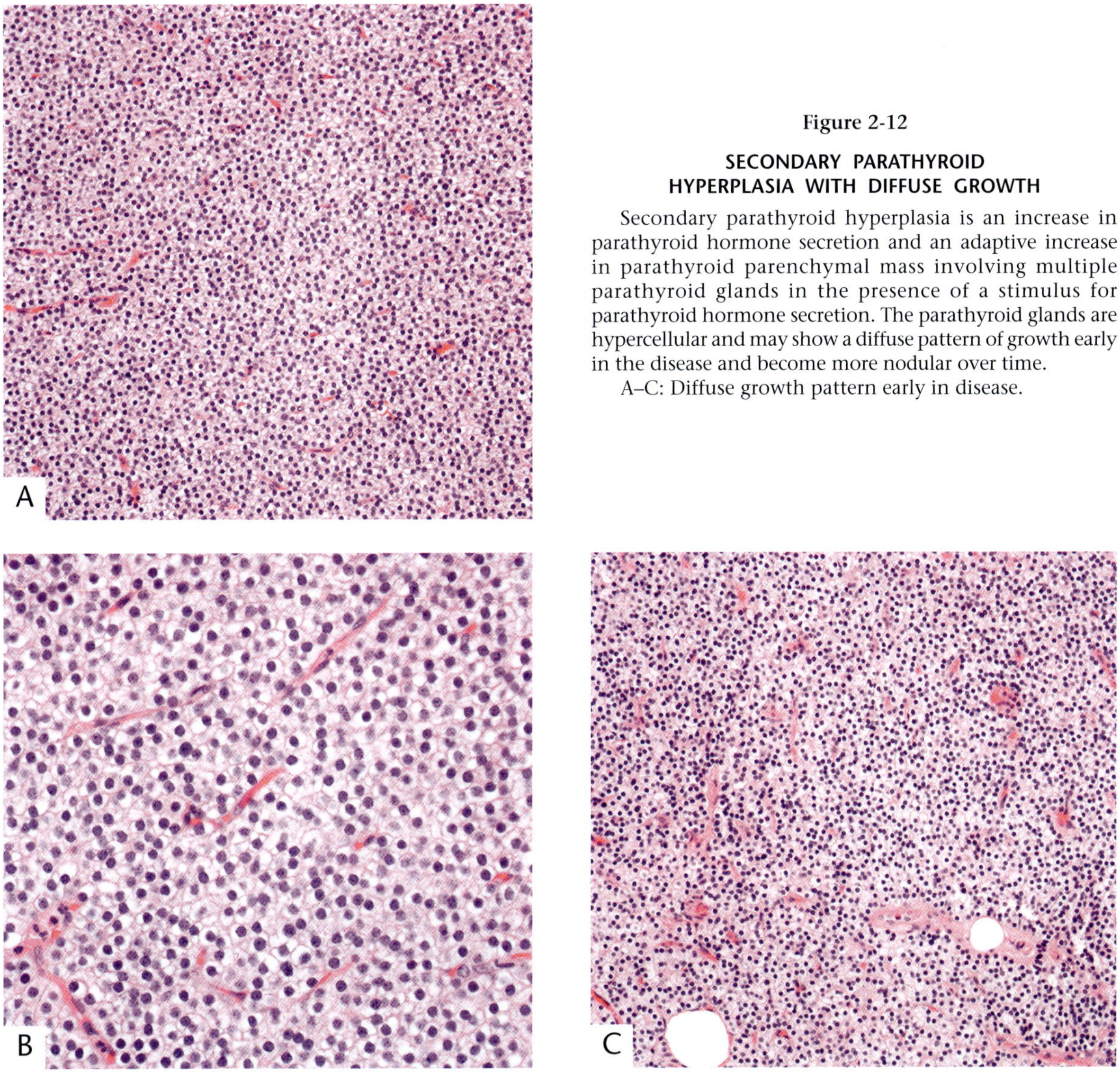

Figure 2-12

SECONDARY PARATHYROID HYPERPLASIA WITH DIFFUSE GROWTH

Secondary parathyroid hyperplasia is an increase in parathyroid hormone secretion and an adaptive increase in parathyroid parenchymal mass involving multiple parathyroid glands in the presence of a stimulus for parathyroid hormone secretion. The parathyroid glands are hypercellular and may show a diffuse pattern of growth early in the disease and become more nodular over time.

A–C: Diffuse growth pattern early in disease.

of three glands, leaving a vascularized remnant of the fourth gland, or total parathyroidectomy with autotransplantation of a portion of parathyroid gland into the neck or forearm. The residual or autotransplanted tissue, however, may become hyperplastic.

TERTIARY HYPERPARATHYROIDISM

Definition. *Tertiary hyperparathyroidism* is the development of autonomous parathyroid hyperfunction in the setting of known secondary hyperparathyroidism.

Clinical Findings. The development of autonomous parathyroid function, with hypersecretion of parathyroid hormone resulting in hypercalcemia, may be caused by a change in the set point of the calcium-sensing receptor. Patients have a history of secondary hyperparathyroidism, usually due to chronic renal failure. After renal transplant, the parathyroid glands may not return to normal, and both nodular and diffuse hyperplasia may be seen. In one study, cases with diffuse hyperplasia had an average parathyroid weight of 0.7 ±0.4 g, while those with nodular hyperplasia were 1.4 ±0.7 g (243). The parathyroid glands may become resistant to calcimimetic drugs despite normalized serum calcium. Hypercalcemia due to hyperparathyroidism may also develop after renal transplantation. This may be transient and resolve spontaneously, or may persist.

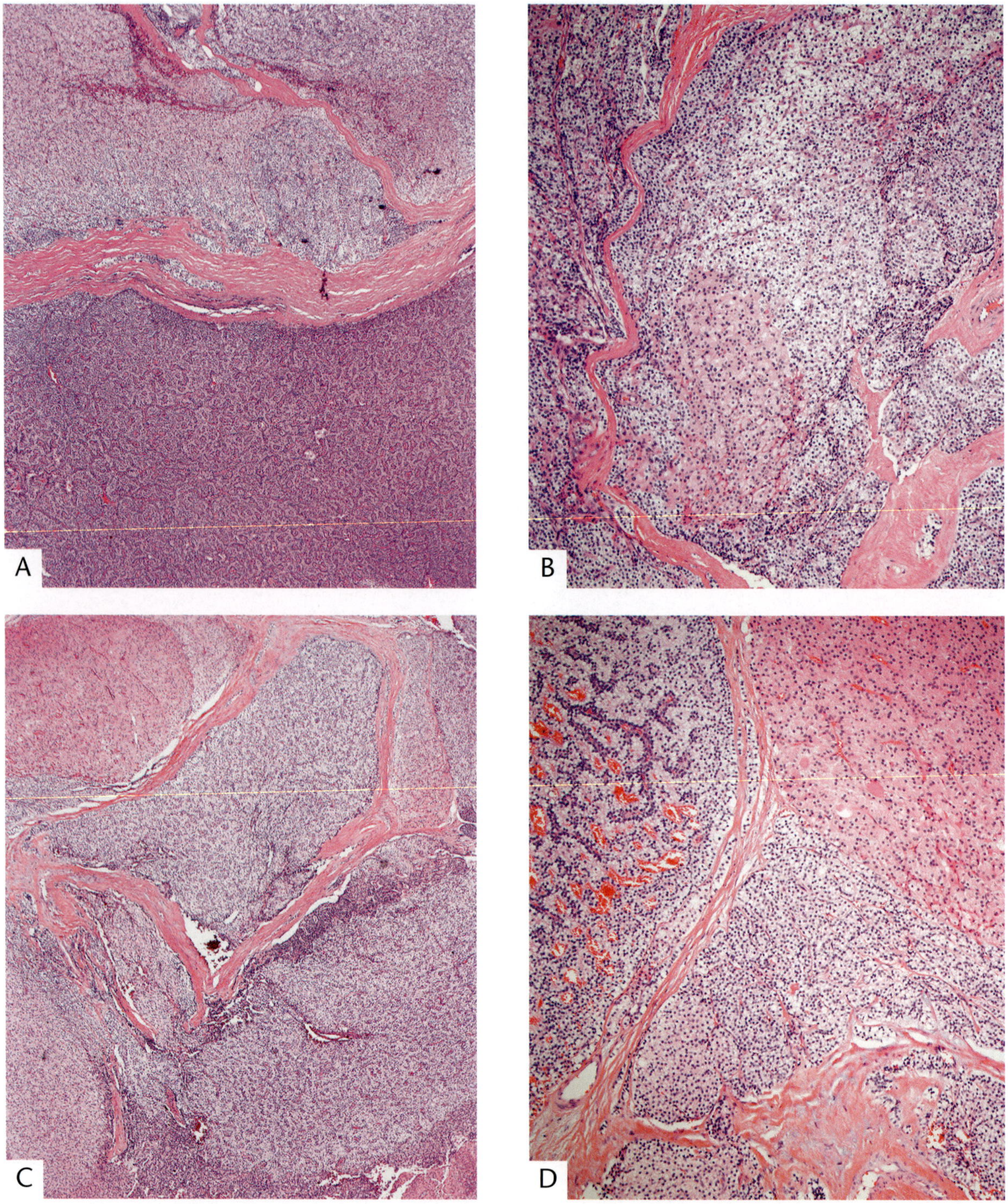

Figure 2-13

SECONDARY PARATHYROID HYPERPLASIA

Although the parathyroid glands in secondary hyperparathyroidism may show diffuse growth early in disease, they become more nodular over time.

A,B: Nodular growth of secondary parathyroid hyperplasia.

C,D: Hypercellular parathyroid showing nodular growth over time, with nodules of chief, oncocytic, and transitional cells, which may become prominent.

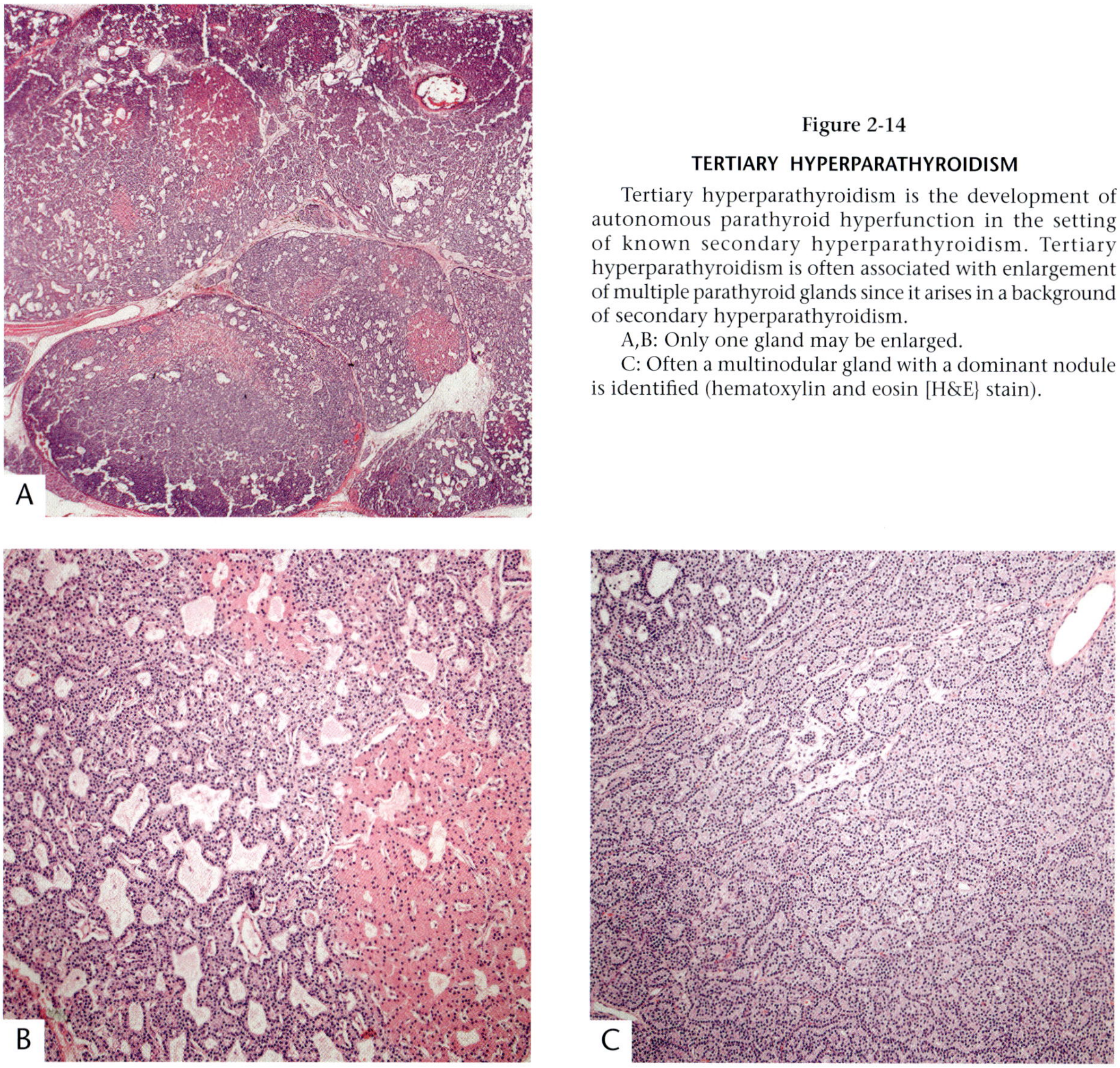

Figure 2-14

TERTIARY HYPERPARATHYROIDISM

Tertiary hyperparathyroidism is the development of autonomous parathyroid hyperfunction in the setting of known secondary hyperparathyroidism. Tertiary hyperparathyroidism is often associated with enlargement of multiple parathyroid glands since it arises in a background of secondary hyperparathyroidism.

A,B: Only one gland may be enlarged.

C: Often a multinodular gland with a dominant nodule is identified (hematoxylin and eosin [H&E} stain).

Gross Findings. Tertiary hyperparathyroidism is often associated with enlargement of multiple parathyroid glands since it arises in a background of secondary hyperparathyroidism, but only one gland may be enlarged. Often, a multinodular gland with a dominant nodule is identified. The superior parathyroid glands may be larger than the inferior glands. In up to one third of cases the disease involves ectopic parathyroid glands.

Microscopic Findings. The histologic findings in tertiary hyperparathyroidism may be identical to those of primary or secondary hyperplasia (fig. 2-14). Tertiary and secondary hyperplasia have more prominent internodular diffuse growth than primary hyperplasia. The glands are hypercellular and show an increase in chief and oxyphil cells and a decrease in adipocytes. In a review by Krause and Hedinger (243) of 128 parathyroid glands from 41 patients with tertiary hyperparathyroidism, 39 of 41 cases (95 percent) showed marked hyperplasia. These hyperplastic glands were composed predominantly of chief cells, but also had oxyphil cells and a marked increase (10 to 40 fold) in parathyroid mass. Adipocytes comprised less than 5 percent of the glands. In patients who had received renal transplants, the hyperplastic glands were moderately enlarged and showed a diffuse pattern of hyperplasia. Individuals treated with dialysis had markedly enlarged hyperplastic parathyroid glands with a nodular growth pattern. Each nodule consisted of one cell type. Although the histologic features differed, the calcium levels were similar. Two of the 41 (5 percent) cases of tertiary hyperparathyroidism were due to parathyroid adenomas (243).

Individual parathyroid glands in the setting of tertiary hyperplasia may show markedly atypical architectural and cytologic features that can mimic malignancy. Therefore, in the setting of tertiary hyperplasia in the setting of tertiary hyperplasia a very high threshold for the diagnosis of parathyroid carcinoma is recommended, for example, unequivocal vascular space invasion or spread outside the neck.

Immunohistochemical Findings. Chromogranin A, parathyroid hormone, and keratins are immunopositive in the parathyroid tissues. Parathyroid tissues are negative for TTF1 and thyroglobulin.

Differential Diagnosis. Tertiary parathyroid hyperplasia occurs in the setting of secondary parathyroid hyperplasia, but persists after renal transplantation or occurs after renal transplantation. Thus, there is a particular clinical setting for the disease. The parathyroid glands in tertiary hyperplasia are usually larger than in primary or secondary hyperplasia.

Treatment. Tertiary hyperparathyroidism may be treated surgically.

PARATHYROMATOSIS

Definition. *Parathyromatosis* is characterized by multiple nodules of hyperfunctioning parathyroid tissue outside the anatomic confines of the parathyroid glands (244).

Clinical Findings. Parathyromatosis is a rare cause of hyperparathyroidism. Parathyromatosis may be caused by hyperfunction of developmental rests in the soft tissue of the neck and mediastinal tissue left behind in development (19). Developmental or ontogenous parathyromatosis, often referred to as *primary parathyromatosis*, occurs when these developmental rests become hyperplasic in the setting of primary or secondary parathyroid hyperplasia. *Secondary* or *postsurgical parathyromatosis* is caused by inadvertent autotransplantation or spillage/seeding of parathyroid tissue during previous parathyroid surgery, including along an endoscopic track (245,246). The median time from prior surgery is approximately 6 years in otherwise healthy individuals and 3 years in the setting of end-stage renal disease (247). Parathyromatosis has also been reported following spontaneous rupture of a parathyroid adenoma (248). Fine-needle aspiration is not thought to be a significant cause of parathyromatosis (249), and can be used diagnostically to identify nodules of parathyromatosis (250).

Parathyromatosis is more common in women than men (247). Most individuals present in the fifth to sixth decade, with a recent review showing median age at presentation of 47 years (range, 18 to 75 years) (247). Although parathyromatosis may occur as persistence or recurrence of primary hyperparathyroidism, most cases complicate secondary hyperparathyroidism.

Gross Findings. Small nodules of hyperpercellular parathyroid tissue are identified in fibrous or fibrofatty tissue or skeletal muscle. The nests of cells are identified in the lower neck, superior mediastinum, or forearm (if site of autotransplantation).

Microscopic Findings. Multiple nests of hyperfunctioning parathyroid tissue are scattered throughout fibrous or fibrofatty tissue or skeletal muscle (fig. 2-15). Parathyromatosis occurring in the setting of prior surgery is associated with prominent fibrosis, invasive growth into skeletal muscle and soft tissue, and mitotic activity (fig. 2-16) (19).

Immunohistochemical Findings. Parathyroid tissue is positive for the neuroendocrine markers chromogranin A and synaptophysin, keratin, and parathyroid hormone. It is negative for TTF1 and thyroglobulin. In an immunohistochemical study of 16 cases of parafibromatosis, galectin-3 was positive in 3 of 16 (18.7 percent) parathyromatosis cases and in 14 of 15 (93.3 percent) parathyroid carcinomas (251).Five of 15 (33.3 percent) parathyroid carcinomas showed loss of Rb expression, but no loss of Rb expression was seen in any of the parathyromatosis cases. Five of 16 (31.3 percent) parathyroid carcinomas showed complete loss of parafibromin, but none of the parathyromatosis cases showed loss of parafibromin. A high Ki-67 proliferative index was seen in 9 of 15 (60 percent) parathyroid carcinomas and in only 1 of 15 (6.7 percent) parathyromatosis cases (251).

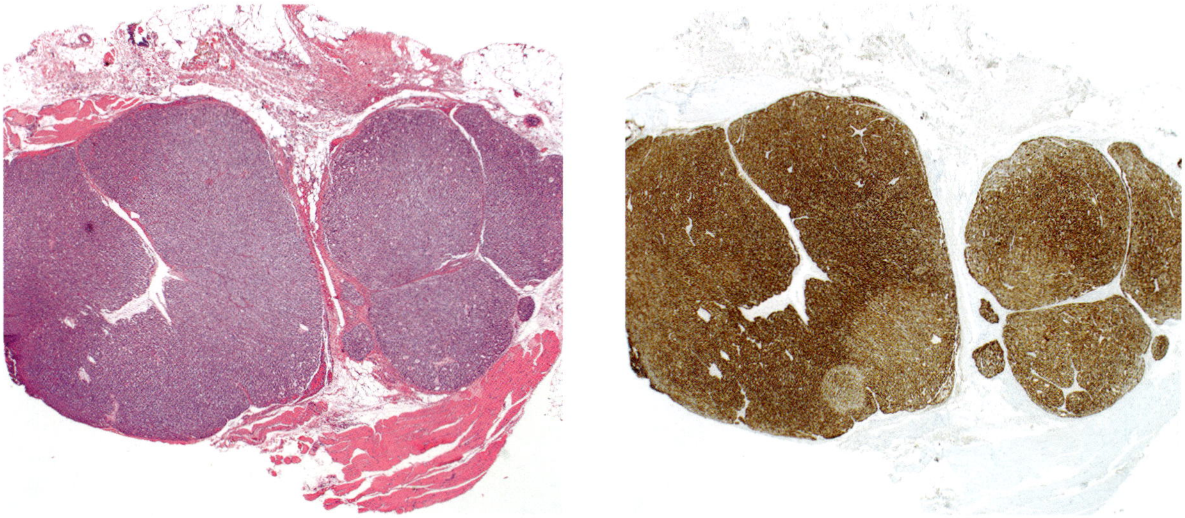

Figure 2-15

PARATHYROMATOSIS

Parathyromatosis is characterized by multiple nodules of hyperfunctioning parathyroid tissue outside the anatomic confines of the parathyroid glands.

Left: Multiple nests of hyperfunctioning parathyroid tissue are scattered throughout fibrous or fibrofatty tissue or skeletal muscle. Right: Parathyroid hormone immunostain shows diffuse positivity in the nodules of parafibromatosis.

Differential Diagnosis. Parathyromatosis must be differentiated from parathyroid carcinoma. Parathyroid carcinoma is usually associated with markedly elevated serum calcium, often 14 mg/dL or higher, which is generally higher than that seen in parathyromatosis (252). Parathyroid carcinoma is more common in men, while parathyromatosis is more common in women (247,252). A palpable neck mass and hoarseness usually only occur in parathyroid carcinoma (252). Rather than a single lesion, parathyromatosis may be multiple small lesions (252). Parathyroid carcinoma shows vascular or perineural invasion, infiltrative growth, and usually prominent mitotic figures. Although desmoplasia is usually not seen in primary parathyromatosis, fibrosis, mitotic activity, and invasive-appearing growth into skeletal muscle and soft tissue can be seen in parathyromatosis patients who have had prior surgery (19). Immunohistochemical studies including Rb, galectin-3, parafibromin, and Ki-67, show some differences between parathyromatosis and parathyroid carcinoma (251).

Treatment. Parathyromatosis may be treated surgically, but medical management with calcimimetics and bisphosphonates is a useful alternative (247,253).

HYPOPARATHYROIDISM

Hypoparathyroidism is insufficient or absent parathyroid hormone production or inability of the hormone to function in target tissues, resulting in hypocalcemia and hyperphosphatemia. The most common cause of hypoparathyroidism is surgical removal of parathyroid glands or damage to the parathyroid glands during anterior neck surgery. Approximately 75 percent of cases of hypoparathyroidism are due to neck surgery and 25 percent are due to other causes (254). Many patients with damage to the parathyroid glands recover parathyroid function, and their hypoparathyroidism is transient, but some have permanent hypoparathyroidism.

In addition to surgical removal or damage to parathyroid glands, Wilson disease, irradiation to the neck, hemachromatosis, and other disorders can compromise the function of the parathyroid glands (254). Other causes of hypoparathyroidism include developmental defects, autoimmune disorders, autoantibodies against the CASR receptor, defective regulation of parathyroid hormone secretion and receptor defects, 22q11.2 deletion (see below), and mitochondrial DNA abnormalities. Hypoparathyroidism is seen

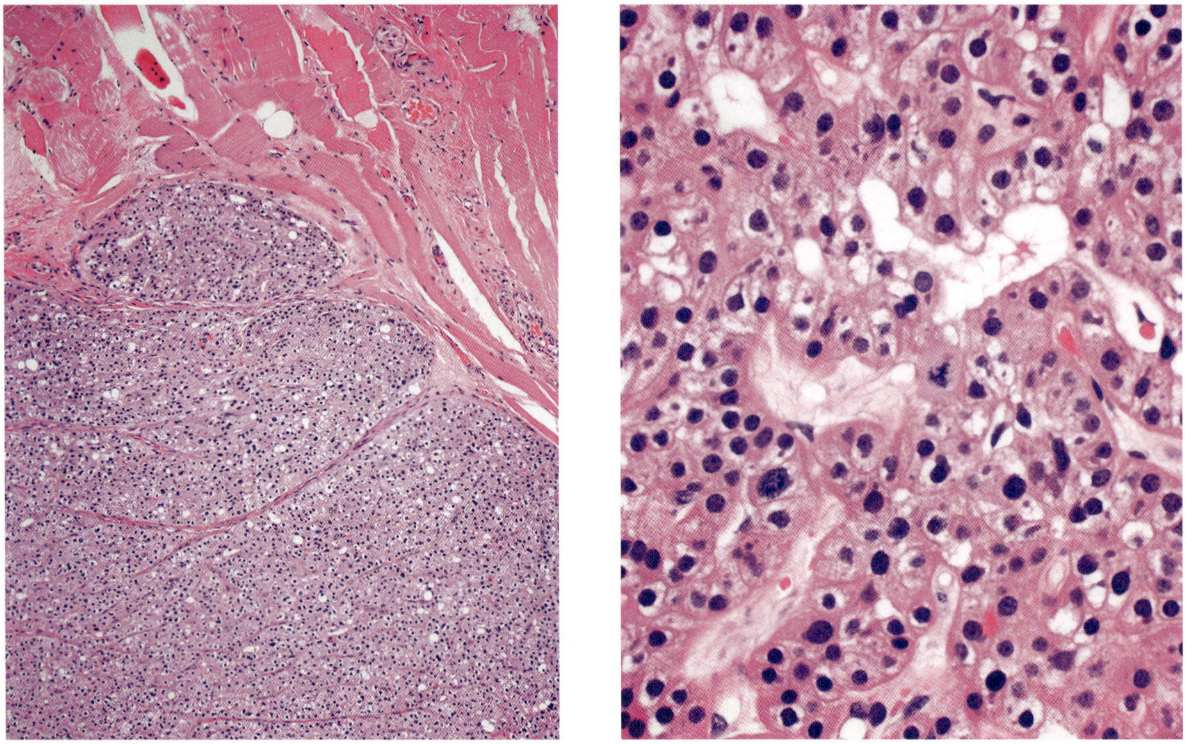

Figure 2-16

PARATHYROMATOSIS

Left: Parathyromatosis occurring in the setting of prior surgery is associated with prominent fibrosis and invasive growth into skeletal muscle.

Right: Extension into soft tissue with mitotic activity.

in a variety of disorders, sporadic and hereditary (Table 2-3). The genetic basis of most cases of sporadic hypoparathyroidism remains unknown; more is known about the pathogenesis of some types of hereditary hypoparathyroidism.

The most common microdeletion syndrome in humans, 22q11.2, occurs in approximately 1 in 4,000 births, arises de novo as a deletion or translocation, or is inherited in an autosomal dominant manner (255–259). Individuals with 22q11.2 deletion syndrome have variable phenotypic expressions. 22q11.2 deletion is present in the majority of individuals with DiGeorge syndrome as well as in velocardiofacial syndrome and conotruncal anomaly face syndrome, and in some patients with Opitz G/BBB and Cayler cardiofacial syndromes (258). DiGeorge syndrome is characterized by congenital aplasia of the thymus (T-cell deficiencies) and parathyroid gland, cardiac abnormalities, characteristic facies, and short stature (260,261). Velocardiofacial syndrome is an autosomal dominant disorder characterized by neonatal hypocalcemia, T-cell dysfunction, cleft palate, cardiac and ocular anomalies, and learning disabilities.

Due to the numerous syndromic names to describe the phenotypic features, classic chromosome 22q11.2 deletion is described according to genetic nomenclature (chromosome 22q11.2 deletion syndrome), and the clinical phenotypes are described using syndromic nomenclature (259). As stated, these syndromes involve hypocalcemia due to hypoparathyroidism, cardiac defects, facial dysmorphism, skeletal and renal abnormalities, impaired thymic development with immune (T-cell) deficiencies, autoimmune disorders, low platelet counts, ocular abnormalities, abnormal/cleft palate, speech difficulties, learning disabilities, psychiatric disorders, and possibly increased risk of malignancy (258,262–267). The *TBX1* gene within the deleted region is thought to be important in cardiac

Table 2-3

HYPOPARATHYROID DISORDERS WITH KNOWN GENETIC ABNORMALITIES

Syndrome	Genetic Abnormality
DiGeorge syndrome	Sporadic 1.5 to 3.0 hemizygous deletion of 22q11.2 (defects due to haploinsufficiency of *TBX1* gene)
Velocardiofacial syndrome	Autosomal dominant 1.5 to 3.0 hemizygous deletion of 22q11.2 (defects due to haploinsufficiency of *TBX1* gene)
Kearns-Sayre syndrome	Mitochondrial DNA abnormalities
Kenny-Caffey syndrome	Autosomal dominant or autorecessive, 1q42-q43, mutations in gene encoding tubulin-specific chaperone E *(TBCE)*
Autosomal dominant hypoparathyroidism	Autosomal dominant *CASR* receptor mutation
Sporadic idiopathic hypoparathyroidism	Sporadic *CASR* receptor mutation
Autoimmune polyglandular syndrome type 1	Autoantigens to *CASR* Autosomal recessive Mutations in the *APECED* (autoimmune polyendocrinopathy-candidiasis-ectodermal dystrophy) or *AIRE* (autoimmune regulator) gene on chromosome 21q22.3
Familial isolated hypoparathyroidism	*CASR* and *PTH* mutations
Autosomal recessive hypoparathyroidism	*PTH* point mutation
Jansen chondrodystrophy	Activating *PTHr1* mutation Autosomal dominant Defect in type 1 parathyroid hormone receptor
Blomstrand chondrodystrophy	Inactivating *PTHr1* mutation Autosomal recessive Defect in type 1 parathyroid hormone receptor
Pseudohypoparathyroidism type 1a (Albright hereditary osteodystrophy)	Autosomal dominant Inactivating mutation in stimulatory guanine nucleotide binding protein (*GNAS1* gene) (*GNAS1* mutation [Gsalpha1 protein of the adenyl cyclase complex])
Pseudohypoparathyroidism type 1b	Defects in methylation of in Stimulatory Guanine Nucleotide Binding Protein (GNAS1 gene [Gsalpha1 protein of the adenyl cyclase complex])
Pseudohypoparathyroidism type 2	Defective cAMP-dependent protein kinase

development, and haploinsufficiency of this gene is associated with some of the features in 22q11.2 deletion syndrome (259,268–270). Numerous genes are involved in the deleted region on 2q11.2, but the specific genes resulting in disease manifestation are controversial (259).

There are many other causes of hypoparathyroidism. These include hereditary forms, and some with known chromosomal abnormalities or patterns of inheritance such as autoimmune polyendocrinopathy candidiasis ectodermal dystrophy (APECED) (autoimmune polyglandular syndrome type 1), autosomal dominant hypocalcemia 1 and 2, Bartter syndrome type 5, CHARGE syndrome, Dubowitz syndrome, hereditary deafness and renal dysplasia syndrome, Kearns-Sayre syndrome, Kenny-Kaffey syndrome, and nonsyndromic isolated hypoparathyroidism.

APECED is an autosomal recessive disorder caused by homozygous or compound heterozygous mutation of the *AIRE* (autoimmune regulator) gene (21q22.3) (271–277). *AIRE* encodes a transcription factor involving the thymus and promotes central tolerance. There are many phenotypic expressions of APECED, but the three characteristic features of the classic phenotype include chronic mucocutaneous candidiasis, hypothyroidism, and adrenal insufficiency (271–276). Two of the features are required for a clinical diagnosis of this disease (271–276). Nevertheless, the diagnosis may be more complicated since some individuals have the classic phenotype and others have the nonclassic phenotype (277). The nonclassic phenotype, unlike the classic phenotype, has an autosomal dominant pattern of inheritance, is associated with heterozygous mutations of *PHD1* (first plant homeodomain) zinc finger of *AIRE,* and may have a milder phenotype, reduced penetrance, and older age at onset (279). Other autoimmune disorders also occur in APECED: over 100 mutations in *AIRE* are associated with this disease, and some are associated with particular populations, but there do not appear to be clear phenotype-genotype correlations (278).

Autosomal dominant hypocalcemia 1 is an autosomal dominant condition involving *CASR* (3q21.1), while autosomal dominant hypocalcemia 2 involves *GNA11* (19p13). Similar to autosomal dominant hypocalcemia 1 and 2, isolated hypoparathyroidism is also nonsyndromic, but can be autosomal dominant (*PTH* 11p15 or *GCMB* 6p24.2), autosomal recessive (*SOX3*, Xq26-27), or X-linked (254,280–284). Bartter syndrome type 5 is an autosomal dominant condition involving *CASR* (3q21.1). CHARGE syndrome, an autosomal dominant disorder associated with *CHD7* and *SEMA3E* (8q12.1-123.2 and 7q21.11) is characterized by coloboma of the eye, heart malformation, choanal atresia, anosmia, retardation of growth and development, ear and genital abnormalities, and deficiency of gonadotropin.

Dubowitz syndrome has an autosomal recessive pattern of inheritance, but the underlying genetic abnormalities are unknown. In addition to hypoparathyroidism, individuals with Dubowitz syndrome have short stature, microcephaly and abnormal facies, and mental retardation. Hereditary deafness and renal dysplasia syndrome is an autosomal dominant condition involving haploinsufficiency of *GATA3* (10p14), which encodes a transcriptional regulator involved in the development of parathyroid glands, kidneys, and inner ear (285–288). This disorder is characterized by hypoparathyroidism, deafness, and renal dysplasia.

Kearns-Sayre syndrome is a rare mitochondrial DNA disorder with prominent ophthalmic features (progressive external ophthalmoplegia, pigmentary retinopathy), cardiac conduction abnormalities, increased cerebrospinal fluid protein, cerebellar ataxia, muscle weakness, hearing loss, short stature, kidney and renal problems, and endocrinopathies such as diabetes and hypoparathyroidism (289–296). Even with characteristic clinical features, individuals thought to have Kearns-Sayre syndrome may also undergo muscle biopsy and genetic testing to confirm the diagnosis (294).

Kenny-Caffey syndrome can have an autosomal dominant or autosomal recessive mode of inheritance (297). Kenny-Caffey syndrome type 1 is associated with a *TBCE* mutation (1q42.3). It is characterized by hypoparathyroidism, short stature, bone abnormalities, hyperopia, and calcification of basal ganglia (297). Kenny-Caffey syndrome type 2 is an autosomal recessive disorder caused by *FAM111A* mutation (11q12.1) and characterized by hypoparathyroidism, short stature, and impaired bone development (298–300).

Pseudohypoparathyroidism type 1a is an autosomal dominant condition caused by an inactivating *GNAS1* (20q13.11) mutation. It is characterized by short stature and bone abnormalities (301). *GNAS* encodes a protein that interacts with parathyroid hormone receptor type 1. Pseudohypoparathyroidism type 1b is caused by defective methylation of GNAS1 and is associated with hypocalcemia with renal parathyroid hormone resistance.

PARATHYROID CYSTS

Parathyroid cysts are rare. They may be incidental findings or clinically apparent as a mass that may be confused with a thyroid nodule (302). Small cysts are usually incidental findings at autopsy while larger cysts may be clinically apparent. In a report of 2,505 parathyroidectomies and 22,009 thyroidectomies, 38 nonfunctional parathyroid cysts were identified (303). Parathyroid cysts account for less than 1 percent of parathyroid lesions (303–305).

Cystic change in a parathyroid tumor, such as a cystic parathyroid adenoma, is more common than a true parathyroid cyst. Unfortunately,

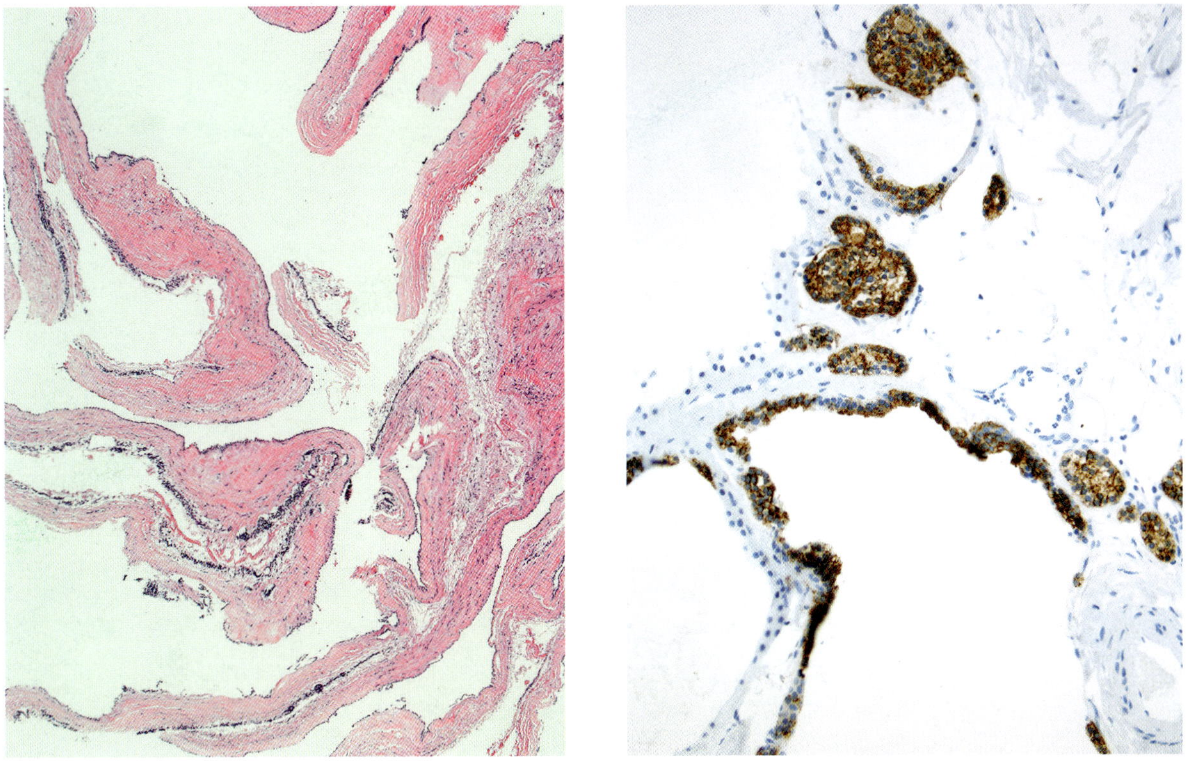

Figure 2-17

PARATHYROID CYST

Parathyroid cysts are rare and may be incidental findings or clinically apparent as a mass that is confused with a thyroid nodule.

Left: Parathyroid cysts are lined by flattened parathyroid epithelial cells that are positive for parathyroid hormone, keratin, and neuroendocrine markers.

Right: This parathyroid cyst shows retained cytoplasmic expression of PTH.

many of these lesions are inappropriately referred to as "functional parathyroid cyst" in the literature. Parathyroid cysts are generally nonfunctional. Parathyroid adenomas or carcinomas can be cystic or associated with a parathyroid cyst and be functional (306). Parathyroid tumors showing cystic degeneration or cystic change are more appropriately referred to as such, rather than a true parathyroid cyst.

Nonfunctional cysts come to clinical attention because of mass effect and may or may not be associated with compressive symptoms, or as incidental findings in radiographic imaging (303). Clinically apparent parathyroid cysts are usually large, some measuring greater than 8 cm (307). The diagnosis is generally made on the basis of histology (fig. 2-17) or needle aspiration of the cyst fluid. Cyst fluid with high parathyroid hormone levels confirms the diagnosis (308,309).

Parathyroid cysts are thought to be nonfunctional remnants of the third and fourth branchial pouches (310). Other etiologies have been suggested, including fluid accumulation within the parathyroid gland, microcysts coalescing into macrocysts, developmental abnormalities, and remnants of Kursteiner canals (311). Parathyroid cysts are usually identified in association with the parathyroid glands, often the lower parathyroid glands, but can be intrathyroidal or mediastinal (303,312–314). They are often located along the thyroid lobes and rarely occur at ectopic sites (304). Mediastinal parathyroid cysts, when containing a component of thymic tissue, are also referred to as *third pharyngeal pouch cysts.*

Parathyroid cysts may be lined by flattened parathyroid epithelial cells, and parathyroid cells may be present in the cyst wall (fig. 2-18). The cyst wall is usually thin and fibrous, but

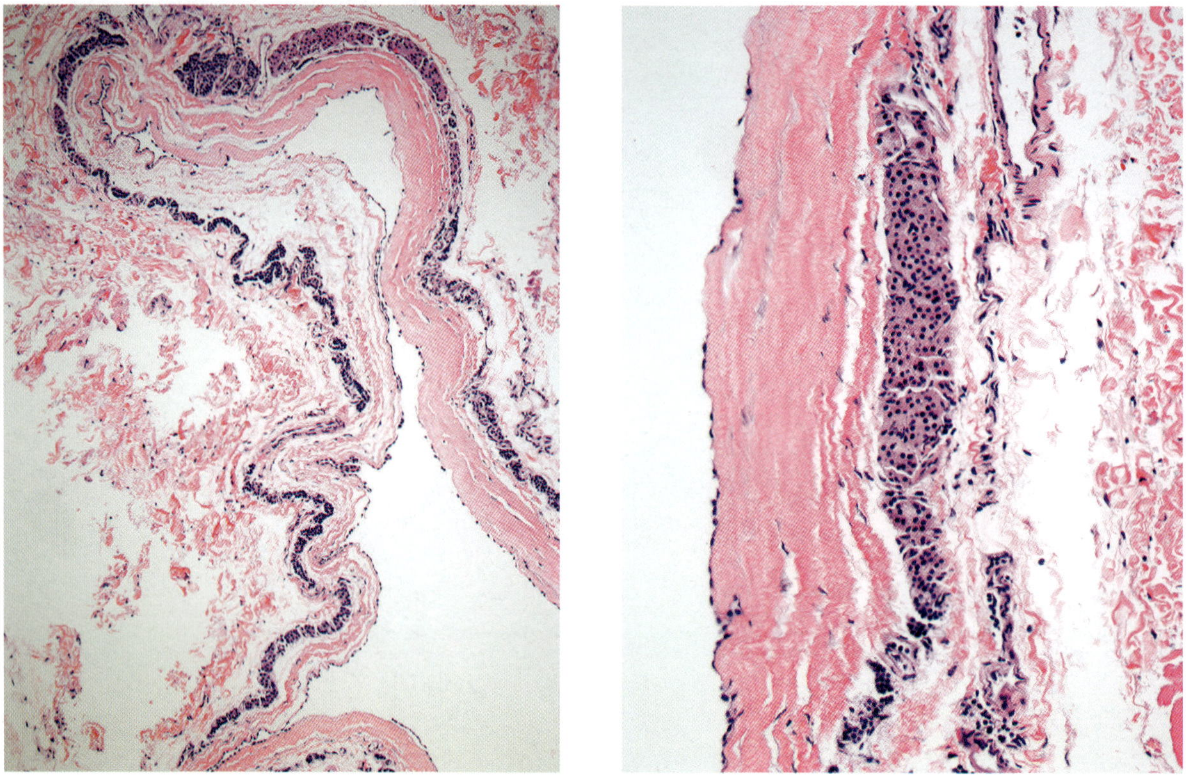

Figure 2-18

PARATHYROID CYST

Left: Parathyroid cysts may have parathyroid cells in the cyst wall.

Right: The cyst wall is usually thin and fibrous, but may also contain other elements such lymphoid, muscular, thymic, salivary, adipose, and mesenchymal tissue.

may also contain lymphoid, muscular, thymic, salivary, adipose, and mesenchymal tissue (303).

PARATHYROIDITIS

Parathyroiditis is an extremely rare entity in diagnostic pathology. It may be autoimmune in nature and associated with hypoparathyroidism or hyperparathyroidism (315–319). When hyperplastic parathyroid glands and parathyroid adenomas are associated with a lymphocytic infiltrate, it is described in the literature as parathyroiditis. However, this may be a nonspecific inflammatory infiltrate rather than an autoimmune process.

The etiology of parathyroiditis is poorly understood. Antibodies to parathyroid tissue have been found in a few patients with idiopathic hypoparathyroidism, although some individuals have other autoimmune disorders and others do not have an antibody identified. Recently, autoantibodies against the calcium sensing receptor were identified in an individual with glucocorticoid responsive lymphocytic parathyroiditis with hypocalciuric hypercalcemia (320). The parathyroid gland histologically shows an infiltrate of lymphocytes, with atrophy of the parathyroid parenchyma and replacement with fibrosis. Although rare in diagnostic pathology, up to 10 percent of parathyroid glands in autopsy series have lymphocytic infiltration without a known parathyroid abnormality in life. Thus, the significance of the histologic finding alone is unclear.

AMYLOID

Amyloid is identified in parathyroid glands in the setting of systemic amyloidosis, but it is also identified in individuals without systemic amyloidosis (fig. 2-19) (321–324). Amyloid is positive for Congo red and thioflavin T, and is often intrafollicular in an involved parathyroid gland (fig. 2-20) (321,323,325).

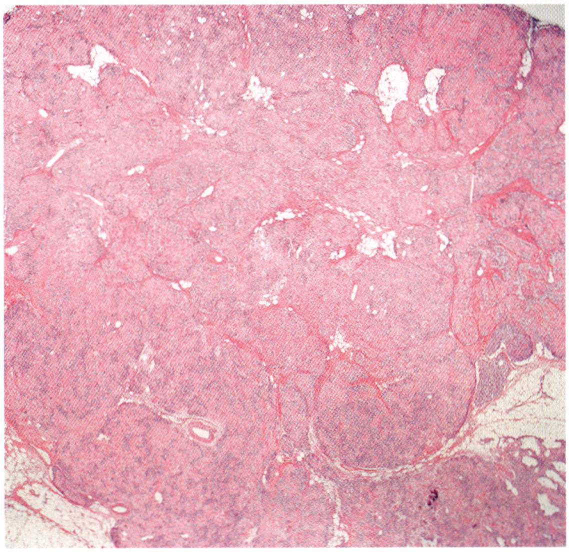

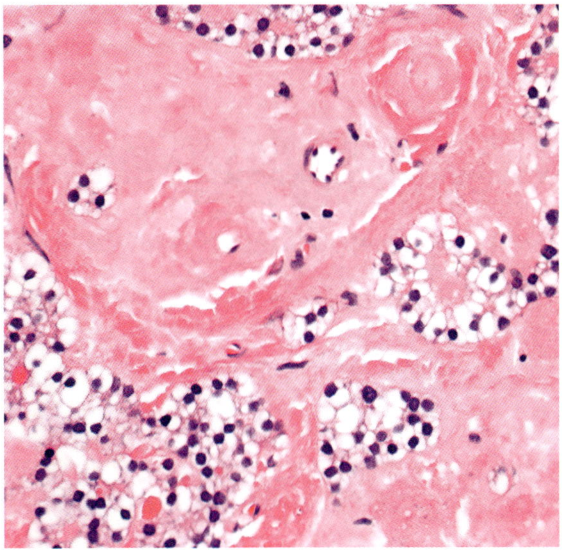

Figure 2-19

PARATHYROID AMYLOID

Amyloid is identified in parathyroid glands in the setting of systemic amyloidosis, but it can also be identified in individuals without systemic amyloidosis.

Left, right: Extensive involvement of the parathyroid gland by amyloid.

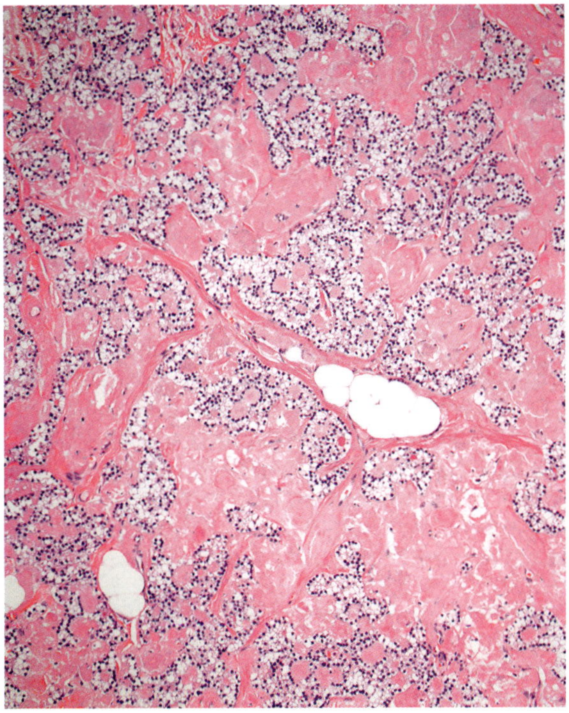

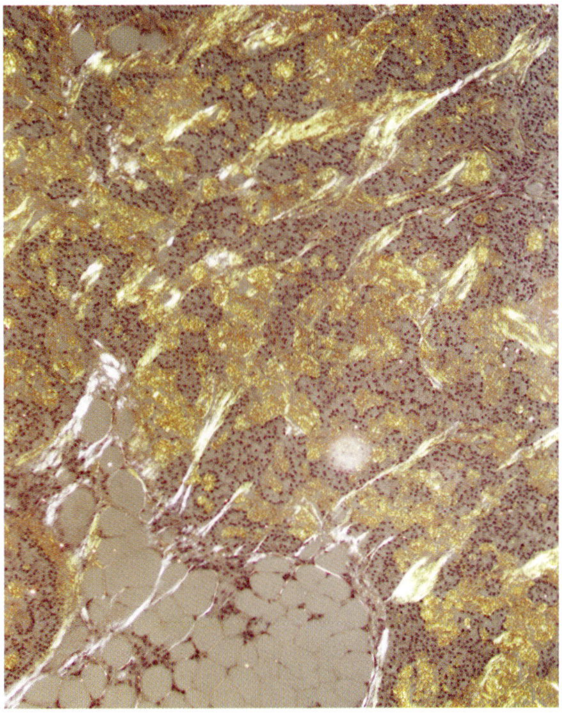

Figure 2-20

PARATHYROID AMYLOID

Left: Parathyroid gland with amyloid stained with H&E.

Right: Congo red shows birefringence in amyloid involving the parathyroid gland.

Amyloid is seen in normal parathyroid glands and pathologic glands such as those involved in hyperparathyroidism (323,325). The amyloid of systemic amyloidosis may not have an intrafollicular distribution, but may be in a prominent periarterial location and in the interstitium around the periphery of the parathyroid gland (323). In an autopsy series, amyloid was identified in 16 percent of surgical cases and 46 percent of autopsied parathyroid glands (323).

GLYCOGEN STORAGE DISEASE

Glycogen storage diseases, such as Pompe disease (type II glycogenosis), can affect many organ systems, including the endocrine organs. Parathyroid glands may be involved (326,327).

REFERENCES

Embryology, Anatomy, Histology

1. Weller GL. Development of the thyroid, parathyroid and thymus glands in man. Contrib Embryol 1933;141:93-139.
2. Hooghe L, Kinnaert P, Van Geertruyden J. Surgical anatomy of hyperparathyroidism. Acta Chir Belg 1992;92:1-9.
3. de Andrade JS, Mangussi-Gomes JP, da Rocha LA, et al. Localization of ectopic and supernumerary parathyroid glands in patients with secondary and tertiary hyperparathyroidism: surgical description and correlation with preoperative ultrasonography and Tc99m-Sestamibi scintigraphy. Braz J Otorhinolaryngol 2014;80:29-34.
4. Gomes EM, Nunes RC, Lacativa PG, et al. Ectopic and extranumerary parathyroid glands location in patients with hyperparathyroidism secondary to end stage renal disease. Acta Cir Bras 2007;22:105-9.
5. Yumita S. Intervention for recurrent secondary hyperparathyroidism from a residual parathyroid gland. Nephrol Dial Transplant 2003;18(Suppl 3):iii62-4.
6. Levin KE, Clark OH. The reasons for failure in parathyroid operations. Arch Surg 1989;124:911-5.
7. Kurman RJ, Prabha AC. Thyroid and parathyroid glands in the vaginal wall: report of a case. Am J Clin Pathol 1973;59:503-7.
8. Lack EE, Delay S, Linnoila RI. Ectopic parathyroid tissue within the vagus nerve. Incidence and possible clinical significance. Arch Pathol Lab Med 1988;112:304-6.
9. Michal M. Ectopic parathyroid within a neck paraganglion. Histopathology 1993;22:85-7.
10. Tsang WY, Chan JK. Ectopic parathyroid within a ganglion, not a paraganglion. Histopathology 1993;23:393.
11. Rubello D, Piotto A, Pagetta C, Pelizzo M, Casara D. Ectopic parathyroid adenomas located at the carotid bifurcation: the role of preoperative Tc-99m MIBI scintigraphy and the intraoperative gamma probe procedure in surgical treatment planning. Clin Nucl Med 2001;26:774-6.
12. Fukumoto A, Nonaka M, Kamio T, et al. A case of ectopic parathyroid gland hyperplasia in the pyriform sinus. Arch Otolaryngol Head Neck Surg 2002;128:71-4.
13. Sahin M, Er C, Unlu Y, Tekin S, Seker M. An ectopic parathyroid gland in the left axillary region: case report. Int Surg 2004;89:6-9.
14. Miura D. Ectopic parathyroid tumor in the sternohyoid muscles: supernumerary gland in a patient with MEN type 1. J Bone Miner Res 2005;20:1478-9.
15. Chang BA, Sharma A, Anderson DW. Ectopic parathyroid adenoma in the soft palate: a case report. J Otolaryngol Head Neck Surg 2016;45:53.
16. Ozgul MA, Seyhan EC, Ozgul G, et al. Endotracheal ectopic parathyroid adenoma mimicking asthma. Respir Med Case Rep 2014;13:28-31.
17. Foroulis CN, Rousogiannis S, Lioupis C, et al. Ectopic paraesophageal mediastinal parathyroid adenoma, a rare cause of acute pancreatitis. World J Surg Oncol 2004;2:41.
18. Noussios G, Anagnostis P, Natsis K. Ectopic parathyroid glands and their anatomical, clinical and surgical implications. Exp Clin Endocrinol Diabetes 2012;120:604-10.
19. Rosai J, DeLellis RA, Carcangiu ML, Frable WJ, Tallini G. Tumors of the thyroid and parathyroid glands. AFIP Atlas of Tumor Pathology, 4th Series, Fascicle 21. Washington DC: American Registry of Pathology; 2014.

20. Akerstrom G, Malmaeus J, Bergstrom R. Surgical anatomy of human parathyroid glands. Surgery 1984;95:14-21.
21. Lappas D, Noussios G, Anagnostis P, Adamidou F, Chatzigeorgiou A, Skandalakis P. Location, number and morphology of parathyroid glands: results from a large anatomical series. Anat Sci Int 2012;87:160-4.
22. Pattou FN, Pellissier LC, Noël C, Wambergue F, Huglo DG, Proye CA. Supernumerary parathyroid glands: frequency and surgical significance in treatment of renal hyperparathyroidism. World J Surg 2000;24:1330-4.
23. Athanasoulis T, Koutsikos J, Korovesis K, Bokos J, Zerva C. Supernumerary parathyroid glands with an unusual ectopic location in recurrent secondary hyperparathyroidism. Nuklearmedizin 2006;45:N4-6.
24. Alexander TH, Beros AD, Orloff LA. Twice-recurrent primary hyperparathyroidism due to parathyroid hyperplasia in an ectopic supernumerary gland. Endocr Pract 2006;12:165-9.
25. Castleman B, Mallory TB. The pathology of the parathyroid gland in hyperparathyroidism: a study of 25 cases. Am J Pathol 1935;11:1-72.
26. Parfitt AM, Wang Q, Palnitkar S. Rates of cell proliferation in adenomatous, suppressed, and normal parathyroid tissue: implications for pathogenesis. J Clin Endocrinol Metab 1998;83:863-9.
27. Grimelius L, Akerström G, Johansson H, Bergström R. Anatomy and histopathology of human parathyroid glands. Pathol Annu 1981;16(Pt 2):1-24.
28. Gilmour JR, Martin WJ. Weight of the parathyroid glands. J Path Bact 1937;44:431-62.
29. Ghandur-Mnaymneh L, Cassady J, Hajianpour MA, Paz J, Reiss E. The parathyroid gland in health and disease. Am J Pathol 1986;125:292-9.
30. Erdheim J. Zur normalen und patghologischen Histologie der Glandula thyreoidea, parathyreoidea und Hypophysis. Beitr z path Anat u z allg Pathol 1903;33:158-236.
31. Kurokawa K. Histological studies of normal and pathological human parathyroid glands. Japan M World 1925;5:241-51.
32. Saffos RO, Rhatigan RM. Intracellular lipid in parathyroid glands. Hum Pathol 1979;10:483-5.
33. Dekker A, Dunsford HA, Geyer SJ. The normal parathyroid gland at autopsy: the significance of stromal fat in adult patients. J Pathol 1979;128:127-32.
34. Dufour DR, Wilkerson SY. The normal parathyroid revisited: percentage of stromal fat. Hum Pathol 1982;13:717-21.
35. Akerstrom G, Grimelius L, Johansson H, Lundqvist H, Pertoft H, Bergströmet R. The parenchymal cell mass in normal human parathyroid glands. Acta Pathol Microbiol Scand A 1981;89:367-75.
36. Erickson LA, Jin L, Papotti M, Lloyd RV. Oxyphil parathyroid carcinomas: a clinicopathologic and immunohistochemical study of 10 cases. Am J Surg Pathol 2002;26:344-9.
37. Tomita T. Immunocytochemical staining patterns for parathyroid hormone and chromogranin in parathyroid hyperplasia, adenoma, and carcinoma. Endocr Pathol 1999;10:145-56.
38. Maeda T, Ashie T, Kikuiri K, Takakura M, Ise T, Shimamotoet K. A case of calcitonin-producing parathyroid adenoma with primary hyperparathyroidism. Jpn J Med 1989;28:640-6.
39. Williamson BR, Carey RM, Innes DJ, et al. Poorly differentiated lymphocytic lymphoma with ectopic parathormone production: visulization of metastatic calcification by bone scan. Clin Nucl Med 1978;3:382-4.
40. Holtz G, Johnson TR Jr, Schrock ME. Paraneoplastic hypercalcemia in ovarian tumors. Obstet Gynecol 1979;54:483-7.
41. Goldberg RS, Pilcher DB, Yates W. The aggressive surgical management of hypercalcemia due to ectopic parathormone production. Cancer 1980;45:2652-4.
42. Max MH, Lawrence PH. Ectopic secretion of parathyroid hormone. Surg Gynecol Obstet 1980;150:411-8.
43. Patel S, Rosenthal JT. Hypercalcemia in carcinoma of prostate. Its cure by orchiectomy. Urology 1985;25:627-9.
44. Buller R, Taylor K, Burg AC, Berman ML, DiSaiaet PJ. Paraneoplastic hypercalcemia associated with adenosquamous carcinoma of the endometrium. Gynecol Oncol 1991;40:95-8.
45. Nakajima K, Tamai M, Okaniwa S. Humoral hypercalcemia associated with gastric carcinoma secreting parathyroid hormone: a case report and review of the literature. Endocr J 2013;60:557-62.
46. Koyama Y, Ishijima H, Ishibashi A, et al. Intact PTH-producing hepatocellular carcinoma treated by transcatheter arterial embolization. Abdom Imaging 1999;24:144-6.
47. Chen L, Dinh TA, Haque A. Small cell carcinoma of the ovary with hypercalcemia and ectopic parathyroid hormone production. Arch Pathol Lab Med 2005;129:531-3.
48. Marchiori E, Pelizzo MR, Herten M, Townsend DM, Rubello D, Boschin IM. Specifying the molecular pattern of sporadic parathyroid tumorigenesis-the Y282D variant of the GCM2 gene. Biomed Pharmacother 2017;92:843-8.
49. Nonaka D. Study of parathyroid transcription factor Gcm2 expression in parathyroid lesions. Am J Surg Pathol 2011;35:145-51.
50. Guan B, Welch JM, Sapp JC, et al. GCM2-Activating mutations in familial isolated hyperparathyroidism. Am J Hum Genet 2016;99:1034-44.

51. Oliveira AM, Tazelaar HD, Myers JL, Erickson LA, Lloyd RV. Thyroid transcription factor-1 distinguishes metastatic pulmonary from well-differentiated neuroendocrine tumors of other sites. Am J Surg Pathol 2001;25:815-9.
52. Agoff SN, Lamps LW, Philip AT, et al. Thyroid transcription factor-1 is expressed in extrapulmonary small cell carcinomas but not in other extrapulmonary neuroendocrine tumors. Mod Pathol 2000;13:238-42.
53. Kaufmann O, Dietel M. Thyroid transcription factor-1 is the superior immunohistochemical marker for pulmonary adenocarcinomas and large cell carcinomas compared to surfactant proteins A and B. Histopathology 2000;36:8-16.
54. Folpe AL, Gown AM, Lamps LW, et al. Thyroid transcription factor-1: immunohistochemical evaluation in pulmonary neuroendocrine tumors. Mod Pathol 1999;12:5-8.
55. Fabbro D, Di LC, Stamerra O, Beltrami CA, Lonigro R, Damante G. TTF-1 gene expression in human lung tumours. Eur J Cancer 1996;32A:512-7.
56. Bisceglia M, Ragazzi M, Galliani CA, Lastilla G, Rosai J. TTF-1 expression in nephroblastoma. Am J Surg Pathol 2009;33:454-61.
57. Nonaka D, Tang Y, Chiriboga L, Rivera R, Ghosseinet R. Diagnostic utility of thyroid transcription factors Pax8 and TTF-2 (FoxE1) in thyroid epithelial neoplasms. Mod Pathol 2008;21:192-200.
58. Graham AD, Williams AR, Salter DM. TTF-1 expression in primary ovarian epithelial neoplasia. Histopathology 2006;48:764-5.
59. Pratt D, Afsar N, Fetsch AM, et al. Re-evaluating TTF-1 immunohistochemistry in diffuse gliomas: expression is clone dependent and associated with tumor location. Clin Neuropathol 2017;36:263-71.
60. Ni YB, Tsang JY, Shao MM, et al. TTF-1 expression in breast carcinoma: an unusual but real phenomenon. Histopathology 2014;64:504-11.
61. Reis HG, Metz CH, Baba HA, Bornfeld N, Schmid KW, Metz KA. [TTF-1 (8G7G3/1) positive colon adenocarcinoma: diagnostic implications.] Pathologe 2011;32:349-51. [German]
62. Hakim SA, Youssef NS. Diagnostic utility of thyroid transcription factor-1 in ovarian carcinoma and its relationship with clinicopathologic prognostic parameters. Appl Immunohistochem Mol Morphol 2017;25:237-43.
63. Matoso A, Singh K, Jacob R, et al. Comparison of thyroid transcription factor-1 expression by 2 monoclonal antibodies in pulmonary and nonpulmonary primary tumors. Appl Immunohistochem Mol Morphol 2010;18:142-9.
64. Dettmer M, Kim TE, Jung CK, Jung ES, Lee KY, Kang CS. Thyroid transcription factor-1 expression in colorectal adenocarcinomas. Pathol Res Pract 2011;207:686-90.
65. Laury AR, Perets R, Piao H, et al. A comprehensive analysis of PAX8 expression in human epithelial tumors. Am J Surg Pathol 2011;35:816-26.
66. Ozcan A, Shen SS, Hamilton C, et al. PAX 8 expression in non-neoplastic tissues, primary tumors, and metastatic tumors: a comprehensive immunohistochemical study. Mod Pathol 2011;24:751-64.
67. Han SI, Tsunekage Y, Kataoka K. Gata3 cooperates with Gcm2 and MafB to activate parathyroid hormone gene expression by interacting with SP1. Mol Cell Endocrinol 2015;411:113-20.
68. Grigorieva IV, Thakker RV. Transcription factors in parathyroid development: lessons from hypoparathyroid disorders. Ann N Y Acad Sci 2011;1237:24-38.
69. Grigorieva IV, Mirczuk S, Gaynor KU, et al. Gata3-deficient mice develop parathyroid abnormalities due to dysregulation of the parathyroid-specific transcription factor Gcm2. J Clin Invest 2010;120:2144-55.
70. Miettinen M, McCue PA, Sarlomo-Rikala M, et al. GATA3: a multispecific but potentially useful marker in surgical pathology: a systematic analysis of 2500 epithelial and nonepithelial tumors. Am J Surg Pathol 2014;38:13-22.
71. Erickson LA, Papouchado B, Dimashkieh H, Zhang S, Nakamura N, Lloyd RV. Cdx2 as a marker for neuroendocrine tumors of unknown primary sites. Endocr Pathol 2004;15:247-52.
72. Heath H 3rd, Hodgson SF, Kennedy MA. Primary hyperparathyroidism. Incidence, morbidity, and potential economic impact in a community. N Engl J Med 1980;302:189-93.
73. Griebeler ML, Kearns AE, Ryu E, Hathcock MA, Melton LJ 3rd, Wermers RA. Secular trends in the incidence of primary hyperparathyroidism over five decades (1965-2010). Bone 2015;73:1-7.
74. Wermers RA, Khosla S, Atkinson EJ, et al. Incidence of primary hyperparathyroidism in Rochester, Minnesota, 1993-2001: an update on the changing epidemiology of the disease. J Bone Miner Res 2006;21:171-7.
75. Yeh MW, Ituarte PH, Zhou HC, et al. Incidence and prevalence of primary hyperparathyroidism in a racially mixed population. J Clin Endocrinol Metab 2013;98:1122-9.

Familial/Hereditary, Sporadic Hyperparathyroidism

76. Thakker RV. Genetics of parathyroid tumours. J Intern Med 2016;280:574-83.
77. Agarwal SK, Burns AL, Sukhodolets KE, et al. Molecular pathology of the MEN1 gene. Ann N Y Acad Sci 2004;1014:189-98.
78. Agarwal SK, Kester MB, Debelenko LV, et al. Germline mutations of the MEN1 gene in familial multiple endocrine neoplasia type 1 and related states. Hum Mol Genet 1997;6:1169-75.

79. Chandrasekharappa SC, Guru SC, Manickam P, et al. Positional cloning of the gene for multiple endocrine neoplasia-type 1. Science 1997;276:404-7.
80. Pellegata NS, Quintanilla-Martinez L, Siggelkow H, et al. Germ-line mutations in p27Kip1 cause a multiple endocrine neoplasia syndrome in rats and humans. Proc Natl Acad Sci U S A, 2006;103:15558-63.
81. Thakker RV. Multiple endocrine neoplasia type 1 (MEN1) and type 4 (MEN4). Mol Cell Endocrinol 2014;386:2-15.
82. Alrezk R, Hannah-Shmouni F, Stratakis CA. MEN4 and CDKN1B mutations: the latest of the MEN syndromes. Endocr Relat Cancer 2017;24:T195-208.
83. Agarwal SK. The future: genetics advances in MEN1 therapeutic approaches and management strategies. Endocr Relat Cancer 2017;24:T119-34.
84. Uchino S, Noguchi S, Nagatomo M, et al. Absence of somatic RET gene mutation in sporadic parathyroid tumors and hyperplasia secondary to uremia, and absence of somatic Men1 gene mutation in MEN2A-associated hyperplasia. Biomed Pharmacother 2000;54(Suppl 1):100s-3s.
85. Uchino S, Noguchi S, Sato M, et al. Screening of the Men1 gene and discovery of germ-line and somatic mutations in apparently sporadic parathyroid tumors. Cancer Res 2000;60:5553-7.
86. Marx SJ, Agarwal SK, Kester MB, et al. Multiple endocrine neoplasia type 1: clinical and genetic features of the hereditary endocrine neoplasias. Recent Prog Horm Res 1999;54:397-9.
87. Brandi ML, Gagel RF, Angeli A, et al. Guidelines for diagnosis and therapy of MEN type 1 and type 2. J Clin Endocrinol Metab 2001;86:5658-71.
88. Friedman E, Sakaguchi K, Bale AE, et al. Clonality of parathyroid tumors in familial multiple endocrine neoplasia type 1. N Engl J Med 1989;321:213-8.
89. Verges B, Boureille F, Goudet P, et al. Pituitary disease in MEN type 1 (MEN1): data from the France-Belgium MEN1 multicenter study. J Clin Endocrinol Metab 2002;87:457-65.
90. Marini F, Falchetti A, Luzi E, et al. Multiple endocrine neoplasia type 1 (MEN1) syndrome. In: Riegert-Johnson DL, Boardman LA, Hefferon T, et al., eds. Cancer syndromes. Bethesda: National Center for Biotechnology Information; 2009.
91. Benya RV, Metz DC, Venzon DJ, et al. Zollinger-Ellison syndrome can be the initial endocrine manifestation in patients with multiple endocrine neoplasia-type I. Am J Med 1994;97:436-44.
92. Carty SE, Helm AK, Amico JA, et al. The variable penetrance and spectrum of manifestations of multiple endocrine neoplasia type 1. Surgery 1998;124:1106-14.
93. Gibril F, Schumann M, Pace A, Jensen RT. Multiple endocrine neoplasia type 1 and Zollinger-Ellison syndrome: a prospective study of 107 cases and comparison with 1009 cases from the literature. Medicine (Baltimore) 2004;83:43-83.
94. Vasen HF, Lamers CB, Lips CJ. Screening for the multiple endocrine neoplasia syndrome type I. A study of 11 kindreds in The Netherlands. Arch Intern Med 1989;149:2717-22.
95. Anlauf M, Perren A, Kloppel G. Endocrine precursor lesions and microadenomas of the duodenum and pancreas with and without MEN1: criteria, molecular concepts and clinical significance. Pathobiology 2007;74:279-84.
96. Skogseid B, Larsson C, Lindgren PG, et al. Clinical and genetic features of adrenocortical lesions in multiple endocrine neoplasia type 1. J Clin Endocrinol Metab 1992;75:76-81.
97. Marini F, Falchetti A, Del Monte F, et al. Multiple endocrine neoplasia type 1. Orphanet J Rare Dis 2006;1:38.
98. Piecha G, Chudek J, Wiecek A. Multiple endocrine neoplasia type 1. Eur J Intern Med 2008;19:99-103.
99. Falchetti A, Marini F, Luzi E, Tonelli F, Brandi ML. Multiple endocrine neoplasms. Best Pract Res Clin Rheumatol 2008;22:149-63.
100. Vierimaa O, Ebeling TM, Kytola S, et al. Multiple endocrine neoplasia type 1 in Northern Finland; clinical features and genotype phenotype correlation. Eur J Endocrinol 2007;157:285-94.
101. Al-Salameh A, Cadiot G, Calender A, et al. Clinical aspects of multiple endocrine neoplasia type 1. Nat Rev Endocrinol 2021;17:207-24.
102. Kouvaraki MA, Lee JE, Shapiro SE, et al. Genotype-phenotype analysis in multiple endocrine neoplasia type 1. Arch Surg 2002;137:641-7.
103. Tsukada T, Yamaguchi K, Kameya T. The MEN1 gene and associated diseases: an update. Endocr Pathol 2001;12:259-73.
104. Ki Wong F, Burgess J, Nordenskjold M, Larsson C, Tean Teh B. Multiple endocrine neoplasia type 1. Semin Cancer Biol 2000;10:299-312.
105. Lemmens I, Van de Ven WJ, Kas K, et al. Identification of the multiple endocrine neoplasia type 1 (MEN1) gene. The European Consortium on MEN1. Hum Mol Genet 1997;6:1177-83.
106. Shen HC, Rosen JE, Yang LM, et al. Parathyroid tumor development involves deregulation of homeobox genes. Endocr Relat Cancer 2008;15:267-75.
107. Crabtree JS, Scacheri PC, Ward JM, et al. A mouse model of multiple endocrine neoplasia, type 1, develops multiple endocrine tumors. Proc Natl Acad Sci U S A 2001;98:1118-23.

108. Loffler KA, Biondi CA, Gartside M, et al. Broad tumor spectrum in a mouse model of multiple endocrine neoplasia type 1. Int J Cancer 2007;120:259-67.
109. Mohr H, Pellegata NS. Animal models of MEN1. Endocr Relat Cancer 2017;24:T161-77.
110. Bertolino P, Tong WM, Galendo D, Wang ZQ, Zhang CX. Heterozygous Men1 mutant mice develop a range of endocrine tumors mimicking multiple endocrine neoplasia type 1. Mol Endocrinol 2003;17:1880-92.
111. Harding B, Lemos MC, Reed AA, et al. Multiple endocrine neoplasia type 1 knockout mice develop parathyroid, pancreatic, pituitary and adrenal tumours with hypercalcaemia, hypophosphataemia and hypercorticosteronaemia. Endocr Relat Cancer 2009;16:1313-27.
112. Guo SS, Sawicki MP. Molecular and genetic mechanisms of tumorigenesis in multiple endocrine neoplasia type-1. Mol Endocrinol 2001;15:1653-64.
113. Scarpelli D, D'Aloiso L, Arturi F, et al. Novel somatic MEN1 gene alterations in sporadic primary hyperparathyroidism and correlation with clinical characteristics. J Endocrinol Invest 2004;27:1015-21.
114. Arnold A, Shattuck TM, Mallya SM, et al. Molecular pathogenesis of primary hyperparathyroidism. J Bone Miner Res 2002;17(Suppl 2):N30-6.
115. Newey PJ, Nesbit MA, Rimmer AJ, et al. Whole-exome sequencing studies of nonhereditary (sporadic) parathyroid adenomas. J Clin Endocrinol Metab 2012;97:E1995-2005.
116. Haven CJ, Howell VM, Eilers PH, et al. Gene expression of parathyroid tumors: molecular subclassification and identification of the potential malignant phenotype. Cancer Res 2004;64:7405-11.
117. Haven CJ, van Puijenbroek M, Tan MH, et al. Identification of MEN1 and HRPT2 somatic mutations in paraffin-embedded (sporadic) parathyroid carcinomas. Clin Endocrinol (Oxf) 2007;67:370-6.
118. Erickson LA, Jalal SM, Harwood A, Shearer B, Jin L, Lloyd RV. Analysis of parathyroid neoplasms by interphase fluorescence in situ hybridization. Am J Surg Pathol 2004;28:578-84.
119. Hsi ED, Zukerberg LR, Yang WI, Arnold A. Cyclin D1/PRAD1 expression in parathyroid adenomas: an immunohistochemical study. J Clin Endocrinol Metab 1996;81:1736-9.
120. He X, Wang L, Yan J, Yuan C, Witze ES, Hua X. Menin localization in cell membrane compartment. Cancer Biol Ther 2016;17:114-22.
121. Guru SC, Goldsmith PK, Burns AL, et al. Menin, the product of the MEN1 gene, is a nuclear protein. Proc Natl Acad Sci U S A 1998;95:1630-4.
122. Cao Y, Liu R, Jiang X, et al. Nuclear-cytoplasmic shuttling of menin regulates nuclear translocation of {beta}-catenin. Mol Cell Biol 2009;29:5477-87.
123. Dreijerink KM, Timmers HT, Brown M. Twenty years of menin: emerging opportunities for restoration of transcriptional regulation in MEN1. Endocr Relat Cancer 2017;24:T135-45.
124. Francis J, Lin W, Rozenblatt-Rosen O, Meyerson M. The menin tumor suppressor protein is phosphorylated in response to DNA damage. PLoS One 2011;6:e16119.
125. Gallo A, Agnese S, Esposito I, Galgani M, Avvedimento VE. Menin stimulates homology-directed DNA repair. FEBS Lett 2010;584:4531-6.
126. Hendy GN, Kaji H, Canaff L. Cellular functions of menin. Adv Exp Med Biol 2009;668:37-50.
127. Wu T, Hua X. Menin represses tumorigenesis via repressing cell proliferation. Am J Cancer Res 2011;1:726-39.
128. Agarwal SK, Guru SC, Heppner C, et al. Menin interacts with the AP1 transcription factor JunD and represses JunD-activated transcription. Cell 1999;96:143-52.
129. Chen G, Jingbo A, Wang M, et al. Menin promotes the Wnt signaling pathway in pancreatic endocrine cells. Mol Cancer Res 2008;6:1894-907.
130. Wu G, Yuan M, Shen S, et al. Menin enhances c-Myc-mediated transcription to promote cancer progression. Nat Commun 2017;8:15278.
131. Hendy GN, Kaji H, Sowa H, Lebrun JJ, Canaff L. Menin and TGF-beta superfamily member signaling via the Smad pathway in pituitary, parathyroid and osteoblast. Horm Metab Res 2005;37:375-9.
132. Lebrun JJ. Activin, TGF-beta and menin in pituitary tumorigenesis. Adv Exp Med Biol 2009;668:69-78.
133. Sowa H, Kaji H, Hendy GN, et al. Menin is required for bone morphogenetic protein 2- and transforming growth factor beta-regulated osteoblastic differentiation through interaction with Smads and Runx2. J Biol Chem 2004;279:40267-75.
134. Dreijerink KM, Mulder KW, Winkler GS, Hoppener JW, Lips CJ, Timmers HT. Menin links estrogen receptor activation to histone H3K4 trimethylation. Cancer Res 2006;66:4929-35.
135. Ellard S, Hattersley AT, Brewer CM, Vaidya B. Detection of an MEN1 gene mutation depends on clinical features and supports current referral criteria for diagnostic molecular genetic testing. Clin Endocrinol (Oxf) 2005;62:169-75.
136. Klein RD, Salih S, Bessoni J, Bale AE. Clinical testing for multiple endocrine neoplasia type 1 in a DNA diagnostic laboratory. Genet Med 2005;7:131-8.

137. Lemos MC, Thakker RV. Multiple endocrine neoplasia type 1 (MEN1): analysis of 1336 mutations reported in the first decade following identification of the gene. Hum Mutat 2008;29:22-32.
138. Pardi E, Borsari S, Saponaro F, et al. Mutational and large deletion study of genes implicated in hereditary forms of primary hyperparathyroidism and correlation with clinical features. PLoS One 2017;12:e0186485.
139. Thakker RV, Newey PJ, Walls GV, et al. Clinical practice guidelines for multiple endocrine neoplasia type 1 (MEN1). J Clin Endocrinol Metab 2012;97:2990-3011.
140. Kimura T, Yoshimoto K, Tanaka C, et al. Obvious mRNA and protein expression but absence of mutations of the RET proto-oncogene in parathyroid tumors. Eur J Endocrinol 1996;134:314-9.
141. Hofstra RM, Landsvater RM, Ceccherini I, et al. A mutation in the RET proto-oncogene associated with multiple endocrine neoplasia type 2B and sporadic medullary thyroid carcinoma. Nature 1994;367:375-6.
142. Eng C, Clayton D, Schuffenecker I, et al. The relationship between specific RET proto-oncogene mutations and disease phenotype in multiple endocrine neoplasia type 2. International RET mutation consortium analysis. JAMA 1996;276:1575-9.
143. Marsh DJ, Andrew SD, Eng C, et al. Germline and somatic mutations in an oncogene: RET mutations in inherited medullary thyroid carcinoma. Cancer Res 1996;56:1241-3.
144. Marsh DJ, Learoyd DL, Andrew SD, et al. Somatic mutations in the RET proto-oncogene in sporadic medullary thyroid carcinoma. Clin Endocrinol (Oxf) 1996;44:249-57.
145. Raue F, Frank-Raue K. Multiple endocrine neoplasia type 2: 2007 update. Horm Res 2007;68(Suppl 5):101-4.
146. Machens A, Dralle H. Multiple endocrine neoplasia type 2 and the RET protooncogene: from bedside to bench to bedside. Mol Cell Endocrinol 2006;247:34-40.
147. de Krijger RR. Endocrine tumor syndromes in infancy and childhood. Endocr Pathol 2004;15:223-6.
148. Phay JE, Moley JF, Lairmore TC. Multiple endocrine neoplasias. Semin Surg Oncol 2000; 18:324-32.
149. Iler MA, King DR, Ginn-Pease ME, O'Dorisio TM, Sotos JF. Multiple endocrine neoplasia type 2A: a 25-year review. J Pediatr Surg 1999;34:92-7.
150. Marsh DJ, Mulligan LM, Eng C. RET proto-oncogene mutations in multiple endocrine neoplasia type 2 and medullary thyroid carcinoma. Horm Res 1997;47:168-78.
151. Oishi S, Sato T, Takiguchi-Shirahama S, Nakamura Y. Mutations of the RET proto-oncogene in multiple endocrine neoplasia type 2A (Sipple's syndrome). Endocr J 1995;42:527-36.
152. Mulligan LM, Marsh DJ, Robinson BG, et al. Genotype-phenotype correlation in multiple endocrine neoplasia type 2: report of the International RET Mutation Consortium. J Intern Med 1995;238:343-6.
153. Mulligan LM, Eng C, Attie T, et al. Diverse phenotypes associated with exon 10 mutations of the RET proto-oncogene. Hum Mol Genet 1994;3:2163-7.
154. Wells SA Jr, Asa SL, Dralle H, et al. Revised American Thyroid Association guidelines for the management of medullary thyroid carcinoma. Thyroid 2015;25:567-610.
155. Mulligan LM, Eng C, Healey CS, et al. Specific mutations of the RET proto-oncogene are related to disease phenotype in MEN 2A and FMTC. Nat Genet 1994;6:70-4.
156. Schuffenecker I, Virally-Monod M, Brohet R, et al. Risk and penetrance of primary hyperparathyroidism in multiple endocrine neoplasia type 2A families with mutations at codon 634 of the RET proto-oncogene. Groupe d'etude des tumeurs a calcitonine. J Clin Endocrinol Metab 1998;83:487-91.
157. Machens A, Dralle H. Genotype-phenotype based surgical concept of hereditary medullary thyroid carcinoma. World J Surg 2007;31:957-68.
158. Fritz A, Walch A, Piotrowska K, et al. Recessive transmission of a multiple endocrine neoplasia syndrome in the rat. Cancer Res 2002;62:3048-51.
159. Piotrowska K, Pellegata NS, Rosemann M, Fritz A, Graw J, Atkinson MJ. Mapping of a novel MEN-like syndrome locus to rat chromosome 4. Mamm Genome 2004;15:135-41.
160. Lee M, Pellegata NS. Multiple endocrine neoplasia type 4. Front Horm Res 2013;41:63-78.
161. Carpten JD, Robbins CM, Villablanca A, et al. HRPT2, encoding parafibromin, is mutated in hyperparathyroidism-jaw tumor syndrome. Nat Genet 2002;32:676-80.
162. Hobbs MR, Pole AR, Pidwirny GN, et al. Hyperparathyroidism-jaw tumor syndrome: the HRPT2 locus is within a 0.7-cM region on chromosome 1q. Am J Hum Genet 1999;64:518-25.
163. Hendy GN, Cole DE. Genetic defects associated with familial and sporadic hyperparathyroidism. Front Horm Res 2013;41:149-65.
164. Iacobone M, Barzon L, Porzionato A, et al. Parafibromin expression, single-gland involvement, and limited parathyroidectomy in familial isolated hyperparathyroidism. Surgery 2007;142:984-91.

165. Simonds WF, Robbins CM, Agarwal SK, Hendy GN, Carpten JD, Marx SJ. Familial isolated hyperparathyroidism is rarely caused by germline mutation in HRPT2, the gene for the hyperparathyroidism-jaw tumor syndrome. J Clin Endocrinol Metab 2004;89:96-102.
166. Teh BT, Farnebo F, Twigg S, et al. Familial isolated hyperparathyroidism maps to the hyperparathyroidism-jaw tumor locus in 1q21-q32 in a subset of families. J Clin Endocrinol Metab 1998;83:2114-20.
167. Kelly TG, Shattuck TM, Reyes-Mugica M, et al. Surveillance for early detection of aggressive parathyroid disease: carcinoma and atypical adenoma in familial isolated hyperparathyroidism associated with a germline HRPT2 mutation. J Bone Miner Res 2006;21:1666-71.
168. Szabo J, Heath B, Hill VM, et al. Hereditary hyperparathyroidism-jaw tumor syndrome: the endocrine tumor gene HRPT2 maps to chromosome 1q21-q31. Am J Hum Genet 1995;56:944-50.
169. Gill AJ, Lim G, Cheung VK, et al. Parafibromin-deficient (HPT-JT type, CDC73 mutated) parathyroid tumors demonstrate distinctive morphologic features. Am J Surg Pathol 2019;43:35-46.
170. Shattuck TM, Valimaki S, Obara T, et al. Somatic and germ-line mutations of the HRPT2 gene in sporadic parathyroid carcinoma. N Engl J Med 2003;349:1722-9.
171. Krebs LJ, Shattuck TM, Arnold A. HRPT2 mutational analysis of typical sporadic parathyroid adenomas. J Clin Endocrinol Metab 2005;90:5015-7.
172. Tan MH, Morrison C, Wang P, et al. Loss of parafibromin immunoreactivity is a distinguishing feature of parathyroid carcinoma. Clin Cancer Res 2004;10:6629-37.
173. Gill AJ, Clarkson A, Gimm O, et al. Loss of nuclear expression of parafibromin distinguishes parathyroid carcinomas and hyperparathyroidism-jaw tumor (HPT-JT) syndrome-related adenomas from sporadic parathyroid adenomas and hyperplasias. Am J Surg Pathol 2006;30:1140-9.
174. Juhlin C, Larsson C, Yakoleva T, et al. Loss of parafibromin expression in a subset of parathyroid adenomas. Endocr Relat Cancer 2006;13:509-23.
175. Cetani F, Ambrogini E, Viacava P, et al. Should parafibromin staining replace HRTP2 gene analysis as an additional tool for histologic diagnosis of parathyroid carcinoma? Eur J Endocrinol 2007;156:547-54.
176. Juhlin CC, Larsson C, Yakoleva T, et al. Parafibromin immunoreactivity: its use as an additional diagnostic marker for parathyroid tumor classification. Endocr Relat Cancer 2007;14:501-12.
177. Lin L, Czapiga M, Nini L, Zhang JH, Simonds WF. Nuclear localization of the parafibromin tumor suppressor protein implicated in the hyperparathyroidism-jaw tumor syndrome enhances its proapoptotic function. Mol Cancer Res 2007;5:183-93.
178. Marcocci C, Cetani F. Parafibromin as a tool for the diagnosis of parathyroid tumors. Adv Anat Pathol 2008;15:179-80.
179. Guarnieri V, Battista C, Muscarella LA, et al. CDC73 mutations and parafibromin immunohistochemistry in parathyroid tumors: clinical correlations in a single-centre patient cohort. Cell Oncol (Dordr) 2012;35:411-22.
180. Rozenblatt-Rosen O, Hughes CM, Nannepaga SJ, et al. The parafibromin tumor suppressor protein is part of a human Paf1 complex. Mol Cell Biol 2005;25:612-20.
181. Yart A, Gstaiger M, Wirbelauer C, et al. The HRPT2 tumor suppressor gene product parafibromin associates with human PAF1 and RNA polymerase II. Mol Cell Biol 2005;25:5052-60.
182. Lin L, Zhang JH, Panicker LM, Simonds WF. The parafibromin tumor suppressor protein inhibits cell proliferation by repression of the c-myc proto-oncogene. Proc Natl Acad Sci U S A 2008;105:17420-5.
183. Zhang C, Kong D, Tan MH, et al. Parafibromin inhibits cancer cell growth and causes G1 phase arrest. Biochem Biophys Res Commun 2006;350:17-24.
184. Jo JH, Chung TM, Youn H, Yoo JY. Cytoplasmic parafibromin/hCdc73 targets and destabilizes p53 mRNA to control p53-mediated apoptosis. Nat Commun 2014;5:5433.
185. Wei J, Lian H, Zhong B, Shu HB. Parafibromin is a component of ifn-gamma-triggered signaling pathways that facilitates JAK1/2-mediated tyrosine phosphorylation of STAT1. J Immunol 2015;195:2870-8.
186. Kikuchi I, Takahashi-Kanemitsu A, Sakiyama N, et al. Dephosphorylated parafibromin is a transcriptional coactivator of the Wnt/Hedgehog/Notch pathways. Nat Commun 2016;7:12887.
187. Warner JV, Nyholt DR, Busfield F, et al. Familial isolated hyperparathyroidism is linked to a 1.7 Mb region on chromosome 2p13.3-14. J Med Genet 2006;43:e12.
188. Wassif WS, Moniz CF, Friedman E, et al. Familial isolated hyperparathyroidism: a distinct genetic entity with an increased risk of parathyroid cancer. J Clin Endocrinol Metab 1993;77:1485-9.
189. El Lakis M, Nockel P, Guan B, et al. Familial isolated primary hyperparathyroidism associated with germline GCM2 mutations is more aggressive and has lesser rate of biochemical cure. Surgery 2018;163:31-4.

190. Marx SJ. Hyperparathyroid and hypoparathyroid disorders. N Engl J Med 2000;343:1863-75.
191. Villablanca A, Wassif WS, Smith T, et al. Involvement of the MEN1 gene locus in familial isolated hyperparathyroidism. Eur J Endocrinol 2002;147:313-22.
192. Teh BT, Kytola S, Farnebo F, et al. Mutation analysis of the MEN1 gene in multiple endocrine neoplasia type 1, familial acromegaly and familial isolated hyperparathyroidism. J Clin Endocrinol Metab 1998;83:2621-6.
193. Yoshimoto K, Endo H, Tsuyuguchi M, et al. Familial isolated primary hyperparathyroidism with parathyroid carcinomas: clinical and molecular features. Clin Endocrinol (Oxf) 1998;48:67-72.
194. Bergman L, The B, Cardinal J, et al. Identification of MEN1 gene mutations in families with MEN 1 and related disorders. Br J Cancer 2000;83:1009-14.
195. Howell VM, Cardinal JW, Richardson AL, Gimm O, Robinson BG, Mars DJ. Rapid mutation screening for HRPT2 and MEN1 mutations associated with familial and sporadic primary hyperparathyroidism. J Mol Diagn 2006;8:559-66.
196. Cardoso L, Stevenson M, Thakker RV. Molecular genetics of syndromic and non-syndromic forms of parathyroid carcinoma. Hum Mutat 2017;38:1621-48.
197. Warner J, Epstein M, Sweet A, et al. Genetic testing in familial isolated hyperparathyroidism: unexpected results and their implications. J Med Genet 2004;41:155-60.
198. Cetani F, Pardi E, Ambrogini E, et al. Genetic analyses in familial isolated hyperparathyroidism: implication for clinical assessment and surgical management. Clin Endocrinol (Oxf) 2006;64:146-52.
199. Mizusawa N, Uchino S, Iwata T, et al. Genetic analyses in patients with familial isolated hyperparathyroidism and hyperparathyroidism-jaw tumour syndrome. Clin Endocrinol (Oxf) 2006;65:9-16.
200. Stokes VJ, Nielsen MF, Hannan FM, Thakker RV. Hypercalcemic disorders in children. J Bone Miner Res 2017;32:2157-70.
201. Attie MF, Gill JR, Stock JL, et al. Urinary calcium excretion in familial hypocalciuric hypercalcemia. Persistence of relative hypocalciuria after induction of hypoparathyroidism. J Clin Invest 1983;72:667-76.
202. Hannan FM, Thakker RV. Calcium-sensing receptor (CaSR) mutations and disorders of calcium, electrolyte and water metabolism. Best Pract Res Clin Endocrinol Metab 2013;27:359-71.
203. Hannan FM, Babinsky VN, Thakker RV. Disorders of the calcium-sensing receptor and partner proteins: insights into the molecular basis of calcium homeostasis. J Mol Endocrinol 2016;57:R127-42.
204. Aida K, Koishi S, Inoue M, Nakazato M, Tawata M, Onaya T. Familial hypocalciuric hypercalcemia associated with mutation in the human Ca(2+)-sensing receptor gene. J Clin Endocrinol Metab 1995;80:2594-8.
205. Pearce SH, Williamson C, Kifor O, et al. A familial syndrome of hypocalcemia with hypercalciuria due to mutations in the calcium-sensing receptor. N Engl J Med 1996;335:1115-22.
206. Janicic N, Soliman E, Pausova Z, et al. Mapping of the calcium-sensing receptor gene (CASR) to human chromosome 3q13.3-21 by fluorescence in situ hybridization, and localization to rat chromosome 11 and mouse chromosome 16. Mamm Genome 1995;6:798-801.
207. Cole DE, Janicic N, Salisbury SR, Hendy GN. Neonatal severe hyperparathyroidism, secondary hyperparathyroidism, and familial hypocalciuric hypercalcemia: multiple different phenotypes associated with an inactivating Alu insertion mutation of the calcium-sensing receptor gene. Am J Med Genet 1997;71:202-10.
208. Nesbit MA, Hannan FM, Howles SA, et al. Mutations affecting G-protein subunit alpha11 in hypercalcemia and hypocalcemia. N Engl J Med 2013;368:2476-86.
209. Aida K, Koishi S, Tawata M, Onaya T. Molecular cloning of a putative Ca(2+)-sensing receptor cDNA from human kidney. Biochem Biophys Res Commun 1995;214:524-9.
210. Hauache OM. Extracellular calcium-sensing receptor: structural and functional features and association with diseases. Braz J Med Biol Res 2001;34:577-84.
211. Hellman P, Carling T, Rask L, Akerstrom G. Pathophysiology of primary hyperparathyroidism. Histol Histopathol 2000;15:619-27.
212. Thakker RV. Diseases associated with the extracellular calcium-sensing receptor. Cell Calcium 2004;35:275-82.
213. Hannan FM, Nesbit MA, Zhang C, et al. Identification of 70 calcium-sensing receptor mutations in hyper- and hypo-calcaemic patients: evidence for clustering of extracellular domain mutations at calcium-binding sites. Hum Mol Genet 2012;21:2768-78.
214. Kifor O, Moore FD Jr, Delaney M, et al. A syndrome of hypocalciuric hypercalcemia caused by autoantibodies directed at the calcium-sensing receptor. J Clin Endocrinol Metab 2003;88:60-72.
215. De Luca F, Baron J. Molecular biology and clinical importance of the Ca(2+)-sensing receptor. Curr Opin Pediatr 1998;10:435-40.

216. Hosokawa Y, Pollak MR, Brown EM, Arnold A. Mutational analysis of the extracellular Ca(2+)-sensing receptor gene in human parathyroid tumors. J Clin Endocrinol Metab 1995;80:3107-10.
217. Castleman B, Schantz A, Roth S. Parathyroid hyperplasia in primary hyperparathyroidism: a review of 85 cases. Cancer 1976;38:1668-75.
218. Bendz H, Sjodin I, Toss G, Berglund K. Hyperparathyroidism and long-term lithium therapy—a cross-sectional study and the effect of lithium withdrawal. J Intern Med 1996;240:357-65.
219. Rao SD, Frame B, Miller MJ, Kleerekoper M, Block MA, Parfitt AM. Hyperparathyroidism following head and neck irradiation. Arch Intern Med 1980;140:205-7.

Primary Chief, Clear Cell Hyperplasia

220. Black WC, Haff RC. The surgical pathology of parathyroid chief cell hyperplasia. Am J Clin Pathol 1970;53:565-79.
221. Akerstrom G, Bergstrom R, Grimelius I, et al. Relation between changes in clinical and histopathological features of primary hyperparathyroidism. World J Surg 1986;10:696-702.
222. Tominaga Y, Grimelius L, Johansson H, et al. Histological and clinical features of non-familial primary parathyroid hyperplasia. Pathol Res Pract 1992;188:115-22.
223. Fallon MD, Haines JQ, Teitelbaum SL. Cystic parathyroid gland hyperplasia—hyperparathyroidism presenting as a neck mass. Am J Clin Pathol 1982;77:104-7.
224. Clark OH. Hyperparathyroidism due to primary cystic parathyroid hyperplasia. Arch Surg 1978; 113:748-50.
225. Erickson LA, Jin L, Wollan P, Thompson GB, van Heerden A, Lloyd RV. Parathyroid hyperplasia, adenomas, and carcinomas: differential expression of p27Kip1 protein. Am J Surg Pathol 1999;23:288-95.
226. Alo PL, Visca P, Mazzaferro S, et al. Immunohistochemical study of fatty acid synthase, Ki67, proliferating cell nuclear antigen, and p53 expression in hyperplastic parathyroids. Ann Diagn Pathol 1999;3:287-93.
227. Lazaris AC, Tseleni-Balafouta S, Papathomas T, et al. Immunohistochemical investigation of angiogenic factors in parathyroid proliferative lesions. Eur J Endocrinol 2006;154:827-33.
228. Albright F, Castleman BE, Chrchill EB. Hyperparathyroidism due to diffuse hyperplasia of all parathyroid glands rather than an adenoma of one. Clinical studies on three such cases. Arch Intern Med 1934;54:315-29.
229. Cope O, Keynes WM, Roth SI, Castleman B. Primary chief-cell hyperplasia of the parathyroid glands: a new entity in the surgery of hyperparathyroidism. Ann Surg 1958;148:375-88.
230. Tisell LE, Hedman I, Hansson G. Clinical characteristics and surgical results in hyperparathyroidism caused by water-clear cell hyperplasia. World J Surg 1981;5:565-71.
231. Dorado AE, Hensley G, Castleman B. Water clear cell hyperplasia of parathyroid: autopsy report of a case with supernumerary glands. Cancer 1976;38:1676-83.
232. Lloyd RV, Douglas BR, Young WF Jr. Endocrine diseases. Atlas of Nontumor Pathology, First Series, Fascicle 1. Washington DC: American Registry of Pathology; 2002:315.

Secondary, Tertiary Hyperparathyroidism

233. Schlosser K, Veit JA, Witte S, et al. Comparison of total parathyroidectomy without autotransplantation and without thymectomy versus total parathyroidectomy with autotransplantation and with thymectomy for secondary hyperparathyroidism: TOPAR PILOT-Trial. Trials 2007;8:22.
234. Pozzoni P, Del Vecchio L, Pontoriero G, Di Filippo S, Locatelli F. Long-term outcome in hemodialysis: morbidity and mortality. J Nephrol 2004;17(Suppl 8):S87-95.
235. Llach F, Velasquez Forero F. Secondary hyperparathyroidism in chronic renal failure: pathogenic and clinical aspects. Am J Kidney Dis 2001;38(Suppl 5):S20-33.
236. Nagaba Y, Heishi M, Tazawa H, Tsukamoto Y, Kobayashi Y. Vitamin D receptor gene polymorphisms affect secondary hyperparathyroidism in hemodialyzed patients. Am J Kidney Dis 1998;32:464-9.
237. Rahamimov R, Silver J. The molecular basis of secondary hyperparathyroidism in chronic renal failure. Isr J Med Sci 1994;30:26-31.
238. Hayes CW, Conway WF. Hyperparathyroidism. Radiol Clin North Am 1991;29:85-96.
239. Stanbury SW. Vitamin D and hyperparathyroidism: the Lumleian Lecture 1981. J R Coll Physicians Lond 1981;15:205-9, 212-7.
240. Arseculeratne G, Evans AT, Morley SM. Calciphylaxis—a topical overview. J Eur Acad Dermatol Venereol 2006;20:493-502.
241. Matsuoka S, Tominaga Y, Uno N, et al. Calciphylaxis: a rare complication of patients who required parathyroidectomy for advanced renal hyperparathyroidism. World J Surg 2005;29:632-5.
242. Wilmer WA, Magro CM. Calciphylaxis: emerging concepts in prevention, diagnosis, and treatment. Semin Dial 2002;15:172-86.
243. Krause MW, Hedinger CE. Pathologic study of parathyroid glands in tertiary hyperparathyroidism. Hum Pathol 1985;16:772-84.

Parathyromatosis

244. Reddick RL, Costa JC, Marx SJ. Parathyroid hyperplasia and parathyromatosis. Lancet 1977; 1549.
245. Fitko R, Roth SI, Hines JR, Roxe DM, Cahill EI. Parathyromatosis in hyperparathyroidism. Hum Pathol 1990;21:234-7.
246. Sharma S, Dey Pranab, Gude G, Saikia UN. Parathyromatosis-A rare occurrence along the endoscopic tract detected on fine needle aspiration cytology. Diagn Cytopathol 2016;44:1125-7.
247. Hage MP, Salti I, El-Hajj Fuleihan G. Parathyromatosis: a rare yet problematic etiology of recurrent and persistent hyperparathyroidism. Metabolism 2012;61:762-75.
248. Sim IW, Farrell S, Grodski S, Jung C, Ng KW. Parathyromatosis following spontaneous rupture of a parathyroid adenoma: natural history and the challenge of management. Intern Med J 2013;43:819-22.
249. Kendrick ML, Charboneau JW, Curlee KJ, van Heerden A, Farley DR. Risk of parathyromatosis after fine-needle aspiration. Am Surg 2001;67:290-4.
250. Baloch ZW, Fraker D, LiVolsi VA. Parathyromatosis as cause of recurrent secondary hyperparathyroidism: a cytologic diagnosis. Diagn Cytopathol 2001;25:403-5.
251. Fernandez-Ranvier GG, Khanafshar E, Tacha D, et al. Defining a molecular phenotype for benign and malignant parathyroid tumors. Cancer 2009;115:334-44.
252. Fernandez-Ranvier GG, Khanafshar E, Tacha D, et al. Parathyroid carcinoma, atypical parathyroid adenoma, or parathyromatosis? Cancer 2007;110:255-64.
253. Unbehaun R, Lauerwald W. Successful use of cinacalcet HCl in a patient with end-stage renal failure and refractory secondary hyperparathyroidism due to parathyromatosis. Clin Nephrol 2007;67:188-92.

Hypoparathyroidism

254. Abate EG, Clarke BL. Review of hypoparathyroidism. Front Endocrinol (Lausanne) 2016;7:172.
255. Fung WL, Butcher NJ, Costain G, et al. Practical guidelines for managing adults with 22q11.2 deletion syndrome. Genet Med 2015;17:599-609.
256. Brauner R, Le Harivel de Gonneville A, Kindermans C, et al. Parathyroid function and growth in 22q11.2 deletion syndrome. J Pediatr 2003;142:504-8.
257. Cuneo BF, Driscoll DA, Gidding SS, Langman CB. Evolution of latent hypoparathyroidism in familial 22q11 deletion syndrome. Am J Med Genet 1997;69:50-5.
258. Cuneo BF. 22q11.2 deletion syndrome: DiGeorge, velocardiofacial, and conotruncal anomaly face syndromes. Curr Opin Pediatr 2001;13:465-72.
259. McDonald-McGinn DM, Sullivan KE. Chromosome 22q11.2 deletion syndrome (DiGeorge syndrome/velocardiofacial syndrome). Medicine (Baltimore) 2011;90:1-18.
260. Kretschmer R, Say B, Brown D, Rosen FS. Congenital aplasia of the thymus gland (DiGeorge's syndrome). N Engl J Med 1968;279:1295-301.
261. Kirkpatrick JA Jr, DiGeorge AM. Congenital absence of the thymus. Am J Roentgenol Radium Ther Nucl Med 1968;103:32-7.
262. Al-Jenaidi F, Makitie O, Grunebaum E, Sochett E. Parathyroid gland dysfunction in 22q11.2 deletion syndrome. Horm Res 2007;67:117-22.
263. McLean-Tooke A, Spickett GP, Gennery AR. Immunodeficiency and autoimmunity in 22q11.2 deletion syndrome. Scand J Immunol 2007;66:1-7.
264. Lambert MP, Arulselvan A, Schott A, et al. The 22q11.2 deletion syndrome: cancer predisposition, platelet abnormalities and cytopenias. Am J Med Genet A 2018;176:2121-7.
265. Morsheimer M, Brown Whitehorn TF, Heimall J, Sullivan KE. The immune deficiency of chromosome 22q11.2 deletion syndrome. Am J Med Genet A 2017;173:2366-72.
266. Stevens T, van der Werff Ten Bosch J, De Rademaeker M, Van Den Bogaert A, van den Akker M. Risk of malignancy in 22q11.2 deletion syndrome. Clin Case Rep 2017;5:486-90.
267. Van L, Boot E, Bassett AS. Update on the 22q11.2 deletion syndrome and its relevance to schizophrenia. Curr Opin Psychiatry 2017;30:191-6.
268. Yagi H, Furutani Y, Hamada H, et al. Role of TBX1 in human del22q11.2 syndrome. Lancet 2003;362:1366-73.
269. Baldini A. DiGeorge syndrome: the use of model organisms to dissect complex genetics. Hum Mol Genet 2002;11:2363-9.
270. Paylor R, Glaser B, Mupo A, et al. Tbx1 haploinsufficiency is linked to behavioral disorders in mice and humans: implications for 22q11 deletion syndrome. Proc Natl Acad Sci U S A 2006;103:7729-34.
271. Chen QY, Lan MS, She JX, Maclaren NK. The gene responsible for autoimmune polyglandular syndrome type 1 maps to chromosome 21q22.3 in US patients. J Autoimmun 1998;11:177-83.
272. Obermayer-Straub P, Manns MP. Autoimmune polyglandular syndromes. Baillieres Clin Gastroenterol 1998;12:293-315.
273. Maclaren N, Chen QY, Kukreja A, Marker J, Zhang CH, Sun ZS. Autoimmune hypogonadism as part of an autoimmune polyglandular syndrome. J Soc Gynecol Investig 2001;8(Suppl):S52-4.

274. Kogawa K, Kudoh J, Nagafuchi S, et al. Distinct clinical phenotype and immunoreactivity in Japanese siblings with autoimmune polyglandular syndrome type 1 (APS-1) associated with compound heterozygous novel AIRE gene mutations. Clin Immunol 2002;103(Pt 1):277-83.
275. Toonkel R, Levine M, Gardner L. Erythropoietin-deficient anemia associated with autoimmune polyglandular syndrome type I. Am J Hematol 2004;75:84-8.
276. Guo CJ, Leung PS, Zhang W, Ma X, Gershwin ME. The immunobiology and clinical features of type 1 autoimmune polyglandular syndrome (APS-1). Autoimmun Rev 2018;17:78-85.
277. De Martino L, Capalbo D, Improda N, et al. Novel findings into AIRE genetics and functioning: clinical implications. Front Pediatr 2016;4:86.
278. Capalbo D, Mazza C, Giordano R, et al. Molecular background and genotype-phenotype correlation in autoimmune-polyendocrinopathy-candidiasis-ectodermal-distrophy patients from Campania and in their relatives. J Endocrinol Invest 2012;35:169-73.
279. Oftedal BE, Hellesen A, Erichsen MM, et al. Dominant mutations in the autoimmune regulator aire are associated with common organ-specific autoimmune diseases. Immunity 2015;42:1185-96.
280. Arnold A, Horst SA, Gardella TJ, Baba H, Levine MA, Kronenberg HM. Mutation of the signal peptide-encoding region of the preproparathyroid hormone gene in familial isolated hypoparathyroidism. J Clin Invest 1990;86:1084-7.
281. Trump D, Dixon PH, Mumm S, et al. Localisation of X linked recessive idiopathic hypoparathyroidism to a 1.5 Mb region on Xq26-q27. J Med Genet 1998;35:905-9.
282. Ding C, Buckingham B, Levine MA. Familial isolated hypoparathyroidism caused by a mutation in the gene for the transcription factor GCMB. J Clin Invest 2001;108:1215-20.
283. Baumber L, Tufarelli C, Patel S, et al. Identification of a novel mutation disrupting the DNA binding activity of GCM2 in autosomal recessive familial isolated hypoparathyroidism. J Med Genet 2005;42:443-8.
284. Maret A, Ding C, Kornfield SL, Levine MA. Analysis of the GCM2 gene in isolated hypoparathyroidism: a molecular and biochemical study. J Clin Endocrinol Metab 2008;93:1426-32.
285. Belge H, Dahan K, Cambier JF, et al. Clinical and mutational spectrum of hypoparathyroidism, deafness and renal dysplasia syndrome. Nephrol Dial Transplant 2017;32:830-7.
286. Bilous RW, Murty G, Parkinson DB, et al. Brief report: autosomal dominant familial hypoparathyroidism, sensorineural deafness, and renal dysplasia. N Engl J Med 1992;327:1069-74.
287. Van Esch H, Devriendt K. Transcription factor GATA3 and the human HDR syndrome. Cell Mol Life Sci 2001;58:1296-300.
288. Van Esch H, Groenen P, Nesbit MA, et al. GATA3 haplo-insufficiency causes human HDR syndrome. Nature 2000;406:419-22.
289. Isotani H, Fukumoto Y, Kawamura H, et al. Hypoparathyroidism and insulin-dependent diabetes mellitus in a patient with Kearns-Sayre syndrome harbouring a mitochondrial DNA deletion. Clin Endocrinol (Oxf) 1996;45:637-41.
290. Maceluch JA, Niedziela M. The clinical diagnosis and molecular genetics of Kearns-Sayre syndrome: a complex mitochondrial encephalomyopathy. Pediatr Endocrinol Rev 2006;4:117-37.
291. Ashrafzadeh F, Ghaemi N, Ahondian J, Toosi MB, Elmi S. Hypoparathyroidism as the first manifestation of Kearns-Sayre syndrome: a case report. Iran J Child Neurol 2013;7:53-7.
292. Berio A, Piazzi A. Multiple endocrinopathies (growth hormone deficiency, autoimmune hypothyroidism and diabetes mellitus) in Kearns-Sayre syndrome. Pediatr Med Chir 2013;35:137-40.
293. Ho J, Pacaud D, Rakic M, Khan A. Diabetes in pediatric patients with Kearns-Sayre syndrome: clinical presentation of 2 cases and a review of pathophysiology. Can J Diabetes 2014;38:225-8.
294. Khambatta S, Nguyen DL, Beckman TJ, Wittich CM. Kearns-Sayre syndrome: a case series of 35 adults and children. Int J Gen Med 2014;7:325-32.
295. Kabunga P, Lau AK, Phan K, et al. Systematic review of cardiac electrical disease in Kearns-Sayre syndrome and mitochondrial cytopathy. Int J Cardiol 2015;181:303-10.
296. Ho J, Pacaud D, Khan A. Kearns-Sayre syndrome is a rare cause of diabetes. Can J Diabetes 2016;40:110-1.
297. Parvari R, Hershkovitz E, Grossman N, et al. Mutation of TBCE causes hypoparathyroidism-retardation-dysmorphism and autosomal recessive Kenny-Caffey syndrome. Nat Genet 2002;32:448-52.
298. Unger S, Gorna MW, Le Bechec A, et al. FAM111A mutations result in hypoparathyroidism and impaired skeletal development. Am J Hum Genet 2013;92:990-5.
299. Isojima T, Doi K, Mitsui J, et al. A recurrent de novo FAM111A mutation causes Kenny-Caffey syndrome type 2. J Bone Miner Res 2014;29:992-8.

300. Abraham MB, Dong Li, Tang D, et al. Short stature and hypoparathyroidism in a child with Kenny-Caffey syndrome type 2 due to a novel mutation in FAM111A gene. Int J Pediatr Endocrinol 2017;2017:1.
301. Shapira H, Friedman E, Mouallem M, Farfel Z. Familial Albright's hereditary osteodystrophy with hypoparathyroidism: normal structural Gs alpha gene. J Clin Endocrinol Metab 1996; 81:1660-2.

Parathyroid Cysts

302. McGoon DC, Cooley DA. Parathyroid cyst: an unusual cervical tumor; report of three cases. Surgery 1951;30:725-32.
303. Ippolito G, Palazzo FF, Sebag F, Sierra M, De Micco C, Henry JF. A single-institution 25-year review of true parathyroid cysts. Langenbecks Arch Surg 2006;391:13-8.
304. Fustar Preradovic L, Danic D, Dzodic R. Small nonfunctional parathyroid cysts: single institution experience. Endocr J 2017;64:151-6.
305. Witherspoon J, Lewis M. Parathyroid cysts: a clinical and radiological challenge. Br J Hosp Med (Lond) 2012;73:108-9.
306. Vazquez, FJ, Aparicio LS, Gallo CG, Diehl M. Parathyroid carcinoma presenting as a giant mediastinal retrotracheal functioning cyst. Singapore Med J 2007;48:e304-7.
307. Zhang XU, Yuan JH, Feng LU, Shan DQ, Wu JF, Liu ST. Giant non-functional parathyroid cyst: a case report. Oncol Lett 2016;11:2237-40.
308. Katz AD, Dunkleman D. Needle aspiration of nonfunctioning parathyroid cysts. Arch Surg 1984;119:307-8.
309. Kodama T, Obara T, Fujimoto Y, Ito Y, Yashiro T, Hirayama A. Eleven cases of nonfunctioning parathyroid cyst—significance of needle aspiration in diagnosis and management. Endocrinol Jpn 1987;34:769-77.
310. Troster M, Chiu HF, McLarty TD. Parathyroid cysts: report of a case with ultrastructural observations. Surgery 1978;83:238-42.
311. Shields TW, Immerman SC. Mediastinal parathyroid cysts revisited. Ann Thorac Surg 1999;67:581-90.
312. Capezzone M, Morabito E, Bellitti P, Giannasio P, de Santis D, Bruno R. Ectopic intrathyroidal nonfunctioning parathyroid cyst. Endocr Pract 2007;13:56-8.
313. Kato H, Kanematsu M, Kiryu T, et al. Nonfunctional mediastinal parathyroid cyst: imaging findings in two cases. Clin Imaging 2008;32:310-3.
314. Ahmad MM, Almohaya M, Almalki MH, Aljohani N. Intrathyroidal parathyroid cyst: an unusual neck mass. Clin Med Insights Endocrinol Diabetes 2017;10:1179551417698135.

Parathyroiditis

315. Van de Casseye M, Gepts W. Case report: primary (autoimmune?) parathyroiditis. Virchows Arch A Pathol Pathol Anat 1973;361:257-61.
316. Furuto-Kato S, Matsukura S, Ogata M, et al. Primary hyperparathyroidism presumably caused by chronic parathyroiditis manifesting from hypocalcemia to severe hypercalcemia. Intern Med 2005;44:60-4.
317. Chetty R, Forder MD. Parathyroiditis associated with hyperparathyroidism and branchial cysts. Am J Clin Pathol 1991;96:348-50.
318. Boyce BF, Doherty VR, Mortimer G. Hyperplastic parathyroiditis—a new autoimmune disease? J Clin Pathol 1982;35:812-4.
319. Sinha SN, McArdle JP, Shepherd JJ. Hyperparathyroidism with chronic parathyroiditis in a multiple endocrine neoplasia patient. Aust N Z J Surg 1993;63:981-2.
320. Song L, Liu L, Miller RT, et al. Glucocorticoid-responsive lymphocytic parathyroiditis and hypocalciuric hypercalcemia due to autoantibodies against the calcium-sensing receptor: a case report and literature review. Eur J Endocrinol 2017;177:K1-6.

Amyloid

321. Lieberman A, DeLellis RA. Intrafollicular amyloid in normal parathyroid glands. Arch Pathol 1973;95:422-3.
322. Koelmeyer TD. Generalised amyloidosis with involvement of the parathyroids: case report. N Z Med J 1977;85:372-3.
323. Anderson TJ, Ewen SW. Amyloid in normal and pathological parathyroid glands. J Clin Pathol 1974;27:656-63.
324. Keven K, Oztas E, Aksoy H, Duman N, Erbay B, Ertürk S. Polyglandular endocrine failure in a patient with amyloidosis secondary to familial Mediterranean fever. Am J Kidney Dis 2001;38:E39.
325. Leedham PW, Pollock DJ. Intrafollicular amyloid in primary hyperparathyroidism. J Clin Pathol 1970;23:811-7.

Glycogen Storage Disease

326. Hui KS, Williams JC, Borit A, Rosenberg HS. The endocrine glands in Pompe's disease. Report of two cases. Arch Pathol Lab Med 1985;109:921-5.
327. Soejima K, Landing BH, Roe TF, Swanson VL. Pathologic studies of the osteoporosis of Von Gierke's disease (glycogenosis 1a). Pediatr Pathol 1985;3:307-19.

3 THYROID GLAND

NORMAL THYROID GLAND

Embryology

The thyroid gland begins to develop between weeks 2 and 3 of gestation, and development is completed by week 11 (1). The gland develops from three structures: one median anlage and two lateral anlagen. The median anlage arises from a portion of endoderm in the ventral midline of the primitive pharynx that later gives rise to the foramen cecum at the junction of the anterior two thirds and posterior one third of the tongue (2). The anlage then migrates caudally along the thyroglossal duct and passes through or in front of the hyoid bone before reaching its final position in front of the trachea in the anterior neck. The thyroglossal duct then atrophies, although the duct may persist, become cystic, and possibly give rise to neoplasms in remnant thyroid tissue within its wall.

The lateral anlage develops during weeks 5 to 7 of fetal life from the so-called "fourth-fifth" branchial pouch, which contains the ultimobranchial body. The calcitonin-secreting cells (C cells), which are derived from the neural crest (3,4), migrate to the ultimobranchial bodies and are subsequently incorporated into the thyroid gland. Around the seventh week of embryonic life, fusion of the medial anlage and both lateral anlagen occurs in the upper lateral aspect of the gland; C cells predominate in this area. Around 9 weeks before term, parathyroid IV separates from the ultimobranchial bodies before they regress. Solid cell nests composed of collections of stratified epithelial cells, with focal mucin production and cyst formation, are believe to be remnants of the ultimobranchial body and may be seen in up to 30 percent of adult thyroid glands (5–7).

Follicles appear within the thyroid gland at the beginning of the second month of gestation and most follicles are formed by the end of the fourth month of gestation (2). Even though it is the first endocrine gland to form, the fetal thyroid does not produce thyroid hormones until 18 to 20 weeks of gestation. Therefore, for the first half of gestation, the fetus is dependent on maternal thyroid hormones that cross the placenta (8).

Anatomy

The thyroid gland consists of two lobes connected by an isthmus. It lies between the anterior borders of the sternocleidomastoid muscles in the anterior triangle of the neck, in front of the trachea and caudal to the cricothyroid membrane and the thyroid cartilage, with the isthmus overlying the second to fourth tracheal rings (fig. 3-1) (2). Each lobe has a pointed superior pole and a blunt inferior pole. The right lobe may be longer than the left, and the isthmus varies in width between individuals. A pyramidal lobe, which is a vestigial remnant of the thyroglossal duct, is observed in up to 50 percent of individuals as an extension of thyroid tissue from the isthmus toward the hyoid bone.

The normal weight of the adult thyroid gland ranges from 15 to 25 g and varies with age, sex, functional status of the gland, hormonal status, nutritional status, and iodine intake (5,9). The gland is generally larger and heavier in women than in men, and changes during pregnancy and the menstrual cycle, increasing up to 50 percent during the early secretory phase of the cycle (9–11).

The thyroid gland has a thin fibrous capsule and septa that divide the gland incompletely into lobules. Incidental nodules may be present in about 10 percent of adult thyroid glands (12). The different types of nodules present are discussed later in this chapter. The parathyroid glands are usually adjacent to the posterior surface of the thyroid gland, while the recurrent laryngeal nerves run in the cleft between the

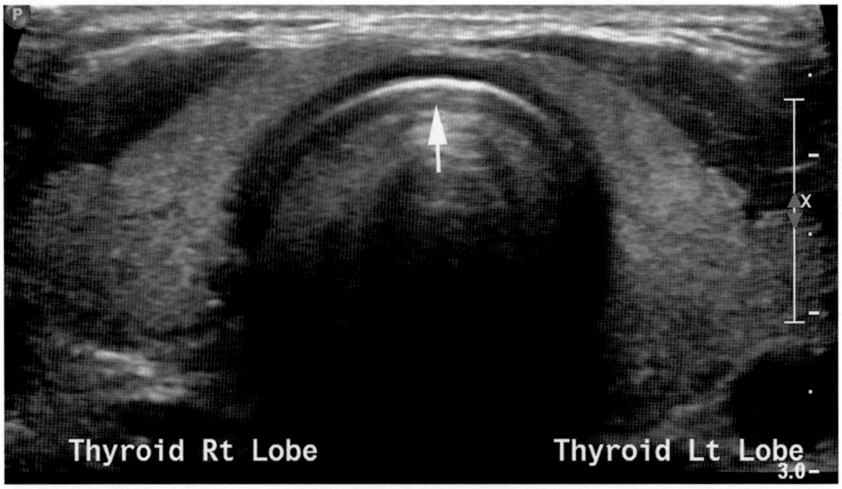

Figure 3-1

NORMAL ADULT THYROID GLAND

Ultrasound of the normal thyroid gland shows the homogeneous nature of the parenchyma. The gland surrounds the anterior aspect of the air-filled trachea (arrow) on this transverse view.

trachea and esophagus medially. Enlargement of the thyroid gland can interfere with the normal functions of the recurrent laryngeal nerves and lead to alterations in phonation.

The thyroid gland receives its blood supply via the superior and inferior thyroid arteries. The superior thyroid artery arises as the first branch of the external carotid artery, or occasionally more centrally, from the common carotid. As the artery approaches the gland, it splits into anterior and posterior branches. The inferior thyroid artery is a branch of the thyrocervical trunk that divides into an inferior branch going to the lower pole and a superior branch that travels on the posterior aspect of the thyroid lobe and may contribute to vascularization of the superior parathyroid gland.

Venous outflow includes the internal jugular, the brachiocephalic, and sometimes the anterior jugular veins. The lymphatic drainage of the thyroid is multidirectional and extensive. It drains initially into central compartment lymph nodes (level VI), which include lymph nodes situated both anterior and posterior to the recurrent laryngeal nerve. The superior poles and the isthmus drain into the Delphian or prelaryngeal lymph nodes (level VI) and jugular lymph nodes (levels II-IV). The lateral part of the lobe drains along the middle thyroid vein through the central compartment and toward the jugular lymph nodes. The inferior part of the lobe drains toward pretracheal/paratracheal lymph nodes (level VI), mediastinal lymph nodes (level VII), and lower jugular lymph nodes (level IV). In recent years, it has become common practice to remove the central compartment lymph nodes in patients with thyroid cancer (2).

The thyroid gland receives sympathetic nerve supply from the superior, middle, and inferior cervical sympathetic ganglia, as well as parasympathetic nerve supply from the superior laryngeal nerve and the recurrent laryngeal nerve. In addition to the parasympathetic component of the autonomic nervous system, adrenergic nerve fibers that end near the follicles may also play a role in regulating thyroid hormone secretion (14,15).

Histology

The thyroid gland is divided into lobules, each composed of 20 to 40 follicles supplied by a branch of the thyroid artery. Each follicle can range from 50 to 500 µm, with an average size of 200 µm (fig. 3-2). The shape of individual follicles is quite variable. Occasionally, collections of small follicles project into the lumen of a larger follicle (known as Sanderson polster). This finding is more common in cases of hyperplasia.

Follicles are lined by a single layer of thyroid follicular epithelial cells that surround the central colloid. Colloid is an amorphous liquid substance that serves as a reservoir for concentrated periodic acid–Schiff (PAS)-positive thyroglobulin. The tinctorial quality of the colloid may vary with the activity of the follicles, i.e., active follicles usually have weakly eosinophilic and flocculent colloid while inactive follicles tend to have more eosinophilic colloid (fig. 3-3). Birefringent calcium oxalate crystals are often found in the colloid of normal or diseased thyroid glands (17).

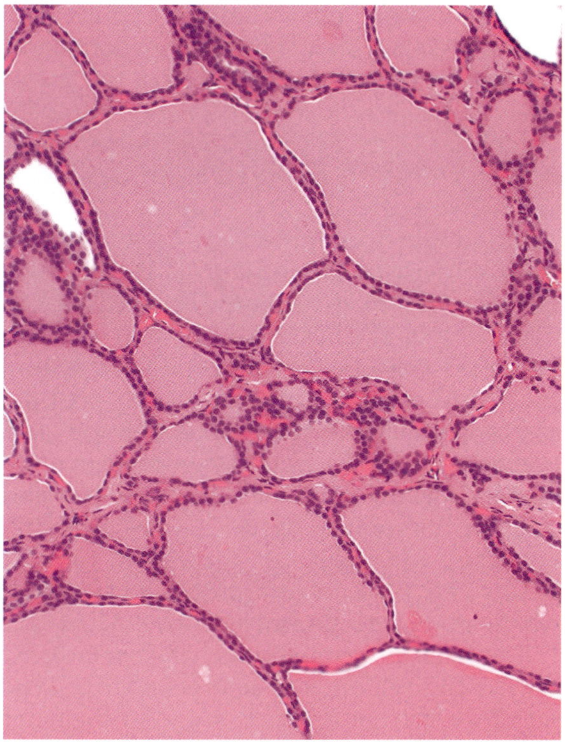

Figure 3-2

NORMAL ADULT THYROID GLAND

Histologic appearance of normal thyroid gland. The follicles are roughly round and lined by cuboidal cells.

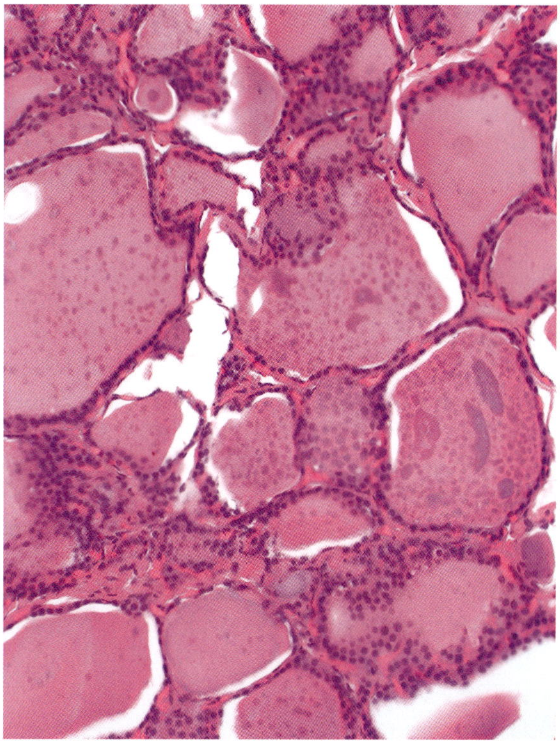

Figure 3-3

NORMAL ADULT THYROID GLAND

Thyroid gland from a 70-year-old patient. The colloid appears more flocculent and basophilic than normal. These changes are within normal limits but are more common in older individuals.

Normal thyroid tissue is occasionally seen outside the thyroid gland. In one study of 56 young adults between 20 to 40 years of age, normal thyroid tissue was found outside the gland in 40 (18). Thyroid tissue was present in the skeletal muscle of the neck in six patients. Patients with hyperplastic or inflammatory thyroid disease may have involvement of these ectopic foci as well (18). Benign thyroid nodules may become anatomically separated from the thyroid gland, in which case they are termed "parasitic nodules."

Follicular Cells

Thyroid follicular cells show variations in their shape and size according to the functional status of the gland. Three subtypes have been described, although they represent various expressions of a morphologic spectrum: flattened (endothelioid), cuboidal, and columnar (cylindrical) (19). Cuboidal cells (their height equaling their width) are the most numerous and their major function is to secrete colloid. The rarer columnar cells resorb the thyroglobulin-containing colloid, liberate the active hormones, and excrete these hormones into the blood stream; they may feature an apical cuticle, apical lipid droplets, and one or more basilar vacuoles. Functional polarity is apparent at the level of the follicle and the follicular cell. A single follicle may have flattened cells on one side and cuboidal or low columnar cells on the other.

At the cellular level, all follicular cells demonstrate polarity, with their base resting on the basement membrane and their apex directed toward the lumen of the follicle. The size and position of the nucleus and some components of the cytoplasm may vary considerably. In the resting thyroid, the nucleus is round to oval, is located toward the center of the cell, and usually contains one eccentrically located nucleolus. Its chromatin may be finely granular or clumped. In actively secreting cells, the nucleus is enlarged and basally located. The cytoplasm is usually weakly eosinophilic. In contrast to parathyroid cells, little or no intracytoplasmic glycogen is present.

Ultrastructurally, the follicular cells rest on a basement membrane, approximately 35 to 40 µm in thickness, which separates them from the interstitial stroma. Microvilli are present at the apical portions of the cells (20,21), and their number and length are increased in actively functioning cells (19). Cell membranes of adjacent cells interdigitate in a complex fashion and are joined by junctional complexes toward the apex (21–24). The cytoplasm contains variable amounts of smooth endoplasmic reticulum, small mitochondria, prominent lysosomes, and moderate amounts of rough endoplasmic reticulum. When the number of mitochondria is highly increased, the cell acquires at the light microscopic level an intensely eosinophilic granular cytoplasmic appearance (Hurthle cells).

A wide variety of immunohistochemical markers with various degrees of specificity and diagnostic significance are expressed by the normal adult follicular cells. Thyroglobulin is the most specific marker of normal follicular cells and their tumors. Other markers include thyroid transcription factor-1 (TTF1) and low molecular weight keratin (15,25–28).

Squamous metaplasia may be seen in the thyroid gland with inflammation, neoplastic conditions, and benign nodules (29). Squamous metaplasia may result from stem cell differentiation of inflamed or reparative follicular epithelium (29).

C Cells

Calcitonin-producing C cells (parafollicular cells) are a minor component of the thyroid gland, comprising no more than 0.1 percent of the normal glandular mass (19). They are often difficult to identify on hematoxylin and eosin (H&E)-stained sections, where they appear polygonal, with clear to pale eosinophilic cytoplasm and oval nuclei (fig. 3-4).

C cells are located, individually or in small groups, within thyroid follicles, mostly at the periphery of the follicular wall, within its basement membrane, and without contact with the follicular lumen. They are located predominantly in the middle and upper third of the lateral lobes; the upper and lower poles are essentially devoid of C cells. They are more numerous in neonates, decrease in the adult thyroid gland, and increase after 60 years of age.

Immunohistochemistry for calcitonin can readily identify C cells (fig. 3-4). They are positive for broad-spectrum neuroendocrine markers such as chromogranin A and synaptophysin, as well as for specific peptides including calcitonin, calcitonin gene–related peptide, somatostatin, gastrin-releasing peptide, and bombesin (30–32). Immunostaining may be positive for low molecular weight keratin and carcinoembryonic antigen (CEA). Normal, hyperplastic and neoplastic C cells are variably positive for TTF1, in contrast to follicular cells which are more uniformly positive (33,34).

The main ultrastructural characteristic of C cells is the presence of neuroendocrine-type secretory granules, which range in diameter from 60 to 550 nm (35,36). Two main types of granules have been identified. Type I granules have an average diameter of 280 nm and a moderately electron-dense, finely granular content which is closely applied to the limiting membranes of the granules. Type II granules are smaller (average diameter of 130 nm), with a more electron-dense content, and are separated from the limiting membranes by a small but distinct electron-lucent space. Immunocytochemical studies performed at the ultrastructural level have shown that both type I and II secretory granules contain immunoreactive calcitonin (35,36). In situ hybridization studies have localized calcitonin and calcitonin gene–related peptide mRNA in C cells (37,38).

Solid Cell Nests

Solid cell nests (SCNs) are clusters of epithelial cells interspersed among the follicles (fig. 3-5). They are usually found in the posterolateral or posteromedial portion of the lateral lobes of the gland and probably represent ultimobranchial body rests (39,40). SCNs are common in the normal thyroid gland and can be detected in almost 90 percent of neonatal thyroid glands (19). In one study, SCNs were found in only 3 percent of routinely examined glands but

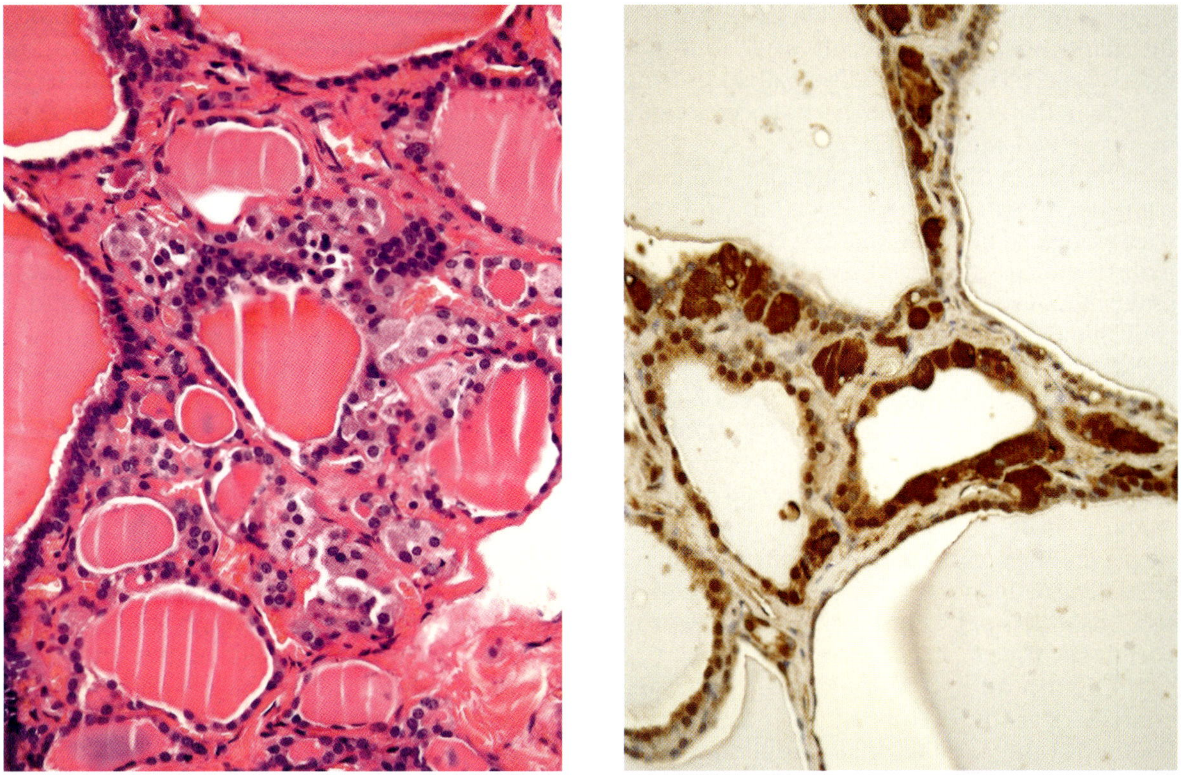

Figure 3-4

NORMAL C CELLS

Left: Normal C cells are inconspicuous on routine hematoxylin and eosin (H&E) staining. In this section, a few clusters of C cells have somewhat pale cytoplasm.

Right: Immunohistochemistry for calcitonin readily identifies C cells.

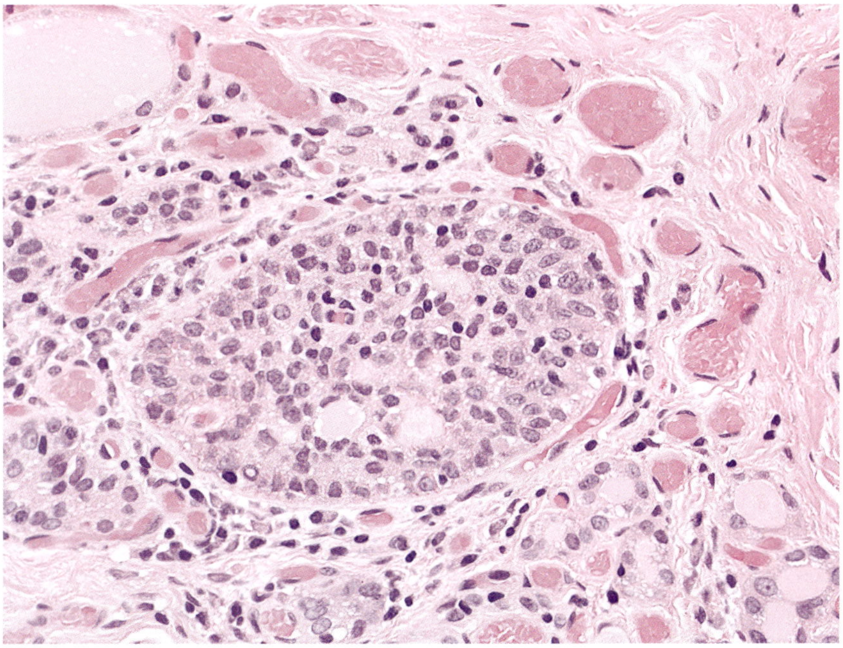

Figure 3-5

SOLID CELL NEST

Solid cell nest in a thyroglossal duct cyst. These ultimobranchial body rests are made up of polygonal to oval cells with oval nuclei and granular chromatin.

in as many as 61 percent of specimens when the gland was blocked serially at 2- to 3-mm intervals (41). They are usually small, measuring 0.1 mm in diameter on average, but occasionally they reach a large size (42).

SCNs have a dual cell population. The main component consists of polygonal to oval cells with elongated nuclei containing granular chromatin and acidophilic cytoplasm. The second cell population is composed of calcitonin-containing cells with clear cytoplasm and round nuclei. SCNs may exhibit squamous differentiation and are sometimes misinterpreted as foci of squamous metaplasia in follicles (29). The SCNs are usually surrounded by stroma and are more or less demarcated by the adjacent thyroid follicles. Adipose tissue, cartilage, or rarely parathyroid tissue, may be present in their vicinity (19). As C cells are usually inconspicuous on H&E-stained sections, SCNs can be used as a marker of the part of the thyroid containing C cells.

Ultrastructurally, the cells of SCNs are characterized by desmosomes, intermediate filaments, and intraluminal cytoplasmic projections. The calcitonin-containing clear cells contain dense-core secretory granules (43,44). Immunohistochemically, the more abundant polygonal cells are positive for high and low molecular weight keratins, galectin-3, CEA, and p63, and are usually negative for TTF1 and thyroglobulin (19). The clear cells express calcitonin and chromogranin (45–47). In contrast to neoplastic C cells, SCNs often demonstrate only weak calcitonin expression, but are more strongly positive for CEA.

Physiology

The main function of the thyroid gland is the production of thyroid hormones, the most important being thyroxin (T4) and triiodothyronine (T3). These hormones regulate metabolism, increase protein synthesis in every tissue of the body, and increase oxygen consumption. Thyroid hormones are particularly important for physical development and for normal maturation of the central and peripheral nervous systems (48).

Ingestion of exogenous iodine from water and food is required for the synthesis of about 100 µg of T4 daily by the thyroid gland (49). Iodine balance is maintained by dietary sources, but many conditions can modify iodine intake, such as geographic location, medications, dietary supplements, and food additives. A series of reactions govern the synthesis and production of thyroid hormone from dietary sources. The first step involves trapping and concentrating iodine from dietary sources and from the recycling of degraded thyroid hormone by the body. A minimum of 60 mg of iodine/day is required for thyroid hormone synthesis and it is estimated that at least 100 mg/day is required to eliminate all signs of iodine deficiency from the population (49).

Once ingested, iodine is absorbed and transported in the extracellular fluid to the thyroid gland, where the intracellular levels are 30 times higher than in the peripheral blood. The active iodide uptake across the basement membrane is mediated by human sodium iodide symporter (hNIS) in a process coupled with the flow of sodium (50). The intrathyroidal iodide is then oxidized to iodine. This reaction is dependent on the action of thyroid peroxidase (TPO) located in the apical membranes of the follicular cells. The results are monoiodotyrosine (MIT) when one iodine molecule is attached or diiodotyrosine (DIT) when two iodine molecules are attached (51,52). The iodotyrosine residues are then condensed to form the biologically active thyroid hormones T4 and T3.

The thyroid gland has both the largest store of hormones among endocrine glands, and the lowest rate of hormonal turnover (1 percent/day). Thyroid hormones are stored in thyroglobulin (Tg), with numerous iodinated tyrosine residues, including biologically active T3 and T4. Tg is a high molecular weight protein that is encoded by a very large gene (more than 260 kilobases) located on chromosome 8 (53). Resorption of Tg takes place through cytoplasmic pseudopodia that engulf minute portions of colloid, which are then drawn into the cell in the form of membrane-bound colloid droplets. These subsequently fuse with lysosomes, and their content is digested by the lysosomal enzymes (54–57). The breakdown products, including T3 and T4, diffuse into the bloodstream, where they are transported primarily by the specific carrier protein thyroxine-binding globulin (TBG). TBG normally transports more than 70 percent of thyroid hormones; approximately 20 percent are carried by transthyretin (prealbumin) and albumin (58). Only a small proportion of circulating thyroid hormones (approximately 0.05 percent of T3 and 0.015 percent of T4) is unbound and, therefore, biologically active (59). Free, circulating, biologically active T3 and T4 are in equilibrium with the hormones bound to the carrier proteins. Intrathyroidal deiodination of the iodotyrosines released during Tg hydrolysis results in free iodine, which is reutilized by the gland for further thyroid hormone synthesis. Tg is present in the plasma of normal individuals at concentrations of up to 50 ng/mL. The source of this Tg is from molecules that have leaked from the thyroid lymphatics. In patients with thyroid cancer, the plasma Tg concentration may be very elevated.

Thyroid biosynthetic and secretory activities are regulated by the blood level of thyroid-stimulating hormone (TSH), a glycoprotein synthesized and secreted by the anterior pituitary gland (60). TSH binds to a specific receptor located on the basolateral surface of the follicular cell membrane, and by activating the adenylate-cyclase pathway, regulates the complex mechanism responsible for T3 and T4 synthesis (61,62). Transcription factors, which are nuclear proteins, are also required for thyroid hormone synthesis and include a thyroid transcription factor, PAX8. These factors regulate the activity of the *Tg* and *TPO* genes in the thyroid gland (63).

Stimulation of the thyroid gland by TSH increases its secretory activity and vascularity, and results in both hypertrophy and hyperplasia of follicular cells, accompanied by reduction of colloid storage. TSH release is in turn regulated by a tripeptide secreted by the hypothalamus, thyrotropin-releasing hormone (TRH). TSH and TRH release are regulated by the circulating levels of free T3 and T4, via a negative feedback on the hypothalamic-pituitary axis (60,64–66).

Calcitonin, which is produced by the C cells, is a 32-amino acid peptide whose main function is the regulation of the level of calcium in the plasma by a feedback mechanism. When calcium plasma levels are increased, calcitonin is released from the thyroid gland, which acts to reduce calcium levels via inhibition of osteoclastic activity and promotion of calcium and phosphate excretion by the kidneys and gastrointestinal tract (67).

Thyroid Adaptation During Normal Pregnancy. The thyroid gland undergoes significant physiologic changes in order to meet the increased metabolic needs during a normal pregnancy (68–70). Serum TBG concentrations rise almost two-fold as a result of estrogenic stimulation of TBG production and TBG sialylation, which results in decreased clearance of TBG (71). To maintain adequate free thyroid hormone concentrations during this period, T4 and T3 production by the thyroid gland must increase. The TBG excess leads to an increase in both serum total, but not free, T4 and T3 concentrations. Levels of total T4 and T3 rise by approximately 50 percent during the first half of pregnancy (70,72), plateauing at approximately 20 weeks of gestation, at which time a new steady state is reached and the overall production rate of thyroid hormones returns to prepregnancy rates.

Human chorionic gonadotrophin (hCG) has weak thyroid-stimulating activity (73). Serum hCG concentrations increase soon after fertilization and peak at 10 to 12 weeks. During this peak, total serum T4 and T3 concentrations increase. Serum free T4 and T3 concentrations increase slightly, usually within the normal range, and serum TSH concentrations are appropriately reduced (73). However, serum TSH concentrations may be transiently low or undetectable in 10 to 20 percent of normal women (74–76).

Iodine requirements are higher in pregnant than in nonpregnant women due both to the increase in maternal T4 production required to maintain maternal euthyroidism and an increase in renal iodine clearance. Severe maternal iodine deficiency during pregnancy results in a reduction in maternal T4 production, inadequate placental transfer of maternal T4, and impairment of fetal neurologic development. Though there is no clinically significant change in the size of the thyroid gland during pregnancy, with iodine deficiency, goiters can develop more easily than in the normal gland because of the increased demand for thyroid hormone. Clinical thyrotoxicosis can develop in patients with hydatidiform moles or choriocarcinoma because of an increase in T3 and T4 (77).

Thyroid Function in Neonates and the Elderly. Serum TSH concentrations rise abruptly to 60 to 80 mU/L within 30 to 60 minutes after delivery in healthy term babies (78). This rise is associated with exposure of the infant to a colder environment and clamping of the umbilical cord. The serum TSH concentration then decreases rapidly to approximately 20 mU/L by 24 hours after delivery and then more slowly to 6 to 10 mU/L at 1 week.

The initial surge in TSH stimulates T4 secretion so that serum total and free T4 concentrations peak at 24 to 36 hours of life (79). Serum T3 concentrations also rise and peak around the same time. The increase in serum T3 is a result of increases in both thyroidal secretion and conversion of T4 to T3 in peripheral tissues. Serum T4 and T3 concentrations are slightly higher in the first year of life but gradually fall to the normal adult range by 2 years of age (80).

With age, the thyroid gland undergoes progressive physiologic fibrosis and atrophy (81), leading to a reduction in thyroid volume. The prevalence of autoantibodies increases with age, reaching up to 20 percent in women over the age of 60 years, and may be partly responsible for the anatomic changes in the thyroid gland (82). Thyroidal iodine uptake decreases with age, leading to decreased T4 secretion in the elderly (82–84). This reduction in T4 secretion is compensated for by a decrease in metabolic clearance of T4 due to decreased 5'deiodinase activity with advanced age (85).

Serum T3 concentrations remain in the normal range into the seventh and eighth decades of life (82). Age-related decline in T3 is demonstrated only in a few studies and only after the age of 90 years (85). This reduction in T3 is also due to decreased T4 synthesis with age and decreased activity of 5'deidoniase.

Peripheral Effects of Thyroid Hormone

Serum free T3 and T4 are available for cellular uptake at any instant in time. In addition, T3 and T4 dissociate from their binding proteins so rapidly that more hormone can become available almost instantaneously. T4 and T3 enter the cells of most organs by carrier-mediated transport and perhaps also by diffusion (86). T3 is also available to cells because it is produced from T4 within them via deiodination in the 5' position, which accounts for about 30 to 40 percent of secreted converted T4.

Some of the locally produced T3 must leave the cells, as evidenced by the observation that serum free (and total) T3 concentrations are near normal in hypothyroid patients taking T4 (levothyroxine) in doses that raise their serum T4 concentrations to normal. However, some of the T3 does not leave, and local production of T3 provides much of the T3 that is bound to T3 nuclear receptors in many tissues. Overall, approximately 90 percent of the extrathyroidal T3 pool of 50 mcg (75 nmoles) is intracellular. The fraction of T3 that is produced locally from T4 and the contribution of locally produced T3 to the amount of T3 bound to its receptors vary substantially from tissue to tissue (64).

The conversion of T4 to T3 may be impaired by various conditions including trauma and certain diseases like the euthyroid sick syndrome. This latter condition is caused by fasting, malnutrition, and hepatic and renal dysfunction, as well as certain drugs such as propylthiouracil, glucocorticoids, propranolol, and amiodarone (87–89).

Nuclear T3 receptors mediate most, if not all, of the physiologic actions of thyroid hormone (93). Cytoplasmic T3 diffuses or is transported into the nucleus, where it binds to the chromatin-localized receptors and exerts its effects. Thyroid hormones stimulate metabolism, increase oxygen consumption, and cause a rise in heat production, cardiac output, and heart rate. They are essential for normal development, growth, and maturation. The acceleration of growth may result from direct action on the cells to increase their rate of division, by acting permissively for other hormones, or by inducing the synthesis of a variety of growth-promoting hormones (56,58,91–95).

DRUG INTERACTIONS WITH THYROID HORMONES

Amiodarone

Amiodarone is a commonly used antiarrythmic drug; however, it is associated with a number of side effects, including thyroid dysfunction. The effects of amiodarone on thyroid function are divided into those that are intrinsic properties of the drug, and those that are due to its high iodine content (96–101). The clinical effects of amiodarone on thyroid function in any individual are dependent upon the underlying status of that individual's thyroid gland.

Direct toxic effects of amiodarone on the thyroid gland are responsible for the side effects seen in patients with no underlying thyroid disease and include: inhibition of outer ring 5'-monodeiodination of T4, thus decreasing T3 production; blocking T3-receptor binding to nuclear receptors; decreasing expression of some thyroid hormone-related genes; and a direct toxic effect on the thyroid resulting in a destructive thyroiditis. In contrast, side effects due to iodine excess are usually seen in patients with underlying thyroid disease and thus defects in the autoregulation of iodine. Patients with autoimmune thyroid disease are more likely to develop amiodarone-induced hypothyroidism in Hashimoto disease (102) and amelioration of Graves hyperthyroidism. Patients with areas of autonomous function within a nodular goiter do not autoregulate iodine, and the addition of more substrate may result in excessive thyroid hormone synthesis and thyrotoxicosis (103).

The clinical manifestations and diagnosis of amiodarone-associated hypothyroidism are similar to those of hypothyroidism from any cause. Hypothyroidism and hypothyroid symptoms may develop as soon as 2 weeks or as late as 39 months after the initiation of amiodarone therapy (104). *Amiodarone-induced thyrotoxicosis* (AIT) is categorized into two types. Type I AIT is typically seen in patients with preexisting multinodular goiter or latent Graves disease, in whom the excess iodine from amiodarone provides increased substrate, resulting in enhanced thyroid hormone production (105). In type II AIT, the hyperthyroidism is a destructive thyroiditis that results in excess release of T4 and T3, without increased hormone synthesis. It typically occurs in patients without underlying thyroid disease and is caused by a direct toxic effect of amiodarone on thyroid follicular epithelial cells (106–108).

The histologic findings in the thyroid gland with type I AIT reflect the underlying disease process. For example, in an individual with a predisposition to Graves disease, amiodarone may precipitate a thyrotoxicosis with a morphology identical to Graves. The morphology of type II AIT is well described and presents as a unique destructive and involutional thyroiditis characterized by the triad of overdistension of follicles by colloid, atrophy and flattening of the follicular lining cells, and infiltration by foamy macrophages (fig. 3-6). In addition to the large follicles lined predominantly by flat epithelial cells containing abundant colloid, and intrafollicular aggregates of CD68-positive macrophages associated with follicular disruption, there may be some interstitial fibrosis (109). Occasional follicular epithelial cells exhibit vacuolar degeneration. Multinucleated giant cells are also identified at sites of follicular damage. The thyroid glands from patients with amiodarone-induced hypothyroidism may show similar features.

Ultrastructural studies have shown lysosomes in the disrupted vacuolated cells, with prominent lamellar configurations (100). It is thought that the release of colloid from the destroyed follicles may lead to increased thyroid hormone levels and hyperthyroidism.

Tyrosine Kinase Inhibitors

Oral tyrosine kinase inhibitors (TKIs) (e.g., lenvatinib, selpercatinib, sunitinib, sorafenib, and imatinib) are commonly used for the treatment of a range of malignancies including gastrointestinal stromal tumors, renal cell carcinoma, hepatocellular carcinoma, and chronic myeloid leukemia; however, they can cause hypothyroidism (110,111). Hypothyroidism occurs in patients with previously normal thyroid function and may be preceded by a destructive thyroiditis and transient suppression of TSH (112). In patients with preexisting hypothyroidism, thyroid hormone requirements may increase (110,113,114). Sorafenib treatment is associated with type 3 deiodination, which increases the metabolism of T4 and T3 (115). This finding may explain, in part, the increased thyroid hormone requirements in patients taking these drugs.

Of all the TKIs, sunitinib appears to be most frequently associated with thyroid dysfunction, with subclinical hypothyroidism reported in 30 to 80 percent of patients (116). It is possible that TKIs like sunitinib that broadly target tyrosine kinases, specifically vascular endothelial growth factor (VEGF) receptors, are more likely to cause hypothyroidism by causing capillary regression and thyroid ischemia. Other possible mechanisms include a destructive thyroiditis (112,117), impaired iodine uptake (118), reduced thyroid peroxidase activity (119), or decreased clearance of TSH (120).

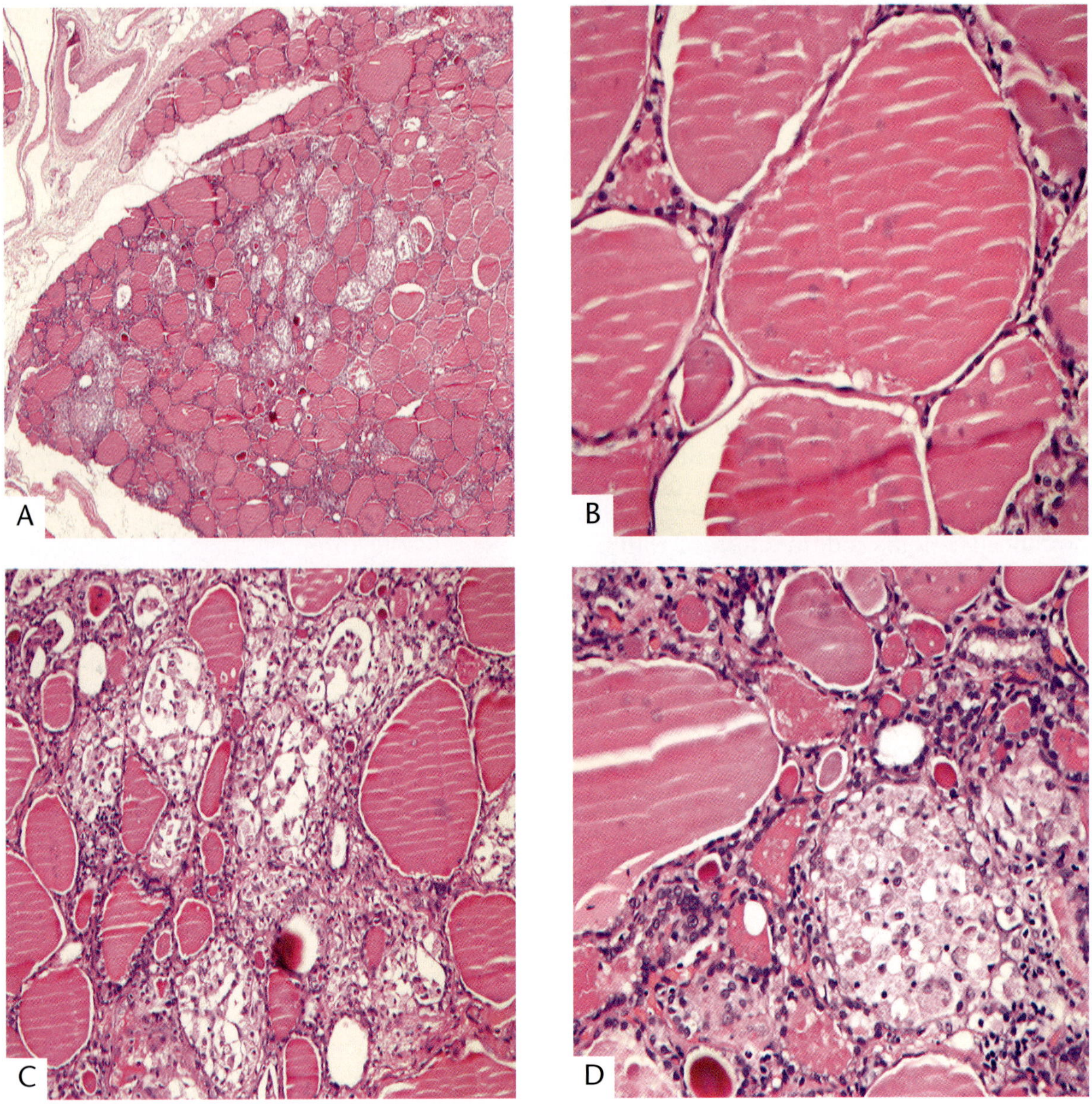

Figure 3-6

AMIODARONE TOXICITY

Amiodarone-induced thyroid injury is characterized by the triad of overdistension of follicles by colloid (A), atrophy of the follicular cells lining the follicles (B), and infiltration of the thyroid gland by foamy macrophages (C,D).

Immune Checkpoint Inhibitors

Immune checkpoint inhibitor (ICI) therapy is now standard of care in oncology for a range of malignancies including melanoma, lung cancer, head and neck cancers, and renal cell carcinoma (121). ICIs include agents targeting cytotoxic T-lymphocyte–associated antigen 4 (CTLA-4) and programmed cell death 1 (PD-1) receptors (e.g., nivolumab, pembrolizumab, ipilimumab), and are now known to be associated with clinically significant endocrinopathies, the most common of which are hypophysitis and hypothyroidism (122). The reported rate of thyroid dysfunction due to PD-1 inhibitors is as high as 21 percent in patients taking pembrolizumab and 29 percent in patients taking nivolumab (123–125). In a systematic review and meta-analysis that included 7,551 patients in 38 randomized trials, the overall incidence of clinically significant endocrinopathies is approximately 10 percent in patients treated with ICIs (126). Studies with long-term follow-up suggest that there is usually no resolution of hypothyroidism once it has developed in these patients (127).

As immunotherapy-related thyroiditis is usually diagnosed biochemically and managed medically, these cases rarely come to the attention of the surgical pathologist unless they are incidental findings in a thyroid gland removed for other reasons. Nonetheless, the histologic features of immunotherapy-induced thyroiditis are distinctive, and characterized by CD8-positive T-cell–mediated destruction of follicles with secondary colloid depletion (123,128). In time, this may lead to secondary atrophy which exaggerates the underlying lobular architecture of the thyroid gland and gives the impression of a firm nodular gland (fig. 3-7). Non-necrotizing granulomas have been described in one case of nivolumab-induced thyroiditis (129).

Drugs that Affect Thyroid Hormone Metabolism

Drugs that increase the metabolism of T4 and T3 include phenobarbital, rifampicin, phenytoin, and carbamazepine. As a result, T4-treated hypothyroid patients may need a higher dose when treated with any of these drugs. Conversely, the doses of these drugs may need adjustment because their metabolism varies according to thyroid status.

Phenobarbital and rifampicin augment the rate of deiodination of T4 and T3, principally by stimulating the hepatic drug-metabolizing enzyme system (130–132). In normal individuals, the hypothalamic-pituitary system compensates for an increase in hormone by augmenting thyroid hormone production and secretion. However, patients with subclinical or overt hypothyroidism cannot augment thyroid hormone production and secretion, and their hypothyroidism is exacerbated.

The interactions of phenytoin and carbamazepine with the thyroid gland are more complex (133,134). These drugs both augment the rate of thyroid hormone metabolism (similar to phenobarbital and rifampin) (98) and displace thyroid hormones from the serum-binding proteins, principally TBG (135). As a result, serum total and free T4 concentrations decrease by approximately 40 percent; the decrease in serum T3 is smaller, but TSH concentrations remain within the normal range. Phenytoin can also induce thyroiditis, possibly by inducing an autoimmune reaction with the thyroid gland as the target organ (136,137).

Miscellaneous Drugs and Chemicals

A wide spectrum of drugs can inhibit thyroid hormone synthesis. They may affect any aspect of the thyroid hormone system, including TSH secretion and thyroidal production of T4 and T3, their transport in serum, and their metabolism (138).

Drugs that inhibit TSH secretion include glucocorticoids (139,140), dobutamine (141), dopamine (142,143), somatostatin analogs such as octreotide (144,145), and the retinoid X receptor ligand bexarotene (146). Glucocorticoids appear to suppress release of TSH through inhibition of TRH in the hypothalamus. However, chronic high-dose glucocorticoids or Cushing syndrome do not appear to cause clinically significant central hypothyroidism (147). Similarly, somatostatin analogs suppress serum TSH, likely through direct effects on pituitary thyrotropes, but these effects are primarily transient and do not appear to cause clinically significant central hypothyroidism (148,149).

A few observational studies have suggested that metformin can lower serum TSH levels (150). One study demonstrated this effect only in type 2 diabetics who also had hypothyroidism, but not in patients with normal thyroid function (151). This effect may occur through altered free T4 levels in patients who are hypothyroid (152), but the exact mechanism is not known.

Drugs that inhibit thyroid iodine transport include thiocyanate and perchlorate, which had been used clinically in the past to treat hypertension and hyperthyroidism, respectively. Many drugs interfere with thyroid hormone binding in the serum, either by increasing or decreasing serum TBG concentrations. Among those that raise serum TBG, the most important are estrogen or selective estrogen receptor modulators (SERMs, including oral contraceptives

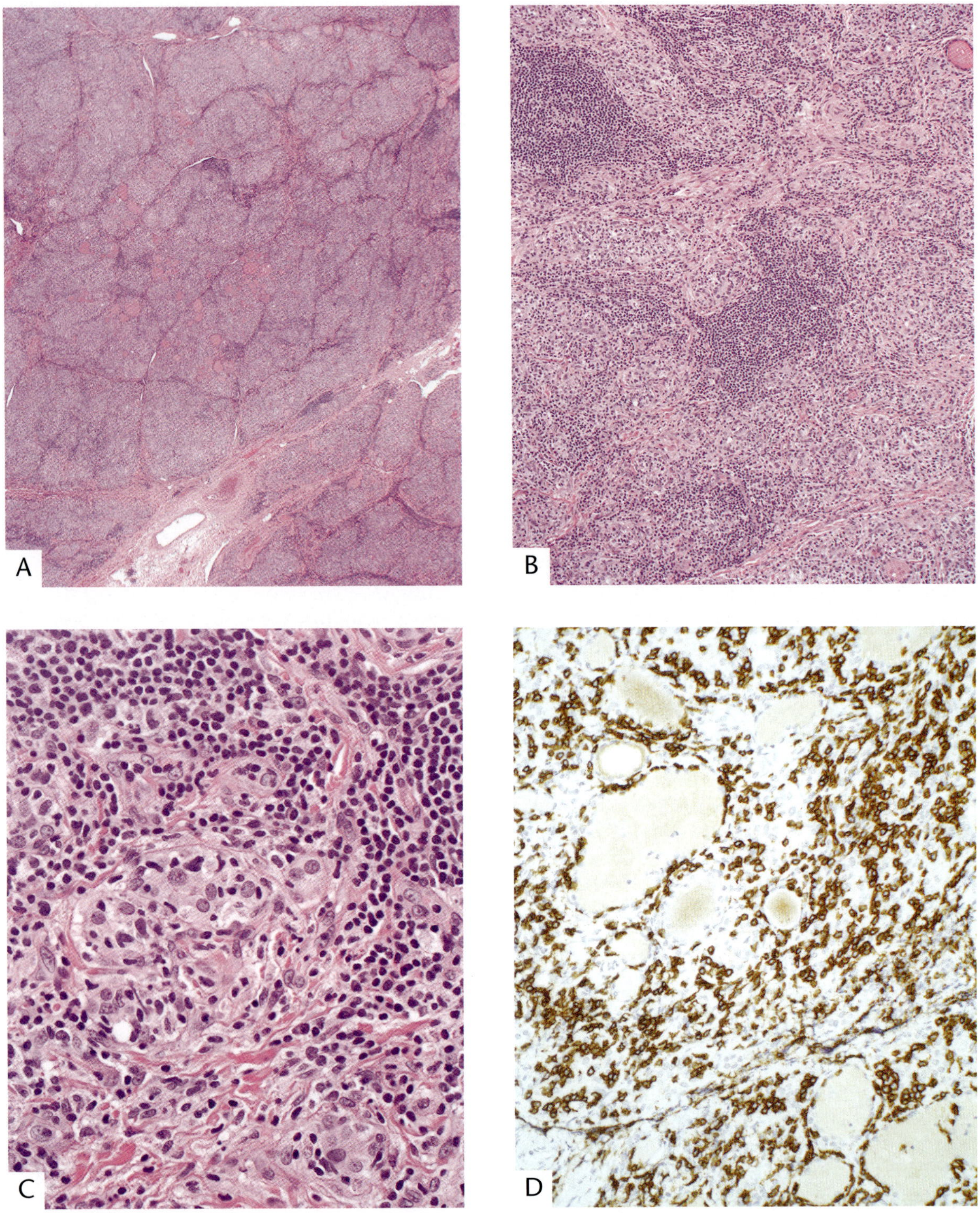

Figure 3-7

IMMUNOTHERAPY-INDUCED THYROIDITIS

This patient was being treated with nivolumab for metastatic melanoma. The thyroid gland has a nodular appearance at low power (A), which is caused by T-cell–mediated destruction of the follicles and secondary colloid depletion (B,C). The majority of the T cells express CD8 by immunohistochemistry (D).

and tamoxifen), methadone, and fluorouracil. Drugs that lower serum TBG include androgens, anabolic steroids, and glucocorticoids.

Several other drugs, such as salicylates, fenclofenac, and furosemide, lower serum T4 and T3 concentrations by blocking hormone binding to TBG (153–156). Heparin may also cause an acute and transient increase in serum free T4 concentrations by stimulating lipoprotein lipase which displaces T4 from binding proteins (157). Other agents that inhibit thyroid hormone binding and coupling reactions include thioamides such as propylthiouracil, methimazole, and phenylbutazone.

Lithium may cause hypothyroidism. A serum lithium concentration in the therapeutic range results in an increase in thyroid gland size in approximately 50 percent of patients and causes mild hypothyroidism in approximately 20 percent (158,159). It inhibits organic binding reactions as well as thyroid hormone release, but its mechanism of action is not well understood (159,160). Some of the patients with hypothyroidism have antithyroid antibodies.

Certain natural foods also contain antithyroid chemicals. Members of the genus Brassica, including cabbage, rutabaga, turnips, and kohlrabi, contain thiocyanates (161). The vegetable cassava contains a cyanogenic glycoside, linamarin, which is metabolized to thiocyanate, so ingestion of cassava can lead to the development of goiters in areas of the world with endemic iodine deficiency.

Abnormal Amounts of Iodine. Abnormalities in thyroid function can result from increased or decreased amounts of dietary iodine. Iodine excess can lead to decreased yields of organic iodine or a relative blockage of organic iodine binding (Wolff-Chaikoff effect) (162). An increase in the iodine-trapping mechanism or impairment of organic iodine formation can lead to goiter or hypothyroidism in patients without Graves disease or in those patients who have not received radioiodine therapy. Treatment with iodine leads to decreased vascularity and hyperplasia of the thyroid gland as is seen with Graves disease before surgical resection (98).

Decreased amounts of iodine in the diet lead to a rapid decrease in serum thyroid hormone (137) and a compensatory increase in TSH. Although T4 levels are low, T3 synthesis is maintained during the development of a goiter (163). With iodine deficiency, other compensatory changes occur including an increase in the conversion of monoiodotyrosine to diiodotyrosine. TSH stimulates thyroid follicular cell replication, leading to the development of a goiter (164,165).

HEREDITARY AND DEVELOPMENTAL DISORDERS

Transcription Factor Deficiencies

Transcription factors regulate gene activity by interacting with nuclear DNA. They control the development and regulation of specific genes in adults: the activity of transcription factors is considered the principal switch that regulates gene expression (166). Thyroid follicular cells simultaneously express four genes that encode the transcription factors homeobox protein Nkx-2.1 (NKX2-1, also known as thyroid transcription factor 1 [TTF1]), forkhead box protein E1 (FOXE1, also known as thyroid transcription factor 2 [TTF2]), paired box protein Pax8 (PAX8), and hematopoietically-expressed homeobox protein Hhex (HHEX), collectively known as the thyroid transcription factors (166–176).

Each of the TTFs is expressed in several other tissue types, but they are only expressed together in epithelial thyroid follicular cells, where they regulate the expression of important proteins such as thyroglobulin, thyroid peroxidase, and the TSH receptor (177,178). Perturbations in levels of expression of the TTFs that result from mutations and/or epigenetic modifications can lead to the development of several clinically relevant conditions. Disruptions to thyroid morphogenesis in humans result in phenotypes such as thyroid agenesis, ectopia, hypoplasia, and hemiagenesis, that together are known as *thyroid dysgenesis* (179,180), which in some cases is accompanied by thyroid dysfunction (169,181,182). Findings from several studies suggest that defects in the functions of TTFs contribute to thyroid dysgenesis; however, the molecular pathways and mechanisms associated with these phenotypes remain unknown (167).

Among these disorders, primary congenital hypothyroidism is the most prevalent, with an incidence of approximately 1 per 3,500 newborns (179,183). Congenital hypothyroidism may be caused by disruptions in thyroid hormone biosynthesis (a condition known as dyshormonogenesis), but in many cases results from thyroid dysgenesis due to *PAX8* mutations (181). Mutations in the *NKX2-1/TTF1* gene have also been described in patients with thyroid dysgenesis (184–186). However, because *NKX2-1* is also expressed in lung and the ventral diencephalon, thyroid dysgenesis in patients with a mutation in this gene is usually also associated with brain and/or lung manifestations (170). *Brain-lung-thyroid syndrome* is characterized by congenital hypothyroidism, respiratory distress, and benign hereditary chorea; however, all three phenotypes are not always present in a given patient and the severity of each phenotype can vary (187–189).

Changes in the expression of genes that encode TTFs and/or sequence variations in these genes have also been implicated in a number of thyroid and nonthyroid tumors. The precise mechanisms that drive each of these processes are still largely unknown (190).

Genetic Disorders of the TSH Receptor Gene

The thyroid-stimulating hormone receptor (TSHR) is expressed on the plasma membrane of thyroid epithelial cells and is central to the regulation of thyroid growth and function (191,192). It is encoded by 10 exons on chromosome 14q31 and is 58 kb in length (193). Several splice variants of the receptor have been described (194). The receptor is coupled to the alpha subunit of the guanine-nucleotide-binding protein stimulatory unit (Gsα), which activates adenylate cyclase and increases the accumulation of cyclic adenosine monophosphate (AMP). Activation of TSHR leads to the initiation of multiple signaling pathways, thyroid hormone synthesis and secretion, and cell proliferation and survival (195).

There are over 30 activating and 13 inactivating mutations in the *TSHR* gene (196). Activating germline mutations lead to an increase in the constitutive activity of the receptor and result in thyroid gland hyperplasia and hyperthyroidism. The mutations are transmitted in an autosomal dominant fashion and result in the clinical condition known as *hereditary toxic thyroid hyperplasia* (also known as *Leclère disease*), which is characterized by hyperthyroidism with a variable age of onset, hyperplastic goiter of variable size but with steady growth, and absence of clinical or biological stigmata of autoimmunity (166,197–200). Toxic thyroid hyperplasia has also been described in children born from unaffected parents (172,201–204). The specific mutations in these cases usually differ from those seen in hereditary cases (205).

Inactivating mutations in the *TSHR* gene are associated with a syndrome of "resistance to TSH" (205). These mutations are inherited in a recessive fashion and result in decreased action of TSH, leading to reduced T4 and T3 synthesis and secretion, with a compensatory increase in TSH secretion. The absence of goiter and the presence of thyroid hypoplasia in these patients are compatible with the dominant role of TSH on the growth of the thyroid gland (206). *TSHR* gene mutations rarely cause severe congenital hypothyroidism (207). The magnitude of serum TSH elevation and the thyroid status of the patients vary with the severity of the functional impairment of the mutant TSH receptor. Approximately half of patients with inactivating *TSHR* mutations have partial impairment of function, while one third can maintain a euthyroid state by an appropriate increase in serum TSH levels (208).

Aplasia and Hypoplasia

Thyroid aplasia (complete absence of thyroid tissue) and *thyroid hypoplasia* (less than the normal amount of thyroid tissue) are part of a spectrum of embryological abnormalities of thyroid gland development known as *thyroid dysgenesis*. Aplasia (or agenesis) and hypoplasia are the most common causes of congenital hypothyroidism. In North America, permanent abnormalities leading to hypothyroidism occur in 1 in 3,000 to 4,000 live births (209). Some infants may have hemiagenesis or aplasia of one lobe of the thyroid gland; this more commonly affects the left lobe and is not associated with functional defects.

Mutations in several genes that regulate thyroid gland development have been reported as rare causes of thyroid dysgenesis (210). These mutations may also be associated with congenital anomalies in other tissues. For example, mutations in the *PAX8* gene causes anomalies of the urogenital tract (211,212); mutations in the *TTF2* gene causes cleft palate, spiky hair,

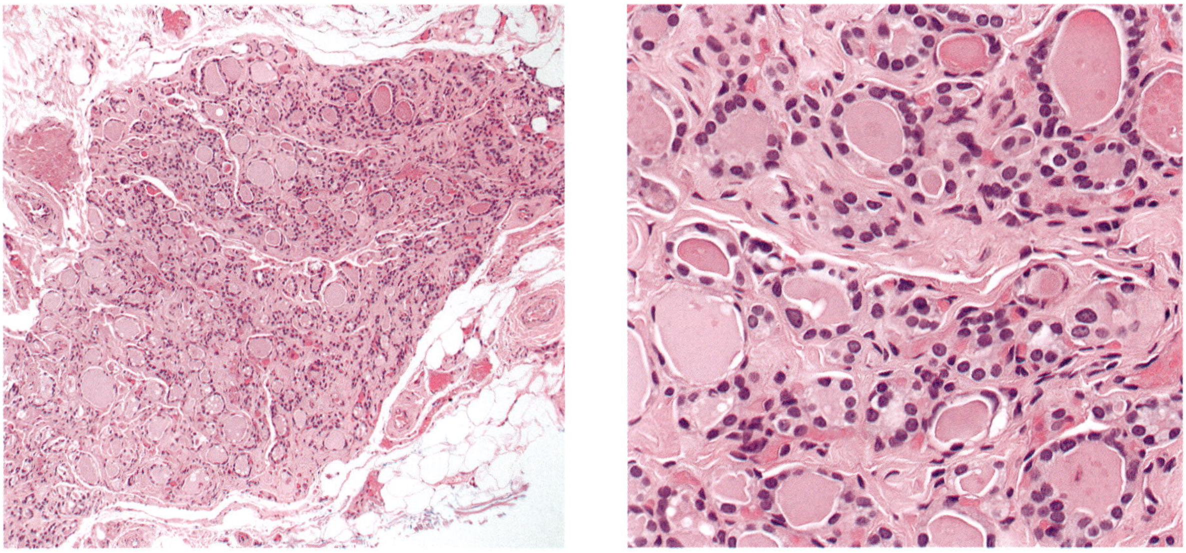

Figure 3-8

ECTOPIC THYROID TISSUE

Left: Ectopic thyroid tissue in a thyroglossal duct remnant shows colloid follicles with cuboidal epithelium. Right: This tissue may be mistaken for metastatic thyroid carcinoma; however, the cytologic features are benign.

and bilateral choanal atresia (Bamforth-Lazarus syndrome) (213,214); and mutations in the *NKX2-1 (TTF1)* gene causes anomalies of the lungs and central nervous system (215,216). Patients with DiGeorge syndrome in which there is arrested development of the parathyroid glands and thymus associated with the third and fourth branchial pouches also have arrested thyroid C-cell development (217,218).

Aberrant (Ectopic) Thyroid Tissue

Thyroid ectopia is defined as functioning thyroid tissue found anywhere other than the usual anatomic location of the thyroid gland. Ectopic thyroid is usually located along the normal path of thyroid gland descent, but rarely has been reported in the mediastinum, heart, esophagus, diaphragm, inguinal region, or porta hepatis (219–225). Along the thyroglossal tract, it may be localized in a lingual, sublingual, supralymphoid, or intrathyroid location. Ectopic thyroid tissue is derived from abnormalities in migration of the medial anlage and hence typically does not contain C cells.

Failure of thyroid gland descent occurs in approximately 1 in 200,000 normal individuals and 1 in 6,000 patients with thyroid disease (226). The true incidence of thyroid ectopia is not known due to the asymptomatic nature of some ectopic thyroid tissue.

The most common site of ectopic thyroid is the tongue. Lingual thyroid tissue results from failure of the medial anlage to descend from the pharynx so that it remains at the base of the tongue (3-15). This may be the only thyroid tissue in some patients and surgical excision can lead to hypothyroidism.

The wall of a thyroglossal duct cyst is the second most common site for ectopic thyroid tissue (227). The ectopic tissue is usually found in the form of small groups of follicles and is present in 25 to 65 percent of cysts examined histologically (fig. 3-8) (228). Up to 2 percent of patients presenting with what appears to be a thyroglossal duct cyst have an ectopic thyroid gland (229).

The presence of normal-appearing thyroid tissue in lymph nodes most commonly represents metastatic papillary carcinoma to lymph nodes (230–233), especially if the thyroid tissue is present within the nodes lateral to the jugular vein (231). However, it is now increasingly accepted in rare cases that normal-appearing thyroid tissue in lateral lymph nodes or salivary gland may represent ectopic thyroid rather than metastatic carcinoma (232), and this has been supported by molecular testing (234).

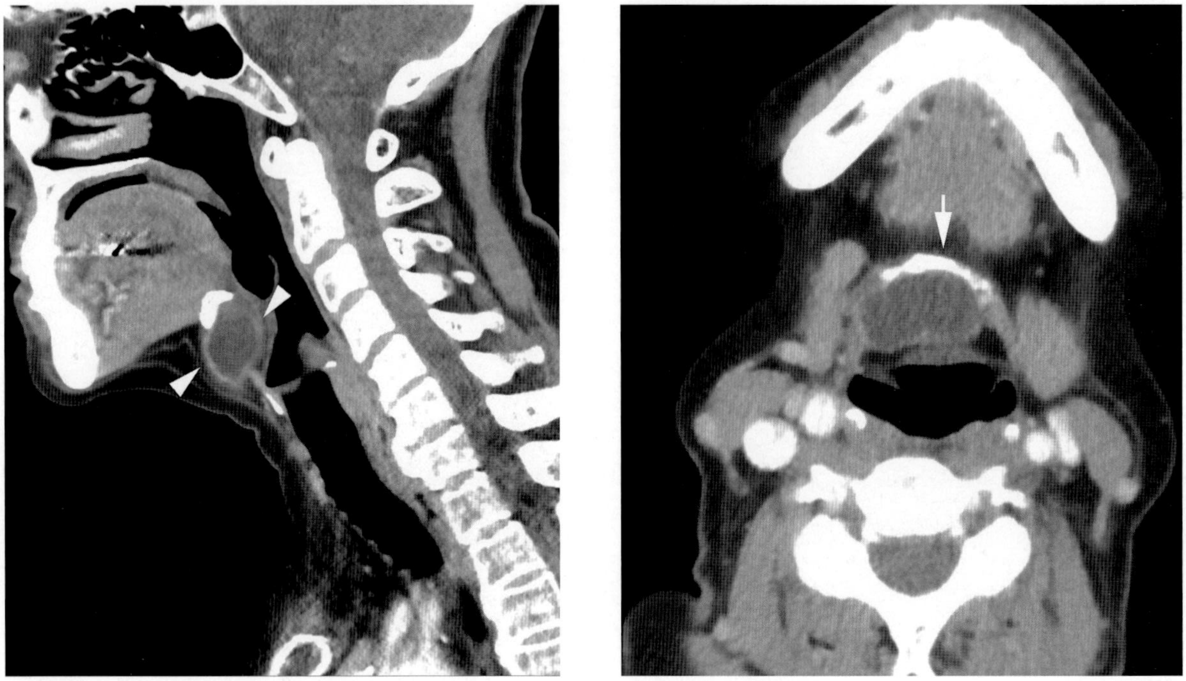

Figure 3-9

THYROGLOSSAL DUCT CYST

Contrast-enhanced sagittal (left) and axial (right) computerized tomography (CT) scans show a fluid-density structure in the soft tissues of the anterior neck (arrowheads). The cyst is intimately associated with the hyoid bone (arrow) as is characteristic of a thyroglossal duct cyst.

Thyroglossal Duct Cysts

Definition. A *thyroglossal duct cyst* is the cystic expansion of a remnant of the thyroglossal duct tract (235,236).

General Features. The median anlage of the thyroid gland forms at the foramen cecum at the posterior tongue and descends in the midline of the neck as the thyroglossal tract to the position of the normal thyroid gland in the base of the neck. The tract usually atrophies and disappears by the 10th week of gestation. Portions of the tract and remnants of thyroid tissue associated with it may persist at any point between the tongue and the thyroid gland.

The stimulus for the cystic expansion of the thyroglossal duct tract remnant to form a cyst is not known. One suggestion is that lymphoid tissue associated with the tract hypertrophies at the time of a regional infection, thereby occluding the tract and resulting in cyst formation (237). Many cystic remnants of the thyroglossal tract are never detected clinically; a postmortem study of 200 adults found a 7 percent incidence of thyroglossal duct cysts (238).

Clinical Features. Patients with thyroglossal duct cysts present with a midline upper neck mass that is cystic. Most patients are children or adolescents, although up to one third are aged 20 years or older (235,236). The mass usually causes no symptoms but may be slightly tender. The cyst may occur anywhere along the thyroglossal duct tract, from the foramen cecum at the base of the tongue to the level of the suprasternal notch. In most cases, the cyst is at or just below the hyoid bone, adjacent to the thyrohyoid membrane (fig. 3-9) (235). Cysts are usually within 2 cm of the midline (235) but can occasionally be more lateral. The cysts move upward with swallowing or protrusion of the tongue. If they become infected, a fistula may develop (239–242).

Gross Findings. Cysts may vary greatly in size, but most are between 1 and 4 cm in diameter. They are unilocular or multilocular,

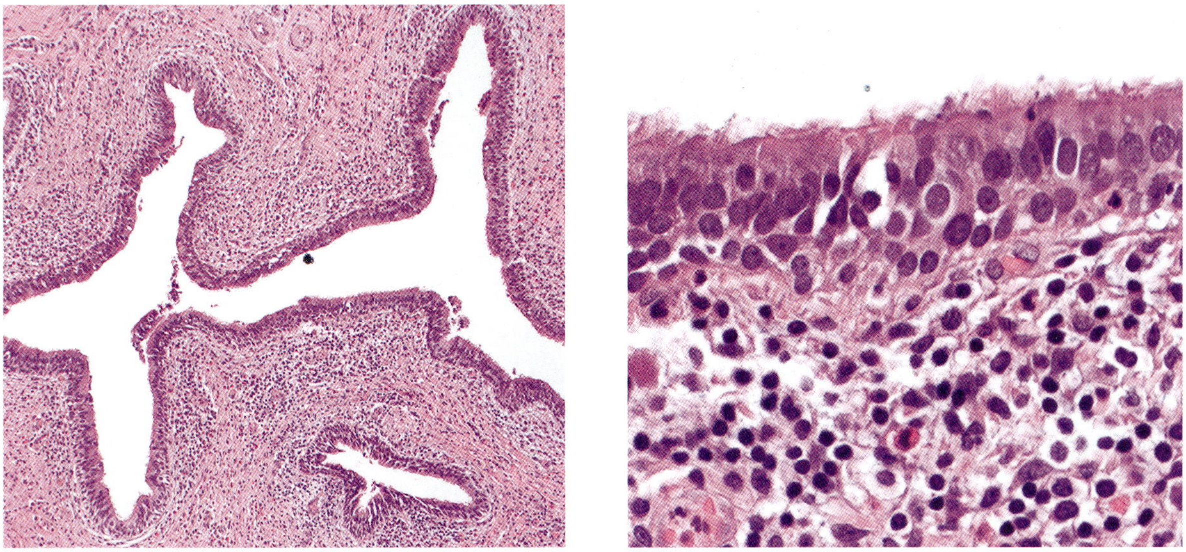

Figure 3-10

THYROGLOSSAL DUCT CYST

Left: Histologic section shows the cyst lined by epithelium, with lymphocytic inflammation in the cyst wall. Right: Ciliated respiratory epithelium lining a thyroglossal duct cyst.

with a rounded, smooth external surface. Cyst contents include clear mucinous or viscous fluid or gelatinous material having a broad range of colors (clear, yellow, tan, red/brown, gray/white) and degrees of opacity. Infected cysts contain purulent material. The cysts are often firmly embedded in the surrounding strap muscles and soft tissues of the neck, without a surrounding plane of cleavage. A sinus tract or fistula may have openings onto the skin or pharynx. The hyoid bone is usually part of the surgical specimen and should be examined for microscopic remnants of the cyst within the bone or immediately adjacent to it.

Microscopic Findings. The cysts are lined by various types of epithelium depending on their site (243). Ciliated pseudostratified columnar (respiratory) epithelium is seen in cysts in the lower neck, while nonkeratinizing squamous epithelium is more common higher in the neck and can also occur as a metaplastic phenomenon in cysts that are inflamed (fig. 3-10). Stratified cuboidal epithelium is seen in cysts at the level of the hyoid bone. Cysts often contain combinations of different epithelial linings.

Secondary inflammation and infection are common, and may result in denudation of the epithelial lining, fibrosis, and granulation tissue formation. In the appropriate clinical setting, a diagnosis of inflamed, fibrotic thyroglossal duct cyst can be made even in the absence of an epithelial lining.

Thyroid follicles are seen in the cyst wall in 30 to 60 percent of cases (328). They are usually present in small irregular groups, sometimes with changes of hyperplasia or even neoplasia. Other tissues that are occasionally seen in the cyst wall are mucus salivary-type glands (244), skin adnexal structures (245), cartilage (246), and bone (247).

Differential Diagnosis. Thyroglossal duct cysts are the most common cause of a midline neck mass, followed by dermoid cysts. In contrast to thyroglossal duct cysts, dermoid and epidermoid cysts tend to be superficial and less tethered to the underlying structures and hyoid bone. These cysts are lined by squamous epithelium and are filled with keratin. Dermoid cysts have skin appendages in their wall.

Branchial cleft cysts may be confused clinically with thyroglossal duct cysts. However, they are usually more laterally located and are more frequently associated with a sinus tract or fistula. They are lined by squamous or ciliated columnar epithelium and have lymphoid follicles in their wall.

Cystic degeneration of a colloid nodule within a multinodular goiter may also enter the differential diagnosis. The midline location and distinct lining with lymphoid follicles are helpful features in establishing the diagnosis of thyroglossal duct cyst.

Treatment and Prognosis. The treatment of thyroglossal duct cyst is surgical resection of the entire cyst in continuity with the midportion of the hyoid bone and a core of tissue from the hyoid upward toward the foramen cecum, an operation known as the Sistrunk procedure (248). This procedure ensures that the entire thyroglossal tract is removed in order to prevent recurrence. Infection is the most common complication of thyroglossal duct cysts and should be treated initially with antibiotics, followed by definitive surgery once the infection has resolved.

The incidence of primary carcinoma of the thyroglossal duct is less than 1 percent in all age groups, the majority of which are papillary thyroid carcinomas (249–251), although other types of thyroid carcinoma have been reported (228,252). Medullary thyroid carcinomas have not been reported because of the absence of C cells in these cysts (253). Most thyroglossal duct cancers occur in adults, although they have occasionally been reported in children under 10 years of age (228). Studies suggest that among patients with thyroglossal duct cyst carcinomas, 11 to 33 percent also have carcinoma within the thyroid gland (249). These tumors may metastasize to lymph nodes, as with other papillary thyroid carcinomas.

Most cases of thyroglossal duct cyst carcinoma are adequately treated by resection with the Sistrunk procedure, with a reported cure rate of 95 percent (249,254). Thyroidectomy may be performed in cases where a primary thyroid malignancy cannot be excluded or when regional lymph nodes are involved (255). Thyroglossal duct cyst carcinoma has an exceedingly low mortality rate.

Parasitic Nodule

Definition. *Parasitic nodule* is a nodule of thyroid tissue that is anatomically separate from the main thyroid gland, which is usually multinodular.

Clinical Features. The parasitic thyroid nodule is an infrequently reported condition, although it may be seen in different clinical settings (256, 257). Conditions labeled as aberrant lateral thyroid tissue and some cases diagnosed as metastatic thyroid carcinoma usually are parasitic nodules. The parasitic nodule probably results from a colloid or hyperplastic nodule located outside the thyroid gland. It becomes enlarged and migrates laterally in the neck. In some cases, it is connected to the thyroid gland by a thin fibrous strand of vascular tissue, while in other cases it is completely separate and obtains its blood supply from the surrounding tissues.

Gross Findings. The nodules vary from 1 to over 4 cm in size. A connection to the thyroid gland by a fibrovascular pedicle is often apparent at surgery with careful dissection.

Microscopic Findings. The histologic appearance of a parasitic nodule is usually that of normal thyroid tissue with dilated colloid-filled and/or hyperplastic follicles. Comparison to the orthotopic thyroid gland shows similar histologic features.

In cases in which there is also Hashimoto thyroiditis, the presence of many lymphoid cells may simulate a lymph node, especially in a small biopsy specimen. The pseudo-ground glass nuclei that are sometimes seen in Hashimoto thyroiditis can simulate papillary carcinoma, and this can be a pitfall in the histologic evaluation of parasitic nodules (258,259).

Differential Diagnosis. The differential diagnosis of parasitic nodule includes metastatic papillary and follicular carcinomas. It also includes mechanical implantation of thyroid tissue into the soft tissues of the neck from previous surgery or accidental trauma. The presence of suture material along with fibrosis may be a clue to previous surgery (260,261). This condition may be similar to the parathyromatosis that may occur after surgical implantation of parathyroid tissue in the neck (259).

Treatment. Particularly in the setting of Graves disease, multiple recurrent parasitic nodules may become a local control problem requiring repeated surgery or radioactive iodine (259).

MISCELLANEOUS CONDITIONS

Radiation Changes

External radiation to the thyroid gland can result in a variety of complications that are dose-related. When radioiodine is administered, hypothyroidism is common, and the incidence increases with time (262–267). Exposure to

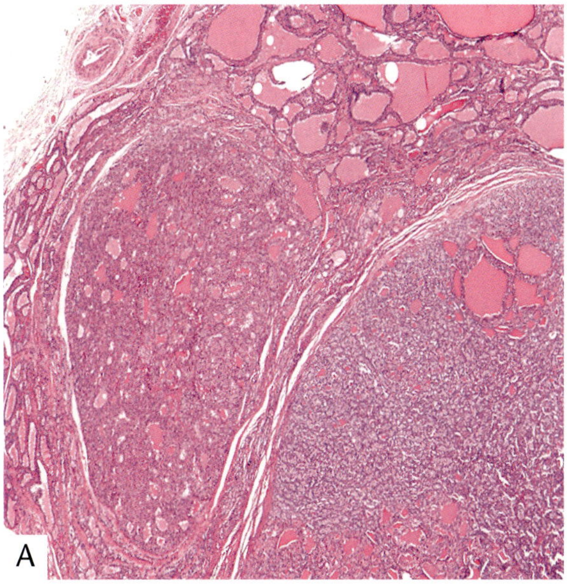

Figure 3-11

RADIATION-INDUCED CHANGES TO THE THYROID GLAND

A: Radiation exposure may drive the development of multiple separate microfollicular nodules, many of which show varying degrees of nuclear atypia.

B: A typical feature of radiation exposure is the presence of scattered individual cells with bizarre nuclear atypia.

C: Some postradiation nodules are composed of cells with powdery to clear nuclear chromatin, in areas with progressively more marked nuclear atypia forming a morphological continuum with the nuclear features of papillary carcinoma.

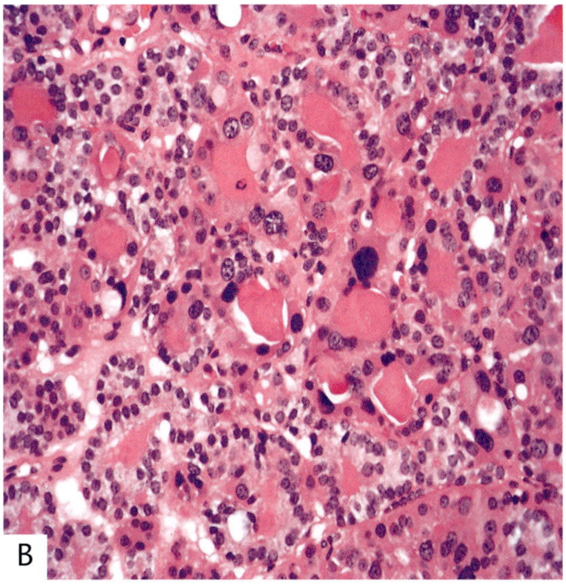

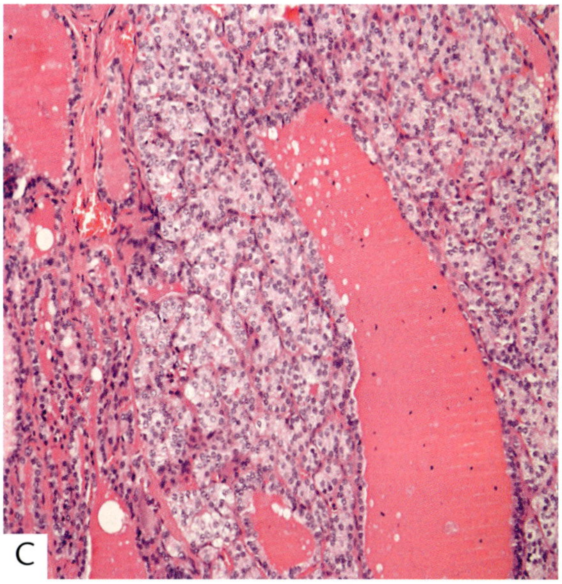

external X ray or internal radioactive iodine at low to moderate doses is associated with the development of thyroid nodules and follicular adenomas (268). These changes are typically seen in thyroid tissue after radiation of the tonsils or other structures adjacent to the thyroid gland. Acute changes in the thyroid gland include follicular disruption, focal hemorrhagic necrosis, and neutrophilic infiltration (266). In one study, there was a decrease in radioactive uptake and serum thyroxine levels between 2 days and 2 months after thyroid radiation (262).

Patients irradiated as treatment for lymphomas may develop striking changes in the thyroid gland that include hypercellularity, cytologic atypia with hyperchromasia, increased nodule size and pleomorphism, fibrosis, and chronic thyroiditis (fig. 3-11) (262). Patients receiving radioactive iodine may have variable changes in the thyroid gland: the cytologic atypia and hypercellularity seen with external irradiation may occur or the changes may be more subtle (264,267). Months or years after radioiodine treatment, the gland is grossly shrunken and fibrotic, and histologically shows fibrosis, follicular atrophy, oncocytic and squamous metaplasia, lymphocytic infiltration, and nuclear abnormalities. Vascular change (intimal thickening and sclerosis of arterial walls, often with inflammatory cell cuffing) is characteristic of radiation damage (269). Although it is well documented that external radiation can lead to an increased incidence of thyroid cancer (270,271), the relationship between treatment with radioactive iodine and thyroid tumor development is less certain.

Langerhans Cell Histiocytosis

Previously considered a reactive phenomenon, but now recognized as a true neoplasm, rare cases of *Langerhans cell histiocytosis* (LCH) involving the thyroid gland have been reported in both adults and children (272,273). Males account for 80 percent of thyroid LCH pediatric cases, while in adults, women are affected more commonly than men (274,275). Patients present with a solitary nodule, a nodular goiter, diffuse goiter, or lymphocytic thyroiditis. Most adults are euthyroid, but some children with severe systemic disease have hypothyroidism and goiter that may precede the involvement of other organs by several months.

Macroscopically, the thyroid gland in LCH is frequently nodular. The cut surface is pale, with areas of hemorrhage. Histologically, there is either diffuse or focal infiltration of the gland by histiocytes, eosinophils, and lymphocytes. Destruction of thyroid follicles may be seen. The histiocytes stain with antibodies to S-100 protein and CD1a (276,277), and ultrastructurally contain Birbeck granules (272,273,278).

Sinus Histiocytosis with Massive Lymphadenopathy (Rosai-Dorfman Disease)

Another clonal histiocytic neoplasm previous thought to be reactive but now recognized as a true neoplasm commonly associated with mutations in the MAPK pathway is *sinus histiocytosis with massive lymphadenopathy (Rosai-Dorfman disease)*. Rare cases have been reported in the thyroid gland (278–280). Thyroid involvement may represent an extension from adjacent lymph nodes. In one reported case, the clinical picture was similar to that of subacute thyroiditis (280).

The microscopic appearance of the thyroid gland is similar to that in other extranodal sites, with numerous histiocytes containing large vesicular nuclei and abundant clear cytoplasm. Some histiocytes show emperipolesis.

Plasma Cell Granuloma

Plasma cell granuloma (PCG) is a pseudotumor-like condition characterized by a polyclonal proliferation of plasma cells. Historically, PCG was considered within the spectrum of inflammatory myofibroblastic tumor (IMT) and some controversy still exists as to whether it truly represents a distinct entity or should be recognized as a morphologic variant of IMT (281). The thyroid gland is rarely affected by PCG (281).

Histologically, PCG is characterized by sheets and clusters of infiltrating mature plasma cells embedded in a fibroblastic stroma. Some of the plasma cells contain Russell bodies (aggregates of immunoglobulin). Lymphocytes, histiocytes, and rare neutrophils and eosinophils may be seen (282,283). Immunohistochemistry and in situ hybridization show that the plasma cells are polyclonal and express both kappa and lambda light chains.

PCG is not associated with *ALK* gene rearrangements, an important distinction since clinically and morphologically it may mimic IMT. Further studies are required to distinguish PCG from other non-neoplastic plasma cell-rich lesions of the thyroid gland such as Riedel thyroiditis and IgG4-related disease.

Scleroderma

Scleroderma (progressive systemic sclerosis) is a connective tissue disorder characterized by sclerosis of the subcutaneous tissue and multiple visceral organs and the presence of autoantibodies to nuclear and other tissue components. The cardinal manifestation of scleroderma of the thyroid gland is diffuse goiter (284). Hypothyroidism and serum antithyroid peroxidase antibodies are common (285).

Histologic examination of the thyroid gland usually shows fibrosis with a chronic lymphocytic infiltrate, even among patients who are euthyroid (286). The fibrosis is probably due to scleroderma, although it could be a result of coexistent thyroid autoimmune disease since many patients also have serologic evidence of autoimmune thyroiditis. Rare cases of scleroderma develop in patients with Graves disease (287).

METABOLIC DISEASES

Metabolic diseases involving the thyroid gland are rare (288–290).

Glycogenosis

A number of inborn errors of glucose and glycogen metabolism result from mutations in genes for virtually all of the proteins involved in glycogen synthesis, degradation, and regulation. Those disorders that result in abnormal storage of glycogen are known as *glycogen storage diseases*. The liver and skeletal muscle are most affected by glycogen storage diseases. Involvement of the thyroid gland has been reported in a few cases (289).

Pompe disease (glycogen storage disease II) is caused by acid alpha-glucosidase deficiency and results in accumulation of glycogen within the lysosome in all tissues (291). The disease is associated with muscle weakness, cardiomegaly, respiratory failure, and death. One study reported accumulation of glycogen in the thyroid gland of patients with Pompe disease (289).

Cystinosis

Cystinosis is a lysosomal storage disease characterized by an accumulation of cystine in different organs and tissues (289), leading potentially to severe organ dysfunction (292–297). The disease is inherited in an autosomal recessive fashion and is caused by mutations in the *CTNS* gene on chromosome 17p13 (298).

Hypothyroidism usually appears between 8 and 12 years of age and develops in approximately half of untreated patients (299). It is detected by the demonstration of high plasma TSH levels, which occur before the thyroxine concentration begins to fall (300). Symptoms of hypothyroidism are uncommon, but this problem may contribute to growth impairment.

The presence of pituitary thyrotroph hyperplasia indicates that the thyroid condition is primary, with secondary hyperplasia of pituitary cells (288). Microscopically, thc gland shows follicular atrophy with fibrosis and decreased amounts of colloid. There are focal areas of papillary hyperplasia with dilated follicles and focal acute inflammation (288). Frozen section examination of the thyroid tissue shows cystine crystals, but these are usually not seen in fixed tissue sections.

Lipidoses

Lipid storage diseases, or the *lipidoses*, are a group of inherited metabolic disorders in which large amounts of lipid accumulate in neurons, autonomic ganglia, muscle, and other cells. Thyroid gland involvement has been reported in some cases (301–303).

Follicular epithelial cells have yellow, autofluorescent pigmented granules which represent lipofuscin. Ultrastructural examination of the thyroid gland shows the osmophilic characteristic deposits in the cytoplasm of follicular cells as is seen in the brain, muscle, and other tissues. Some patients with Batten-Spielmeyer-Vogt disease have hypothyroidism and there is decreased or absent thyroid peroxidase activity (301).

Iron Pigment Accumulation

Hereditary hemochromatosis and *hemosiderosis* due to iron overload from repeated red blood cell transfusions result in excessive iron accumulation in multiple tissues, including the pituitary and thyroid glands, causing parenchymal damage and fibrosis (304–307). Thyroid dysfunction occurs slowly as iron accumulates in the thyroid or pituitary gland. Pituitary dysfunction (due to iron deposition in the pituitary thyrotrophs), with resultant central hypothyroidism, is more common than primary hypothyroidism caused by iron deposition in the thyroid gland (308). Thyroid infiltration is not typically associated with thyroid enlargement (306,309). Tissue deposition of iron is confirmed by the characteristic low intensity signals on T1- and T2-weighted magnetic resonance imaging (MRI) (307).

Macroscopically, the gland may appear dark brown to rust colored due to significant iron accumulation. Golden brown granular pigments are seen in follicular epithelial cells, causing lymphocytic inflammation and follicular cell destruction (304). Occasionally, increased fibrosis is present.

Minocycline-Associated Changes

The accumulation of a dark brown to black pigment in the thyroid gland occurs secondarily to the administration of minocycline and related tetracyclines. The striking changes in the thyroid gland that occur with this condition have led to the designation "black thyroid." This was first

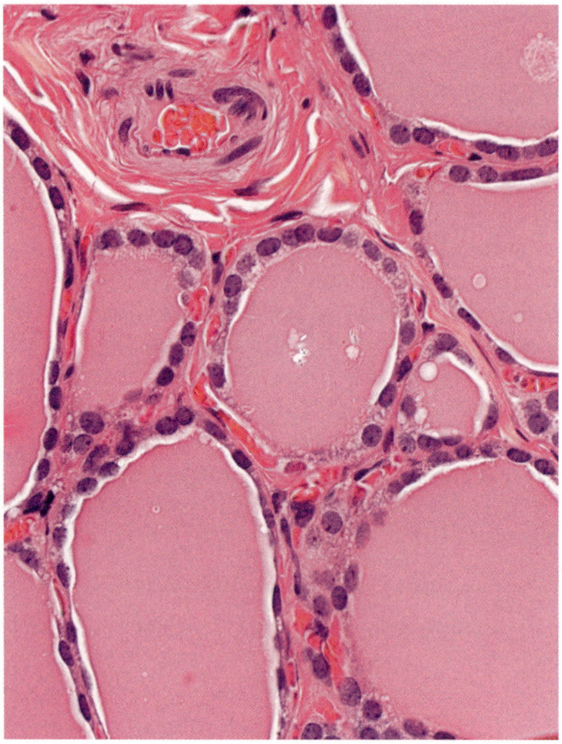

Figure 3-12

CALCIUM OXALATE IN COLLOID

Refractile calcium oxalate crystals are readily found in the colloid of thyroid tissues.

observed in animals given minocycline and has been well documented in humans (310–316).

Patients have been prescribed long-term minocycline for a variety of conditions including acne, pyelonephritis, and pulmonary infections. The minocycline-associated pigmentation is thought to be lipofuscin as well as other associated unknown compounds (313). Ultrastructural studies have shown that there is an accumulation of pigments in the lysosomes resulting from a lipid-drug complex. The impaired lysosomal function leads to pigment accumulation (310,314).

Patients are typically asymptomatic with normal thyroid function. The changes are usually noted incidentally in the pathologic evaluation after thyroidectomy for other unrelated thyroid pathology.

Gross examination shows a diffusely black thyroid gland; in contrast, benign and malignant thyroid neoplasms are often more pale than the background normal thyroid gland (317). Histologically, the follicular epithelial cells contain amorphous aggregates of brown-black pigment in the apical portion of the cytoplasm, which may be present in the colloid as well (318–320). Follicular cell hyperplasia and hypertrophy have been reported in experimental animals given minocycline, but not in humans. Electron microscopy shows the subcellular localization of electron-dense pigment with lipid within the thyrocyte lysosomes (317,318).

Minocycline-associated thyroid pigmentation is an incidental finding of no clinical significance. There is no reported increased risk of neoplastic development or thyroid dysfunction.

Crystal Deposits in the Thyroid Gland

Anisotropic *crystals of calcium oxalate* may be present within the colloid in normal adult thyroid glands, particularly in older or less active glands (fig. 3-12). Crystals may be seen in ordinary light, but are more easily identified under polarized light (315,321–324). Their shape varies from rhomboid to irregular plaques and there is considerable variation in their size (324).

In one autopsy study, calcium oxalate crystals were found in 88 percent of nodular goiters, 60 percent of follicular adenomas, 33 percent of follicular carcinomas, and only 5 percent of papillary carcinomas (325). A heavy deposit of these crystals was seen almost exclusively in benign conditions. The overall prevalence is 85 percent of all thyroid glands examined (325).

The number of crystals appears to increase with age; this, together with the observation that they have been found more frequently in colloid with low positivity for TGB, has prompted the suggestion that they result from variations in colloid and calcium concentration in the gland secondary to a low functional state of the thyroid gland (323–325).

In one study, the number of crystals was markedly elevated in glands with subacute thyroiditis, where they were found in the giant cells, in remnants of colloid, and in the thyroid stroma (326). In the same study, crystals were identified only rarely in thyroid glands with chronic thyroiditis or glandular hyperplasia. Oxalate crystals are also seen in the thyroid in large numbers in patients undergoing dialysis for renal failure (321). Scully et al. (327) have used the presence of calcium oxalate crystals in struma ovarii as one of the features for identifying thyroid tissue. Similarly, their identification at the time of frozen section within a follicular structure can be useful in distinguishing thyroid from parathyroid tissue (328).

Teflon (Polytef) in the Thyroid Gland

Teflon (polytetrafluoroethylene) is a crystal-like material. Between the 1960s and 1990s, intracordal injection of teflon was used to treat dysphonia in patients with unilateral vocal cord paralysis due to recurrent nerve injury. Subsequently, accumulation of teflon was reported the thyroid gland and cervical tissues (329–331). Teflon produces a granulomatous foreign body-type reaction with fibrosis that results in improved phonation in these patients. Rarely, overinjection and misplaced injection causes direct migration from the vocal cord through the cricothyroid membrane, resulting in foreign-body type granulomas containing birefringent particles within the thyroid gland (332).

INFECTIOUS AND GRANULOMATOUS DISEASES

Acute Thyroiditis

Definition. *Acute thyroiditis* is characterized by acute (neutrophilic) inflammation of the thyroid gland, almost always caused by infection.

General Features. Acute thyroiditis is rare and is almost always a result of infection, although it may also be encountered shortly following radiation exposure (333–335). Acute (infectious) thyroiditis is caused by many different organisms, which reach the thyroid gland either via hematogenous spread, usually in an immunocompromised patient, or via direct spread due to trauma or fistula (336). Spread of infection from a fistula arising in the pyriform sinus adjacent to the larynx is probably the most common cause of suppurative thyroiditis, and it usually occurs in children (333,337,338).

Causative organisms include Gram-positive bacteria (*Staphylococcus, Streptococcus*), and Gram-negative bacteria especially in the setting of neck trauma (339,340). Fungal organisms including *Aspergillus, Cryptococcus, Candida,* and *Coccidioides* have also been reported in acute thyroiditis (333,337,338,341–344). Viral infections such as rubella (345) and cytomegalovirus (336) have been documented. The latter is usually seen in the setting of immunosuppression, such as in patients with human immunodeficiency virus (HIV) or acquired immunodeficiency syndrome (AIDS). In this setting, the infectious thyroiditis tends to be more chronic (347–349).

Clinical Features. Acute infectious thyroiditis is characterized by the sudden onset of neck pain and tenderness, which is usually unilateral and accompanied by fever, chills, and other symptoms and signs of infection. Some patients have dysphagia or hoarseness. On physical examination, most patients have a unilateral neck mass, which may be fluctuant. Thyroid function tests are usually normal, but thyrotoxicosis may be present (335). Thyroid ultrasound can differentiate between subacute thyroiditis (diffuse heterogeneity and low intensity vascular flow) and infectious thyroiditis (abscess).

Patients with more chronic thyroid infections often have bilateral disease. Thyroid pain and tenderness are less prominent than with acute infections, and some patients have hypothyroidism.

Gross Findings. Grossly, the gland often appears normal, but may be slightly enlarged and diffusely softened, with areas of purulence.

Microscopic Findings. Infiltration with neutrophils, areas of microabscesses, necrotic foci, and vasculitis may be present. Careful search of the H&E-stained sections may show clusters of bacteria or fungal organisms. Special stains are helpful for identifying the etiologic organism. Viral infections are identified by intranuclear or cytoplasmic inclusions, or by specific cytopathic changes. Immunohistochemical or in situ hybridization studies are important when a viral etiology is suspected. In immunosuppressed individuals, multiple infectious organisms may be present, often with a paucity of inflammatory cells.

Differential Diagnosis. The differential diagnosis includes other conditions with neutrophilic infiltrates such as subacute thyroiditis (discussed in next section) and ischemic necrosis.

Treatment and Prognosis. After the diagnosis is made by serologic tests, histologic examination, special stains and culture, drainage of the thyroid gland and treatment with the appropriate agents, such as antibiotics and antifungal or antiviral agents, is indicated. Most patients with bacterial thyroiditis recover from the infection and the thyroid function stabilizes unless there is overwhelming sepsis or other underlying conditions. Surgical incision and drainage may be required in patients who do not respond to percutaneous drainage and systemic antibiotic therapy. With appropriate management, mortality from bacterial thyroiditis is rare (339,350,351).

Fungal thyroiditis is very uncommon. It is most frequently seen as an autopsy finding in immunosuppressed patients with widespread systemic fungal infection.

Granulomatous (Subacute) Thyroiditis

Definition. *Granulomatous thyroiditis*, also referred to as *subacute thyroiditis, nonsuppurative thyroiditis,* or *de Quervain thyroiditis,* is a rare entity that is thought to be caused by a systemic viral infection or postviral inflammatory process.

General Features. Granulomatous thyroiditis affects women more often than men (3 to 5:1) (352–355). There is an association of subacute thyroiditis with certain viral epidemics including mumps, Coxsackie adenovirus, measles, and influenza (356–358). Antibodies to different viruses have been detected in some patients. There is a strong association with human leukocyte antigen (HLA) Bw35 (359–361). It is hypothesized that the disorder results from a subclinical viral infection that provides an antigen, either of viral origin or resulting from virus-induced host tissue damage, that uniquely binds to HLA-B35 molecules on macrophages. The resulting antigen-HLA-B35 complex activates cytotoxic T lymphocytes that then damage thyroid follicular cells because the cells have partial structural similarity to the infection-related antigen. Unlike autoimmune thyroid disease, however, the immune reaction is not self-perpetuating, so the process is self-limiting.

Clinical Features. The most common presenting symptom is neck pain, reported in 96 percent of patients in one study (362). The onset may be sudden or gradual, and may be preceded by an upper respiratory tract infection. The pain may be limited to the region of the thyroid gland or radiate to the upper neck, jaw, throat, upper chest, and ears. Fever, fatigue, malaise, anorexia, and myalgia are common (353,363–366).

The thyroid gland is typically slightly or moderately, diffusely or asymmetrically enlarged, and is nearly always tender. Approximately half of patients have symptoms and signs of hyperthyroidism, but the neck pain and tenderness usually dominate the illness, and the diagnosis should not routinely be made in their absence. It is important to recognize that subacute thyroiditis may involve the thyroid gland in an asymmetric manner, and may therefore give the clinical impression of a solitary painful nodule.

Clinically, three phases of the disease are recognized: a hyperthyroid, a hypothyroid, and a recovery phase (356,367). In the hyperthyroid phase there is destruction of the follicles by the inflammatory process, usually subsiding within 2 to 8 weeks. Iodine uptake is usually low in this phase. The phase of hypothyroidism occurs after a significant portion of the gland is destroyed and there is decreased capacity to synthesize thyroid hormone. Antithyroid antibodies may be detected in the serum during this phase (364,365). After several weeks to months, patients usually recover and become euthyroid.

A few patients with subacute thyroiditis present without pain, and there may be confusion with silent or painless thyroiditis in such cases (364). Some patients with painless subacute thyroiditis may present with thyromegaly associated with fever and weight loss. Laboratory studies usually show an elevated erythrocyte sedimentation rate and low radioiodine uptake (361,368).

Gross Findings. The thyroid gland is usually asymmetrically enlarged and firm. Cut sections show a firm, tan gland with irregular white-tan nodules, ranging from a few millimeters to a few centimeters in size.

Cytologic Findings. Cytologic features include follicular cells admixed with neutrophils, lymphocytes, histiocytes, and multinucleated giant cells containing foamy cytoplasm and up to 50 nuclei per cell. The follicular cells may be arranged in sheets, nests, or as single units. Colloid may be present within acini or as an isolated finding.

Microscopic Findings. In the active phase, the thyroid gland is infiltrated by multinucleated giant cells, lymphocytes, plasma cells, histiocytes, areas of acute inflammation, and a variable degree of fibrosis. During the early or

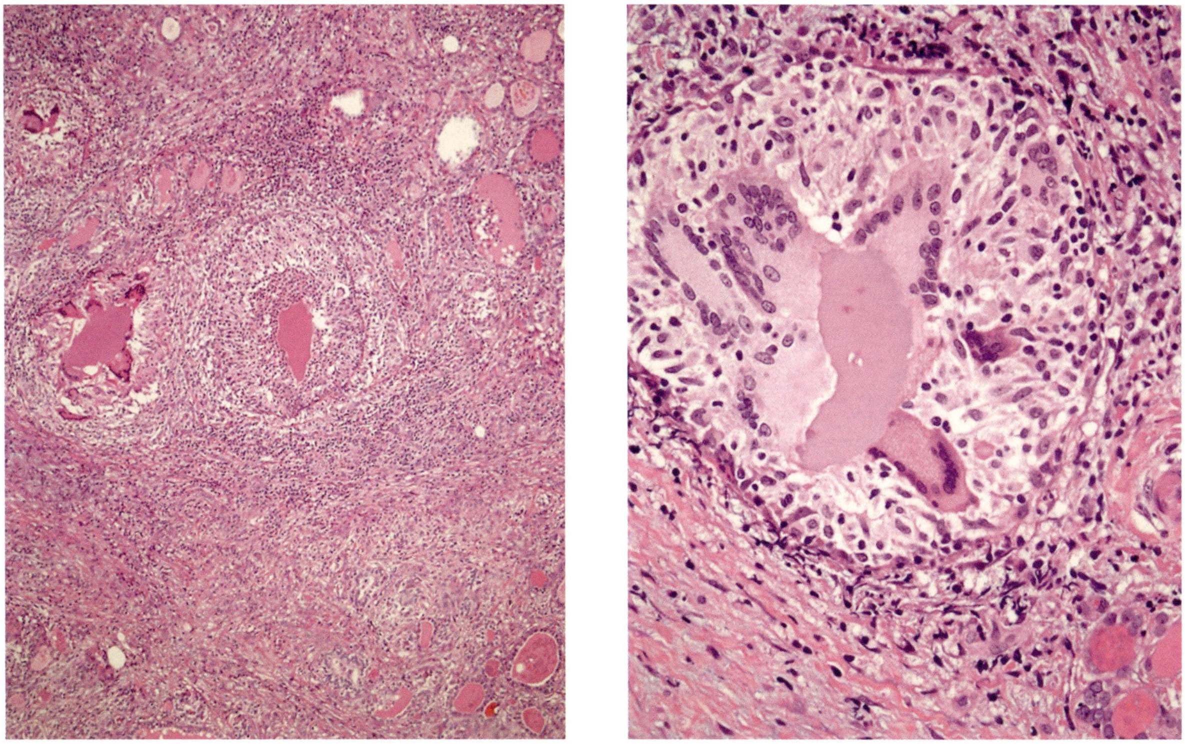

Figure 3-13

GRANULOMATOUS THYROIDITIS

Left, right: The granulomas give the appearance of actively engulfing the colloid within nodules, an essentially pathognomonic feature of granulomatous thyroiditis.

hyperthyroid phase, there is disruption of the follicular cells and depletion of colloid, which correlates with clinical hyperthyroidism from the contribution of the escaping colloid and thyroid hormone. There may be acute inflammation with microabscess formation. A few multinucleated giant cells may be present at this stage.

In the hypothyroid phase, the follicular epithelium may disappear. There is a mixed inflammatory infiltrate of histiocytes, multinucleated giant cells, lymphocytes, and plasma cells. A rim of histiocytes and giant cells replaces the follicular epithelium, imparting a granulomatous appearance. Characteristically, the multinucleated giant cells engulf colloid, a feature which, if widespread, is virtually pathognomonic for subacute thyroiditis (fig. 3-13).

During the recovery phase, there is regeneration of follicles and a central fibrotic reaction. Failure to appreciate that fibrosis often dominates in the recovery phase may lead to misdiagnosis unless a careful search is undertaken for the characteristic granulomas engulfing colloid (fig. 3-14) (369,370).

Treatment and Prognosis. Treatment of patients with subacute thyroiditis should be directed at providing relief for thyroid pain and tenderness, and ameliorating symptoms of hyperthyroidism, if present. For pain relief, aspirin or a nonsteroidal anti-inflammatory drug (NSAID) is adequate in most cases. Prednisone can be used for patients with severe pain. Beta-blockers (such as propranolol or atenolol) may be used to treat the symptoms of thyrotoxicosis (371). Thyroid function tests should be monitored every 2 to 8 weeks to confirm resolution of hyperthyroidism, detection of hypothyroidism, and subsequent normalization of thyroid function.

Most patients experience a complete and spontaneous recovery and a return to normal thyroid function; nevertheless, some patients have permanent hypothyroidism. In a follow-up study of 160 patients with subacute thyroiditis seen at the Mayo Clinic, 15 percent eventually

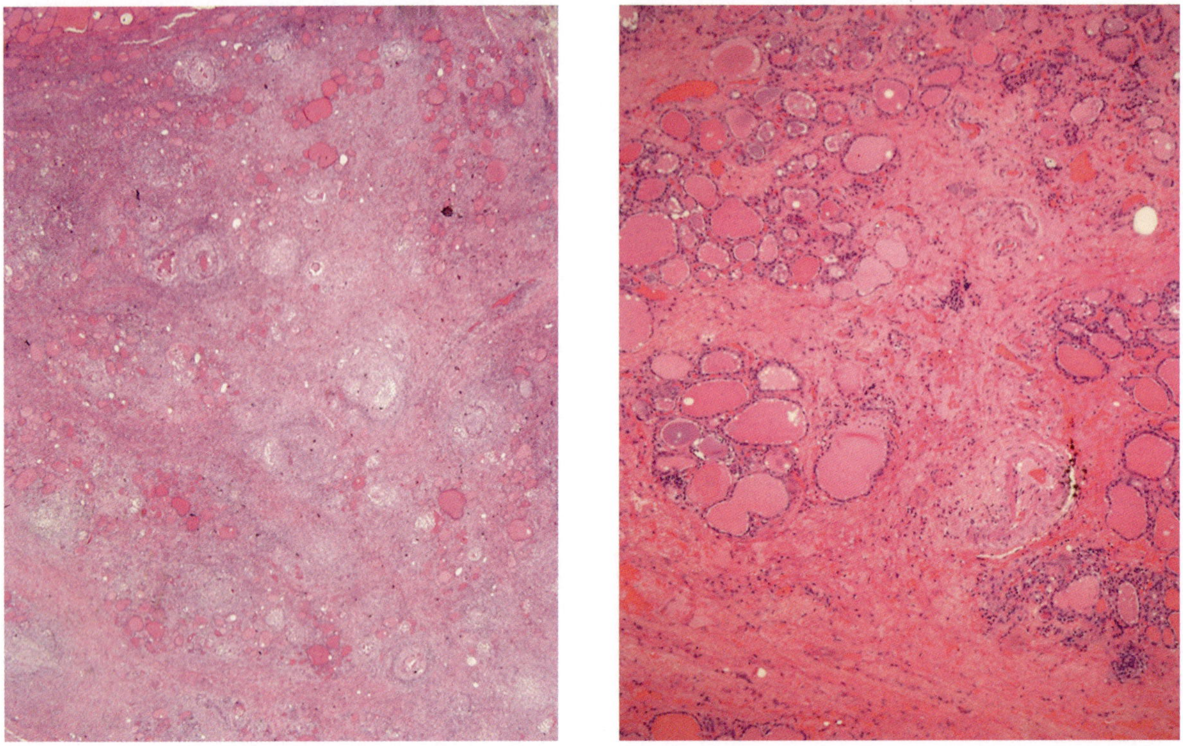

Figure 3-14

GRANULOMATOUS THYROIDITIS

Left: In florid active granulomatous thyroiditis, the granulomatous inflammation is widespread and not easily overlooked. Right: In the resolving ('burnt out') phase, the predominant low-power appearance is atrophy of the thyroid gland with patchy interstitial fibrosis which exaggerates the lobulated architecture of the thyroid. In these cases, a careful search of multiple sections may be required to identify the characteristic granulomas engulfing colloid.

developed permanent hypothyroidism requiring levothyroxine therapy (362).

Other Granulomatous Thyroiditides

Tuberculous Thyroiditis. Tuberculous thyroiditis, whether primary or secondary, is an extremely rare disease (372), even in countries with a high prevalence of tuberculosis. A number of factors are believed to confer relative immunity to the thyroid gland from tuberculosis, including the gland's well-developed capsule and high iodine content (373), the bactericidal properties of colloid (374), the increased physiologic activity of phagocytes in hyperthyroidism, and the possible antitubercular role of thyroid hormones (375).

The infection is associated with caseating granulomas as are seen at other sites of tuberculous infection (376,377). Tuberculosis in the thyroid gland is most commonly associated with disseminated disease. Immunosuppressed patients are more susceptible to the development of tuberculous thyroiditis.

Fungal Thyroiditis. Fungal infections of the thyroid gland are uncommon. *Aspergillus* is the most common cause of fungal thyroiditis (378). Immunocompromised patients, such as those with HIV, leukemia, autoimmune disease, and those on immunosuppressive treatment after organ transplant, are particularly at risk for fungal thyroiditis. Most patients with fungal thyroiditis have disseminated fungal infection and, in many instances, the diagnosis is made at autopsy in patients without clinical manifestations or laboratory evidence of thyroid dysfunction (378). Cases may be associated with an acute necrotizing reaction or very little reaction. Occasionally, granulomatous inflammation occurs with the fungal reaction (379).

Postoperative Necrotizing Granulomas. Postoperative necrotizing granulomas of the

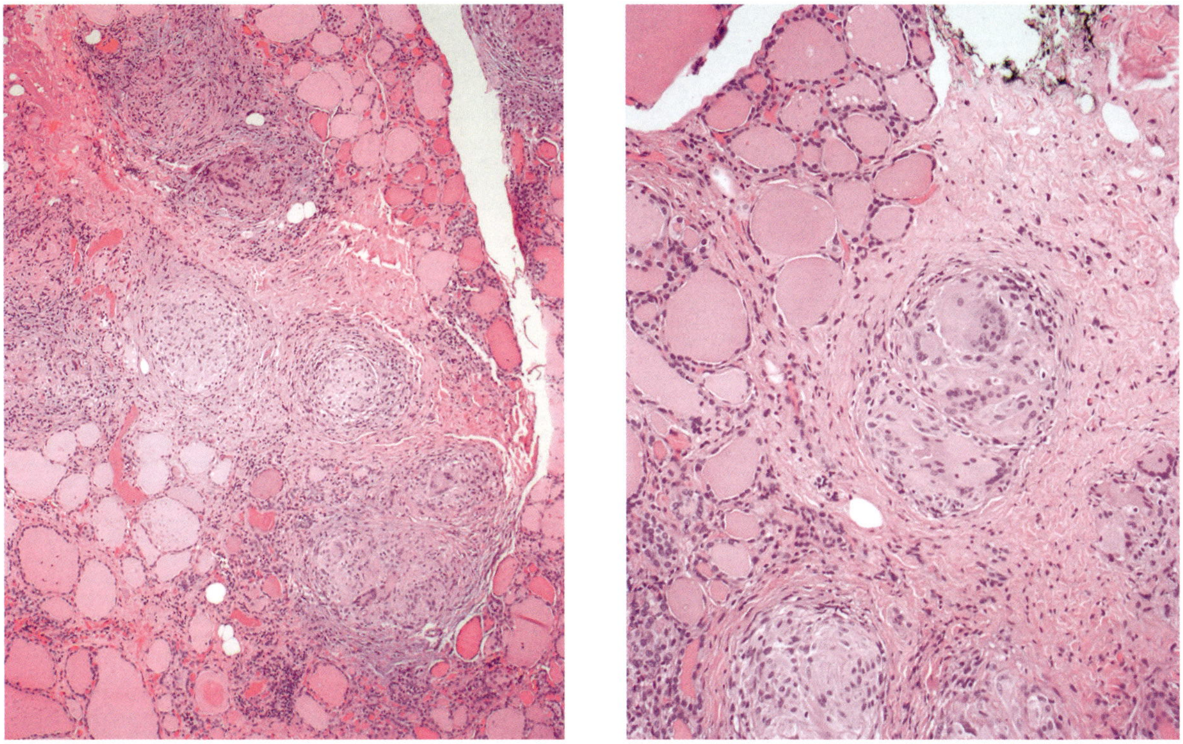

Figure 3-15

SARCOIDOSIS

Left, right: In sarcoidosis involving the thyroid gland, the granulomas are characteristically located in the collagenous connective tissue of the interstitium. This contrasts strongly with the granulomas of granulomatous subacute (de Quervains) thyroiditis, which are centered on the follicles and actively engulf colloid.

thyroid gland have been reported in isolated cases (380). These are similar to the postoperative granulomas seen in other organs, such as the bladder and prostate gland. They are thought to result from a response to tissue necrosis caused by diathermy, possibly related to altered collagen (381). The clinical history and absence of organisms by special stains or culture help support the diagnosis.

Granulomatous Vasculitis. Granulomatous vasculitides, such as granulomatosis with polyangiitis (formerly known as Wegener granulomatosis), rarely secondarily involve the thyroid gland (382). Hypersensitivity vasculitis can occur as a reaction to phenytoin (383).

Sarcoidosis. Sarcoidosis is a multisystem disorder of unknown etiology that can affect any organ system, including the thyroid gland (371,384–387). In postmortem studies, the thyroid gland is affected in approximately 4 to 5 percent of patients with sarcoidosis (387). Sarcoidosis can cause diffuse goiter or rarely, solitary or multiple thyroid nodules (386,388,389) that may resemble a malignancy, especially when associated with cervical lymphadenopathy (390).

The histologic features are similar to those of sarcoidosis elsewhere and consist of noncaseating granulomas with epithelioid histiocytes, multinucleated giant cells, and lymphocytic infiltration. In contrast to subacute and palpation thyroiditis, the granulomas in sarcoidosis are usually located in the interstitium rather than being centered on colloid (fig. 3-15). If there is no history of sarcoidosis, special stains for fungi and acid-fast organisms should be performed to exclude infection.

Sarcoidosis typically presents as an incidental finding. It is common for sarcoidal granulomas to be identified in lymph nodes, particularly level 6 or level 7 nodes removed during treatment for papillary thyroid carcinoma. If well-formed epithelioid granulomas are identified in these

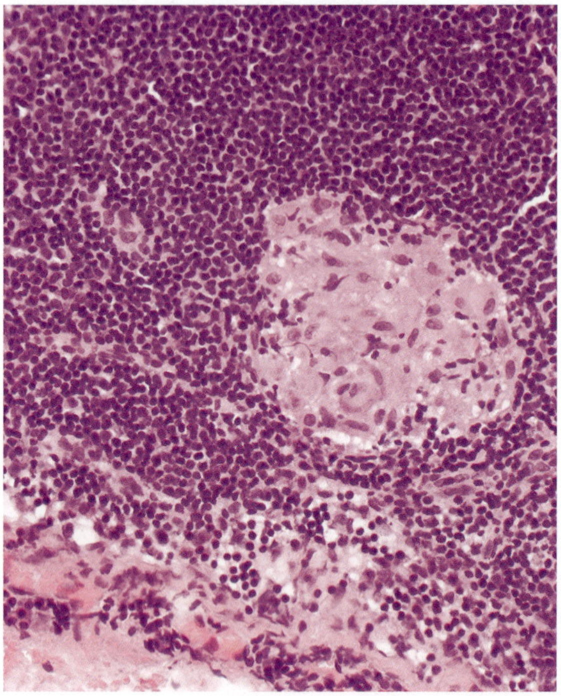

Figure 3-16

SARCOIDOSIS

If well-formed epithelioid granulomas are identified as an incidental finding in perithyroidal or level 6/7 lymph nodes, the possibility of subclinical sarcoidosis should be suggested.

lymph nodes (fig. 3-16), even in the absence of a clinical history, the possibility of subclinical sarcoidosis should be suggested.

Palpation Thyroiditis. Palpation thyroiditis (multifocal granulomatous folliculitis) is an iatrogenic condition caused by vigorous palpation of the thyroid gland. It is found in 85 to 95 percent of surgically resected glands (391) and probably represents the thyroid's response to minor trauma. It is also caused by physical examination or trauma to the neck. Patients present with transient neck pain and tenderness and transient hyperthyroidism (392–396).

Occasionally, gross areas of hemorrhage are present, but there are usually no other macroscopic abnormalities. The histologic features of palpation thyroiditis include multiple isolated follicles or small groups of follicles that show partial or circumferential loss of epithelium and replacement by inflammatory cells, predominantly histiocytes (fig. 3-17).

Palpation thyroiditis may be confused with sarcoidosis or granulomatous inflammation. However, in sarcoidosis, the granulomas are usually interstitial. The unique appearance of palpation thyroiditis and the absence of necrosis allow distinction from granulomatous

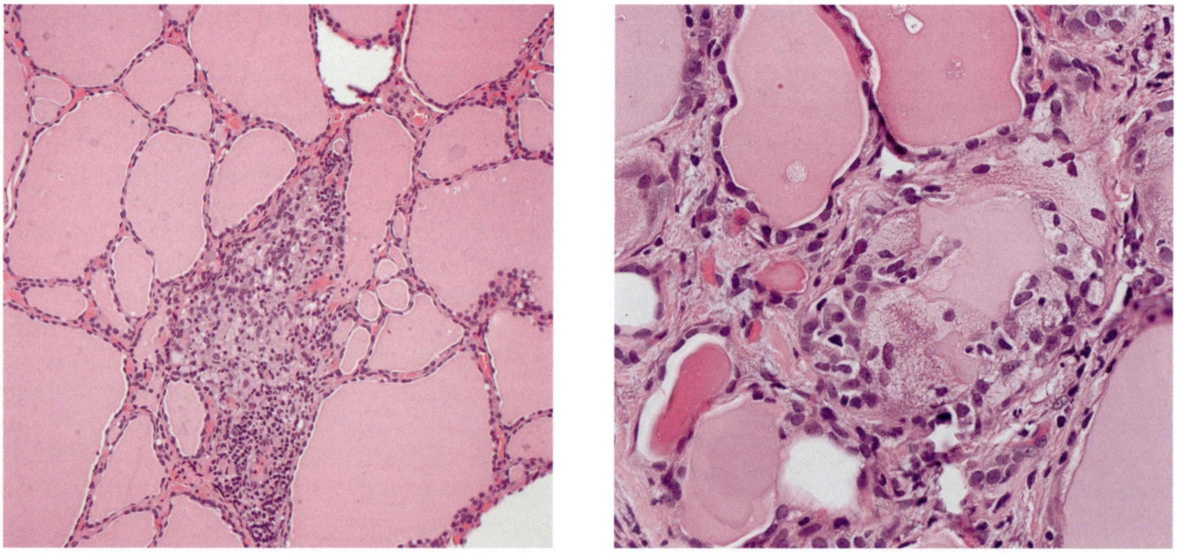

Figure 3-17

PALPATION THYROIDITIS

Left: Palpation thyroiditis is characterized by ill-defined and poorly cohesive groups of histiocytes within follicles rather than true well-formed epithelioid granulomas.

Right: At high power, the characteristically disorganized nature of the histiocytes in palpation thyroiditis is evident.

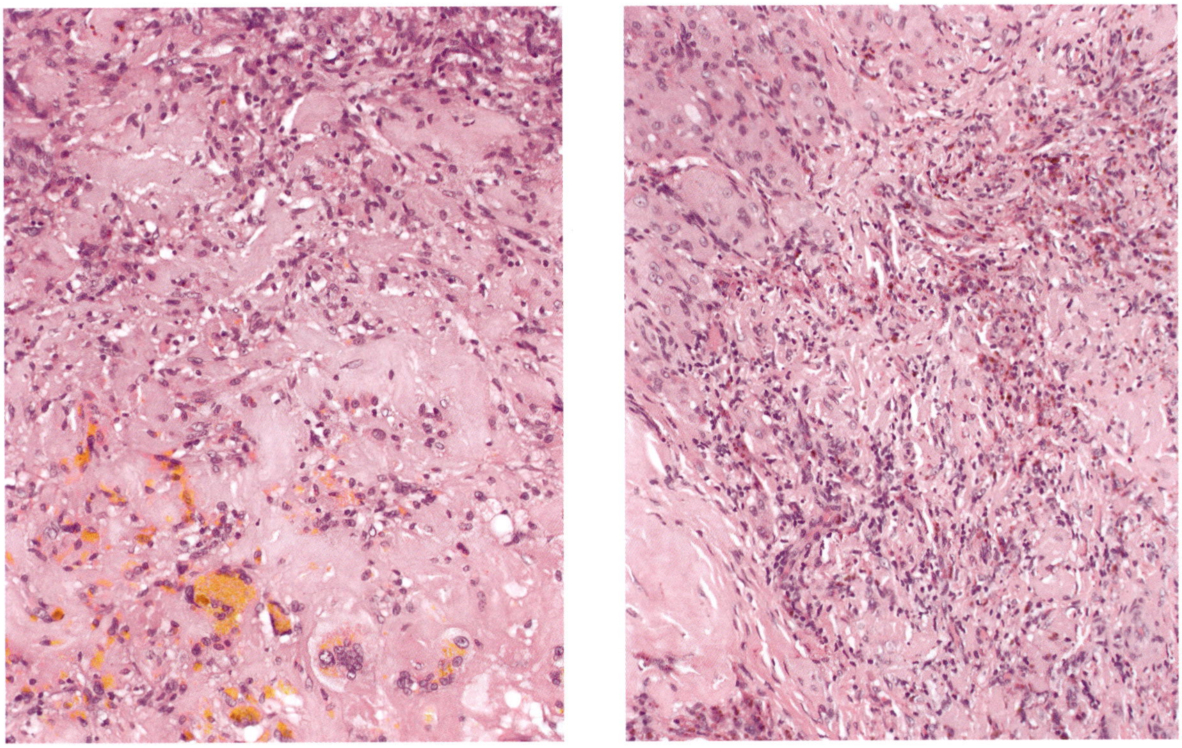

Figure 3-18

FINE-NEEDLE ASPIRATION BIOPSY CHANGES

Left, right: Fine-needle aspiration biopsy changes in a thyroid gland include fibrosis, chronic inflammation, and hemosiderin deposition. These changes may simulate some of the changes seen in the fibrous variant of Hashimoto thyroiditis or in thyroid carcinoma.

infectious diseases such as tuberculous or fungal thyroiditis (fig. 3-17). The acute inflammatory cells seen in subacute thyroiditis, along with the granulomas, allows distinction from palpation thyroiditis, which is not associated with acute inflammation. Palpation thyroiditis is a self-limited condition of no clinical significance, and no specific treatment is required.

Alterations Following Fine-Needle Aspiration Biopsy. Fine-needle aspiration (FNA) of the thyroid gland is a common diagnostic procedure. One of the consequences of this procedure is histologic alteration and distortion of the thyroid tissue, potentially altering the histologic appearance of the resected lesion (fig. 3-18). In some cases, benign lesions are mistakenly interpreted as malignant due to artefactual displacement of epithelium mimicking invasive growth (397–399). A history of previous FNA or histologic clues suggesting previous FNA, such as infarction, hemorrhage and hemosiderosis, inflammation, and early fibrosis, should lead to a cautious interpretation of unusual histologic features. In the modern era, small thyroid nodules, which are particularly prone to FNA-induced infarction, are being increasingly biopsied and may cause confusion. Review of preoperative cytology samples may permit a definitive diagnosis.

AUTOIMMUNE THYROID DISEASE

Autoimmune thyroid disease encompasses a broad spectrum of clinical and morphologic entities that share certain features suggesting their autoimmune etiology. The spectrum includes diffuse toxic goiter (Graves disease) associated with hyperthyroidism on the one hand and lymphocytic thyroiditis (Hashimoto disease) associated with hypothyroidism on the other. A number of other conditions, such as atrophic autoimmune thyroiditis (primary myxedema) and Riedel thyroiditis, are also included in this category.

Autoimmune thyroid disease is a polygenetic disease in which susceptibility genes and environmental factors act together to initiate both cellular and humoral immune responses against the thyroid gland (400,401). Genome-wide screening and linkage analyses have identified several chromosomal foci that are linked to autoimmune thyroid disease. Sequence variants in the following loci have been implicated: *HT-1* (chromosome 13q33) and *HT-2* (chromosome 12q22) for Hashimoto thyroiditis and *GD-1* (chromosome 14q31), *GD-2* (chromosome 20q11.2), and *GD-3* (chromosome Xq21) for Graves disease (402). A number of other genes have also been proposed as susceptibility or immunoregulatory genes. The exact mechanism of involvement of environmental factors in autoimmune thyroid disease is still not clearly understood; these factors include a combination of dietary iodine, medication, and infection (404).

Hashimoto Thyroiditis

Definition. *Hashimoto thyroiditis (chronic autoimmune thyroiditis)* is characterized clinically by gradual thyroid failure, with or without goiter formation, due to autoimmune-mediated destruction of the thyroid gland. Nearly all patients have elevated circulating thyroid antibodies, including thyroid peroxidase and thyroglobulin autoantibodies.

General Features. Hashimoto thyroiditis is primarily a disease of middle-aged women, with a sex ratio of approximately 7 to 1 (405,406). It is the most common cause of hypothyroidism in iodine-sufficient areas of the world and the most common cause of sporadic goiter in children (407). Nearly all patients with Hashimoto thyroiditis have high serum concentrations of antibodies to thyroglobulin (Tg) and thyroid peroxidase (TPO). Hashimoto thyroiditis may coexist with other autoimmune diseases including Sjögren syndrome, pernicious anemia, chronic active hepatitis, adrenal insufficiency, diabetes mellitus, and Graves disease. An association of HLA-DR3 and HLA-DR5 with the atrophic and goitrous forms of Hashimoto disease has been reported (408).

There is now good evidence for a genetic susceptibility to Hashimoto thyroiditis and for autoimmune thyroid disease in general (409). Hashimoto thyroiditis clusters in families, sometimes alone and sometimes in combination with Graves disease (410). Up to 5 percent of first-degree relatives of patients with the condition have antithyroid antibodies, which are inherited as a dominant trait (408,411,412). There is a strong association with certain HLA alleles such as DR3 and an association with the presence of arginine at position 74 (p.Arg74) in the DR3 amino acid-binding pocket (413). Interestingly, patients from Japan with HLA-DR2 and HLA-DQ1 have a protective effect against autoimmune thyroid disease (411,414). There is a high prevalence of autoimmune thyroid disease in patients with Down syndrome, familial Alzheimer disease, and Turner syndrome (415–418).

A number of precipitating factors for Hashimoto thyroiditis have been postulated, including infection, stress, sex steroids, pregnancy, iodine intake, and radiation exposure. Mild iodine deficiency is associated with a lower prevalence of Hashimoto thyroiditis and hypothyroidism, while excessive intake is associated with a higher prevalence (406,419). Similarly, iodine-containing drugs, such as amiodarone, can also precipitate autoimmune thyroiditis (420). Other drugs associated with elevated levels of antithyroid antibodies and hypothyroidism are lithium (421) and interferon-alpha (411,422).

During pregnancy, there is a marked increase in CD4/CD25-positive regulatory T cells, which lead to diminished function of both T and B cells and ensure maternal tolerance of the fetus (423). The rebound from this immunosuppression in the postnatal period is thought to contribute to the development of postpartum Hashimoto thyroiditis. In addition, pregnancy-associated immune changes are associated with a shift to Th2 T cells and a shift in cytokine profiles (424). Approximately 20 percent of patients with postpartum thyroiditis go on to develop classic Hashimoto thyroiditis in later years (425). Although no infection is known to cause or even to be closely associated with Hashimoto thyroiditis in humans (404), certain viral infections have been shown to induce thyroiditis in experimental animal studies (426).

Chronic autoimmune thyroiditis is a common condition. At autopsy, about 45 percent of women and 20 percent of men in the United States and the United Kingdom have focal thyroiditis (1 to 10 foci/cm^2) and 5 to 15 percent of women and 1 to 5 percent of men have severe thyroiditis (more than 40 foci/cm^2) (427). The prevalence of elevated titers of autoantibodies for thyroid disease increases with age and may reach as high as 33 percent in women after age 70 (428,429).

Pathogenesis. The pathogenesis of autoimmune thyroiditis involves activation of the CD4 or helper T cells targeted at thyroid antigens (430). The current major hypotheses for all autoimmune diseases include those known as "molecular mimicry" and "bystander activation." According to the molecular mimicry proposal, Hashimoto disease is caused by the immune response to a foreign antigen such as a virus that is structurally similar to an endogenous substance (431–436). Serologic evidence of recent viral and bacterial infections has been reported (404,436), but the link between the infection and initiation of the autoimmune disease has not been conclusively documented. According to the bystander activation hypothesis, the arrival of a thyroid cell virus or activated nonspecific lymphocytes within the thyroid gland causes the local release of cytokines, which in themselves may activate resident local thyroid-specific T cells (437,438).

An alternative hypothesis is that HLA class II antigens are present on thyroid follicular cells in patients with Hashimoto thyroiditis, but not normal subjects. Expression of these molecules on thyroid follicular cells can be induced by interferon-gamma and other products of T cells, when the T cells are activated (e.g., by a viral infection) (439). Thyroid cells expressing major histocompatibility complex (MHC) class II molecules would be able to present antigens, either foreign or self, to T cells, thereby activating them (431,440). This view is supported by the observation that thyroid cells in patients with autoimmune thyroiditis, in contrast to normal thyroid cells, express HLA-DR, HLA-DPd, and HLA-DQ proteins which are required for CD4 antigen presentation (410,441,442). Adhesion molecules such as intercellular adhesion molecule (ICAM) have also been identified on the thyroid epithelial cells from patients with Hashimoto thyroiditis (443).

Once the helper T cells are activated, they stimulate B cells, which are recruited into the thyroid gland and produce antibodies (440). The principal antigen targets are Tg, TPO, and TSH receptor. Two types of TSH receptor antibodies have been recognized. TSH receptor antibodies of the stimulating variety were first identified in the serum of patients with Graves disease, but have since been detected in the serum of some patients with Hashimoto thyroiditis (444,445). Despite the presence of these stimulating TSH receptor antibodies, they are unable to stimulate much thyroid hormone synthesis because of the thyroid cell destruction (446). Blocking-type TSH receptor antibodies, in addition to autoantibodies to TPO and Tg, are sometimes found in patients with overt and subclinical autoimmune thyroiditis, most commonly of the atrophic variety. These blocking-type antibodies prevent TSH from binding to its receptors in thyroid epithelial cells.

Ultimately, thyroid cell death is the central pathologic phenomenon in Hashimoto disease. Normal thyroid epithelial cells express a variety of death receptors, including Fas. Activation of the Fas ligand-Fas signaling system by cytokine stimulation from antigen-presenting cells and Th1 cells (e.g., interleukin-1) could contribute to the follicular cell destruction characteristic of Hashimoto thyroiditis (447).

Clinical and Radiologic Features. Patients with autoimmune thyroiditis can present with a goiter, hypothyroidism, or both. The mean age at diagnosis is 59 years for women and 58 years for men (415,428). Children rarely develop autoimmune thyroiditis before age 5, however, autoimmune thyroiditis may account for about 40 percent of goiters in adolescents (448). Patients often have a goiter, but compression of the trachea or recurrent laryngeal nerves is uncommon. Patients often complain of a feeling of tightness in the neck. Rare patients have thyroid pain and tenderness (449,450), particularly if there is rapid thyroid swelling, and such patients may even require surgical relief. Typically, the thyroid is symmetrically enlarged and has a firm bosselated surface. The thyroid gland may be asymmetric on physical examination and may be mistaken for a solitary nodule or multinodular goiter, especially in glands with fibrosis.

While hypothyroidism is the characteristic functional abnormality, the inflammatory process early in the course may involve enough apoptosis to cause thyroid follicular disruption and thyroid hormone release, resulting in transient hyperthyroidism, sometimes referred to as "hashitoxicosis" (451). Rare patients cycle between hypothyroidism and hyperthyroidism, perhaps secondary to

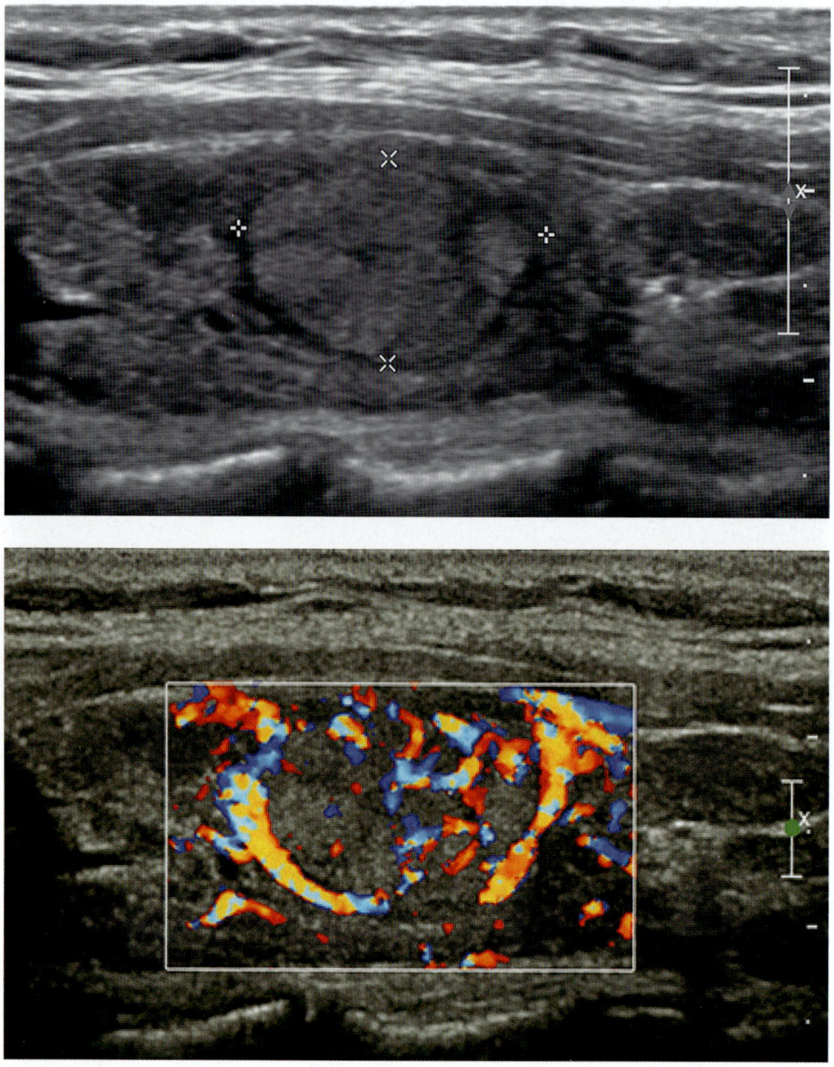

Figure 3-19

HASHIMOTO THYROIDITIS

Top: Longitudinal ultrasound image of the thyroid gland shows a diffusely enlarged gland with a heterogeneous nodular appearance.

Bottom: The color Doppler image shows diffusely increased blood flow to the gland.

alternating production of TSH receptor blocking and stimulating antibodies (452,453).

Thyroid imaging is usually not necessary in patients with autoimmune thyroiditis. However, if it is done for an unusual case, the radioiodine scan can be very misleading because the uptake pattern may be similar to that of Graves disease and multinodular goiter: a hot or cold nodule (454). The uptake of radioiodine is usually normal or elevated in patients with autoimmune thyroiditis with goiter, even when the patient is hypothyroid; in contrast, patients with subacute thyroiditis or silent thyroiditis usually have a low radioiodine uptake (455). Ultrasonographic studies show an enlarged gland with a hypoechogenic pattern in the majority of patients (fig. 3-19) (456,457).

Laboratory Findings. Laboratory tests show thyroid-specific autoantibodies in the serum. Antithyroglobulin antibodies are present in about 60 percent of patients with diffuse goiter, hypothyroidism, or both, and antibodies to TPO are present in 95 percent (458). The titers tend to be higher in patients with the atrophic form of autoimmune thyroiditis. Antibodies to the TSH receptor are found in a small number of Hashimoto patients (approximately 15 percent) and may be of the blocking or stimulating variety (446,459).

Gross Findings. The thyroid gland is firm and symmetrically enlarged, weighing between 25 and 250 g (460). The surface of the gland is bosselated or irregular due to accentuation of the normal lobulation by interlobular fibrosis. The gland may

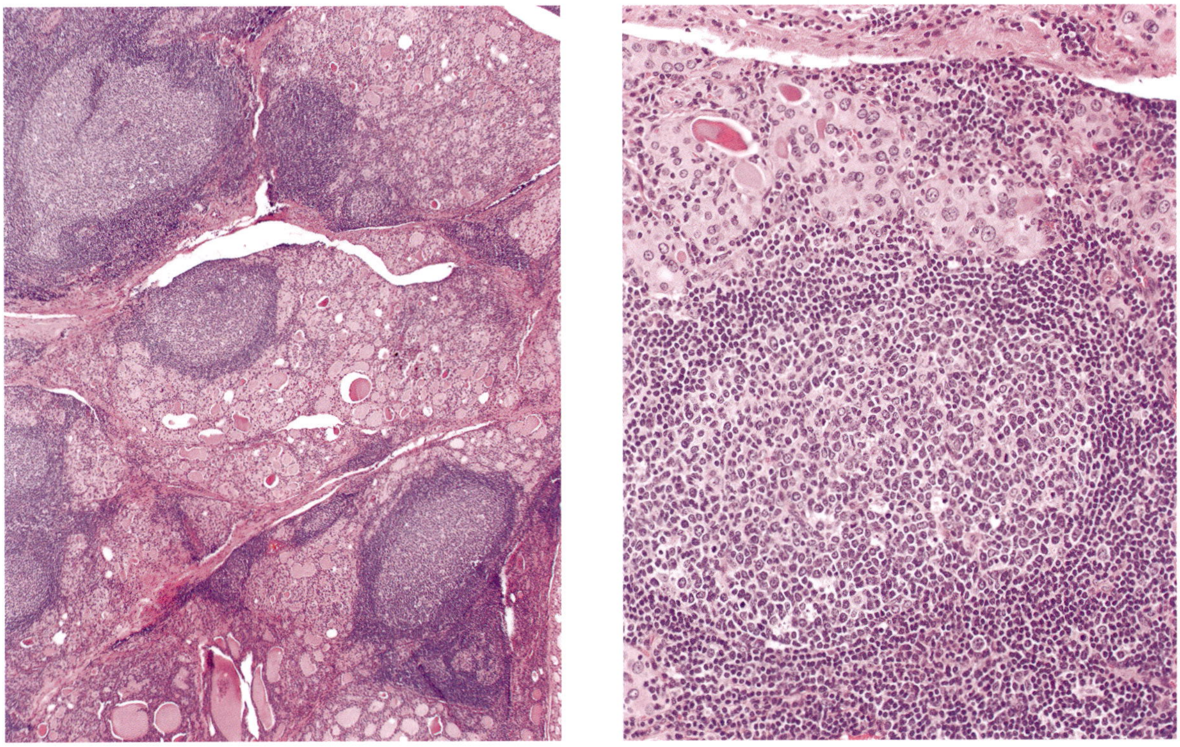

Figure 3-20

HASHIMOTO THYROIDITIS

Lymphoid follicles with prominent germinal centers (left) and a background of lymphocytes and plasma cells (right) associated with Hurthle cell change.

be very fibrotic, especially in older patients, suggesting malignancy on gross examination. Patients with atrophic autoimmune thyroiditis have small thyroid glands with no evidence of a goiter.

In Hashimoto thyroiditis, the enlargement is usually symmetric, but a midline pyramidal lobe may be prominent. Cut sections often reveal accentuated lobulation with increased fibrosis. The gland is tan-yellow rather than the usual red-brown because of the abundant lymphoid tissue.

Cytologic Findings. The aspiration biopsy is usually cellular and consists of Hurthle cells and lymphocytes. Hurthle cells are larger than follicular cells, contain more abundant granular cytoplasm, and usually appear as sheets or small groups of cells or as single cells. Lymphocytes are usually abundant. Plasma cells, neutrophils, and macrophages may also be seen.

Microscopic Findings. Histologic examination shows sheets of lymphocytes and plasma cells, often with prominent germinal centers (fig. 3-20). The thyroid follicles are small and atrophic, and the colloid appears dense or may be absent. There is usually fibrosis in the background. The follicular cells are metaplastic and include oncocytic (Hurthle cell) (fig. 3-21), clear, and squamous cell types (fig. 3-22). The nuclei of the follicular epithelial cells often show clearing, enlargement, and overlapping; this reactive nuclear change can be mistaken for foci of papillary thyroid carcinoma. In cases with marked nuclear clearing and atypia, a useful clue to distinguish thyroiditis from papillary carcinoma is the gradual transition from atypical areas (most prominent in the areas of active inflammation) to bland follicular cells, which contrast to the abrupt transition in cases of neoplasia. On rare occasions, a few multinucleated giant cells are present in the thyroid gland, but they do not occur in the same histologic background as subacute thyroiditis. The inflammatory cells may extend to the adjacent skeletal muscle, especially if there is preexisting intermingling

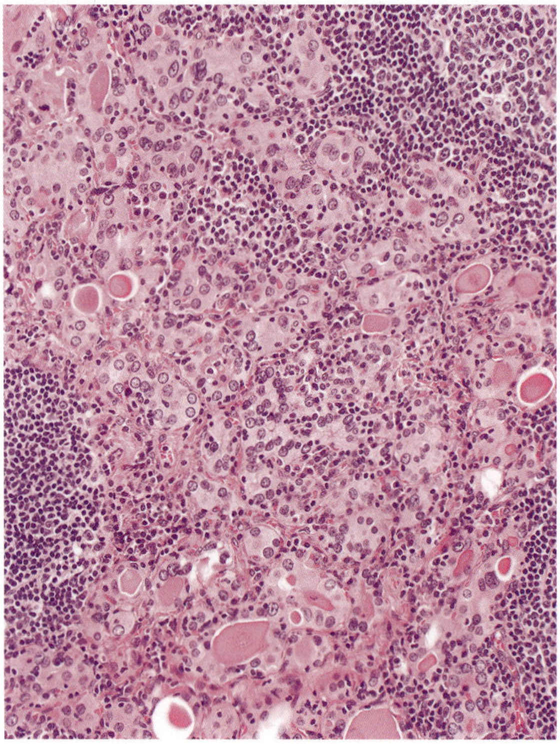

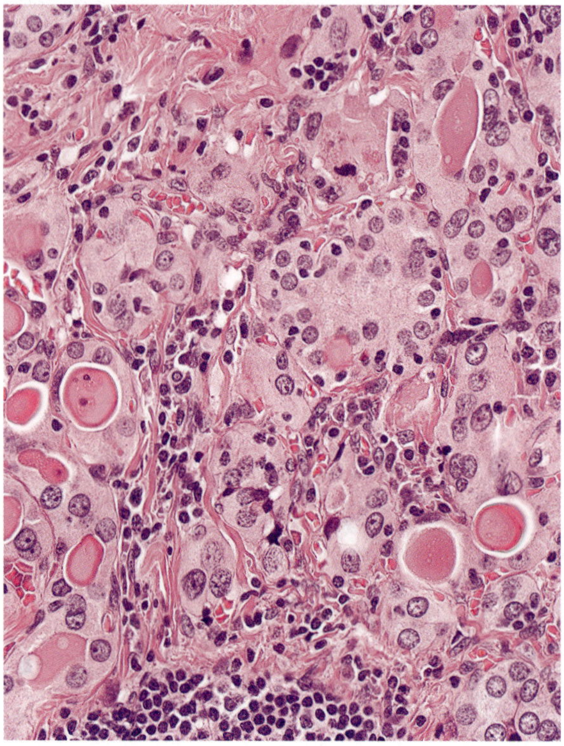

Figure 3-21

HASHIMOTO THYROIDITIS

Hurthle cell (oxyphil cell) metaplasia is usually a prominent feature in Hashimoto thyroiditis.

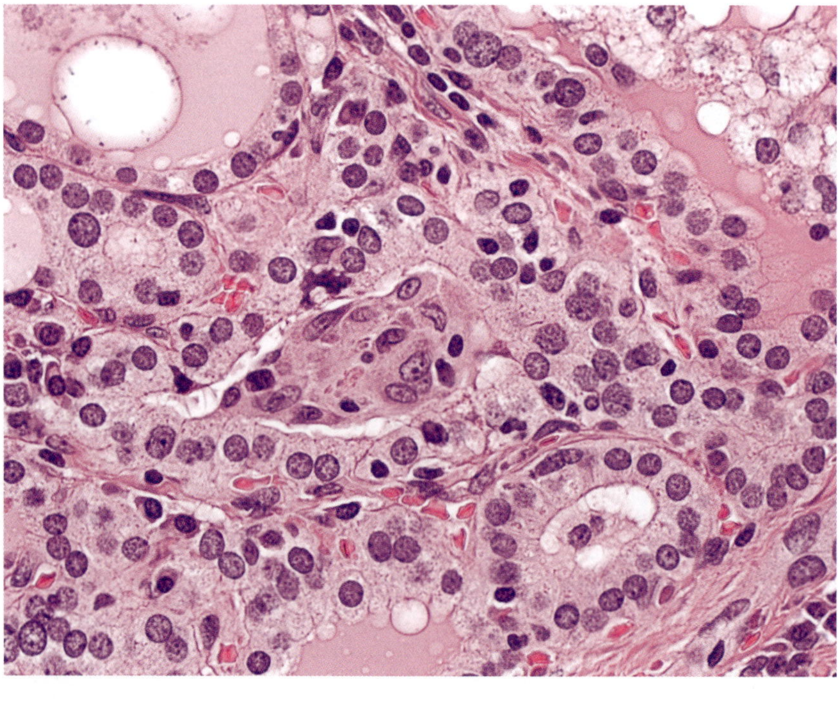

Figure 3-22

HASHIMOTO THYROIDITIS

Focal squamous metaplasia of the follicular cells is common in longstanding Hashimoto thyroiditis.

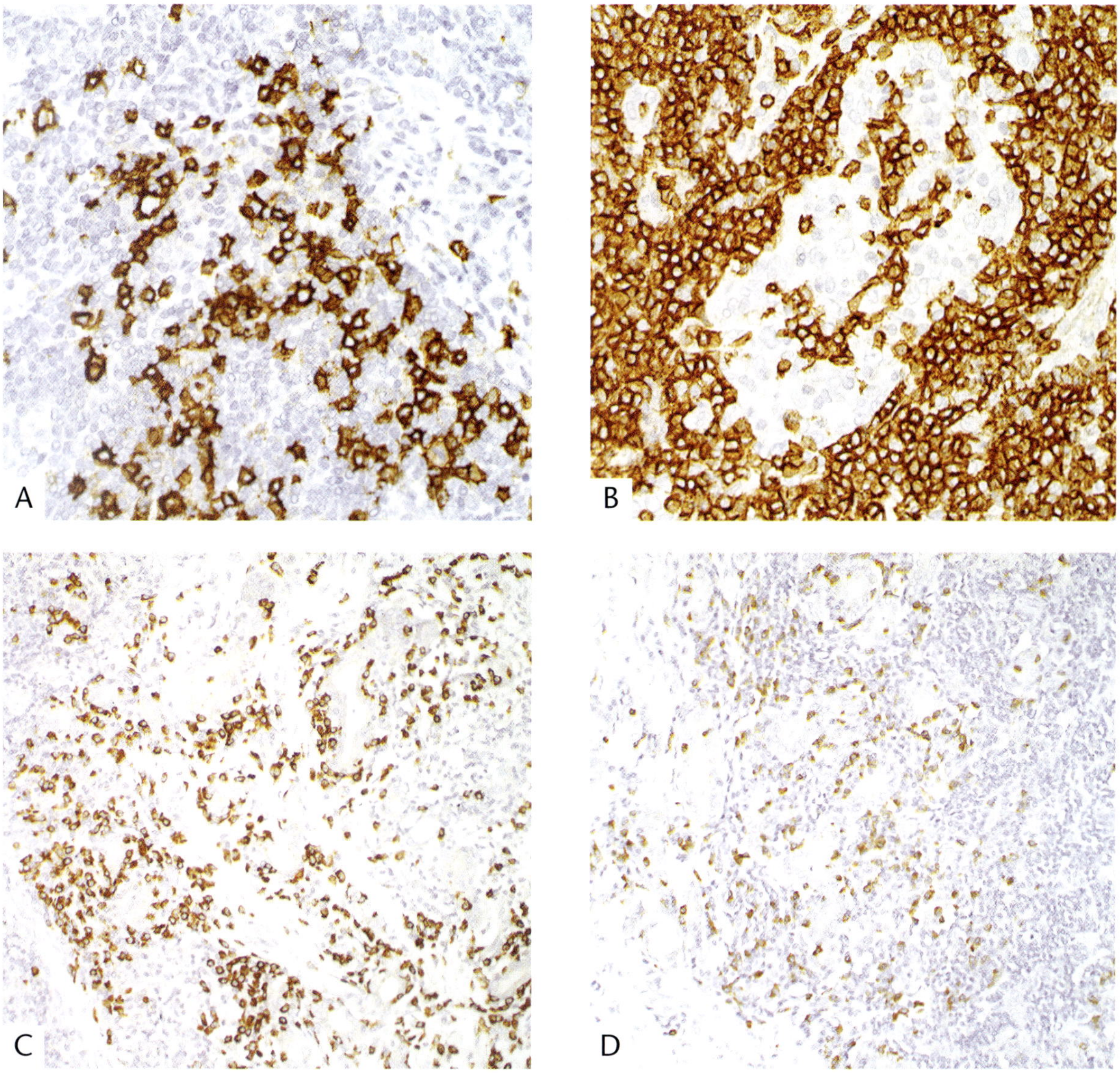

Figure 3-23

HASHIMOTO THYROIDITIS

A mixed B (CD20-positive) cell (A) and T (CD3-positive) cell infiltrate (B) is present. The T cells infiltrate the follicular epithelium (B). Immunostaining for kappa (C) and lambda (D) light chains of immunoglobulin shows polyclonal B cells.

of skeletal muscle and thyroid follicles, resulting in adherence of the gland to the surrounding neck structures during surgery. The perithyroid lymph nodes are usually enlarged and show reactive follicular hyperplasia.

Immunohistochemical Findings. Immunohistochemical staining with B- and T-cell antibodies show that the lymphocytic infiltrate is composed of a mixture of B and T cells in an almost 1 to 1 ratio (fig. 3-23) (461). The plasma cells have a spectrum of immunoreactivity for heavy chains (IgG, IgM, IgA) as well as polyclonal kappa and lambda light chains. Many of the T cells are CD4-positive helper cells with HLA-DR2 expression, suggesting activated T cells (408,442). CD4-positive cells are widely distributed in the thyroid gland and are present between follicles, around blood vessels, and in lymphoid tissue. CD8-positive suppressor/cytotoxic T cells are admixed with the CD4-positive

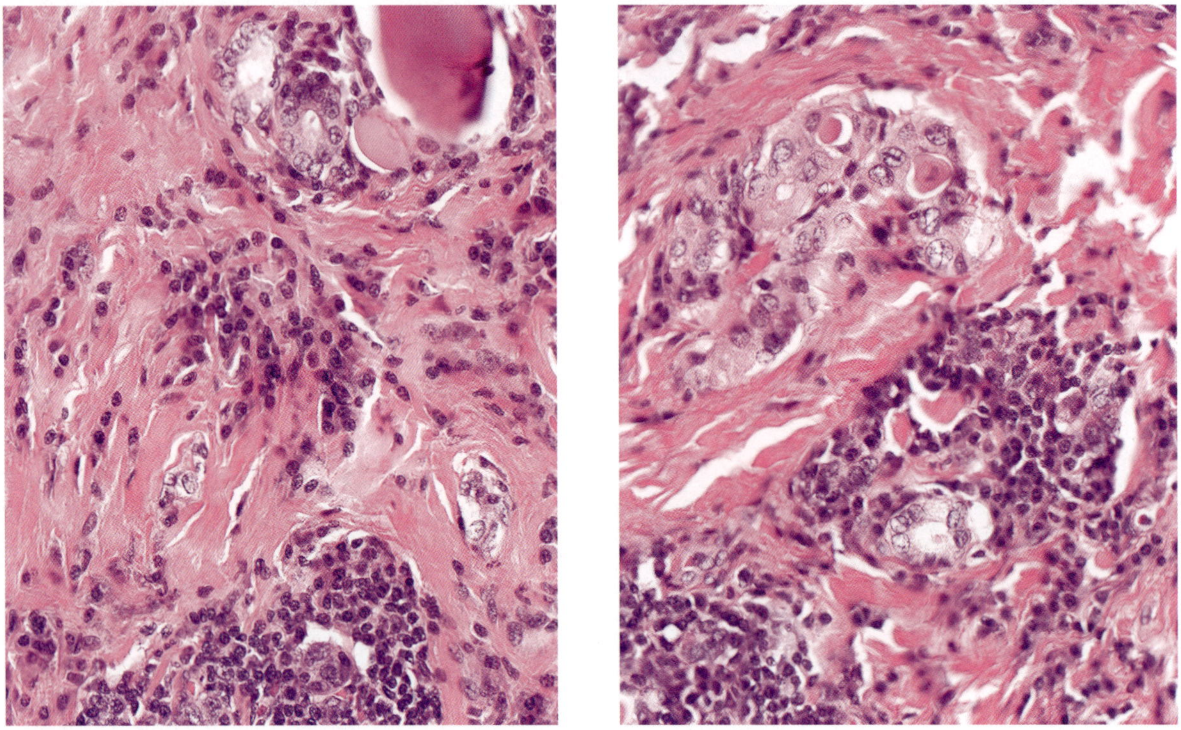

Figure 3-24

FIBROUS VARIANT OF HASHIMOTO THYROIDITIS

Left: There is a marked fibrotic reaction, admixed with lymphocytes and plasma cells in the background.

Right: The fibrotic reaction compresses and traps some of the Hurthle cells. This may be mistaken for a carcinoma, particularly if there is nuclear clearing. Careful examination of the pattern of growth and the gradual transition to bland epithelium usually allows confident diagnosis.

and B cells in the follicles, and are present in the extracellular spaces formed by invaginations of the epithelial cell membranes (408). The suppressor T cells have cytolytic activity in vitro, so they may contribute to the follicular damage seen in this disease. The B cells are usually of the immunoglobulin kappa subclass.

Ultrastructural Findings. Ultrastructurally, the Hurthle cells contain many large mitochondria with decreased volumes of other organelles including rough endoplasmic reticulum and Golgi complexes (462). The T cells can be observed between epithelial cells. Duplication and rupture of the follicular basement membrane has also been noted (402,405,463).

Variants of Hashimoto Thyroiditis. *Fibrous Variant.* The fibrous (or fibrosing) variant accounts for 10 to 13 percent of all cases of Hashimoto disease (402). It occurs in a slightly older age group, and the patients present with marked hypothyroidism and a large symptomatic goiter. The disorder usually requires surgical treatment to relieve the local symptoms of dysphagia or dyspnea. Laboratory studies show a markedly elevated antithyroglobulin antibody titer and elevated serum TSH levels.

The gland is usually larger and more fibrotic than the usual Hashimoto gland. Histologic examination shows preservation of the lobulated pattern of the normal thyroid gland. Atrophic follicular cells with degenerative changes and broad areas of dense keloid-like fibrosis are present (fig. 3-24). Features of usual Hashimoto thyroiditis, with Hurthle cells and lymphoplasmacytic infiltrates with prominent germinal centers, are readily seen; however, there is more extensive squamous metaplasia of the follicular epithelium. The pathogenesis of this variant is unknown, but Katz and Vickery (464) suggest that the fibrous variant may develop by progression from the usual type of Hashimoto thyroiditis.

Fibrous Atrophy Variant. Some patients with or without a history of Hashimoto thyroiditis present with a small fibrotic gland associated with hypothyroidism, or "idiopathic myxedema." The gland is very small and weighs between 1 and 6 g. Histologically, there is extensive destruction of the thyroid parenchyma with minimal preservation of the follicular architecture (fig. 3-25). Extensive fibrosis and lymphoplasmacytic infiltrates are features reminiscent of the fibrous variant of Hashimoto thyroiditis. These patients, who are usually elderly, often have high titers of antithyroid antibodies and severe hypothyroidism (465). The histologic features are similar to those of the fibrous variant of Hashimoto thyroiditis (464), but the thyroid gland is much smaller.

Juvenile Variant. The juvenile variant is a poorly defined form of chronic lymphocytic thyroiditis that has been described in younger patients (466,467). Some patients present with hyperthyroidism that progresses to hypothyroidism with time. The glands show the typical lymphoplasmacytic infiltrate; however, oxyphilia and fibrosis are less prominent in children, and hyperplasia of epithelial cells may be marked. Follicular atrophy is uncommon.

Differential Diagnosis. The differential diagnosis of Hashimoto thyroiditis includes thyroid gland carcinoma, lymphoma of the thyroid, Riedel disease, Graves disease, and "hashitoxicosis" or "hyperthyroiditis" (467). The extensive fibrosis that is seen in some cases of Hashimoto thyroiditis, especially in the fibrous variant which may be adherent to the thyroid gland at surgery, may suggest a carcinoma. The histologic preservation of the normal lobulated architecture, the lack of architectural and cytologic features of papillary or other carcinomas, and the histologic features of Hashimoto thyroiditis allow distinction. Patients with Hashimoto thyroiditis are at increased risk of papillary carcinoma and B-cell lymphoma of marginal or diffuse large B type (468,469). In addition, patients with Hashimoto disease may be susceptible to the development of plasmacytomas within the thyroid gland (470). A peculiar variant of mucoepidermoid carcinoma, known as sclerosing mucoepidermoid carcinoma with eosinophilia, has also been recognized in patients with Hashimoto disease (402). Whether there is an increase in the incidence of papillary thyroid carcinoma is still debated.

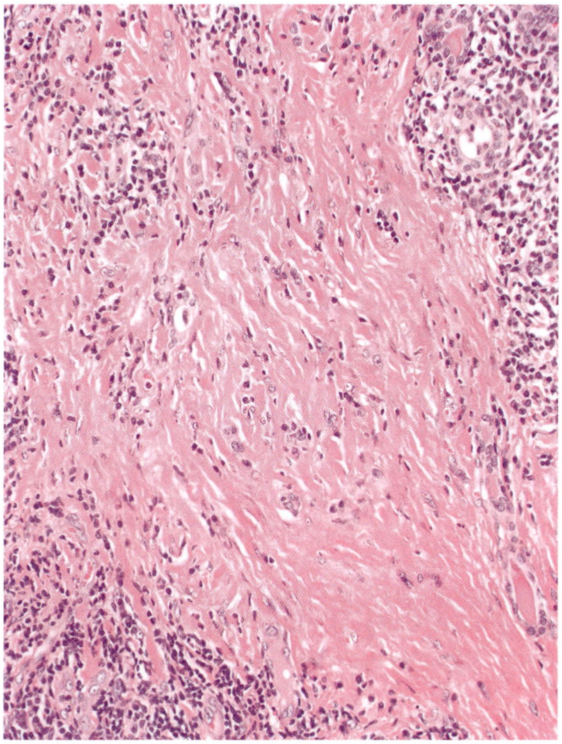

Figure 3-25

FIBROUS ATROPHY IN CHRONIC HASHIMOTO THYROIDITIS

There is extensive fibrosis, and atrophic follicles are entrapped in the fibrotic areas.

Fine-needle aspiration biopsy of the thyroid gland before surgical excision of a lobe may lead to hemorrhage, chronic inflammation, and fibrosis. These features may simulate carcinoma or the fibrous variant of Hashimoto thyroiditis.

A condition described as peritumor thyroiditis may be confused with Hashimoto thyroiditis, especially in small biopsy specimens. In this condition, there are lymphocytic and mixed lymphoplasmacytic infiltrates at the periphery of a thyroid neoplasm, especially an infiltrative papillary carcinoma (471). Adequate sampling of the lesion should lead to the correct diagnosis.

Hashimoto thyroiditis, especially the fibrous variant, may be difficult to distinguish from Riedel thyroiditis (discussed below). The adherence of the thyroid gland to neck structures and the extensive fibrosis of the gland contribute to this confusion. Riedel thyroiditis is associated with more extreme fibrosis in which fibrous tissue extends into nerve, fat, and muscle as well as the parathyroid glands, and is often associated with vasculitis with intimal proliferation and destruction of the media, features not commonly seen in Hashimoto thyroiditis. In contrast to Hashimoto thyroiditis, where the inflammatory and fibrotic process is confined to the gland, in Riedel thyroiditis the fibrosing process is either centered outside the thyroid gland or extends outside the gland.

The dense lymphoplasmacytic infiltrate in Hashimoto thyroiditis may simulate a lymphoma. This situation is made more difficult by the fact that most thyroid lymphomas arise in a setting of Hashimoto thyroiditis (472,473). The mixed lymphoplasmacytic infiltrate seen in the latter may also be present in thyroid lymphoma. The absence of sheets of atypical lymphoid cells allows distinction between these two conditions. Immunophenotyping for B- and T-cell markers, and clonality and gene rearrangement studies can readily allow distinction between the lymphoid hyperplasia of Hashimoto thyroiditis and the monoclonal B-cell proliferation of most thyroid lymphomas. Woolner et al. (467,474) previously described a lymphoma-like thyroiditis that may be challenging for the pathologist to distinguish from thyroid lymphoma. Some of these cases undoubtedly represent mucosa-associated lymphoid tissue (MALT) lymphomas of the thyroid gland as described by Isaacson (475) and later by other investigators (476).

Hashimoto thyroiditis may share some histologic features with Graves disease, and a condition with features of both Graves disease and Hashimoto thyroiditis has been described (451). These patients have glands with lymphoplasmacytic infiltrates, Hurthle cell changes, and follicular atrophy or hyperplasia. Clinical and laboratory studies, along with the usual histologic features, allow separation of classic Hashimoto thyroiditis and Graves disease.

Treatment and Prognosis. Because Hashimoto thyroiditis is usually asymptomatic and the goiter is small, many patients do not require treatment (477). When hypothyroidism is present, treatment with T4 is indicated, with the dose adjusted to normalize the serum TSH. Successful treatment reverses all the symptoms and signs of hypothyroidism, although some neuromuscular and psychiatric symptoms may not disappear for several months. Up to 24 percent of patients with hypothyroidism due to chronic autoimmune thyroiditis who are treated with T4 for more than 1 year remain euthyroid when the drug is withdrawn (478). This may be related to the disappearance of TSH-blocking antibodies (478,479).

Various complications can develop in patients with Hashimoto thyroiditis. Over-replacement with T4 can cause subclinical or even overt hyperthyroidism. In rare cases, patients develop Graves disease, possibly caused by the presence of different autoantibodies that include TSH-stimulating and TSH-blocking antibodies (480).

Thyroid lymphoma is a serious complication of Hashimoto thyroiditis. In one study of women in Japan followed for an average of 8 years, 0.1 percent developed thyroid lymphoma, a prevalence rate that was 80-fold higher than expected (473). Lymphomas arising in patients with Hashimoto thyroiditis are usually of B-cell phenotype and are usually confined to the thyroid gland (481). Treatment with radiotherapy alone or with chemotherapy leads to 5-year survival rates of 13 to 92 percent, depending on the grade of the tumor (482).

In pregnant women, two patterns of postpartum thyroid dysfunction are recognized: postpartum thyroiditis and postpartum exacerbation of chronic lymphocytic thyroiditis (405). Postpartum thyroiditis is characterized by transient hyperthyroidism, which may be followed by hypothyroidism that may be transient or permanent. Postpartum exacerbation of Hashimoto thyroiditis is characterized by postpartum progression of autoimmune destruction. It may cause a transient or permanent increase in thyroid hormone requirements. In one study, more than 50 percent of women with Hashimoto thyroiditis required an increase in their pregestational T4 dose in the postpartum period (483).

Graves Disease

Definition. *Graves disease* is an autoimmune condition characterized by hyperthyroidism, goiter, thyroid eye disease (Graves orbitopathy), and occasionally, a dermopathy known as pretibial or localized myxedema.

General Features. Graves disease is estimated to affect 2 to 3 percent of the general population and is the most common cause of spontaneous hyperthyroidism in patients younger than 40 years of age (484,485). There is abundant epidemiologic evidence for a genetic susceptibility to Graves disease and chronic autoimmune thyroiditis (486–488). There is an increased incidence of Graves disease among family members of affected individuals, with the sibling risk ratio exceeding 10.0 (489). The development of Graves disease is strongly associated with HLA-B8 and HLA-DR3 haplotypes (490).

The main autoantigen in Graves disease is the TSH receptor (TSHR), which is expressed primarily in the thyroid gland but also in adipocytes, fibroblasts, bone cells, and a variety of additional sites (491,492). TSHR is a G-protein coupled receptor with seven transmembrane-spanning domains. TSH, acting via TSHR, regulates thyroid growth, hormone production, and hormone secretion. Factors that contribute to TSHR presentation as a target for the immune system in humans are not well understood, but are considered to be primarily factors that build on a state of enhanced genetic susceptibility combined with a failure of immune tolerance. Such susceptibility may be translated by variable expression of TSHR on thymic epithelial cells, which is important in determining self-tolerance (493,494).

TSHR antibodies are specific for autoimmune thyroid disease, especially Graves disease, in contrast to antibodies to thyroglobulin and thyroid peroxidase which occur in the "normal" population and in patients with Hashimoto thyroiditis. Almost all patients with Graves disease have detectable TSHR antibodies when measured by sensitive assays (495–497). The stimulating variety of TSHR antibodies, like TSH, stimulate the synthesis and activity of the sodium-iodide symporter, explaining the increased uptake of iodide by thyroid tissue in Graves disease in the absence of TSH (498). In contrast, the blocking variety of TSHR antibodies block the binding and action of TSH and, therefore, can cause hypothyroidism. Blocking TSHR antibodies are found in 10 to 15 percent of patients with Hashimoto thyroiditis. Some patients with Graves disease have a mixture of both stimulating and blocking TSHR antibodies, and the clinical presentation may depend upon a balance between them.

Pathogenesis. The primary pathophysiology of Graves disease is related to TSHR antibodies of the immunoglobulin G1 (IgG1) subclass, which are driven by interferon-gamma (499,500). TSHR is present in the binding pocket of the HLA molecule on the surface of the antigen-presenting cells, resulting in T-cell activation. Patients with Graves disease also have reduced levels of circulatory suppressor/cytotoxic or CD8-positive cells, suggesting that a lack of suppressor/cytotoxic T cells may be responsible for the breakdown of immune tolerance in Graves disease (501,502). A possible etiologic factor is infection, which induces MHC class II expression in experimental animals. Hepatitis C infection is a well-recognized precipitator of autoimmune thyroid disease when treated with interferon therapy, although, most commonly, thyroiditis develops rather than Graves disease (503).

Various forms of stress have been suggested as etiologic stimuli in Graves disease. Some research suggests that stress results in general immune suppression because of the effects of glucocorticoids and corticotropin-releasing hormone on the immune system (504,505); the immune system overcompensates for the stress, leading to autoimmune disease.

The role of gonadal steroid hormones in the initiation of Graves disease has been suggested because of the higher incidence in women (approximately 8 to 1), and the development of the disease in times of changes in gonadal steroid hormone production, such as puberty, pregnancy, and menopause. In men, Graves disease occurs at a later age, is more severe, and is frequently associated with ophthalmopathy (484).

Ophthalmopathy and Dermopathy. The main autoantigen in Graves disease, TSHR, appears to be closely aligned with the insulin-like growth factor 1 (IGF1) receptor. Some muscle cells and fibroblasts also express HLA class II antigens, as seen on thyroid cells in patients with autoimmune thyroid disease, suggesting that they can present antigen to T cells and resident dendritic cells, thereby serving to perpetuate if not initiate the pathologic process (506–509). TSHR antibodies and activated T cells also play an important role in the pathogenesis of *Graves orbitopathy* by activating retro-ocular fibroblast and adipocyte TSHR and IGF1 receptors and initiating a retro-orbital inflammatory environment (510). This is supported by the observation that high titers of TSHR antibodies correlate with the presence and severity of Graves orbitopathy.

The volume of both the extraocular muscles and retro-ocular connective tissue is increased due to fibroblast proliferation, inflammation, and accumulation of hydrophilic glycosaminoglycans (GAG), mostly hyaluronic acid (511,512). GAG secretion by fibroblasts is increased by thyroid-stimulating antibodies and activated T cells (via cytokine secretion), implying that both B- and T-cell activation are integral to this process. The accumulation of hydrophilic GAG in turn leads to fluid accumulation, muscle swelling, and an increase in pressure within the orbit. These changes, together with retro-ocular adipogenesis, displace the eyeball forward, leading to extraocular muscle dysfunction and impaired venous drainage.

Graves dermopathy, also known as *pretibial myxedema,* results from accumulation in the dermis of GAG, especially hyaluronic acid, secreted by fibroblasts under the stimulation of local cytokines. The cytokines arise from lymphocytic infiltration, which is best seen in early lesions. The resulting characteristic pathologic changes are mucinous edema and fragmentation of dermal collagen fibers, with deposition of acid mucopolysaccharides (hyaluronic acid) in the papillary and reticular dermis, with subsequent extension into deeper tissues (513,514).

Clinical and Radiologic Features. Graves disease occurs at all ages, with a peak in the third and fourth decades. Its incidence is higher in women (2 percent) than in men (0.2 percent). Patients typically present with a diffuse goiter and symptoms of thyrotoxicosis, sometimes accompanied by orbitopathy and skin manifestations. The most common symptoms include heat intolerance, sweating, weight loss, palpitations, tremor, and hair loss.

Men tend to develop Graves disease at an older age than women, and the degree of thyroid hyperfunction is often more severe, but symptoms may also be less severe. Muscle weakness and eye disease are more likely to develop in men (484).

Imaging studies with radioactive iodine show a diffuse increase in tracer uptake in the thyroid gland, indicating increased de novo synthesis of thyroid hormone rather than inflammation or destruction of thyroid tissue with release of preformed hormone into the circulation (515,516). This increase is not unique to Graves disease, since other conditions, including Hashimoto thyroiditis, can lead to a similar imaging pattern (515,516).

Laboratory Findings. The serum TSH level is very low, while serum T3 and T4 levels are elevated. Serum T3 concentration is usually more elevated than T4. The presence of TSHR antibodies of the stimulating subtype in the serum is diagnostic of Graves disease (484), but this does not always correlate with the clinical state. The thyroid radioactive iodine uptake values (RAIU) are invariably increased in patients with Graves disease. The RAIU is measured 24 hours after the isotope is given, since it usually reaches a plateau at this time. However, in Graves disease, the RAIU usually peaks early, and therefore, this test is not as diagnostically accurate as others, such as determination of TSH and free T4 levels (484).

Gross Findings. The thyroid gland is usually diffusely enlarged and symmetric, and weighs between 50 and 150 g. It glistens and has a prominent vascular pattern. The cut surface shows beefy, deep red parenchyma.

Cytologic Findings. Cytologic examination shows sheets of follicular cells, with rings that are formed by the peripherally situated basal nuclei, and marginal vacuoles of variable sizes with discrete vacuoles adjacent to the nuclei. These vacuoles can also be found in aspirates from patients without hyperthyroidism.

Microscopic Findings. Untreated Graves disease has a striking low-power appearance (fig. 3-26). Diffuse hyperplasia of the follicular cells causes the follicles to take on very irregular convoluted and serrated contours. There is extensive colloid scalloping. The lobular architecture is retained, and there is prominent vascular congestion and follicular hyperplasia. The follicular cells are columnar with enlarged nuclei. Papillary infoldings of the follicular epithelium are also seen (fig. 3-27). The follicular cells can demonstrate marked nuclear clearing, which in combination with papillary hyperplasia, can be mistaken for papillary carcinoma (517). In addition to the absence of classic nuclear features of papillary carcinoma, the diffuse nature of the process is an important clue to the diagnosis.

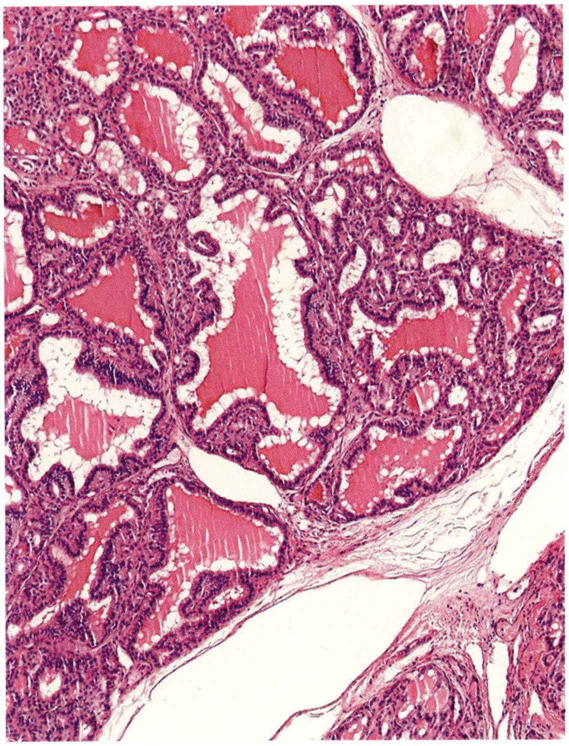

Figure 3-26

UNTREATED GRAVES DISEASE

The follicles have convoluted and irregular contours. There is extensive colloid scalloping.

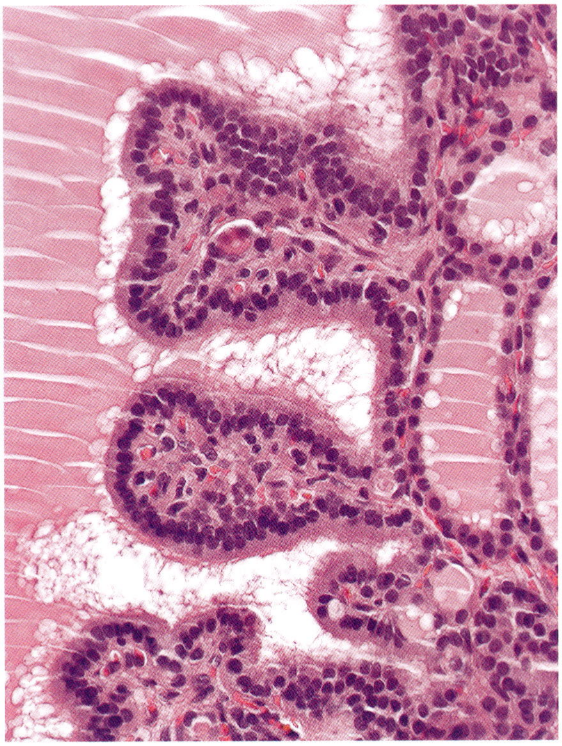

Figure 3-27

GRAVES DISEASE

There may be such marked papillary hyperplasia that the diagnosis of papillary thyroid carcinoma is considered. Careful attention to the nuclear features helps avoid this pitfall.

Colloid is decreased in the untreated gland and when present, there is a decreased follicular lumen and prominent "scalloping" or vacuoles in the colloid adjacent to the apex of the follicular cells. The "scalloping" is a fixation artefact due to decreased amounts of colloid and marked shrinkage during tissue processing.

Lymphocytes are commonly seen in the stroma around follicles. Mixed B and T cells are often present, with B cells in the germinal centers and T cells in a perifollicular location (518). Dendritic cells, which are antigen-producing cells, are usually increased in number (519,520). The dendritic cells and thyrocytes often express adhesion molecules such as CD54, while the mononuclear cells express other adhesion molecules such as CD11a/CD18 (519). In a diffuse goiter removed from a hyperthyroid patient, the absence of stromal lymphocytes should raise the possibility of a secondary form of hyperthyroidism, such as a pituitary lesion (measurements of TSH will resolve this issue).

In cases of treated Graves disease, a different histologic patterns may be seen, depending on the type of treatment (fig. 3-28), and often different degrees of activity are present within the same gland. Treatment with iodine before surgery leads to a decrease in the vascularity of the gland and involution of the follicular epithelium, which becomes cuboidal rather than columnar. There is also an increase in colloid storage in many follicles, which may still retain their convoluted contour (521). If radioactive iodine treatment is subsequently followed by surgery, the histologic changes may be variable, depending on the period of iodine therapy. The histologic features can vary from slight nuclear atypia to extensive follicle cell destruction, fibrosis, and oncocytic metaplasia with marked nuclear pleomorphism (522,523). Treatment with drugs that interfere with thyroid hormone synthesis, such as propylthiouracil, leads to increased hyperplastic changes while treatment

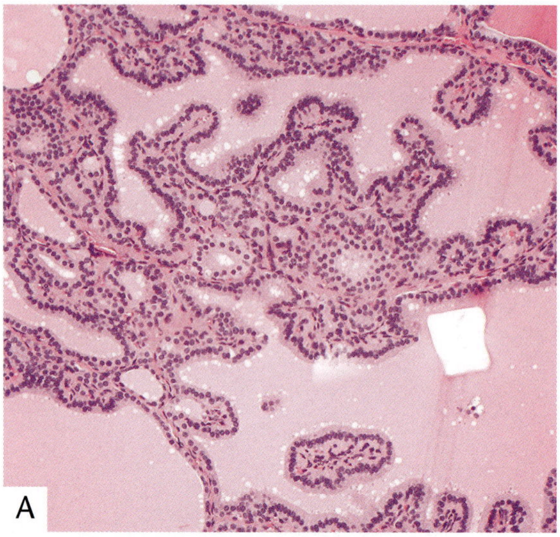

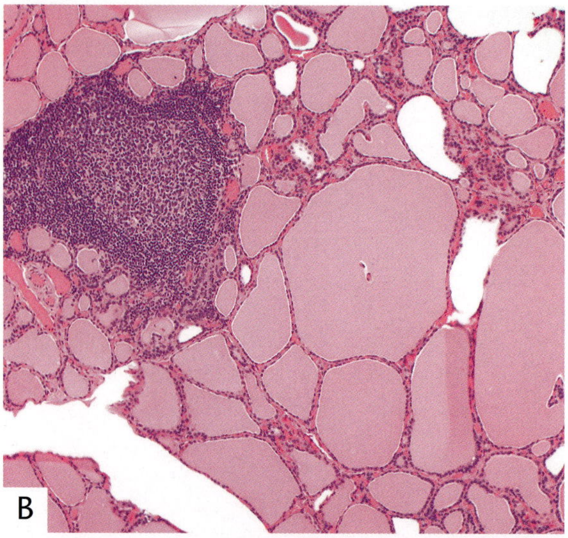

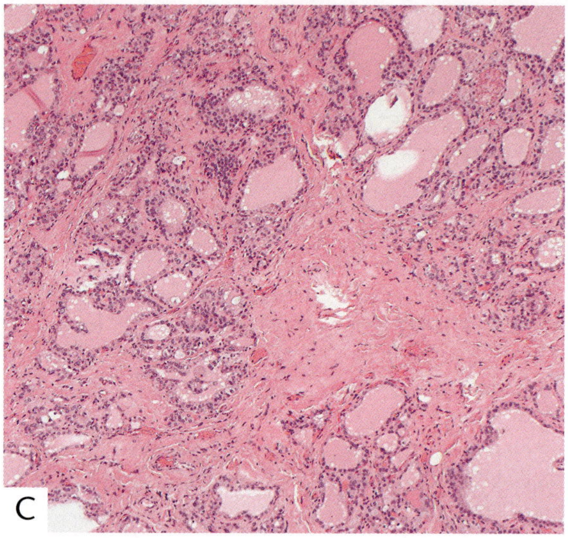

Figure 3-28

TREATED GRAVES DISEASE

The morphologic findings in treated Graves disease are highly variable and dependent on both the duration of therapy and the type of treatment.

A: In most cases, the follicles maintain their irregular contours, but with less marked colloid scalloping. Often the colloid has a watery ink quality.

B: Sometimes the follicles in treated Graves disease give the impression of being overdistended by colloid, often while still maintaining irregular contours.

C: In other cases of treated Graves disease, there is atrophy causing the underlying lobulated architecture of the thyroid gland to be exaggerated.

D,E: Nuclear hyperchromasia and contour irregularity are common features in treated Graves disease. This nuclear atypia, often seen in only some cells dispersed in a random manner throughout the thyroid gland, is quite different from the nuclear atypia seen in papillary thyroid carcinoma and should not suggest malignancy.

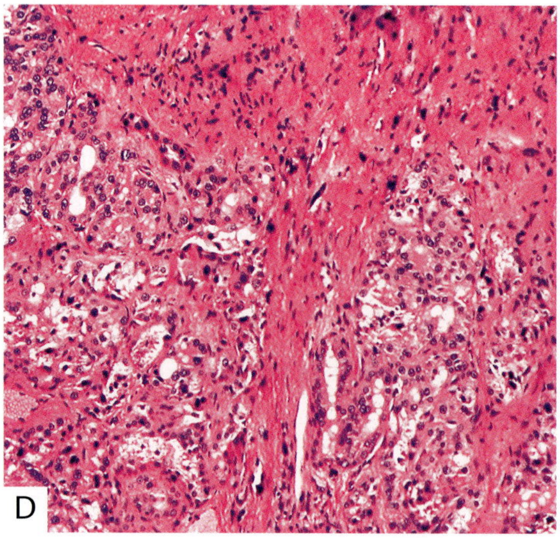

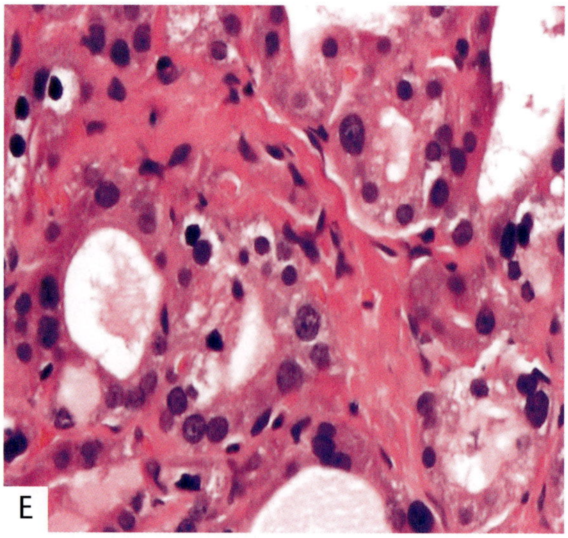

with beta-adrenergic blockers, such as propranolol, does not have any predictable morphologic effect on the thyroid gland (524,525).

Psammoma bodies are occasionally seen in Graves disease without associated papillary carcinoma (526). Some studies have reported an increased incidence of thyroid cancer in patients with Graves disease (527).

The changes of Graves orbitopathy or ophthalmopathy include increases in the volume of orbital tissue resulting in increased mass of the extraocular muscles, an increase in retrobulbar connective tissue secondary to edema, and accumulation of hyaluronic acid and chondroitin sulfates leading to proptosis. Histologically, the muscles are swollen and edematous, with loss of striation, fragmentation of fibers, and a diffuse lymphocytic inflammatory infiltrate (528,529). These changes in the muscle and connective tissues lead to the clinical presentation of exophthalmos.

The histologic changes of pretibial myxedema consist of deposits of mucopolysaccharides in the dermis and thickening of the skin in these areas (530). This develops in 5 to 10 percent of patients with Graves disease.

Ultrastructural Findings. Ultrastructural examination of the thyroid gland in patients with Graves disease shows increased activity of the follicular cells which contain prominent rough endoplasmic reticulum and Golgi complexes in addition to well- developed nucleoli in the enlarged nuclei (531). Other studies have reported immune complex deposits in the basement membrane of the thyroid follicles (532).

Immunohistochemical Findings. Immunohistochemical staining for HLA-DR is positive in the thyrocytes as well as the lymphoid cells in Graves disease (533). Studies of the lymphocytes in the thyroid gland have shown changes in the T cells, with increases in the CD4-positive cells (534). Differences in the ratio of B and T cells in the thyroid gland compared to the peripheral blood have been reported by immunophenotyping studies (535,536).

Differential Diagnosis. The histologic differentiation of Graves disease from papillary thyroid carcinoma can be difficult, since papillae with fibrovascular cores and psammoma bodies may be seen in Graves disease. In addition, the diffuse hyperplasia in Graves disease may involve the skeletal muscle adjacent to the thyroid gland, simulating an invasive papillary carcinoma. However, the absence of enlarged overlapping nuclei, nuclear grooves, and cytoplasmic invaginations into the nucleus helps to make the distinction.

Many other thyroid conditions can lead to hyperthyroidism and may be a problem clinically, but these can be readily distinguished histologically. Hashitoxicosis, or hyperthyroiditis (fig. 3-29), is a disease in which patients have thyrotoxicosis along with the histologic features of epithelial hyperplasia admixed with the changes seen in Hashimoto thyroiditis (519), including lymphocytic infiltration with germinal centers, Hurthle cell changes, and variable follicular atrophy. The clinical and pathologic features distinguish these cases from typical Graves disease.

Neoplasms, including hyperfunctioning follicular adenomas and primary and metastatic follicular carcinomas, are associated with thyrotoxicosis (537–540). The histologic appearance of hyperfunctioning neoplasms is similar to that of nonfunctioning ones, so it is usually easy to distinguish these lesions from Graves disease.

Toxic nodular goiters are frequently associated with hyperthyroidism (541). Grossly, there is a multinodular gland rather than the diffuse hyperplasia of Graves disease. Microscopically, focal areas of toxic goiter are seen in some of the nodules rather than diffusely as in Graves disease. Retrogressive changes are common in multinodular goiters, including fibrosis, recent and old hemorrhage, and calcification, while these changes are uncommon in diffuse hyperplasia.

Other conditions associated with hyperthyroidism but which are usually not a clinical problem in the histologic differential diagnosis are gestational trophoblastic disease (542); iodine-induced hyperthyroidism (543); struma ovarii with hyperthyroidism (520); rapidly growing tumors of the thyroid gland including lymphomas, anaplastic carcinomas, and metastatic tumors to the gland; ectopic production of TSH; TRH resistance to thyroid hormone (534, 544) usually secondary to pituitary resistance (545); and factitious or exogenous thyroid hormone intake (546), including accidental overdose from consumption of contaminated food (532).

Treatment and Prognosis. The therapeutic approach to Graves hyperthyroidism consists of both rapid amelioration of symptoms with a

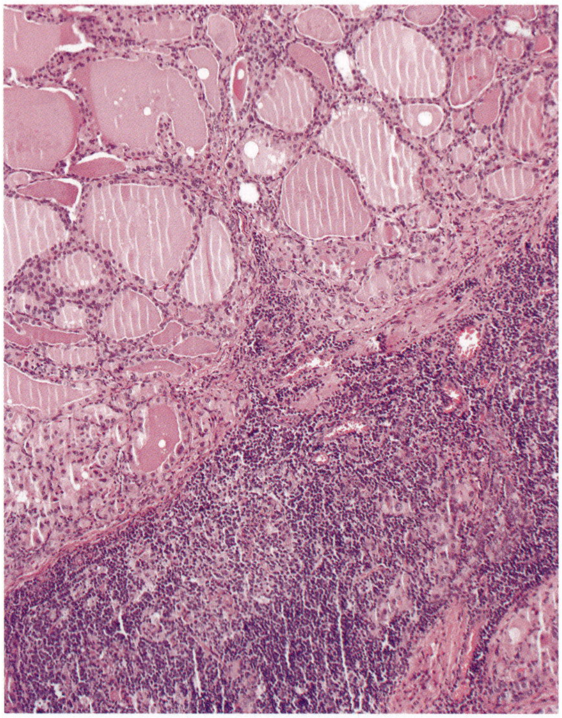

Figure 3-29

HASHITOXICOSIS

Features of both Graves disease and Hashimoto thyroiditis, with lymphoid follicles and germinal centers, are present.

beta-blocker and measures aimed at decreasing thyroid hormone synthesis including administration of antithyroid medication, radioiodine ablation, or surgery (547). The thionamide methimazole is the primary drug used to treat Graves hyperthyroidism (547). It is now almost exclusively used because of its longer duration of action, allowing for once-daily dosing, more rapid efficacy, and lower incidence of side effects. Propylthiouracil is preferred during the first trimester of pregnancy, because of the more significant teratogenic effects of methimazole, and in patients who have minor drug reactions to methimazole who refuse radioiodine or surgery. In countries where methimazole is unavailable, carbimazole is used as an alternative.

Radioiodine is less expensive and has a lower complication rate than surgery and is often preferred as definitive therapy for hyperthyroidism in nonpregnant patients, except in patients with moderate or severe orbitopathy. Patients with significant symptoms of hyperthyroidism, who are older or who have underlying heart disease or other comorbidities, can be treated with a thionamide (in addition to beta-blockers) to restore euthyroidism prior to radioiodine treatment. For patients with mild, well-tolerated hyperthyroidism, there is no need to pretreat with a thionamide, and radioiodine can be given soon after the diagnosis is made. Radioiodine induces extensive tissue damage, resulting in ablation of the thyroid gland within 6 to 18 weeks (548). Approximately 10 to 20 percent of patients fail the first radioiodine treatment and require a second dose. These patients usually have more severe hyperthyroidism or larger goiters.

The disadvantages of radioiodine treatment include a high frequency of late hypothyroidism, which is seen in 30 percent of patients at 5 years and 40 percent at 10 years (484). Also, there is increasing evidence that radioiodine therapy can cause the development or worsening of Graves orbitopathy more often than antithyroid drug therapy or surgery. There is no documented increase in thyroid cancer with this therapeutic approach (484).

The surgical treatment of Graves disease is bilateral subtotal thyroidectomy. Surgery is being used more frequently as first-line treatment in patients with active orbitopathy (549). The remission rates with surgery are superior to those of nonsurgical modalities, and it is primarily indicated in patients who have a large or obstructive goiter; in patients with moderate to severe active orbitopathy who desire definitive therapy for their hyperthyroidism; in pregnant women who are allergic to antithyroid drugs; and in patients who have allergies or poor compliance on antithyroid drugs but refuse radioiodine (550). Surgery is also indicated if there is a coexisting suspicious or malignant thyroid nodule or primary hyperparathyroidism. However, most thyroid nodules associated with Graves disease are benign, in which case surgery is not recommended (551). The frequency of recurrent hyperthyroidism is usually less than 5 percent, but the incidence of permanent hypothyroidism ranges from 4 to 30 percent (484).

Postpartum Thyroiditis

Definition. *Postpartum thyroiditis* is a destructive thyroiditis induced by an autoimmune mechanism within 1 year after parturition.

General Features. The reported prevalence of postpartum thyroiditis varies globally, and ranges from 1 to 17 percent (552–555), with a mean prevalence in the general population of approximately 7 to 8 percent (553,554). Higher rates, up to 25 percent, have been reported in women with type 1 diabetes mellitus (553,554). Postpartum thyroiditis, like painless thyroiditis, is considered a variant form of chronic autoimmune thyroiditis (Hashimoto thyroiditis). The pathologic findings in the two disorders are similar and both are associated with HLA-B and HLA-D haplotypes, suggesting that inherited risk factors are important (556).

Women with postpartum thyroiditis have a relative increase in B cells in the thyroid gland compared to the peripheral blood, and a relative decrease in suppressor T cells, leading to an increase in the ratio of helper to suppressor T cells (557,558). The resulting thyroid inflammation damages the thyroid follicles and activates proteolysis of the thyroglobulin stored within the follicles. The result is unregulated release of large amounts of T4 and T3 into the circulation and subsequent hyperthyroidism. This state lasts only until the stores of thyroglobulin are exhausted because new hormone synthesis ceases, not only because of damage to the thyroid follicular cells, but also because of inhibition of TSH secretion by the increased serum T4 and T3 concentrations. As the inflammation subsides, the thyroid follicles regenerate, and thyroid hormone synthesis and secretion resume. There may be a transient period of hypothyroidism and increased TSH secretion before thyroid secretion normalizes.

Approximately 20 to 30 percent of women with postpartum thyroiditis have the characteristic sequence of hyperthyroidism, which usually begins 1 to 4 months after delivery and lasts 2 to 8 weeks, followed by hypothyroidism, which lasts approximately 2 weeks to 6 months, and then recovery (559,560). However, 20 to 40 percent have only hyperthyroidism and the remaining 40 to 50 percent have only hypothyroidism. In some women, normal endogenous thyroid function is not restored after the initial episode of hypothyroidism (561,562).

The symptoms and signs of hyperthyroidism, when present, are typically mild and consist mainly of fatigue, weight loss, palpitations, heat intolerance, anxiety, irritability, tachycardia, and tremor. Similarly, hypothyroidism is also usually mild, leading to lack of energy, cold intolerance, constipation, sluggishness, and dry skin (560).

Gross Findings. The thyroid gland may be slightly enlarged diffusely without localized nodules.

Microscopic Findings. There is lymphocytic infiltration, with occasional germinal centers, variable disruption and collapse of thyroid follicles, and slight follicular hyperplasia (563–565). During recovery, lymphocytic infiltration is still seen and there may be some fibrosis, but the thyroid follicles are more normal in appearance.

Differential Diagnosis. The histologic differential diagnosis includes Hashimoto thyroiditis. However, because of the absence of chronicity, Hurthle cell metaplasia and fibrosis are not usually seen in postpartum thyroiditis. The histologic features of postpartum thyroiditis are similar to those of silent thyroiditis: lymphocytic infiltrates without significant Hurthle cell metaplasia. By definition, the history of recent pregnancy allows distinction between these two entities.

Treatment and Prognosis. The majority of women with postpartum thyroiditis need no treatment during the hyperthyroid or hypothyroid phases of their disease. Women who have bothersome symptoms of hyperthyroidism can be treated with a beta-blocker, while those with symptomatic hypothyroidism should be treated with T4 (levothyroxine). Most women recover and are euthyroid within 1 year of parturition, although approximately 20 to 40 percent develop permanent hypothyroidism (555,560–562,564,566).

Painless/Silent Thyroiditis

Definition. *Painless/silent thyroiditis* is an autoimmune thyroiditis characterized by transient hyperthyroidism, followed sometimes by hypothyroidism, and then recovery. Synonyms include *transient hyperthyroidism with lymphocytic thyroiditis, atypical subacute thyroiditis,* and *chronic lymphocytic thyroiditis.*

General Features. Painless thyroiditis accounts for 0.5 to 5.0 percent of cases of hyperthyroidism (567,568). It is considered a variant form of Hashimoto thyroiditis (569) and has many similarities with postpartum thyroiditis but by definition, excludes women who have a painless thyroiditis syndrome within 1 year after pregnancy. Many patients with painless thyroiditis have high serum concentrations of antithyroid peroxidase and antithyroglobulin antibodies, many have a family history of thyroid autoimmune disease, and some develop overt chronic autoimmune thyroiditis several years later (570,571). It affects women more frequently than men (569).

Painless thyroiditis is associated with specific HLA haplotypes, most often HLA-DR3, suggestive of an inherited susceptibility (572). The etiology of the condition is uncertain. Factors postulated to initiate the condition include excess iodine intake and various cytokines. A syndrome similar to this disorder occurs in patients treated with interferon-alfa, interleukin-2, lithium, tyrosine kinase inhibitors, and immune checkpoint inhibitors (573). These observations raise the possibility that cytokines released in response to some (subclinical) injury or infection might initiate the disorder. A specific viral etiology has been sought, but not confirmed.

Regardless of the initiating factors, the resulting thyroid inflammation damages thyroid follicles and activates proteolysis of the thyroglobulin stores within them. The result is unregulated release of large amounts of T4 and T3 into the circulation, leading to clinical and biochemical hyperthyroidism. As in postpartum thyroiditis, this state only lasts until the stores of thyroglobulin have been exhausted, because new hormone synthesis ceases, not only because of damage to the thyroid follicular cells but also because of inhibition of TSH secretion by the high serum T4 and T3 concentrations.

As the inflammation subsides, the undamaged or repaired thyroid follicles resume synthesis and secretion of thyroid hormone. There may be a transient period of hypothyroidism and increased TSH secretion before thyroid secretion normalizes again. In a minority of patients, thyroid damage is sufficient to result in permanent hypothyroidism.

Gross Findings. A diffuse goiter in a slightly enlarged thyroid gland is usually present.

Microscopic Findings. The thyroid gland shows a variable lymphocytic infiltrate with preservation of the normal lobular architecture. There may be focal or diffuse thyroiditis. Hurthle cell change is uncommon, but may be seen focally. Fibrosis is usually absent or focal. Follicular destruction is present in all cases and is a useful diagnostic feature. The degree of follicular destruction is variable, and lymphoid follicles are present in about half the cases. During the late recovery phase, there is usually focal thyroiditis but the thyroid follicles are normal.

Immunohistochemical Findings. The studies of Mizukami et al. (574) showed T cells concentrated in the paracortex, with a small percentage in the germinal centers, while B cells predominated in the germinal centers. The lymphoid cells infiltrating between follicles were predominantly T cells (575).

Differential Diagnosis. The main differential diagnosis is postpartum thyroiditis, which has many clinical and histologic similarities to silent painless/thyroiditis. The principal difference is the antecedent pregnancy in the former. The findings differ from those of chronic autoimmune thyroiditis in that there is more follicular disruption but fewer lymphocytes, fewer germinal centers, and less fibrosis in painless thyroiditis (576). Subacute thyroiditis differs clinically and by the presence of multinucleated giant cells.

Treatment and Prognosis. Many patients need no treatment during either the hyperthyroid or hypothyroid phase, because the thyroid dysfunction is rarely severe and is transient. Patients who have symptomatic hyperthyroidism can be treated with a betablocker, while those with sufficient symptoms of hypothyroidism may warrant thyroxine therapy. Although thyroid function abnormalities resolve in the majority of patients, approximately 20 to 30 percent develop permanent hypothyroidism (568,571). Patients may also experience recurrent episodes of thyroiditis, often many years after the initial episode.

Focal Lymphocytic Thyroiditis

Definition. *Focal lymphocytic thyroiditis* is usually discovered incidentally in surgically excised thyroid glands or at autopsy. It is characterized by dense collections of lymphocytes in the interlobular or intralobular fibrous tissue of the thyroid gland. Synonyms include *nonspecific thyroiditis* and *focal autoimmune thyroiditis.*

General Features. Focal lymphocytic thyroiditis is found in 5 to 20 percent of adult autopsies and is most common in elderly women (577–580). It is thought to be related to the addition of iodide to the water and salt supplies in the United States (581). It has been suggested that iodide (iodine) may combine with a protein, act as an antigen, and evoke an immune response localized to the thyroid gland. Some authors believe that focal lymphocytic infiltration of the thyroid represents an immunologic disorder associated with aging (582). The lesions are commonly present in older patients who are asymptomatic for thyroid disease and who have low levels of antithyroid antibodies.

Gross Findings. There are no specific gross findings. The thyroid gland is usually of normal size. Occasionally, nodules are present.

Microscopic Findings. Focal aggregates of lymphocytes in the interlobular or intralobular fibrous tissue are present. Rarely, there is germinal center formation. Hurthle cell change is rarely present. The infiltrate usually spares any adenomatous nodules present. Follicular atrophy or disruption does not usually occur (582). In surgical specimens, this disorder may be associated with nontoxic nodular goiters, adenomas, and carcinomas, as well as in glands resected for Graves disease.

Treatment. Focal thyroiditis is an incidental finding usually identified at the time of surgery for other reasons. It is most likely not an early form of diffuse autoimmune thyroiditis and does not require any treatment.

Riedel Thyroiditis

Definition. *Riedel thyroiditis (Riedel disease, invasive fibrous thyroiditis, Riedel struma)* is a fibrotic process associated with a mononuclear cell inflammatory infiltrate that extends beyond the thyroid gland into the perithyroidal soft tissue.

General Features. Riedel thyroiditis is extremely rare, with an outpatient incidence of 1 in 100,000 and a prevalence of 0.05 percent or less of surgical thyroid diseases (583). Women are four times more likely than men to be affected, and the disease occurs most commonly between the ages of 30 and 50 years (584,585).

The etiology of Riedel thyroiditis is not known. Although the disease is often believed to be part of a systemic disease known as multifocal fibrosclerosis (586), there is limited evidence suggesting that Riedel thyroiditis is associated with autoimmune disease or a primary fibrotic process (587,588). Features supporting an autoimmune mechanism include the presence of mononuclear cells and vasculitis within the fibrous tissue and the high serum antithyroid antibody concentrations seen in many patients (588). However, the elevated thyroid antibodies have also been postulated to be a reaction to antigens released by destroyed thyroid tissue (588), and the disease has been associated with both Hashimoto thyroiditis and Graves disease (585).

Riedel thyroiditis is a primary fibrotic disorder, with fibroblast proliferation induced by cytokines produced by B or T lymphocytes (589). It may be associated with other fibrosing conditions including retroperitoneal fibrosis, fibrosing mediastinitis, sclerosing cholangitis, pancreatitis, lacrimal fibrosis, orbital pseudotumor fibrosis, and tumefactive fibroinflammatory lesions of the head and neck (589,590).

Riedel thyroiditis may also occur as part of immunoglobulin G4 (IgG4)-related systemic disease (591). The hallmarks of IgG4-related disease are a lymphoplasmacytic inflammatory infiltrate of mainly IgG4-positive plasma cells and small lymphocytes accompanied by fibrosis, obliterative phlebitis, and elevated serum levels of IgG4. High numbers of IgG4-positive plasma cells have been identified in thyroidectomy specimens from patients with Riedel thyroiditis (592).

Patients often present with a slowly growing, painless goiter, sometimes associated with anterior neck pressure, dysphagia, hoarseness, dyspnea, or hypoparathyroidism if the process has extended beyond the thyroid gland (593–595). Most patients with Riedel thyroiditis have normal serum thyroid hormones and TSH concentrations at presentation. However, approximately one third experience hypothyroidism due to fibrous infiltration and thyroid parenchymal destruction (584,585,595,596). Up to two thirds of patients have elevated serum thyroid autoantibody concentrations, but the titers are often less than those seen in Hashimoto thyroiditis (595,597).

Gross Findings. The thyroid gland is very hard, with extensive fibrosis involving part or all of the gland. The fibrosis usually extends to the adjacent muscle and soft tissues of the neck. The cut surface is tan-gray, woody, and avascular, with loss of the normal lobulation.

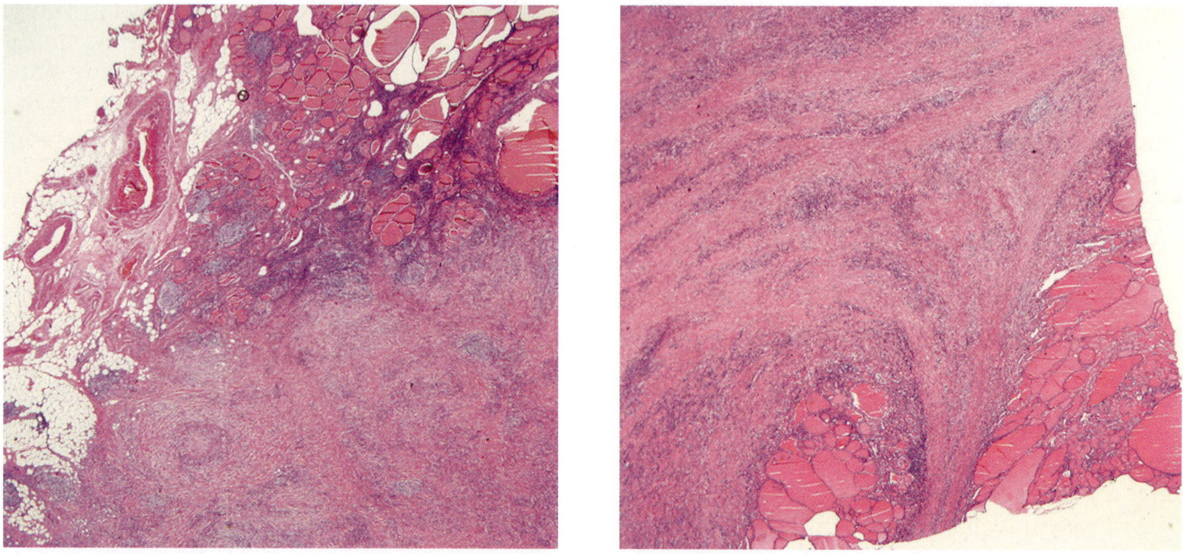

Figure 3-30

RIEDEL THYROIDITIS

Left: An important clue to the diagnosis of Riedel thyroiditis is that the fibrosing process is not confined to the thyroid gland, but rather it extends out of the gland into the adjacent connective tissue.

Right: A characteristic low-power feature is the irregularity of the involvement of the thyroid gland. There are often areas that are completely spared directly adjacent to areas that are obliterated by the fibrosing process.

Microscopic Findings. An important clue to the diagnosis of Riedel thyroiditis is that the fibrosing process is not confined to the thyroid gland, but it extends into the perithyroidal connective tissue (fig. 3-30). In many cases the fibrosis gives the impression of not being centered in the thyroid gland, but rather involving the thyroid together with other structures in the neck. Characteristically, Riedel thyroiditis involves only part of the thyroid gland, with relative sparing of other areas. In involved areas, the normal lobular architecture of the gland is obliterated by dense keloid-like fibrous tissue (fig. 3-31) associated with lymphocytes, plasma cells, a few neutrophils, and eosinophils. This keloid-like fibrosis is unusual in Hashimoto thyroiditis and is useful in the differential diagnosis. The fibrous tissue extends into muscle, nerves, and fat, and entraps blood vessels (fig. 3-32). In approximately 25 percent of cases, the parathyroid glands are also encased (595,598).

There may be an associated vasculitis (predominantly phlebitis) characterized by intimal proliferation and thrombosis (598–600). This occlusive phlebitis results from diffuse infiltration of the walls of small and medium-sized veins by lymphocytes and plasma cells. Adenomas or cysts (degenerative or hemorrhagic) occur within the fibrous mass in approximately 25 percent of cases (584). The relationship between the adenoma and the fibrous reaction is unknown, however, it has been suggested that the fibrous tissue proliferation may be a reaction to the adenoma or its products.

Immunohistochemical Findings. Staining of the plasma cells reveals a predominance of lambda light chains and IgA heavy chains, in contrast to Hashimoto thyroiditis which usually has kappa light chains and IgG heavy chains (601). Analysis of B and T cells shows a similar proportion of CD4-positive helper and CD8-positive suppressor cells, as is seen in Hashimoto thyroiditis (595). Immunofluorescence studies of proteins in human eosinophilic granules show extracellular deposition of major basic protein in Riedel thyroiditis but not in Hashimoto thyroiditis (602). Immunohistochemistry for IgG4 helps identify the subgroup of Riedel thyroiditis patients who have IgG4-related disease.

Differential Diagnosis. Thyroid fibrosis is also seen in Hashimoto thyroiditis and in the diffuse sclerosing variant of papillary thyroid carcinoma. However, the extent of the fibrosis in these disorders is much less and does not extend beyond the thyroid gland as it does in Riedel thyroiditis. Clinical features such as the absence of very high antibody titers as well as histologic features such as the absence of Hurthle cell change and a limited lymphocytic infiltrate, also help distinguish Riedel thyroiditis from Hashimoto disease.

The paucicellular variant of anaplastic thyroid carcinoma may mimic Riedel thyroiditis, especially in small biopsy specimens (604). The presence of marked cellular pleomorphism and atypical mitoses in the tumor cell population should suggest this diagnosis. Elevated serum IgG4 levels support an association with IgG4-related disease in some patients.

Treatment and Prognosis. The treatment of Riedel thyroiditis is aimed at managing both the hypothyroidism if present, and the manifestations related to fibrosclerosis. Hypothyroidism is treated with T4 (levothyroxine), but this treatment rarely has any effect on the goiter or the progressive spread of fibrosclerosis. Surgery is

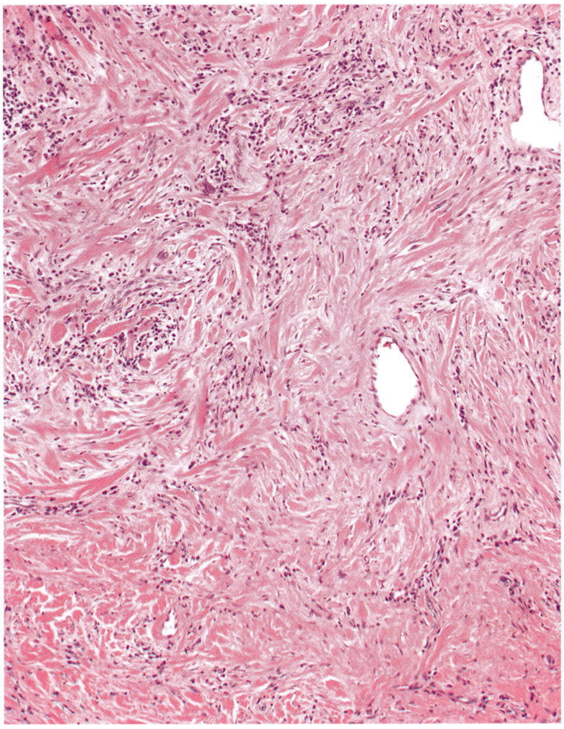

Figure 3-31

RIEDEL THYROIDITIS

The normal lobular pattern is obliterated. Dense "keloid-like" collagen fibers are commonly present.

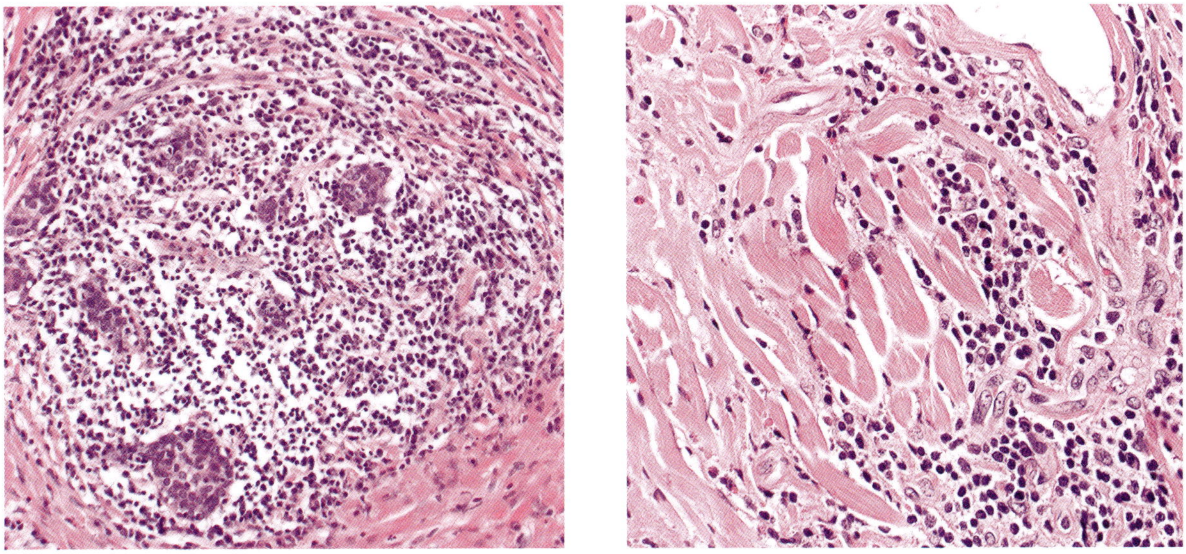

Figure 3-32

RIEDEL THYROIDITIS

Left: A few residual follicles are seen in the fibroinflammatory background.

Right: The infiltrate of lymphocytes and plasma cells, along with the sclerosis, extends beyond the thyroid gland into the adjacent skeletal muscle.

often required to relieve tracheal or esophageal compression or to rule out malignancy (605). The operation should be limited to relieving the obstructive symptoms. Extensive resection (i.e., total thyroidectomy) is not indicated, because of the lack of resection planes and risk of injury to adjacent adhering structures, including the parathyroid glands and recurrent laryngeal nerves. Some medical therapies have been used with variable success. These include glucocorticoids (593,606), tamoxifen (585,607,608), rituximab (609), and mycophenolate mofetil (610). Low-dose radiation therapy has also been used in cases that are refractory to other treatments.

Untreated, Riedel thyroiditis is usually slowly progressive, but may stabilize or even regress spontaneously (590). Mortality is typically attributed to recurrent pneumonia due to bronchial compression and dyspnea.

SIMPLE, MULTINODULAR, DYSHORMONOGENETIC, AND DIFFUSE TOXIC GOITERS

The term *goiter* is nonspecific and refers simply to an enlargement of the thyroid gland. Goiters have a variety of etiologies, ranging from inflammatory diseases, non-neoplastic enlargements, and neoplasms. The presence of a goiter does not reflect the functional activity of the gland, which can range from euthyroid to hyperthyroid and hypothyroid states. Thyroid hyperplasia often results from disturbances of the feedback system in which decreased production of thyroid hormone leads to increased production of TRH from the hypothalamus and TSH from the pituitary. TSH stimulates hyperplasia of the thyroid gland, leading to enlargement of the gland.

Simple and Multinodular Nontoxic Goiter

Definition. *Simple and multinodular nontoxic goiter* is a diffuse enlargement of the thyroid gland with varying degrees of nodularity not associated with hyperthyroidism or hypothyroidism and not resulting from inflammation or neoplasia.

General Features. The incidence of nodular goiter depends on the criteria used to define it, and essentially all thyroid glands show some degree of nodularity histologically. Approximately 2 to 4 percent of the population have a clinical mass. Approximately 10 percent of thyroid glands removed at autopsy contain gross evidence of nodules, usually multiple, and 40 to 50 percent harbor microscopic nodules (611–613). Most simple and multinodular goiters are associated with a euthyroid state, although hyperthyroidism may occur (see Toxic Multinodular Goiter below).

Simple goiters usually weigh more than 40 g, although this may vary with the patient's geographic location. In the pathogenesis of simple and multinodular goiters, it is assumed that there is initially diffuse hyperplasia, which becomes nodular with increasing size, fibrosis, and distortion of the vascular supply of the thyroid gland.

The etiologic stimuli for goiter development are quite variable. Iodine deficiency is the most common cause of goiter worldwide. Low iodine intake leads to reduced T4 and T3 production, which results in increased TSH secretion in an attempt to restore T4 and T3 production to normal. TSH also stimulates thyroid growth; thus, goiter occurs as part of the compensatory response to iodine deficiency. In the United States, where significant iodine deficiency is rare, multinodular goiter, chronic autoimmune (Hashimoto) thyroiditis, and Graves disease are more common causes of goiter. In older adults, multinodular goiter is most common.

Chemically-induced goiters may result from excessive consumption of goitrogenic foods including thiocyanates (found in plant foods such as cassava, cabbage, broccoli, and cauliflower) and perchlorates (found in food containers and food processing equipment), which affect the iodine-trapping mechanisms in the thyroid gland. The thiocarbamides and aniline derivatives, which inhibit the organification of iodine, can also lead to simple goiter. Elements such as iodine and lithium can interfere with the breakdown of thyroglobulin and the release of thyroid hormone.

Drug-induced goiters from compounds such as sulfonamides, phenylbutazones, and phenindione act by inhibiting organification of iodine. Excessive intake of iodine-containing medications, such as amiodarone, may lead to goiters by interfering with thyroglobulin proteolysis, as discussed above. Some natural goitrogens that occur in foods, such as members of the Brassica family, lead to goiters by releasing isothiocyanates, thiocyanates, and thiouracil-like compounds (614–616).

The pathogenesis of simple and multinodular nontoxic goiter is probably related to factors that impair hormone synthesis (617,618). This impairment leads to hypersecretion of TSH, which stimulates thyroid growth and increased thyroid hormone biosynthesis to compensate for the prior impairment. This homeostatic compensation results from the increase in thyroid mass and a greater capacity to produce the required amount of thyroid hormone.

Another theory about the development of simple and multinodular goiters without hyperthyroidism is that the thyroid gland is stimulated by thyroid growth immunoglobulins (TGI) which do not stimulate thyroid adenylate cyclase activity (as TSH autoantibodies do) so that there is no concomitant hyperthyroidism (619–622). The nodularity of the gland in progression from simple to multinodular goiter would be explained by the "marine cycle" in which hyperstimulation of the thyroid by TSH (because of prior decreased hormone production and hypothyroidism) leads to homeostasis; further stimulation leads to focal nodular hyperplasia and this is followed by involutional changes with fibrosis, hemorrhagic changes in blood flow, and dystrophic calcification, leading to more retrogressive changes (623,624).

Molecular genetic analyses have provided some insight into the pathogenesis of adenoma and adenomatous nodules in multinodular goiters. The principle of the Lyon hypothesis is used in these studies, which assumes that during embryonic development in females, one of the two sets of X chromosomes is inactivated, with different X chromosomes inactivated in different cells. There is a random distribution of cells with X inactivation throughout the body (625,626). Many X-linked genes have been used in these studies in which the subject must be a female. In a true neoplasm, the clonal cells express the active X-chromosome marker. A comparison between the normal cells and tumor tissues or nodules shows two bands in the normal tissue and inactivation of one allele in the clonal nodule or tumor. Although the results from different studies have been variable (627–630), when careful morphologic studies are correlated with molecular biological studies, some nodules within multinodular goiters are indeed monoclonal and represent true adenomas. More recent clonality studies have used the androgen receptor on the X chromosome. This technique is more sensitive than earlier X chromosome targets and has the potential to analyze greater than 90 percent of the target female population rather than the 25 percent in earlier studies (631,632). Studies in a few families have implicated a role of heredity in the development of nontoxic goiters (633,634).

Clinical and Radiologic Features. The clinical manifestations of goiter depend upon the presence of thyroid dysfunction and upon the rate of growth of the goiter. Most goiters grow very slowly over many decades, and thus most patients with goiter are asymptomatic. The goiter may first be noted on physical examination or found incidentally on imaging studies performed for other reasons. Some patients have symptoms and biochemical evidence of hypothyroidism or hyperthyroidism. However, most patients with goiter are asymptomatic and biochemically euthyroid.

Patients with large goiters may develop symptoms of obstruction due to progressive compression of the trachea or sudden enlargement (usually accompanied by pain) secondary to hemorrhage into a nodule. The most common symptom in patients with obstructive cervical or substernal goiter is exertional dyspnea, which is present in 30 to 60 percent of patients (635–641). When tracheal compression becomes severe (luminal diameter less than 5 mm), stridor or wheezing occurs at rest (642).

Occasionally, venous engorgement from narrowing of the thoracic inlet may occur. Vocal cord paralysis occasionally results from a large multinodular goiter, although this sign is usually more diagnostic of thyroid carcinoma.

Nontoxic goiters have a female predominance of about 8 to 1. They tend to develop more frequently during adolescence and pregnancy (643). However, since goiters tend to grow slowly, substernal goiters are most commonly discovered during the fifth and sixth decades of life (640,644). Scintigraphic thyroid scanning with radioactive iodine in patients with diffuse or nodular simple goiter often shows a heterogeneous picture (645). Evidence of degenerative changes and calcification may be seen on X ray. The radioactive iodine uptake is usually normal but may be increased due to mild iodine deficiency or other defects. T3 and T4 concentrations are usually normal, although the proportion of T3 to T4 may be increased secondary to defective iodination of thyroglobulin (646). Serum thyroglobulin levels are often elevated.

Gross Findings. Simple goiters are firm and diffusely enlarged, weighing up to 40 g. The cut surface is a uniform amber color with a translucent appearance.

Multinodular goiters range in size from 60 to 1,000 g or more. The gland has a distorted shape and is nodular. Dystrophic calcification is common. Some nodules are partially or completely separated from the gland, constituting sequestered or parasitic nodules. The cut surface shows multiple nodules of varying consistency separated by variable amounts of normal-appearing thyroid tissue. There are areas of nodularity, fibrosis, old and recent hemorrhage, and calcification. The nodules may be variable in size and number, and may be cystic. Some nodules contain a thickened fibrous connective tissue capsule and have the general appearance of follicular neoplasms.

Some multinodular goiters may descend into the anterior mediastinum as substernal or plunging goiters. Extensive hemorrhage in these goiters as part of the degenerative or retrogressive process may lead to tracheal compression and present as a surgical emergency.

Cytologic Findings. Cytologic smears show colloid and a mixed cell population, with few cells in the aspirate, which can include follicular cells, inflammatory cells, and Hurthle cells. Follicular cells are present in sheets of 15 to 100 monolayers, with uniform nuclei. Colloid is present in large globules or as discrete rounded or hexagonal droplets. Hypercellular foci within a multinodular goiter may simulate a follicular neoplasm.

Microscopic Findings. Microscopically, colloid lakes are seen alternating with normal to hyperplastic-appearing foci of thyroid follicles, hemorrhage, siderosis, fibrosis, calcification, and even bone. The follicles vary in size and may be as large as a few millimeters in diameter. They are lined by flattened epithelium with involutional changes. Smaller follicles are lined by epithelial cells that are more columnar (suggestive of more active binding of organic iodine and synthesis of thyroglobulin).

The histology of multinodular goiter parallels the macroscopic features, with nodules consisting of irregularly enlarged follicles of variable size, with distended colloid and flattened epithelium, adjacent to smaller follicles with taller, more active epithelium (fig. 3-33). The larger nodules compress the adjacent parenchyma. Large distended follicles often coalesce to create cystic areas, which may be several millimeters in diameter. Focal areas of papillary hyperplasia with "Sanderson polsters," or a cluster of small follicles protruding into large dilated follicles, may be seen (fig. 3-34). The polster is usually made up of taller, more columnar epithelial cells since the colloid in the cytoplasm adjacent to the follicular cells contains reabsorption vacuoles. There may be evidence of old and recent hemorrhage, with hemosiderin-laden macrophages and ischemic necrosis with lipid-laden macrophages. Cholesterol clefts are usually evident along with extensive irregular fibrosis, variable dystrophic calcification, and ossification.

Hyperplastic nodules with variable degrees of encapsulation are often seen. Some of these nodules may be completely encapsulated and look like adenomas. They are often referred to as *adenomatous* or *adenomatoid nodules* and the goiter as *adenomatous goiter*. Colloid nodules usually refer to the dilated follicles lined by flattened epithelium.

It may be difficult to distinguish on a morphologic basis an adenomatous nodule in multinodular goiter from a true follicular neoplasm arising in this setting, and the distinction may be arbitrary. Clues commonly used to support an adenoma include compression of the adjacent thyroid tissue, a well-formed and complete capsule, and a distinct growth pattern that contrasts with other nodules in the gland. Clonality studies with X-chromosome inactivation have shown that some of these adenomatous nodules are monoclonal (628,629,631). It has been postulated that the hyperplastic process with more rapidly proliferating cells may lead to somatic mutations and neoplastic development in these multinodular goiters (631).

Differential Diagnosis. The differential diagnosis includes other causes of multinodular goiter and neoplasms arising in multinodular

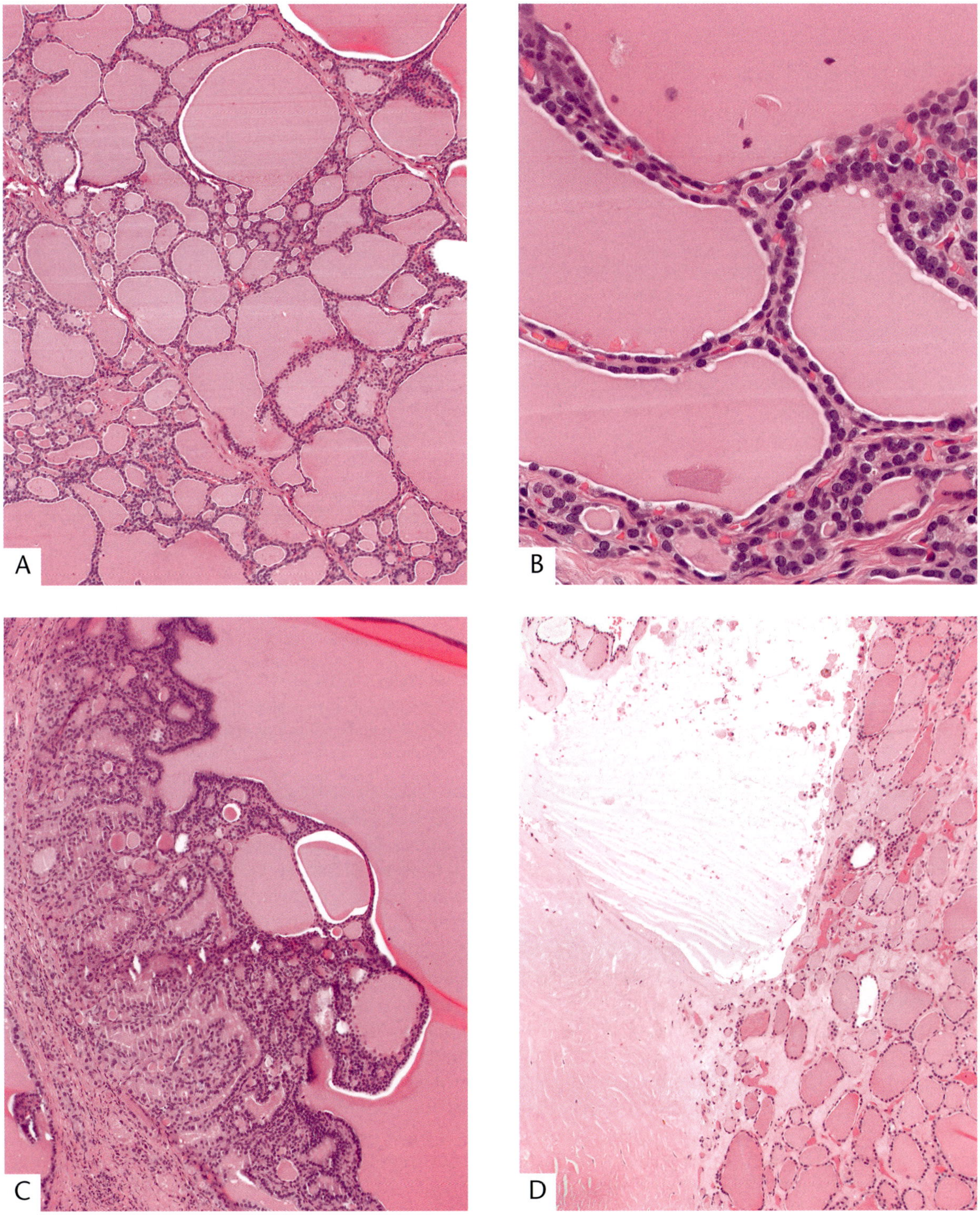

Figure 3-33

MULTINODULAR GOITER

Multiple colloid nodules of variable size (A) and dilated follicles lined by flattened epithelium (B) are present. Other areas show hyperplastic smaller follicles, papillary hyperplasia (C), and retrogressive changes including fibrosis and cystic degeneration (D).

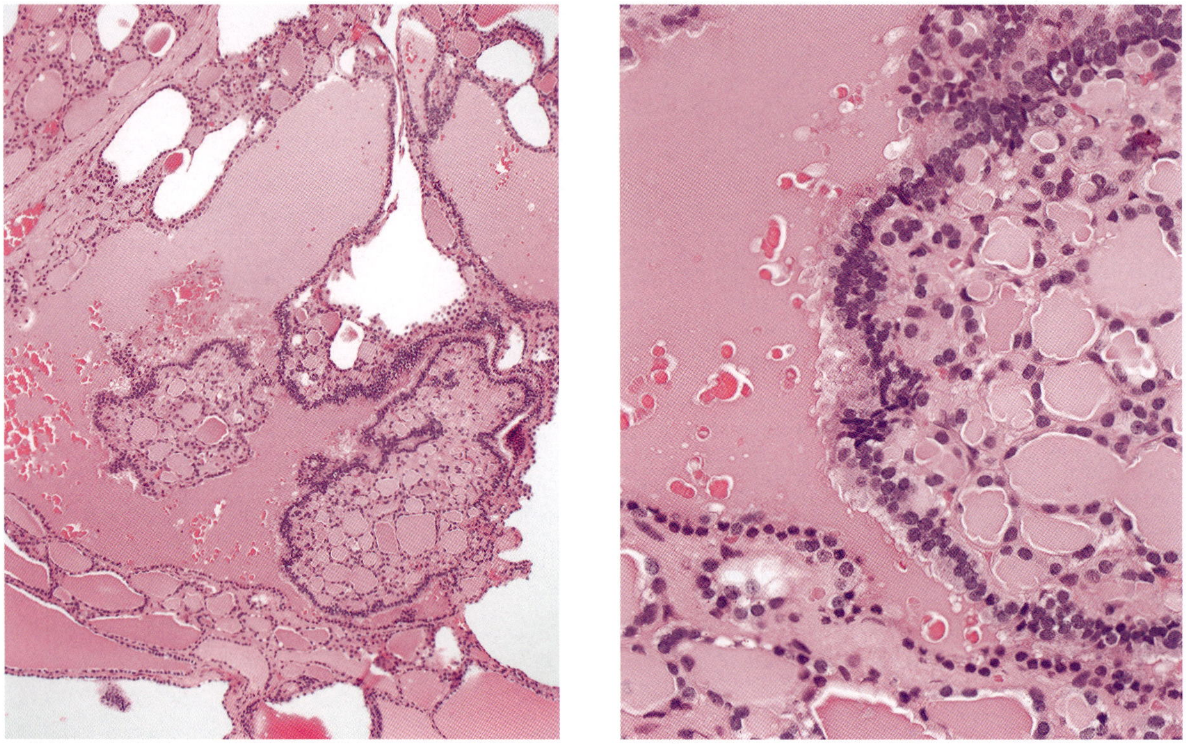

Figure 3-34

MULTINODULAR GOITER

Left: A Sanderson polster is formed by a cluster of small follicles protruding into a large dilated follicle. Right: Higher magnification shows that the polster is lined by more columnar epithelium.

goiter. Multinodular goiters can be confused with dyshormonogenetic goiters, which are caused by thyroid hyperplasia secondary to enzyme defects in thyroid hormone biosynthesis. The cellular appearance of dyshormonogenetic goiter is similar to that of multinodular goiter, but the increased cellularity of the latter is usually focal as opposed to the diffuse involvement in dyshormonogenetic goiter (647).

True follicular neoplasms arising in the setting of multinodular goiter may be difficult to distinguish from a prominent or dominant adenomatous nodule. The distinction between a benign adenoma and an adenomatous nodule in the setting of multinodular goiter is of little clinical significance. However, the distinction between an irregularly contoured nodule and a follicular carcinoma, especially those of the macrofollicular variant, is important. The presence of unequivocal vascular space invasion clearly warrants a diagnosis of carcinoma (648,649). Individual nodules in the setting of multinodular goiter commonly demonstrate contour irregularity which, in different circumstances, may warrant a diagnosis of minimally invasive follicular carcinoma. Features that favor benign adenomatous nodule over carcinoma in the setting of an irregularly contoured nodule include a heterogenous population of small and large follicles; a variably thickened fibrous capsule so that it is most prominent in the area of contour irregularity but minimal or absent elsewhere; the absence of cellular monotony or nuclear atypia (including subtle so called RAS-like atypia); and a similar pattern of growth and morphology in multiple nodules including clearly noninvasive nodules. Although this difficult differential diagnosis may be the cause of significant interobserver disagreement, particularly if only a few slides are reviewed so that the multinodular background is not appreciated, paying careful attention to the pattern of growth elsewhere in the thyroid gland usually resolves the issue. Follicular carcinoma arising in the setting of multinodular goiter is extremely rare (particularly when vascular space invasive carcinomas are excluded), and the prognosis of follicular carcinomas that are negative for vascular space invasion is very good. For this reason, it is extremely unusual for the authors to even consider the diagnosis of minimally invasive follicular carcinoma in the setting of established multinodular goiter unless there is unequivocal vascular space invasion.

Treatment and Prognosis. Asymptomatic, euthyroid patients with benign multinodular goiters do not require any specific treatment, but should be monitored for the development of thyroid dysfunction or for continued growth of the thyroid gland and development of obstructive symptoms. For euthyroid patients with large goiters or goiters that cause obstructive symptoms or cosmetic concerns, total or near-total thyroidectomy is the treatment of choice. Radioiodine therapy is an alternative option for patients who are poor surgical candidates. However, due to high dietary iodine intake in the United States, thyroid iodine uptake is typically low in patients with nontoxic goiter, and therefore, a relatively high dose of radioiodine is required for goiter reduction.

Thyroid hormone suppressive therapy is effective in reducing goiter volume and preventing the development of new nodules in some patients, particularly in patients from regions of the world with borderline or low iodine intake (650–653). However, the efficacy of thyroid hormone suppressive therapy in euthyroid patients with sporadic nontoxic goiter is controversial, and its use has declined due to concerns about potential long-term side effects from the induction of subclinical or overt hyperthyroidism.

The natural history of benign goiter is variable. In up to 20 percent of women and 5 percent of men, the goiter stabilizes or spontaneously regresses over time (654). In other patients, the goiter gradually increases in size over time with the development of additional nodules, compressive symptoms, and cosmetic concerns (655).

Dyshormonogenetic Goiter

Definition. *Dyshormonogenetic goiters* are caused by genetic defects in various pathways involved in thyroid hormone biosynthesis (656).

General Features. Dyshormonogenetic goiter refers to a family of inborn errors of metabolism that lead to defects in the synthesis of thyroid hormone. The prevalence of this disease is 1 in 30,000 to 50,000 live births in Europe and North America (657). It is the second most common cause of permanent congenital hypothyroidism, accounting for 10 to 20 percent of cases (658,659).

Defects in virtually all steps in thyroid hormone biosynthesis and secretion have been described. Various mutations in the *NIS, TPO, DUOX2, TG IYF/DEHAL1,* and *PDS* genes have been primarily implicated (660,661). Most of these mutations are nonsyndromic and familial, and are inherited in an autosomal recessive fashion, except for an occasional autosomal dominantly inherited heterozygous *DUOX2* mutation (662,663). Approximately 500 genetic mutations have been reported thus far (664). The resultant alterations in thyroid gland homeostasis, disturbance of the feedback system, and chronic TSH stimulation lead to enlarged thyroid glands, or goiters.

Thyroid dyshormonogenesis is usually first detected by newborn screening for congenital hypothyroidism (665). Occasionally, patients come to attention when prenatal ultrasonography identifies fetal goiter with no evidence of maternal thyroid disease or antithyroid antibodies. It also may be identified prenatally because of a family history of congenital hypothyroidism with an autosomal recessive pattern of inheritance.

Clinical and Radiologic Features. The clinical presentation depends on the severity of the inborn error. A severe defect leads to neonatal or congenital hypothyroidism, goiter, mental retardation, and growth abnormalities (cretinism) (659). Milder defects present later in life. In a study of 56 patients with dyshormonogenetic goiter, ages ranged from neonates to 52 years, with a median of 16 years at presentation (666). Most goiters (75 percent) develop before 24 years of age. Patients usually present with clinical evidence of a goiter, although sometimes the diagnosis is made at autopsy. Of the 56 cases described above, there were 34 females and 22 males; a family history of goiter or hypothyroidism was present in 11 cases (20 percent). One patient had Pendred syndrome, defined as dyshormonogenetic goiter associated with sensorineural deafness (667,668).

Ultrasound scans at 20 to 22 weeks of gestation can detect fetal thyroid hypertrophy (669). Radionuclide uptake and scans also help in the diagnosis of dyshormonogenetic goiter (658,670). For example, patients with mutations in *NIS,* which result in iodide trapping, show no or very low radioiodine uptake in the thyroid. In iodide organification defects (caused by mutations in *TPO, DUOX2,* and *PDS*), radioiodine uptake is normal, but the perchlorate discharge test is positive because most radioisotope is washed out into the bloodstream instead of accumulating in the thyroid gland (670).

Gross Findings. The thyroid gland is enlarged and multinodular; it can weigh up to 600 g. Fibrous bands encapsulate individual nodules. As with other multinodular goiters, areas of cystic change, fibrosis, and old or recent hemorrhage are present. The cut surface is firm and tan, with nodules of varying size, which may be up to a few centimeters in diameter.

Microscopic Findings. Histologically, the process characteristically affects the entire thyroid gland and no normal thyroid tissue is found. The nodules are hypercellular and show variable architectural patterns including solid, insular, trabecular, and papillary (fig. 3-35). Colloid is usually markedly decreased and in many areas is absent. Some authors have described a characteristic microcystic pattern associated with a thyroglobulin synthetic defect (671). Myxoid change may also be seen. The most common patterns in the nodules are solid and microfollicular. The extensive fibrosis in the internodular tissue may simulate true capsular invasion. The follicular cells typically exhibit severe cytologic atypia, including bizarre and markedly enlarged, hyperchromatic nuclei (659,666). There may be focal areas with clear cytoplasmic change.

Immunohistochemical studies show positive staining for thyroglobulin and negative staining for calcitonin (666). Ultrastructural studies show cells with abundant mitochondria, rough endoplasmic reticulum with dilated cisternae, and tall follicular cells with numerous microvilli (666).

Differential Diagnosis. Usually the diagnosis of dyshormonogenetic goiter is known before surgery, and the primary issue for the surgical pathologist is to confirm or exclude the diagnosis of carcinoma arising in the goiter. This is particularly problematic because many of the nodules are so hypercellular or show such unusual architectures that the diagnosis of poorly differentiated (insular) carcinoma may be considered. It is accepted that this is a challenging differential diagnosis and that there is an increased risk of carcinoma; however, to make the diagnosis of carcinoma arising in dyshormonogenetic goiter, there must be unequivocal vascular invasion, massive extrathyroidal extension, distant metastasis, or architectural and cytologic features of papillary carcinoma (666). Molecular approaches to this differential diagnosis are increasingly being used as it appears that the development of carcinoma requires the accumulation of additional somatic mutations (672).

Treatment and Prognosis. Treatment may be medical (thyroid hormone replacement) or surgical if there is symptomatic enlargement of the thyroid gland. Early treatment is particularly important in severe cases to avoid or diminish mental retardation and growth abnormalities. The prognosis is excellent with early diagnosis and treatment.

Toxic Multinodular Goiter

Definition. *Toxic multinodular goiter* (MNG) is characterized by multiple autonomously functioning nodules, resulting in hyperthyroidism.

General Features. Toxic MNG is the result of focal and/or diffuse hyperplasia of thyroid follicular cells whose functional capacity is independent of regulation by TSH (673). The term usually includes both *toxic MNG (Plummer disease)* and *toxic adenoma* (674,675). Activating somatic mutations of the TSH receptor genes are identified in both toxic adenomas and nodules of toxic MNG (676–678). Activating mutations of the Gs-alpha protein have been identified in toxic adenomas (676,677), but it is uncertain whether they occur in toxic MNG (677). Other mutations are also thought to play a role (679).

Toxic MNG is more common in areas where iodine intake is relatively low (680). In iodine sufficient countries, the prevalence of multinodular goiter is up to 4 percent in adults beyond the age of 30 years (681). The incidence of toxic adenomas is not related to iodine intake.

In some toxic MNGs, hyperthyroidism develops abruptly after exposure to increased amounts of iodine (673). The iodine allows

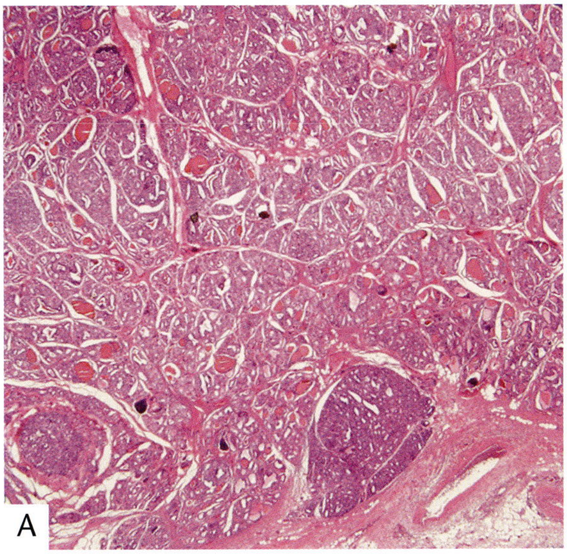

Figure 3-35

DYSHORMONOGENETIC GOITER

A: The entire thyroid gland is abnormal. There are varying areas of hyperplasia, often with an exaggerated lobular architecture.

B: Some of the nodular areas are distinct and give the low-power impression of independent growth.

C: At high power, varying degrees of nuclear atypia are present. In many areas there may be sufficient nuclear atypia to consider the diagnosis of papillary thyroid carcinoma.

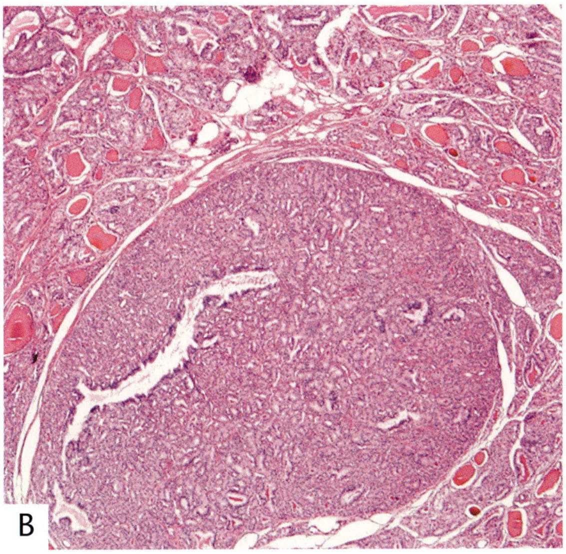

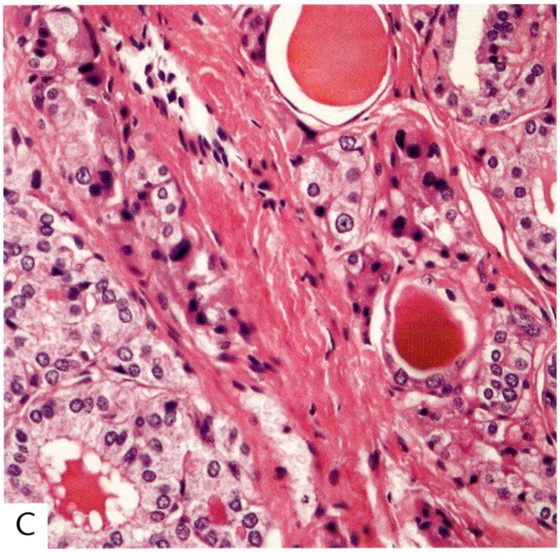

autonomous foci to increase thyroid hormone secretion to high levels, exacerbating chronic mild hyperthyroidism.

Clinical and Radiologic Features. Toxic MNG is much more common in women than in men. It is most often detected as a mass in the neck, but at times, an enlarging gland produces obstructive symptoms. Occasionally, tenderness and a sudden increase in size herald hemorrhage into a cyst. Hyperthyroidism develops in a large proportion of these patients after a few decades, frequently after iodine excess. Rare complications are paralysis of the recurrent laryngeal nerve and Horner syndrome caused by pressure on the superior sympathetic ganglion (682).

Thyroid function tests are normal or show subclinical or overt hyperthyroidism. The overproduction of thyroid hormone in toxic MNG is usually less than in Graves disease. The serum T3 and T4 concentrations may be only slightly increased, while suppressed TSH may be the only early manifestation of this condition (683). Thyroid autoantibodies are usually absent or low. Toxic MNG is rarely associated with eye disease. The principal clinical manifestation in patients with overt hyperthyroidism is cardiovascular, with atrial fibrillation or tachycardia. Weakness and muscle wasting are also common (683).

Ultrasonography may detect nonpalpable nodules, will estimate nodule and goiter size

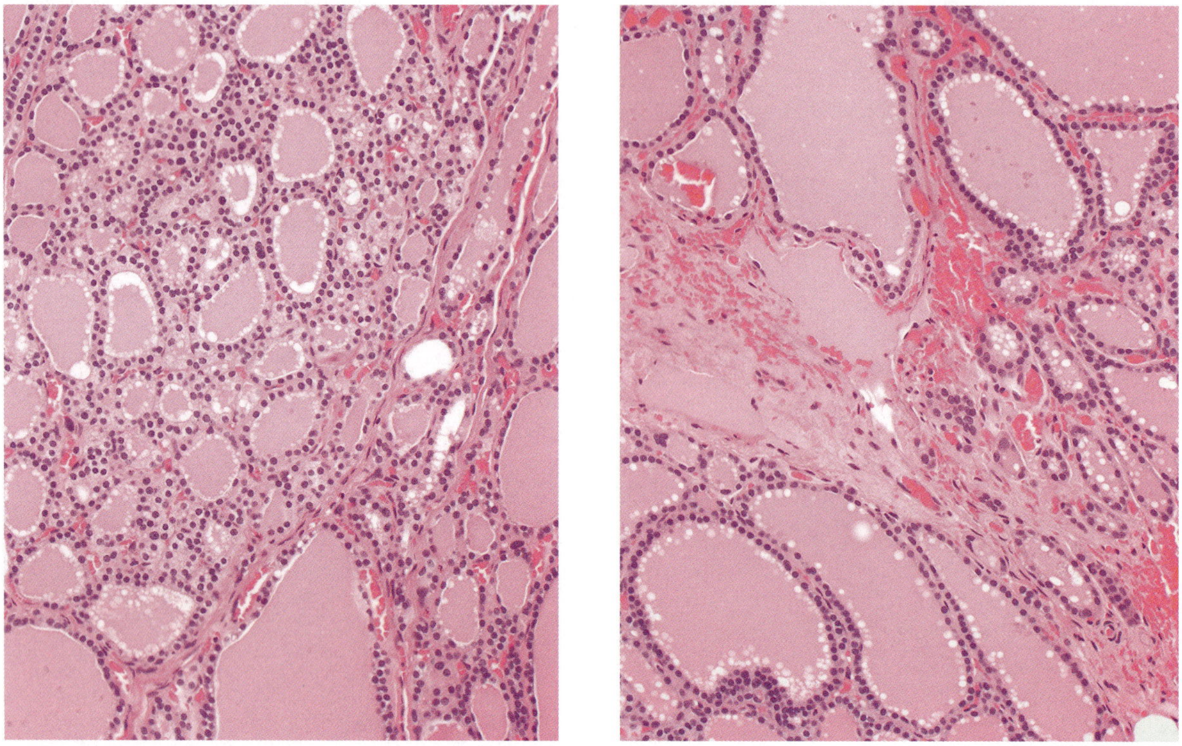

Figure 3-36

TOXIC MULTINODULAR GOITER

Left: Variably hypercellular areas show scalloping of the colloid.
Right: Other areas show dilated follicles, fibrosis, and old hemorrhage in a patient with a toxic goiter.

(volume), and will guide fine needle aspiration biopsy. Hypoechogenicity, microcalcifications, indistinct borders, and increased vascularity may have predictive value in distinguishing malignant from benign nodules. The radioactive iodine uptake is not helpful in establishing the diagnosis because thyrotoxicosis may exist in association with values that are normal or only slightly increased. Imaging studies with radioiodine show localization in one or more discrete nodules of the multinodular gland while iodine accumulation in the remainder of the gland is usually suppressed. TSH administration stimulates iodine uptake in the inactive areas, confirming that the suppression is due to a lack of TSH.

Gross Findings. The thyroid gland is enlarged, with multiple nodules showing fibrosis, old and recent hemorrhage, and dystrophic calcification.

Microscopic Findings. The nodules are variable in appearance. Although it is impossible to morphologically determine whether an individual nodule is hyperfunctioning, hyperactive nodules are often well demarcated and consist of follicles with tall columnar cells and papillary hyperplasia. The other nonfunctional nodules may appear inactive and have degenerative changes as seen in the usual multinodular goiter (fig. 3-36).

Differential Diagnosis. The histologic appearance of a toxic MNG and a nontoxic MNG overlap, so distinguishing these two entities without clinical correlation is impossible. The increased cellularity of individual toxic nodules may resemble dyshormonogenetic goiter, but the changes are focal rather than diffuse. Graves disease is distinguished by the uniformity of changes, with diffuse hyperplasia and absence of nodules in the diffusely enlarged thyroid gland.

Treatment and Prognosis. Treatment of hyperthyroidism due to toxic MNG consists of symptomatic relief with beta-blockers, decreasing the production of thyroid hormone and relieving obstructive symptoms with radioiodine ablation or surgery (684). Patients with symptoms or signs of compression/obstruction, a need for rapid return to euthyroidism, or coexisting thyroid cancer require surgery. In the absence of these indications for surgery, either radioiodine or surgery may be preferred for definitive treatment. Patients who are undergoing surgery should be treated with an antithyroid drug (methimazole or carbimazole) until they are euthyroid. The antithyroid drug should be discontinued immediately postoperatively. Most treated patients have a good prognosis. A worse prognosis is related to untreated hyperthyroidism.

Endemic Goiters

Endemic goiters are those that develop in certain geographic regions, usually secondary to iodine deficiency. They are also termed *iodine deficiency disorder.*

Historically, endemic goiters have developed in geographic areas such as the Great Lakes Region in North America, the mountainous regions of South America, and Western Europe (685,686). The use of the term "iodine deficiency disorder" emphasizes the results of iodine deficiency on mental and physical development (687,688). The incidence of endemic goiter and cretinism decreases markedly when iodine is introduced into the diet as a supplement.

Low iodine intake leads to reduced T4 and T3 production, which results in increased TSH recreation in an attempt to restore T4 and T3 production to normal. TSH also stimulates thyroid growth; thus, goiter occurs as part of the compensatory response to iodine deficiency. There are probably interactions between iodine deficiency and other factors that work permissively to lead to endemic goiters: there are reports that increased calcium and fluoride in the water of iodine-deficient regions contribute to goiter development (689). In some regions of the world, cassava ingestion can interact with iodine deficiency and lead to goiter development (690,691). Contamination of drinking water by sewage has been associated with goiter development in the Himalayas (692). Thus, the interaction of nutritional and environmental factors leading to endemic goiters appears more complex than simply iodine deficiency (690,691).

Endemic goiters are initially diffuse but eventually become nodular because the cells in some of the follicles proliferate more than others, and over time, nodules enlarge and undergo cystic degeneration, hemorrhage, and calcification. In regions of iodine deficiency, children and adolescents generally have diffuse goiters, while adults who have lived in conditions of long-standing iodine deficiency have nodular goiters. The process of thyroid follicular cell replication induced by iodine deficiency increases the chance of developing mutations in the TSH receptor gene, which may lead to constitutive activation of the receptor and thus TSH-independent growth and function (693,694).

The macroscopic and microscopic findings in endemic goiters are similar to those of multinodular goiters, with multiple nodules in a background of degenerative changes, including fibrosis, old and recent hemorrhage, and dystrophic calcification. The histologic picture includes variably sized follicles, with thin follicular cells in dilated colloid follicles admixed with smaller follicles lined by columnar epithelium.

For most patients, iodine-deficiency goiter is only a cosmetic problem. However, in some cases, the goiter may be large enough to cause compression of the trachea or esophagus, or to delay recognition of a coexisting malignancy.

Amyloid Goiter

In the thyroid gland, amyloid is most often detected in the stroma in medullary thyroid cancers. Rarely, it is deposited diffusely in the gland in patients with systemic amyloidosis, sometimes causing goiter and occasionally hypothyroidism (695–697).

Presentation with an *amyloid goiter* (as opposed to its incidental finding in a patient known to have amyloidosis) is rare. The amyloid in the thyroid is usually of the secondary (AA) type, which is associated with chronic infection or inflammation, but primary (AL) amyloid, usually associated with plasma cell abnormalities, may also be present (698,699). Symptomatic amyloid goiter has been reported in 0.04 percent of patients with systemic primary amyloidosis (700).

Patients present with gradual, painless, diffuse, firm thyroid enlargement, occasionally sufficient to cause tracheal compression, sometimes accompanied by chronic hoarseness (700,701). Thyroid ultrasound may show diffuse fatty infiltration. Occasionally, symptomatic

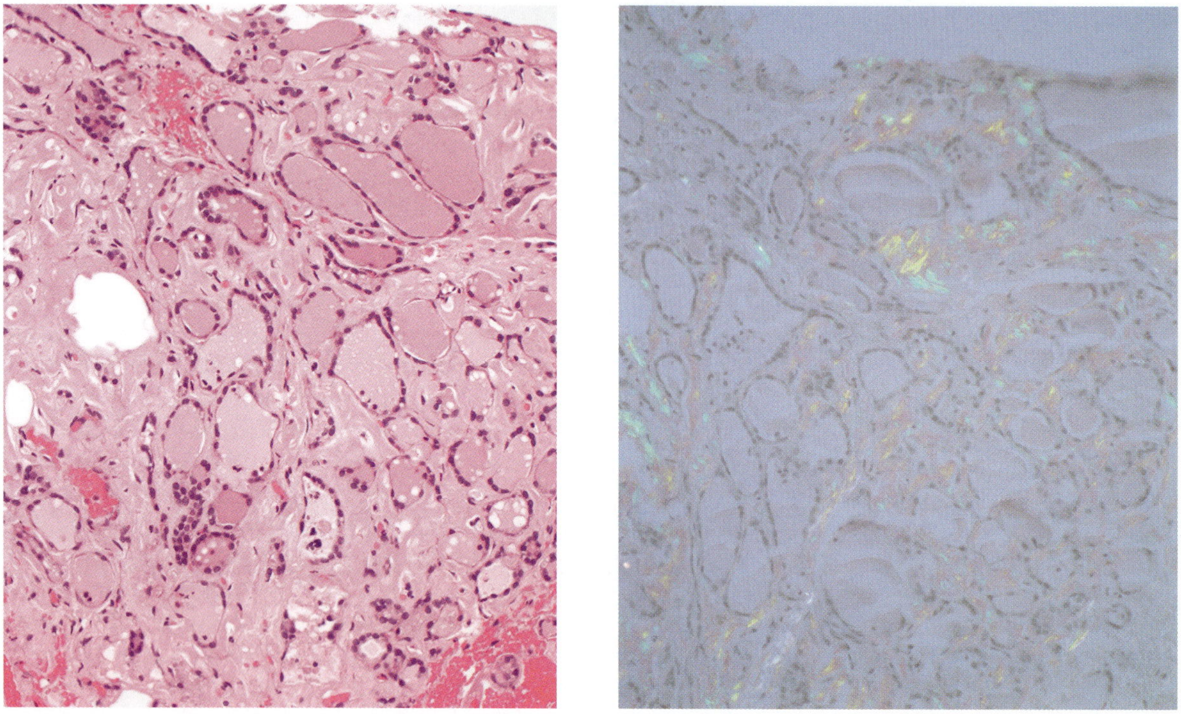

Figure 3-37

AMYLOIDOSIS OF THE THYROID GLAND

Left: H&E-stained section shows deposition of amorphous eosinophilic material surrounding the follicles. Right: Apple-green birefringence after Congo red staining of the amyloid under polarized light.

amyloid goiter is the first manifestation of systemic amyloidosis (702).

The thyroid gland is usually enlarged and has a solid white to tan appearance (703,704). Microscopic examination shows deposition of amorphous eosinophilic fibrillary material between or replacing the thyroid follicles and in the walls of blood vessels (fig. 3-37). Other histologic findings include focal lymphocytic thyroiditis, foreign body giant cells, calcification, and, rarely, focal squamous metaplasia of the follicular epithelium. Focal or diffuse fatty infiltration admixed with the amyloid may be seen (705,706).

Congo red staining shows the characteristic green birefringence of the amyloid deposits under polarized light microscopy. Immunohistochemical staining using antibodies to kappa and lambda light chains, amyloid A protein, and other forms of amyloid can aid in determining the specific amyloid type (695).

The differential diagnosis includes medullary thyroid carcinoma in which the neoplastic cells should be present, amyloidosis associated with multiple myeloma or other plasma cell disorders, and possibly hyalinizing trabecular adenomas. The prognosis of patients with amyloid goiter depends on the underlying disease. Treatment is directed both toward the affected organ and to the specific underlying disease. Hypothyroidism should be corrected, if present, and thyroidectomy may be indicated in patients with obstructive symptoms (701).

C-Cell Hyperplasia

Definition. *C-cell hyperplasia* is an increase in the total mass of C cells within the thyroid gland, usually appearing as multifocal areas and nodules of C cells.

General Features. C cells are often difficult to identify on H&E-stained sections, so calcitonin immunostaining is important for their detection (fig. 3-38). The etiology of C-cell hyperplasia, as well as the definition of this disorder, can be quite variable (707–714). Several different criteria for the diagnosis of C-cell hyperplasia have been proposed, including: clusters of 4 or more C cells (715); greater than 6 C cells per follicle (716); clusters of 20 or more C cells (717); or clusters of at least 50 C cells in one low-power field (707). Regardless of the criteria used, it is clear that when C cells are actively sought by performing immunohistochemistry on otherwise unremarkable thyroid glands, more C cells may be evident than otherwise thought (718).

In a study using autopsy-obtained thyroid glands from 42 adults, it was found that 33 percent (41 percent of males and 15 percent of females) had at least 50 C cells per three 10X fields (720), suggesting that C-cell hyperplasia may be more common than previously thought (707,708,719–721). C-cell hyperplasia occurs more often in men than in women in various studies (720,721).

Some investigators divide C-cell hyperplasia into physiologic and neoplastic types (722,723). Physiologic hyperplasia is observed in neonatal and elderly patients, and is associated with various neoplastic and non-neoplastic conditions including hyperparathyroidism and hypergastrinemia. Neoplastic hyperplasia is associated with multiple endocrine neoplasia (MEN) type 2. It has been suggested that physiologic hyperplasia may be associated with overstimulation by TSH (707,708).

The progression of C-cell hyperplasia to neoplasia is the hallmark of inherited forms of medullary thyroid carcinoma in the setting of familial cancer syndromes associated with *RET* proto-oncogene germline mutations (711,725–730). This is supported by the frequent finding of C-cell hyperplasia in association with tumors from patients with MEN2 syndrome (714). In this model of neoplastic C-cell hyperplasia, the C cell undergoes hypertrophy and focal and diffuse hyperplasia, and then progresses to nodular hyperplasia and medullary thyroid carcinoma (711,725).

The *RET* proto-oncogene, which encodes a member of the tyrosine kinase family of transmembrane receptors, is frequently associated with germline mutations in familial medullary thyroid carcinoma (MTC) and in some sporadic cases of MTC (731–734). Recent studies have shown that RET immunoreactivity is associated with *RET* codon 918 or codon 803 mutations in patients with sporadic MTC (732). However, *RET* mutations and results of immunostaining in C-cell hyperplasia have not been examined.

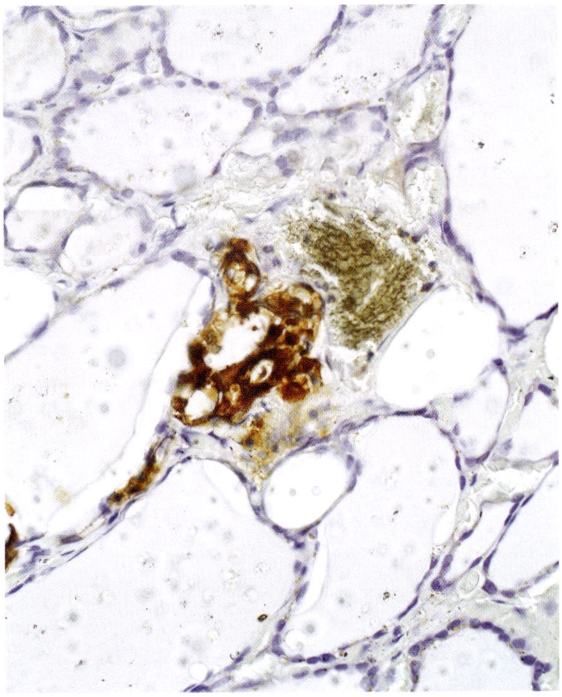

Figure 3-38

C-CELL HYPERPLASIA

Physiologic C-cell hyperplasia shows increased numbers of calcitonin-positive C cells.

Another gene involved in the pathogenesis of MTC is the trk family neurotropin receptor. Of the three members of the trk family, trkB was consistently expressed in C-cell hyperplasia while in normal C cells, only a subset expressed trk receptors (735). TrkB was also reduced in MTC while trkC was increased in these tumors, suggesting that expression of some members of the neurotropin receptor family is turned off or on at various stages in C-cell disease progression (735).

In the era before the widespread availability of genetic testing, the presence of large numbers of C cells displaying a nodular pattern of growth in the non-neoplastic thyroid gland (nodular C-cell hyperplasia) was used as a marker of MEN2. It may still be used in the resource-poor setting to triage patients for genetic testing (730).

As several authors have demonstrated, C-cell hyperplasia is often identified in thyroid glands removed for reasons other than MTC (708,718, 720,736). In a recent study, the number of C cells in 118 completion thyroidectomy specimens from patients without MTC and with no risk factors for

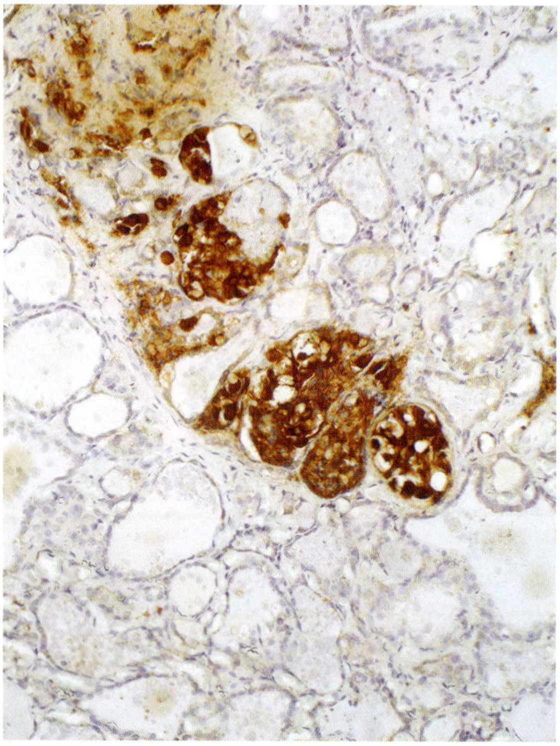

Figure 3-39

C-CELL HYPERPLASIA

C-cell hyperplasia and microscopic medullary thyroid carcinoma in a patient with multiple endocrine neoplasia (MEN) type 2B. After calcitonin immunostaining, the increased numbers of C cells are readily appreciated.

MEN2 was investigated by universal immunohistochemistry; C-cell hyperplasia was found in 4.2 to 30.5 percent of cases depending on the criteria used (718). Therefore, large numbers of C cells sufficient to fulfil criteria for C-cell hyperplasia are a common finding in the general population and do not have a role in triaging patients for genetic testing except in the most resource poor settings. Moreover, there is no evidence that C-cell hyperplasia can progress to MTC outside the setting of familial disease (728).

Gross Findings. There are usually no macroscopic changes associated with C-cell hyperplasia. Rarely, careful sectioning may show small 1- to 2-mm foci in the middle to upper portions of both thyroid lobes.

Microscopic Findings. In the classic case of C-cell hyperplasia, multifocal areas of increased numbers of large amphophilic cells replacing follicular epithelium and also forming nodules that completely replace follicles in the middle and upper regions of the thyroid lobes are seen. Several different diagnostic criteria have been proposed (see above). In the World Health Organization (WHO) 2017 classification, C-cell hyperplasia is defined as greater than 6 to 8 C cells per cluster in several foci with greater than 50 C cells per low-power (10X) field (714).

Immunohistochemical staining is positive for calcitonin, chromogranin A, calcitonin gene-related peptide, and many other peptides in the C cells (727). Ultrastructural studies show a proliferation of C cells within the basement membrane of the follicles, and these are separated from the interstitium by the follicular basal lamina (711,725). There are two types of secretory granules: smaller type II granules, 130 nm in diameter, and larger type I granules, 280 mm. The secretory granules of normal C cells and hyperplastic C cells are similar, while MTC cells have a smaller number of secretory granules. Other distinctive ultrastructural features include a prominent rough endoplasmic reticulum and Golgi complex.

Differential Diagnosis. One of the more difficult diagnostic challenges is the distinction between physiologic and neoplastic C-cell hyperplasia. Demographic data and patient history should help in this distinction. Gross and microscopic findings, such as the presence of a follicular neoplasm, hyperparathyroidism, or hypergastrinemia, suggest a diagnosis of physiologic hyperplasia. Patients with a family history of MEN type 2A or 2B most likely have neoplastic C-cell hyperplasia (fig. 3-39). Histologic features of C-cell hyperplasia associated with MEN2A or MEN2B are cytologic atypia of the C cells, hyperplasia adjacent to the MTC, and bilateral involvement; these features are not seen with physiologic hyperplasia (722). It is not known at this time whether analysis of the *RET* proto-oncogene by molecular or immunohistochemical methods would separate physiologic from preneoplastic C-cell hyperplasia.

In the differential diagnosis of nodular C-cell hyperplasia, it is important to distinguish C-cell nodules from solid cell nests, squamous metaplasia, parathyroid tissue, and thyroid tissue. Solid cell nests represent remnants of the ultimobranchial bodies embedded in the thyroid gland (738). They usually stain for keratin and

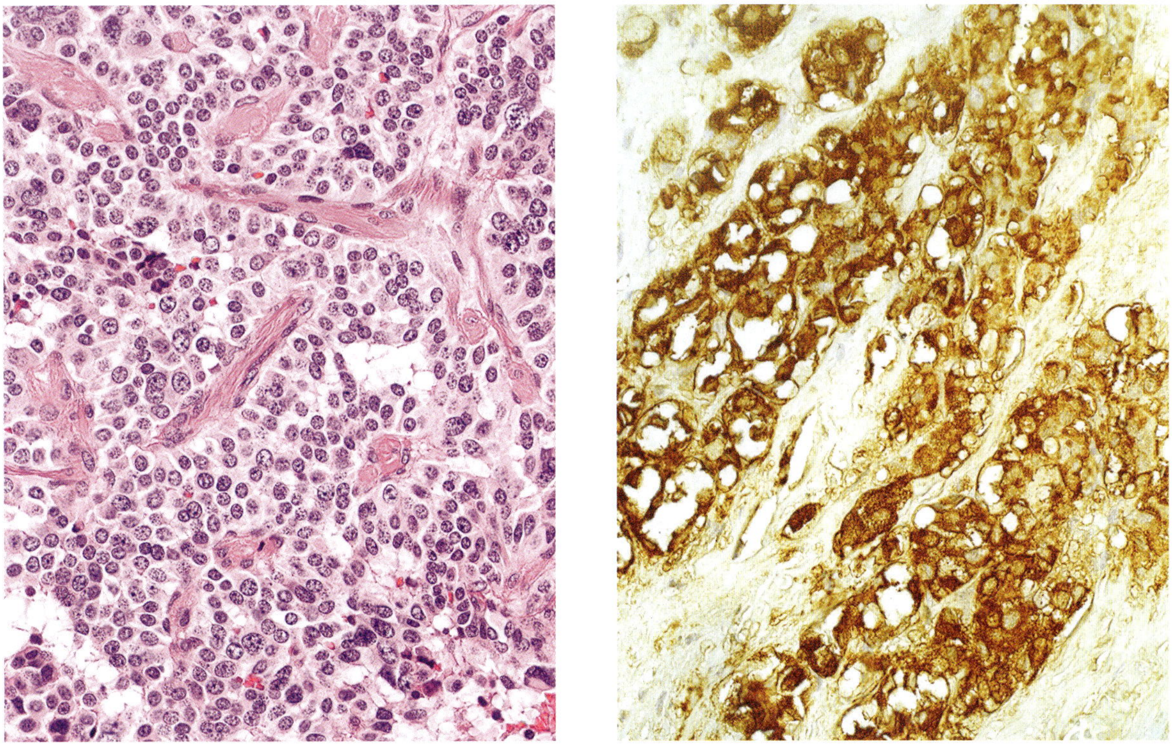

Figure 3-40

MEDULLARY THYROID CARCINOMA

Left: Medullary thyroid carcinoma is composed of spindle and epithelioid cells.
Right: Calcitonin immunostain is strong in the tumor cells.

polyclonal CEA, and may be positive for calcitonin, which makes the distinction even more difficult. The ultrastructural findings of intermediate filaments and cytoplasmic projections that lack dense core secretory granules allow a diagnosis of solid cell nests in difficult cases.

Squamous metaplasia usually occurs with inflammatory conditions such as Hashimoto thyroiditis, including the fibrous variant. These cell nests are more reminiscent of solid cell nests (739). Parathyroid nests and thymic remnants can usually be distinguished from C-cell nests by histologic findings such as distinct and prominent cell membranes in parathyroid cells, the lymphoid elements associated with thymic cells, or immunostaining for calcitonin and parathyroid hormone.

The distinction between nodular C-cell hyperplasia and MTC can also be challenging. The diagnosis of MTC should be made when the C cells have breached the follicular basement membrane (740). A PAS stain or anticollagen IV antibodies can be used to highlight the basement membrane in difficult cases.

Occasionally, nests of thyroid follicular cells may look like foci of nodular C-cell hyperplasia. The nuclear features of neuroendocrine cells are absent in the follicular nests and the cytoplasm does not have the clear to granular basophilic appearance seen in C cells. Staining for calcitonin and thyroglobulin assists in making the distinction.

Treatment and Prognosis. Prophylactic total thyroidectomy is common for patients with confirmed MEN2, and the distinction of C-cell hyperplasia from MTC (fig. 3-40) in this situation helps guide follow-up (741–746).

REFERENCES

Embryology, Anatomy, and Histology

1. Hoyes AD, Kershaw DR. Anatomy and development of the thyroid gland. Ear Nose Throat J 1985;64:318-33.
2. Walls GV, Mihai R. Thyroid gland embryology, anatomy, and physiology. In: Ledbetter DJ, Johnson PR, eds. Endocrine surgery in children. Berlin: Springer Nature; 2018:3-52.
3. Le Douarin N, Fontaine J, Le Lievre C. New studies on the neural crest origin of the avian ultimobranchial glandular cells—interspecies combinations and cytochemical characterization of C-cells based on the uptake of biogenic amine precursors. Histochemistry 1974;38:297-305.
4. Pearse AG, Polak JM. Cytochemical evidence for the neural crest origin of mammalian ultimobranchial C-cells. Histochemie 1971;27:96-102.
5. Mansberger AR Jr, Wei JP. Surgical embryology and anatomy of the thyroid and parathyroid glands. Surg Clin North Am 1993;73:727-46.
6. Ohri AK, Ohri SK, Singh MP. Evidence for thyroid development from the fourth branchial pouch. J Laryngol Otol 1994;108:71-3.
7. Sugiyama S. The embryology of the human thyroid gland including ultimobranchial body and others related. Adv Anat Embryol Cell Biol 1970;44:6-110.
8. De Escobar GM, Obregon MJ, del Rey FR. Maternal thyroid hormones early in pregnancy and fetal brain development. Best Pract Res Clin Endocrinol Metab 2004;18:225-48.
9. Hegedus L, Perrild H, Poulsen LR, et al. The determination of thyroid volume by ultrasound and its relationship to body weight, age, and sex in normal subjects. J Clin Endocrinol Metab 1983;56:260-3.
10. Eales JG. The influence of nutritional state on thyroid function in various vertebrates. Am Zool 1988;20:351-62.
11. Hegedus L, Karstrup S, Rasmussen N. Evidence of cyclic alterations of thyroid size during the menstrual cycle in healthy women. Am J Obstet Gynecol 1986;155:142-5.
12. Brown RA, Al-Moussa M, Beck J. Histometry of normal thyroid in man. J Clin Pathol 1986;39:475-82.
13. LiVolsi VA. Surgical pathology of the thyroid: major problems in pathology, vol. 22. Philadelphia: WB Saunders; 1990.
14. Melander A, Ljunggren JG, Norberg KA, et al. Sympathetic innervation and noradrenaline content of normal human thyroid tissue from fetal, young, and elderly subjects. J Endocrinol Invest 1978;1:175-7.
15. Tice LW, Creveling CR. Electron microscopic identification of adrenergic nerve endings on thyroid epithelial cells. Endocrinology 1975;97:1123-9.
16. Zannini M, Francis-Lang H, Plachov D, DiLauro R. Pax 8, a paired domain containing protein binds to a sequence overlapping the recognition site of a homeodomain and activates transcription from two thyroid-specific promoters. Mol Cell Biol 1992;12:4230-41.
17. Reid JD, Choi CH, Oldroyd NO. Calcium oxalate crystals in the thyroid. Their identification, prevalence, origin, and possible significance. Am J Clin Pathol 1987;87:443-54.
18. Hanson GA, Komorowski RA, Cerletty JM, Wilson SD. Thyroid gland morphology in young adults: normal subjects versus those with prior low-dose neck irradiation in childhood. Surgery 1983;94:984-8.
19. Carcangiu ML. Thyroid. In: Mills SE, ed. Histology for pathologists. Philadelphia: Lippincott Williams & Wilkins 2012;1185-207.
20. Gould VE, Johannesser JV, Sobrinho-Simoes M. The thyroid gland. In: Johannessen JV, ed. Electron microscopy in human medicine, vol. 10. New York: McGraw-Hill; 1981:29-107.
21. Klinck GH, Oertel JE, Winship T. Ultrastructure of normal human thyroid. Lab Invest 1970;22:2-22.
22. Heimann P. Ultrastructure of human thyroid. A study of normal thyroid, untreated and treated diffuse goiter. Acta Endocrinol (Copenh) 1966; 53(Suppl 110):1-102.
23. Ingbar SH. The thyroid gland. In: Wilson JD, Foster DW, eds. Williams textbook of endocrinology, 7th ed. Philadelphia: WB Saunders; 1985:682-815.
24. Kameda Y, Okamoto K, Ito M, Tagawa T. Innervation of the C cells of chicken ultimobranchial glands studied by immunohistochemistry, fluorescence microscopy, and electron microscopy. Am J Anat 1988;182:353-68.
25. Buley ID, Gatter KC, Heryet A, Mason DR. Expression of intermediate filament proteins in normal and diseased thyroid glands. J Clin Pathol 1987;40:136-42.
26. Dockhorn-Dworniczak B, Franke WW, Schroeder S, Czernobilsky B, Gould VE, Bocker W. Pattern of expression of cytoskeletal proteins in human thyroid gland and thyroid carcinomas. Differentiation 1987;35:53-71.

27. Kurata A, Ohta K, Mine M, et al. Monoclonal antihuman thyroglobulin antibodies. J Clin Endocrinol Metab 1984;59:573-9.
28. Stanta G, Carcangiu ML, Rosai J. The biochemical and immunohistochemical profile of thyroid neoplasia. Pathol Annu 1988;23(Pt 1):129-57.
29. LiVolsi VA, Merino MJ. Squamous cells in the human thyroid gland. Am J Surg Pathol 1978;2:133-40.
30. DeLellis RA, Wolfe HJ. The pathobiology of the human calcitonin (C)-cell: a review. Pathol Annu 1981;16:25-52.
31. Scopsi L, Ferrari C, Pilotti S, et al. Immunocytochemical localization and identification of prosomatostatin gene products in medullary carcinoma of human thyroid gland. Hum Pathol 1990;21:820-30.
32. Wolfe HJ, DeLellis RA, Voelkel EF, Tashjian AH Jr. Distribution of calcitonin-containing cells in the normal neonatal human thyroid gland: a correlation of morphology with peptide context. J Clin Endocrinol Metab 1975;41:1076-81.
33. Bejarano PA, Nikiforov YE, Swenson ES, Biddinger PW. Thyroid transcription factor-1, thyroglobulin, cytokeratin 7 and cytokeratin 20 in thyroid neoplasms. Appl Immunohistochem Mol Morphol 2000;8:189-94.
34. Katoh R, Miyagi E, Nakamura N, et al. Expression of thyroid transcription factor-1 in human C-cells and medullary thyroid carcinoma. Hum Pathol 2000;31:386-93.
35. DeLellis RA, Nunnemacher G, Wolfe HJ. C-cell hyperplasia. An ultrastructural analysis. Lab Invest 1977;36:237-48.
36. DeLellis RA, May L, Tashjian AH Jr, Wolfe HJ. C-cell granule heterogeneity in man. An ultrastructural immunohistochemical study. Lab Invest 1978;38:263-9.
37. Hankin RC, Lloyd RV. Detection of messenger RNA in routinely processed tissue sections with biotinylated oligonucleotide probes. Am J Clin Pathol 1989;92:166-71.
38. Zajac JD, Penshow J, Mason T, Tregear G, Coghlan J, Martin TJ. Identification of calcitonin and calcitonin gene-related peptide messenger ribonucleic acid in medullary thyroid carcinomas by hybridization histochemistry. J Clin Endocrinol Metab 1986;62:1037-43.
39. Harach HR. Solid cell nests of the thyroid. J Pathol 1988;155:191-200.
40. Williams ED, Toyn CE, Harach HR. The ultimobranchial gland and congenital thyroid abnormalities in man. J Pathol 1989;159:135-41.
41. Harach HR. Solid cell nests of the human thyroid in early stages of postnatal life. Systematic autopsy study. Acta Anat (Basel) 1986;127:262-4.
42. Fellegara G, Dorji T, Bajimeta MR, Rosai J. Images in pathology. "Giant" solid cell nest of the thyroid: a hyperplastic change? Int J Surg Pathol 2009;17:268-9.
43. Martin V, Martin L, Viennet G, et al. Ultrastructural features of "solid cell nest" of the human thyroid gland: a study of 8 cases. Ultrastruct Pathol 2000;24:1-8.
44. Yamaoka Y. Solid cell nest (SCN) of the human thyroid gland. Acta Pathol Jpn 1973;23:493-506.
45. Cameselle-Teijeiro J, Varela-Durán J, Sambade C, et al. Solid cell nests of the thyroid: light microscopy and immunohistochemical profile. Hum Pathol 1994;25:684-93.
46. Janzer RC, Weber E, Hedinger C. The relation between solid cell nests and C cells of the thyroid gland: an immunohistochemical and morphometric investigation. Cell Tissue Res 1979;197:295-312.
47. Mizukami Y, Nonomura A, Michigishi T, et al. Solid cell nests of the thyroid. A histologic and immunohistochemical study. Am J Clin Pathol 1994;101:186-91.

Physiology

48. Carcangiu ML. Thyroid. In: Mills SE, ed. Histology for pathologists. Philadelphia: Lippincott Williams & Wilkins; 2012:1185-207.
49. Larsen PR, Davies TF, Hay ID. The thyroid gland. In: Wilson JD, Foster DW, Kronenberg HM, Larsen PR, eds. Williams textbook of endocrinology, 9th ed. Philadelphia: WB Saunders; 1998:389-515.
50. Lin JD, Hsueh C, Chao TC, et al. Expression of sodium iodide symporter in benign and malignant human thyroid tissues. Endocr Pathol 2001; 12:15-21.
51. Michalkiewicz M, Huffman LJ, Connors JM, Hedge GA. Alterations in thyroid blood flow induced by varying levels of iodine intake in the rat. Endocrinology 1989;125:54-60.
52. Zannini M, Francis-Lang H, Plachov D, DiLauro R. Pax 8, a paired domain containing protein binds to a sequence overlapping the recognition site of a homeodomain and activates transcription from two thyroid-specific promoters. Mol Cell Biol 1992;12:4230-41.
53. DiLaino R, Obici S, Condliffe D, et al. The sequence of 967 amino acids at the carboxyl end of rat thyroglobulin: location and surroundings of two thyroxine-forming sites. Eur J Biochem 1985;148:7-11.
54. Bjorkman U, Ekholm R, Elmqvist LG, Ericson LE, Melander A, Smeds S. Induced unidirectional transport of protein into the thyroid follicular lumen. Endocrinology 1974;95:1506-17.
55. Ericson LE, Engstrom G. Quantitative electron microscopic studies on exocytosis and endocytosis in the thyroid follicle cell. Endocrinology 1978;103:883-92.

56. Green WI. The physiology of the thyroid gland and its hormones. In: Green WI, ed. The thyroid. New York: Elsevier; 1987:1-46.
57. Ide M. Immunoelectron microscopic localization of thyroglobulin in the human thyroid gland. Acta Pathol Jpn 1984;34:575-84.
58. Sterling K. Thyroid hormone action at the cell level (first of two parts). N Engl J Med 1979;300:117-23.
59. Walls GV, Mihai R. Thyroid gland embryology, anatomy, and physiology. In: Ledbetter DJ, Johnson PR, eds. Endocrine surgery in children. Berlin: Springer Nature; 2018:3-52.
60. Larsen PR. Thyroid-pituitary interaction: feedback regulation of thyrotropin secretion by thyroid hormones. N Engl J Med 1982;306:23-32.
61. Davies T, Marians R, Latif R. The TSH receptor reveals itself. J Clin Invest 2002;110:161-4.
62. Farid NR, Szkudlinski MW. Minireview: structural and functional evolution of the thyrotropin receptor. Endocrinology 2004;145:4048-57.
63. Sinclair AJ, Lonigro R, Civitareale D, Ghibelli L, Di Lauro R. The tissue-specific expression of the thyroglobulin gene requires interaction between thyroid-specific and ubiquitous factors. Eur J Biochem 1990;193:311-8.
64. Larsen PR, Silva JE, Kaplan MM. Relationships between circulating and intracellular thyroid hormones: physiological and clinical implications. Endocr Rev 1981;2:87-102.
65. Pittman JA Jr. Thyrotropin-releasing hormone. Adv Intern Med 1974;19:303-25.
66. Wilber JF. Thyrotropin-releasing hormone: secretion and actions. Ann Rev Med 1973;24:353-64.
67. Austin LA, Heath H. Calcitonin: physiology and pathophysiology. N Engl J Med 1981;304:269-78.
68. Freeman R, Rosen H, Thysen B. Incidence of thyroid dysfunction in an unselected postpartum population. Arch Intern Med 1986;146:1361-4.
69. Glinoer D. The regulation of thyroid function in pregnancy: pathways of endocrine adaptation from physiology to pathology. Endocr Rev 1997;18:404-33.
70. Guillaume J, Schussler GG, Goldman J. Components of the total serum thyroid hormone concentrations during pregnancy: high free thyroxine and blunted thyrotropin (TSH) response to TSH-releasing hormone in the first trimester. J Clin Endocrinol Metab 1985;60:678-84.
71. Ain KB, Mori Y, Refetoff S. Reduced clearance rate of thyroxine-binding globulin (TBG) with increased sialylation: a mechanism for estrogen-induced elevation of serum TBG concentration. J Clin Endocrinol Metab 1987;65:689-96.
72. Burrow GH. Thyroid status in normal pregnancy. J Clin Endocrinol Metab 1990;71:274-5.
73. Ballabio M, Poshychinda M, Ekins RP. Pregnancy-induced changes in thyroid function: role of human chorionic gonadotropin as putative regulator of maternal thyroid. J Clin Endocrinol Metab 1991;73:824-31.
74. Glinoer D, de Nayer P, Bourdoux P, et al. Regulation of maternal thyroid during pregnancy. J Clin Endocrinol Metab 1990;71:276-87.
75. Goodwin TM, Montoro M, Mestman JH, Pekary AE, Hershman JM. The role of chorionic gonadotropin in transient hyperthyroidism of hyperemesis gravidarum. J Clin Endocrinol Metab 1992;75:1333-7.
76. Yeo CP, Khoo DH, Eng PH, Tan HK, Yo SL, Jacob E. Prevalence of gestational thyrotoxicosis in Asian women evaluated in the 8th to 14th weeks of pregnancy: correlations with total and free beta human chorionic gonadotrophin. Clin Endocrinol (Oxf) 2001;55:391-8.
77. Norman RJ, Green-Thompson RW, Jialal I, Soutter WP, Pillay NL, Joubert SM. Hyperthyroidism in gestational trophoblastic neoplasia. Clin Endocrinol 1981;15:395-401.
78. Fisher DA, Odell WD. Acute release of thyrotropin in the newborn. J Clin Invest 1969;48:1670-7.
79. Fisher DA, Brown RS. Thyroid physiology in the perinatal period and during childhood. In: Braverman LE, Utiger RD, eds. Werner's and Ingbar's the thyroid. Philadelphia: Lippincott Williams & Wilkins; 2000:959.
80. Thorpe-Beeston JG, Nicolaides KH, Felton CV, Butler J, McGregor AM. Maturation of the secretion of thyroid hormone and thyroid-stimulatory hormone in the fetus. N Engl J Med 1991;324:532-6.
81. Irvine RE. Thyroid disease in old age. In: Brock Lehurst JC, ed. Text book of pediatric medicine and gerontology. New York: Churchill Livingstone; 1973:435-58.
82. Mariotti S, Franceschi C, Cossarizza A, Pinchera A. The aging thyroid. Endocr Rev 1995;16:686-715.
83. Adler SM, Burman KD. Abnormalities in thyroid function parameters and subclinical thyroid disease in the elderly: a brief review; 2007. http://www.hotthyroidology.com.
84. Hollowell JG, Staehling NW, Hannon WH, et al. Iodine nutrition in the United States. Trends and public health implications: iodine excretion data from National Health and Nutrition Examination Surveys I and III (1971-1974 and 1988-1994) J Clin Endocrinol Metab 1998;83:3401-8.
85. Ajish TP, Jayakumar RV. Gerioatric thyroidology: an update. Indian J Endocrinol Metab 2012;16:542-7.
86. Hennemann G, Docter R, Friesema EC, et al. Plasma membrane transport of thyroid hormones and its role in thyroid hormone metabolism and bioavailability. Endocr Rev 2001;22:451-76.

87. Brent GA, Hershman JM. Thyroxine therapy in patients with severe nonthyroidal illnesses and low serum thyroxine concentration. J Clin Endocrinol Metab 1986;63:1-8.
88. Wartofsky L, Burman KD. Alterations in thyroid function in patients with systemic illness: the "euthyroid sick syndrome." Endocr Rev 1982;3:164-217.
89. Wiersinga WM, Trip MD. Amiodarone and thyroid hormone metabolism. Postgrad Med J 1986;62:909-14.
90. Yen PM. Genomic and nongenomic actions of thyroid hormones. In: Braverman LE, Utiger RD, eds. Werner and Ingbar's the thyroid: a fundamental and clinical text, 9th ed. Philadelphia: Lippincott Williams & Wilkins; 2005:135.
91. Müller MJ, Seitz HJ. Thyroid hormone action on intermediary metabolism. Part I: respiration, thermogenesis and carbohydrate metabolism. Klin Wochenschr 1984;62:11-8.
92. Müller MJ, Seitz HJ. Thyroid hormone action on intermediary metabolism. Part II: lipid metabolism in hyper- and hypothyroidism. Klin Wochenschr 1984;62:49-55.
93. Müller MJ, Seitz HJ. Thyroid hormone action on intermediary metabolism. Part III. Protein metabolism in hyper- and hypothyroidism. Klin Wochenschr 1984;62:97-102.
94. Oppenheimer JH. Thyroid hormone action at the nuclear level. Ann Intern Med 1985;102:374-84.
95. Oppenheimer JH, Samuels HH, eds. Molecular basis of thyroid hormone action. New York: Academic Press; 1983.

Drug Interactions With Thyroid Hormones

96. Gammage MD, Franklyn JA. Amiodarone and the thyroid. Q J Med 1987;238:83-6.
97. Hawthorne GC, Campbell NP, Geddes JS, et al. Amiodarone-induced hypothyroidism. Arch Intern Med 1985;145:1016-9.
98. Martino E, Safran M, Aghini-Lombardi F, et al. Environmental iodine intake and thyroid dysfunction during chronic amiodarone therapy. Ann Intern Med 1984;101:28-34.
99. Hansen JM, Skovsted L, Lauridsen UB, Kirkegaard C, Siersbaek-Nielsen K. The effects of diphenylhydantoin on thyroid function. J Clin Endocrinol Metab 1974;39:785-9.
100. Smyrk TC, Goellner JR, Brennan MD, Carney JA. Pathology of the thyroid in amiodarone-associated thyrotoxicosis. Am J Surg Pathol 1987;11:197-204.
101. Wiersinga WM, Trip MD. Amiodarone and thyroid hormone metabolism. Postgrad Med J 1986;62:909-14.
102. Braverman LE, Ingbar SH, Vagenakis AG, et al. Enhanced susceptibility to iodide myxedema in patients with Hashimoto's disease. J Clin Endocrinol Metab 1971;32:515-21.
103. Stanbury JB, Ermans AE, Bourdoux P, et al. Iodine-induced hyperthyroidism: occurrence and epidemiology. Thyroid 1998;8:83-100.
104. Nademanee K, Piwonka RW, Singh BN, Hershman JM. Amiodarone and thyroid function. Prog Cardiovasc Dis 1989;31:427-37.
105. Fradkin JE, Wolff J. Iodide-induced thyrotoxicosis. Medicine (Baltimore) 1983;62:1-20.
106. Bartalena L, Grasso L, Brogioni S, Aghini-Lombardi F, Braverman LE, Martino E. Serum interleukin-6 in amiodarone-induced thyrotoxicosis. J Clin Endocrinol Metab 1994;78:423-7.
107. Brennan MD, Erickson DZ, Carney JA, Bahn RS. Nongoitrous (type I) amiodarone-associated thyrotoxicosis: evidence of follicular disruption in vitro and in vivo. Thyroid 1995;5:177-83.
108. Lambert M, Unger J, De Nayer P, Brohet C, Gangji D. Amiodarone-induced thyrotoxicosis suggestive of thyroid damage. J Endocrinol Invest 1990;13:527-30.
109. Mete O, Asa SL. Images in endocrine pathology: thyrotoxicosis associated with destructive thyroiditis. Endocr Pathol 2012;23:212-4.
110. Hamnvik OP, Larsen PR, Marqusee E. Thyroid dysfunction from antineoplastic agents. J Natl Cancer Inst 2011;103:1572-87.
111. Jannin A, Penel N, Ladsous M, Vantyghem MC, Cao CD. Tyrosine kinase inhibitors and immune checkpoint inhibitors-induced thyroid disorders. Crit Rev Oncol Hematol 2019;141:23-35.
112. Shinohara N, Takahashi M, Kamishima T, et al. The incidence and mechanism of sunitinib-induced thyroid atrophy in patients with metastatic renal cell carcinoma. Br J Cancer 2011;104:241-7.
113. Brassard M, Neraud B, Trabado S, et al. Endocrine effects of the tyrosine kinase inhibitor vandetanib in patients treated for thyroid cancer. J Clin Endocrinol Metab 2011;96:2741-9.
114. de Groot JW, Zonnenberg BA, Plukker JT, van Der Graaf WT, Links TP. Imatinib induces hypothyroidism in patients receiving levothyroxine. Clin Pharmacol Ther 2005;78:433-8.
115. Abdulrahman RM, Verloop H, Hoftijzer H, et al. Sorafenib-induced hypothyroidism is associated with increased type 3 deiodination. J Clin Endocrinol Metab 2010;95:3758-62.
116. Makita N, Iiri T. Tyrosine kinase inhibitor-induced thyroid disorders: a review and hypothesis. Thyroid 2013;23:151-9.
117. Desai J, Yassa L, Marqusee E, et al. Hypothyroidism after sunitinib treatment for patients with gastrointestinal stromal tumors. Ann Intern Med 2006;145:660-4.

118. Mannavola D, Coco P, Vannucchi G, et al. A novel tyrosine-kinase selective inhibitor, sunitinib, induces transient hypothyroidism by blocking iodine uptake. J Clin Endocrinol Metab 2007;92:3531-4.
119. Wong E, Rosen LS, Mulay M, et al. Sunitinib induces hypothyroidism in advanced cancer patients and may inhibit thyroid peroxidase activity. Thyroid 2007;17:351-5.
120. Verloop H, Smit JW, Dekkers OM. Sorafenib therapy decreases the clearance of thyrotropin. Eur J Endocrinol 2013;168:163-7.
121. Wang D, Gilbert J, Kim YJ. Immunotherapy: who is eligible. Otolaryngol Clin North Am 2018;50:867-74.
122. Alvarez-Sierra D, Marin-Sanchez A, Ruiz-Blazquez P, et al. Analysis of the PD-1/PD-L1 axis in autoimmune thyroid disease: insights into pathogenesis and clues to immunotherapy associated thyroid autoimmunity. J Autoimmun 2019;103:102285.
123. Imblum B, Baloch Z, Fraker D, LiVolsi VA. Pembrolizumab-induced thyroiditis. Endocr Pathol 2019;30:163-7.
124. Osorio JC, Ni A, Chaft JE, et al. Antibody-mediated thyroid dysfunction during T-cell checkpoint blockade in patients with non-small-cell lung cancer. Ann Oncol 2016;28:583-9.
125. Yano S, Ashida K, Nagata H, et al. Nivolumab-induced thyroid dysfunction lacking antithyroid antibody is frequently evoked in Japanese patients with malignant melanoma. BMC Endocr Disord 2018;18:36.
126. Barroso-Sousa R, Barry WT, Garrido-Castro AC, et al. Incidence of endocrine dysfunction following the use of different immune checkpoint inhibitor regimens: a systematic review and meta-analysis. JAMA Oncol 2018;4:173-82.
127. Lee H, Hodi FS, Giobbie-Hurder A, et al. Characterization of thyroid disorders in patients receiving immune checkpoint inhibitory therapy. Cancer Immunol Res 2017;5:1133-40.
128. Zaborowski M, Sywak M, Nylen C, Gill AJ, Chou A. Unique and distinctive histological features of immunotherapy-related thyroiditis. Pathology 2020;52:271-3.
129. Neppl C, Kaderli R, Trepp R, et al. Histology of nivolumab-induced thyroiditis. Thyroid 2018;28:1727-8.
130. Cavlieri RR, Sung LC, Becker CE. Effects of phenobarbital on thyroxine and triiodothyronine kinetics in Graves' disease. J Clin Endocrinol Metab 1973;37:308.
131. Isley WL. Effect of rifampin therapy on thyroid function tests in a hypothyroid patient on replacement L-thyroxine. Ann Intern Med 1987; 107:517-8.
132. Oppenheimer JH, Bernstein G, Surks MI. Increased thyroxine turnover and thyroidal function after stimulation of hepatocellular binding of thyroxine by phenobarbital. J Clin Invest 1968;47:1399-406.
133. Smith PJ, Surks MI. Multiple effects of 5,5'-diphenylhydantoin on the thyroid hormone system. Endocr Rev 1984;5:514.
134. Surks MI, DeFesi CR. Normal serum free thyroid hormone concentrations in patients treated with phenytoin or carbamazepine. A paradox resolved. JAMA 1996;275:1495.
135. Oppenheimer JH, Tavernetti RR. Displacement of thyroxine from human thyroxine binding globulin by analogues of hydantoin. J Clin Invest 1962;41:2213-20.
136. Kuiper JJ. Lymphocytic thyroiditis possibly induced by diphenylhydantoin. JAMA 1969;210:2370-2.
137. Larsen PR, Davies TF, Hay ID. The thyroid gland. In: Wilson JD, Foster DW, Kronenberg HM, Larsen PR, eds. Williams textbook of endocrinology, 9th ed. Philadelphia: WB Saunders; 1998:389-515.
138. Burch HB. Drug effects on the thyroid. N Engl J Med 2019;381:749-61.
139. Brabant G, Brabant A, Ranft U, et al. Circadian and pulsatile thyrotropin secretion in euthyroid man under the influence of thyroid hormone and glucocorticoid administration. J Clin Endocrinol Metab 1987;65:83-8.
140. Brabant G, Prank K, Hoang-Vu C, Hesch RD, von zur Mühlen A. Hypothalamic regulation of pulsatile thyrotopin secretion. J Clin Endocrinol Metab 1991;72:145-50.
141. Lee E, Chen P, Rao H, Lee J, Burmeister LA. Effect of acute high dose dobutamine administration on serum thyrotrophin (TSH). Clin Endocrinol (Oxf) 1999;50:487-92.
142. Agner T, Hagen C, Andersen AN, Djursing H. Increased dopaminergic activity inhibits basal and metoclopramide-stimulated prolactin and thyrotropin secretion. J Clin Endocrinol Metab 1986;62:778-82.
143. Cooper DS, Klibanski A, Ridgway EC. Dopaminergic modulation of TSH and its subunits: in vivo and in vitro studies. Clin Endocrinol (Oxf) 1983;18:265-75.
144. Bertherat J, Brue T, Enjalbert A, et al. Somatostatin receptors on thyrotropin-secreting pituitary adenomas: comparison with the inhibitory effects of octreotide upon in vivo and in vitro hormonal secretions. J Clin Endocrinol Metab 1992;75:540-6.
145. Braverman LE. Iodine induced thyroid disease. Acta Med Austriaca 1990;17(Suppl 1):29-33.

146. Sherman SI, Gopal J, Haugen BR, et al. Central hypothyroidism associated with retinoid X receptor-selective ligands. N Engl J Med 1999;340:1075-9.
147. Haugen BR. Drugs that suppress TSH or cause central hypothyroidism. Best Pract Res Clin Endocrinol Metab 2009;23:793-800.
148. Lamberts SW, Zuyderwijk J, den Holder F, van Koetsveld P, Hofland L. Studies on the conditions determining the inhibitory effect of somatostatin on adrenocorticotropin, prolactin and thyrotropin release by cultured rat pituitary cells. Neuroendocrinology 1989;50:44-50.
149. Murray RD, Kim K, Ren SG, Lewis I, Weckbecker G, Bruns C, Melmed S. The novel somatostatin ligand (SOM230) regulates human and rat anterior pituitary hormone secretion. J Clin Endocrinol Metab 2004;89:3027-32.
150. Vigersky RA, Filmore-Nassar A, Glass AR. Thyrotropin suppression by metformin. J Clin Endocrinol Metab 2006;91:225-7.
151. Cappelli C, Rotondi M, Pirola I, et al. TSH-lowering effect of metformin in type 2 diabetic patients: differences between euthyroid, untreated hypothyroid and euthyroid on l-T4 therapy patients. Diabetes Care 2009;32:1589-90.
152. Isidro ML, Penin MA, Nemina R, Cordido F. Metformin reduces thyrotropin levels in obese, diabetic women with primary hypothyroidism on thyroxine replacement therapy. Endocrine 2007;32:79-82.
153. Faber J, Waetjen I, Siersbaek-Nielsen K. Free thyroxine measured in undiluted serum by dialysis and ultrafiltration: effects of non-thyroidal illness, and an acute load of salicylate or heparin. Clin Chim Acta 1993;223:159-67.
154. McConnell RJ. Abnormal thyroid function test results in patients taking salsalate. JAMA 1992;267:1242-3.
155. Samuels MH, Pillote K, Asher D, Nelson JC. Variable effects of nonsteroidal antiinflammatory agents on thyroid test results. J Clin Endocrinol Metab 2003;88:5710-6.
156. Stockigt JR, Lim CF, Barlow JW, et al. Interaction of furosemide with serum thyroxine-binding sites: in vivo and in vitro studies and comparison with other inhibitors. J Clin Endocrinol Metab 1985;60:1025-31.
157. Jaume JC, Mendel CM, Frost PH, Greenspan FS, Laughton CW. Extremely low doses of heparin release lipase activity into the plasma and can thereby cause artifactual elevations in the serum-free thyroxine concentration as measured by equilibrium dialysis. Thyroid 1996;6:79-83.
158. Perrild H, Hegedüs L, Baastrup PC, Kayser L, Kastberg S. Thyroid function and ultrasonically determined thyroid size in patients receiving long-term lithium treatment. Am J Psychiatry 1990;147:1518-21.
159. Spaulding SW, Burrow GN, Bermudez F, Himmelhoch JM. The inhibitory effect of lithium on thyroid hormone release in both euthyroid and thyrotoxic patients. J Clin Endocrinol Metab 1972;35:905-11.
160. Bocchetta A, Bernardi F, Pedditzi M, et al. Thyroid abnormalities during lithium treatment. Acta Psychiatr Scand 1991;83:193-8.
161. Ermans AM, Delange F, Van der Velden M, Kinthaert J. Possible role of cyanide and thiocyanate in the etiology of endemic cretinism. Adv Exp Med Biol 1972;30:455-86.
162. Wolff J, Chaikoff IL. Plasma inorganic iodide as a homeostatic regulator of thyroid function. J Biol Chem 1948;174:555-64.
163. Abrams GM, Larsen PR. Triiodothyronine and thyroxine in the serum and thyroid glands of iodine-deficient rats. J Clin Invest 1973;52:2522-31.
164. Boyages SC. Clinical review 49: iodine deficiency disorders. J Clin Endocrinol Metab 1993;77:587-91.
165. Boyages SC, Halpern JP. Endemic criticism: toward a unifying hypothesis. Thyroid 1993;3:59-69.

Hereditary and Developmental Disorders

166. Latchman DS. Transcription-factor mutations and disease. N Engl J Med 1996;334:28-33.
167. De Felice M, Di Lauro R. Thyroid development and its disorders: genetics and molecular mechanisms. Endocr Rev 2004;25:722-46.
168. Fabbro D, DiLoreto C, Beltram CA, Belfiore A, DiLauro R, Damante G. Expression of thyroid-specific transcription factors TTF-1 and PAX-8 in human thyroid neoplasms. Cancer Res 1994;54:4744-9.
169. Fagman H, Nilsson M. Morphogenesis of the thyroid gland. Mol Cell Endocrinol 2010; 323:35-54.
170. Fernandez LP, Lopez-Marquez A, Santisteban P. Thyroid transcription factors in development, differentiation and disease. Nat Rev Endocrinol 2015;11:29-42.
171. Guazzi S, Price M, DeFelice M, Damante G, Mattein MG, DiLauro R. Thyroid nuclear factor 1 (TTF-1) contains a homeodomain and displays a novel DNA binding specificity. EMBO J 1990;9:3631-9.
172. Kopp P, Muirhead S, Jourdain N, et al. Congenital hyperthyroidism caused by a solitary toxic adenoma harboring a novel somatic mutation (serine281-->isoleucine) in the extracellular domain of the thyrotropin receptor. J Clin Invest 1997;100:1634-9.

173. Rossi DL, Acebron A, Santisteban P. Function of the homeo and paired domain proteins TTF-1 and Pax-8 in thyroid cell proliferation. J Biol Chem 1995;270:23139-42.
174. Santisteban P, Bernal J. Thyroid development and effect on the nervous system. Rev Endocr Metab Disord 2005;6:217-28.
175. Shimura H, Shimura Y, Ohmuri M, Ikuyama S, Kohn LD. Single-strand DNA-binding proteins and thyroid transcription factor-1 conjointly regulate thyrotropin receptor gene expression. Mol Endocrinol 1995;9:527-39.
176. Zannini M, Francis-Lang H, Plachov D, Di Lauro R. Pax-8, a paired domain containing protein binds to a sequence overlapping the recognition site of a homeodomain and activates transcrip- tion from two thyroid-specific promoters. Mol Cell Biol 1992;12:4230-41.
177. Acebron A, Aza-Blanc P, Rossi DL, Lamas L, Santisteban P. Congenital human thyroglobulin defect due to low expression of the thyroid-specific transcription factor TTF-1. J Clin Invest 1995;96:781-5.
178. Sinclair AJ, Lonigro R, Civitareale D, Ghibelli L, DiLauro R. The tissue-specific expression of the thyroglobulin gene requires intervention between thyroid-specific and ubiquitous factors. Eur J Biochem 1990;193:311-8.
179. Fisher DA, Klein AH. Thyroid development and disorders of thyroid function in the newborn. N Engl J Med 1981;304:702-12.
180. Gruters A, Krude H. Detection and treatment of congenital hypothyroidism. Nat Rev Endocrinol 2011;8:104-13.
181. Park SM, Chatterjee VK. Genetics of congenital hypothyroidism. J Med Genet 2005;42:379-89.
182. Van Vliet G. Development of the thyroid gland: lessons from congenitally hypothyroid mice and men. Clin Genet 2003;63:445-55.
183. Flicek P, Amode MR, Barrell D, et al. Ensembl 2014. Nucleic Acids Res 2014;42:D749-55.
184. Nettore IC, Mirra P, Ferrara AM, et al. Identification and functional characterization of a novel mutation in the NKX2-1 gene: comparison with the data in the literature. Thyroid 2013;23:675-82.
185. Patel NJ, Jankovic J. NKX2-1-Related disorders. In: Adam MP, Ardinger HH, Pagon RA, et al., eds. GeneReviews®[Internet]. Seattle: University of Washington; 2014. https://www.ncbi.nlm.nih.gov/books/NBK185066/.
186. Thorwarth A, Schnittert-Hubener S, Schrumpf P, et al. Comprehensive genotyping and clinical characterisation reveal 27 novel NKX2-1 mutations and expand the phenotypic spectrum. J Med Genet 2014;51:375-87.
187. Carre A, Szinnai G, Castanet M, et al. Five new TTF1/NKX2.1 mutations in brain-lung-thyroid syndrome: rescue by PAX8 synergism in one case. Hum Mol Genet 2009;18:2266-76.
188. Krude H, Schutz B, Biebermann H, et al. Choreoathetosis, hypothyroidism, and pulmonary alterations due to human NKX2-1 haploinsufficiency. J Clin Invest 2002;109:475-80.
189. Willemsen MA, Breedveld GJ, Wouda S, et al. Brain-thyroid-lung syndrome: a patient with a severe multi-system disorder due to a de novo mutation in the thyroid transcription factor 1 gene. Eur J Pediatr 2005;164:28-30.
190. Kimura S. Thyroid-specific transcription factors and their roles in thyroid cancer. J Thyroid Res 2011;2011:710213.
191. Rees Smith B, McLachlan SM, Furmaniak J. Autoantibodies to the thyrotropin receptor. Endocr Rev 1988;9:106-21.
192. Vassart G, Dumont JE. The thyrotropin receptor and the regulation of thyrocyte function and growth. Endocr Rev 1992;13:596-611.
193. Davies TF, Yin X, Latif R. The genetics of the thyroid stimulating hormone receptor: history and relevance. Thyroid 2010;20:727-36.
194. Mitchell PJ, Tijan R. Transcriptional regulation in mammalian cells by sequence-specific DNA binding proteins. Science 1989;245:371-8.
195. Morshed SA, Latif R, Davies TF. Characterization of thyrotropin receptor antibody-induced signaling cascades. Endocrinology 2009;150:519-29.
196. Wonerow P, Neumann S, Gudermann T, Raschke R. Thyrotropin receptor mutations as a tool to understand thyrotropin receptor action. J Mol Med 2001;79:707-21.
197. de Roux N, Misrahi M, Brauner R, et al. Four families with loss of function mutations of the thyrotro- pin receptor. J Clin Endocrinol Metab 1996;81:4229-35.
198. Duprez L, Parma J, van Sande J, et al. Germline mutations in the thyrotropin receptor gene causes non-autoimmune autosomal dominant hyperthyroidism. Nature Genet 1994;7:396-401.
199. Fuhrer D, Holzapfel HP, Wonerow P, Scherbaum WA, Paschke R. Somatic mutations in the thyrotropin receptor gene and not in the Gs alpha protein gene in 31 toxic thyroid nodules. J Clin Endocrinol Metab 1997;82:3885-91.
200. Paschke R, Ludgate M. The thyrotropin receptor in thyroid diseases. N Engl J Med 1997;337:1675-81.
201. Gruters A, Schoneberg T, Biebermann H, et al. Severe congenital hyperthyroidism caused by a germline neo mutation in the extracellular portion of the thyrotropin receptor. J Clin Endocrinol Metab 1998;83:1431-6.
202. Holzapfel HP, Wonerow P, von Petrykowski W, Henschen M, Scherbaum WA, Paschke R. Sporadic congenital hyperthyroidism due to a spontaneous germline mutation in the thyrotropin receptor gene. J ClinEndocrinol Metab 1997;82:3879-84.

203. Karges B, Krause G, Homoki J, Debatin KM, de Roux N, Karges W. TSH receptor mutation V509A causes familial hyperthyroidism by release of interhelical constraints between transmembrane helices TMH3 and TMH5. J Endocrinol 2005;186:377-85.
204. Kopp P, van Sande J, Parma J, et al. Brief report: congenital hyperthyroidism caused by a mutation in the thyrotropin-receptor gene. N Engl J Med 1995;332:150-4.
205. Kleinau G, Vassart G. TSH receptor mutations and diseases. In: Feingold KR, Anawalt B, Boyce A, et al., eds. Endotext [Internet]. South Dartmouth: MDText.com, Inc.; 2000-. https://www.ncbi.nlm.nih.gov/books/NBK279140/.
206. Abramowicz MJ, Duprez L, Parma J, et al. Familial congenital hypothyroidism due to inactivating mutation of the thyrotropin receptor causing profound hypoplasia of the thyroid gland. J Clin Invest 1997;99:3018-24.
207. Ahlbom BD, Yaqoob M, Larsson A, Ilicki A, Anneren G, Wadelius C. Genetic and linkage analysis of familial congenital hypothyroidism: exclusion of linkage to the TSH receptor gene. Hum Genet 1997;99:186-90.
208. Van Sande J, Parma J, Tonacchera M, Swillens S, Dumont J, Vassart G. Somatic and germline mutations of the TSH receptor gene in thyroid diseases. J Clin Endocrinol Metab 1995;80:2577-85.

Aplasia and Hypoplasia

209. Fisher DA, Klein AH. Thyroid development and disorders of thyroid function in the newborn. N Engl J Med 1981;304:702-12.
210. Narumi S, Muroya K, Asakura Y, et al. Transcription factor mutations and congenital hypothyroidism: systematic genetic screening of a population-based cohort of Japanese patients. J Clin Endocrinol Metab 2010;95:1981-5.
211. Carvalho A, Hermanns P, Rodrigues AL, et al. A new PAX8 mutation causing congenital hypothyroidism in three generations of a family is associated with abnormalities in the urogenital tract. Thyroid 2013;23:1074-8.
212. Vilain C, Rydlewski C, Duprez L, et al. Autosomal dominant transmission of congenital thyroid hypoplasia due to loss-of-function mutation of PAX8. J Clin Endocrinol Metab 2001;86:234-8.
213. Baris I, Arisoy AE, Smith A, et al. A novel missense mutation in human TTF-2 (FKHL15) gene associated with congenital hypothyroidism but not athyreosis. J Clin Endocrinol Metab 2006; 91:4183-7.
214. Castanet M, Park SM, Smith A, et al. A novel loss-of-function mutation in TTF-2 is associated with congenital hypothyroidism, thyroid agenesis and cleft palate. Hum Mol Genet 2002;11:2051-9.
215. Doyle DA, Gonzalez I, Thomas B, Scavina M. Autosomal dominant transmission of congenital hypothyroidism, neonatal respiratory distress, and ataxia caused by a mutation of NKX2-1. J Pediatr 2004;145:190-3.
216. Shetty VB, Kiraly-Borri C, Lamont P, Bikker H, Choong CS. NKX2-1 mutations in brain-lung-thyroid syndrome: a case series of four patients. J Pediatr Endocrinol Metab 2014;27:373-8.
217. Acebron A, Aza-Blanc P, Rossi DL, Lamas L, Santisteban P. Congenital human thyroglobulin defect due to low expression of the thyroid-specific transcription factor TTF-1. J Clin Invest 1995;96:781-5.
218. Burke B, Wick M. Thyroid C cells in DiGeorge syndrome. Lab Invest 1986;54:2.

Ectopic Thyroid and Thyroglossal Duct Cysts

219. Baughman RA. Lingual thyroid and lingual thyroglossal tract remnants. A clinical and histopathologic study with review of the literature. Oral Surg Oral Med Oral Pathol 1972;34:781-99.
220. Block MA, Wylie JH, Patton RB, Miller JM. Does benign thyroid tissue occur in the lateral part of the neck? Am J Surg 1966;112:476-81.
221. deSouza FM, Smith PE. Retrosternal goiter. J Otolaryngol 1983;12:393-6.
222. Neinas FW, Gorman CA, Devine KD, Woolner LB. Lingual thyroid. Clinical characteristics of 15 cases. Ann Intern Med 1973;79:205-10.
223. Pollice L, Caruso G. Struma cordis. Ectopic thyroid goiter in the right ventricle. Arch Pathol Lab Med 1986;110:452-3.
224. Rahm J. An unusual heterotopic of thyroid gland tissue. Zbl Allg Pat 1959;99:80-6.
225. Spinner RJ, Moore KL, Gottfried MR, Lowe JE, Sabiston DC Jr. Thoracic intrathymic thyroid. Ann Surg 1994;220:91-6.
226. Di Benedetto V. Ectopic thyroid gland in the submandibular region simulating a thyroglossal duct cyst: a case report. J Pediatr Surg 1997;32:1745-6.
227. Lilley JS, Lomenick JP. Delayed diagnosis of hypothyroidism following excision of a thyroglossal duct cyst. J Pediatr 2013;162:427-8.
228. LiVolsi VA, Perzin KH, Savetsky L. Carcinomas arising in median ectopic thyroid (including thyroglossal duct tissue). Cancer 1974;34:1303-15.
229. Radkowski D, Arnold J, Healy GB, et al. Thyroglossal duct remnants. Preoperative evaluation and management. Arch Otolaryngol Head Neck Surg 1991;117:1378-81.
230. Asp A, Hasbargen J, Blue P, Kidd GS. Ectopic thyroid tissue on thallium/technetium parathyroid scan. Arch Intern Med 1987;147:595-6.

231. LiVolsi VA. Surgical pathology of the thyroid. Major problems in pathology series, vol 22. Philadelphia: WB Saunders Co.; 1990.
232. Meyer JS, Steinberg LS. Microscopically benign thyroid follicles in cervical lymph nodes. Serial section study of lymph node inclusions and entire thyroid gland in 5 cases. Cancer 1969;24:302-11.
233. Moses DC, Thompson NW, Nishiyama RH, Sisson JC. Ectopic thyroid tissue in the neck. Benign or malignant? Cancer 1976;38:361-5.
234. Kakudo K, Shan L, Nakamura Y, Inoue D, Koshiyama H, Sato H. Clonal analysis helps to differentiate aberrant thyroid tissue from thyroid carcinoma. Hum Pathol 1998;29:187-90.
235. Allard RH. The thyroglossal cyst. Head Neck Surg 1982;5:134-46.
236. Brown RL, Azizkhan RG. Pediatric head and neck lesions. Pediatr Clin North Am 1998;45:889-905.
237. Noyek AM, Friedberg J. Thyroglossal duct and ectopic thyroid disorders. Otolaryngol Clin North Am 1981;14:187-201.
238. Ellis PD, van Nostrand AW. The applied anatomy of thyroglossal tract remnants. Laryngoscope 1977;87:765-70.
239. Jain SN. Lingual thyroid. Int Surg 1969;52:320-5.
240. Jaques DA, Chambers RG, Oertel JE. Thyroglossal tract carcinoma. A review of the literature and addition of 18 cases. Am J Surg 1970;120:439-46.
241. Sohn N, Gumport SL, Blum M. Thyroglossal duct carcinoma. N Y State J Med 1974;74:2004-5.
242. Stanley DG, Robinson FW. Thyroid carcinoma in thyroglossal duct : a case report and literature review. Am Surg 1970;36:581-2.
243. Ali AA, Al-Jandan B, Suresh CS, Subaei A. The relationship between the location of thyroglossal duct cysts and the epithelial lining. Head Neck Pathol 2013;7:50-3.
244. Batsakis JG, El-Naggar AK, Luna MA. Thyroid gland ectopias. Ann Otol Rhinol Laryngol 1996;105:996-1000.
245. Phillips PS, Ramsay A, Leighton SE. A mixed thyroglossal cyst. J Laryngol Otol 2004;118:996-8.
246. Tovi F, Barki Y, Maor E. Cartilage within a thyroglossal duct anomaly. Int J Pediatr Otorhinolaryngol 1988;15:205-10.
247. Davis JP, Toma AG, Robinson PJ, Friedmann I. Ossified thyroglossal cyst—is it of embryological significance? J Laryngol Otol 1994;108:168-70.
248. Sistrunk WE. The surgical treatment of cysts of the thyroglossal tract. Ann Surg 1920;71:121-2.
249. Heshmati HM, Fatourechi V, van Heerden JA, Hay JA, Goellner JR. Thyroglossal duct carcinoma: report of 12 cases. Mayo Clin Proc 1997;72:315-9.
250. Peretz A, Leiberman E, Kapelushnik J, Hershkovitz E. Thyroglossal duct carcinoma in children: case presentation and review of the literature. Thyroid 2004;14:777-85.
251. Wood CB, Bigcas JL, Alava I, Bischoff L, Langerman A, Kim Y. Papillary-type carcinoma of the thyroglossal duct cyst: the case for conservative management. Ann Otol Rhinol Laryngol 2018;127:710-6.
252. Nussbaum M, Buchwald RP, Ribner A, Mori K, Litwins J. Anaplastic carcinoma arising from median ectopic thyroid (thyroglossal duct remnant). Cancer 1981;48:2724-8.
253. Stein T, Murugan P, Li F, El Hag MI. Can medullary thyroid carcinoma arise in thyroglossal duct cysts? A search for parafollicular C-cells in 41 resected cases. Head Neck Pathol 2018;12:71-4.
254. Weiss SD, Orlich CC. Primary papillary carcinoma of a thyroglossal duct cyst: report of a case and literature review. Br J Surg 1991;78:87-9.
255. Hans SS, Lee PT, Proctor B. Carcinoma arising in thyroglossal duct remnants. Am Surg 1976; 42:773-7.

Parasitic Nodule

256. Hathaway BM. Innocuous accessory thyroid nodules. Arch Surg 1965;90:222-7.
257. Sisson JC, Schmidt RW, Beierwaltes WH. Sequested nodular goiter. N Engl J Med 1964;270:927-32.
258. Assi A, Sironi M, Di Bella C, Declich P, Cozzi L, Pareschi R. Parasitic nodule of the right carotid triangle. Arch Otolaryngol Head Neck Surg 1996;122:1409-11.
259. Baker LJ, Gill AJ, Chan C, et al. Parasitic thyroid nodules: cancer or not? Endocrinol Diabetes Metab Case Rep 2014;2014:140027.
260. Block MA, Wylie JH, Patton RB, Miller JM. Does benign thyroid tissue occur in the lateral part of the neck? Am J Surg 1966;112:476-81.
261. Klopp CT, Kirson SM. Therapeutic problems with ectopic non-cancerous follicular thyroid tissue in the neck: 18 case reports according to etiological factors. Ann Surg 1966;163:653-64.

Radiation Changes

262. Carr RF, LiVolsi VA. Morphologic changes in the thyroid after irradiation for Hodgkin's and non-Hodgkin's lymphoma. Cancer 1989;64:825-9.
263. Constine LS, Donaldson SS, McDougall IR, Cox RS, Link MP, Kaplan HS. Thyroid dysfunction after radiotherapy in children with Hodgkin's disease. Cancer 1984;53:878-83.
264. Droese M, Kempken K, Schneider ML, Hor G. [Cytologic changes in aspiration biopsy smears from various conditions of the thyroid treated with radioiodine.] Verh Dtsch Ges Pathol 1973; 57:336-8. [German]

265. Favus MJ, Schneider AB, Stachura ME, et al. Thyroid cancer occurring as a late consequence of head-and-neck irradiation. Evaluation of 1,056 patients. N Engl J Med 1976;294:1019-25.
266. Holten I. Acute response of the thyroid to external radiation. Acta Pathol Microbiol Scand Suppl 1983;283:1-111.
267. Kennedy JS, Thomson JA. The changes in the thyroid gland after irradiation with 131I or partial thyroidectomy for thyrotoxicosis. J Pathol 1974;112:65-81.
268. Ron E, Brenner A. Non-malignat thyroid diseases following a wide range of radiation exposures. Radiat Res 2010;174:877-88.
269. Komorowski RA, Hanson GA. Morphologic changes in the thyroid following low-dose childhood radiation. Arch Pathol Lab Med 1977;101:36-9.
270. Fogelfeld L, Wiviott MB, Shore-Freedman E, et al. Recurrence of thyroid nodules after surgical removal in patients irradiated in childhood for benign conditions. N Eng J Med 1989;320:835-40.
271. Schneider AB, Recant W, Pinsky SM, Ryo UY, Bekerman C, Shore-Freedman E. Radiation- induced thyroid carcinoma. Clinical course and results of therapy in 296 patients. Ann Int Med 1986;105:405-12.

Langerhans Cell Histiocytosis and Sinus Histiocytosis with Massive Lymphadenopathy (Rosai-Dorfman Disease)

272. Coode PE, Shaikh MU. Histiocytosis X of the thyroid masquerading as thyroid carcinoma. Hum Pathol 1988;19:239-41.
273. Teja K, Sabio H, Langdon DR, Johanson AJ. Involvement of the thyroid gland in histiocytosis X. Hum Pathol 1981;12:1137-9.
274. Behrens RJ, Levi AW, Westra WH, et al. Langerhans cell histiocytosis of the thyroid: a report of two cases and review of the literature. Thyroid 2001;11:697-705.
275. Foulet-Rogé A, Josselin N, Guyetant S, et al. Incidental langerhans cell histiocytosis of thyroid: case report and review of the literature. Endocr Pathol 2002;13:227-33.
276. Elliott DD, Sellin R, Egger JF, Medeiros LJ. Langerhans cell histiocytosis presenting as a thyroid gland mass. Ann Diagn Pathol 2005;9:267-74.
277. Thompson LD, Wenig BM, Adair CF, et al. Langerhans cell histiocytosis of the thyroid: a series of seven cases and a review of the literature. Mod Pathol 1996;9:145-9.
278. Schofield JB, Alsarjari NA, Davis J, MacLennan KA. Eosinophilic granuloma of lymph nodes associated with metastatic papillary carcinoma of the thyroid. Histopathology 1992;20:181-3.
279. Carpenter RJ III, Banks PM, McDonald TJ, Sanderson DR. Sinus histiocytosis with massive lymphadenopathy (Rosai-Dorfman disease): report of a case with respiratory involvement. Laryngoscope 1978;88:1963-9.
280. Larkin DF, Dervan PA, Munnelly J, Finucane J. Sinus histiocytosis with massive lymphadenopathy simulating subacute thyroiditis. Hum Pathol 1986;17:321-4.

Plasma Cell Granuloma

281. Patil PA, DeLellis RA. Plasma cell granuloma of the thyroid: review of an uncommon entity. Arch Pathol Lab Med 2018;142:998-1005.
282. Holck S. Plasma cell granuloma of the thyroid. Cancer 1981;48:830-2.
283. Yapp R, Linder J, Schenken JR, Karrer FW. Plasma cell granuloma of the thyroid. Hum Pathol 1985;16:848-50.

Scleroderma

284. D'Angelo WA, Fries JF, Masi AT, Shulman LE. Pathologic observations in systemic sclerosis (scleroderma). A study of fifty-eight autopsy cases and fifty-eight matched controls. Am J Med 1969;46:428-40.
285. Antonelli A, Fallahi P, Ferrari SM, et al. Incidence of thyroid disorders in systemic sclerosis: results from a longitudinal follow-up. J Clin Endocrinol Metab 2013;98:E1198.
286. Gordon MB, Klein I, Dekker A, Rodnan GP, Medsger TA Jr. Thyroid disease in progressive systemic sclerosis: increased frequency of glandular fibrosis and hypothyroidism. Ann Intern Med 1981;95:431-5.
287. Nelson DF, Reddy KV, O'Mara RE, Rubin P. Thyroid abnormalities following neck irradiation for Hodgkin's disease. Cancer 1978;42:2553-62.

Metabolic Diseases

288. Chan AM, Lynch MJ, Bailey JD, Ezrin C, Fraser D. Hypothyroidism in cystinosis. A clinical, endocrinologic and histologic study involving sixteen patients with cystinosis. Am J Med 1970;48:678-92.
289. Hui KS, Williams JC, Borit A, Rosenberg HS. The endocrine glands in Pompe's disease. Report of two cases. Arch Pathol Lab Med 1985;109:921-5.
290. Lucky AW, Howley PM, Megyesi K, Spielberg SP, Schulman JD. Endocrine studies in cystinosis: compensated primary hypothyroidism. J Pediatr 1977;91:204-10.
291. Hirschhorn R, Reuser A. Glycogen storage disease type II: acid alpha-glucosidase (acid maltase) deficiency. In: Scriver CR, Beaudet AL, Sly WS, Valle D. The metabolic and molecular bases of inherited disease, 8th ed. New York: McGraw-Hill; 2001:3389.

292. Gahl WA, Thoene JG, Schneider JA. Cystinosis. N Engl J Med 2002;347:111-21.
293. Gahl WA, Bashan N, Tietze F, Bernardini I, Schulman JD. Cystine transport is defective in isolated leukocyte lysosomes from patients with cystinosis. Science 1982;217:1263-5.
294. Gahl WA, Tietze F, Bashan N, Steinherz R, Schulman JD. Defective cystine exodus from isolated lysosome-rich fractions of cystinotic leucocytes. J Biol Chem 1982;257:9570-5.
295. Jonas AJ, Smith ML, Allison WS, et al. Proton-translocating ATPase and lysosomal cystine transport. J Biol Chem 1983;258:11727.
296. Jonas AJ, Smith ML, Schneider JA. ATP-dependent lysosomal cystine efflux is defective in cystinosis. J Biol Chem 1982;257:13185-8.
297. Thoene JG, Oshima RG, Ritchie DG, Schneider JA. Cystinotic fibroblasts accumulate cystine from intracellular protein degradation. Proc Natl Acad Sci U S A 1977;74:4505-7.
298 Town M, Jean G, Cherqui S, et al. A novel gene encoding an integral membrane protein is mutated in nephropathic cystinosis. Nat Genet 1998;18:319-24.
299. Levtchenko E. Endocrine complications of cystinosis. J Pediatr 2017;83S:S5-8.
300. Burke JR, El-Bishti MM, Maisey MN, Chantler C. Hypothyroidism in children with cystinosis. Arch Dis Child 1978;53:947-51.

Lipidosis

301. Armstrong D, Van Warner DE, Neville H, Dimmitt S, Clingan E. Thyroid peroxidase deficiency in Batten-Speilmeyer-Vogt disease. Arch Pathol 1975;99:430-5.
302. Dayan AD, Trickey RJ. Thyroid involvement in juvenile amaurotic family idiocy (Batten's disease). Lancet 1970;2:296-7.
303. Dolman CL, Chang E. Visceral lesions in amaurotic family idiocy with curvilinear bodies. Arch Pathol 1972;94:425-30.

Iron Pigment Accumulation

304. Edwards CQ, Kelly TM, Ellwein G, Kushner JP. Thyroid disease in hemochromatosis. Increased incidence in homozygous men. Arch Intern Med 1983;143:1890-3.
305. el-Reshaid K, Seshadri MS, Hourani H, et al. Endocrine abnormalities in hemodialysis patients with iron overload: reversal with iron depletion. Nutrition 1995;11(Suppl):521-6.
306. Livadas DP, Sofroniadou K, Souvatzoglou A, et al. Pituitary and thyroid insufficiency in thalassaemic haemosiderosis. Clin Endocrinol (Oxf) 1984;20:435-43.
307. Noma S, Konishi J, Morikawa M, Yoshida Y. MR imaging of thyroid hemochromatosis. J Comput Assist Tomogr 1988;12:623-5.
308. Vantyghem MC, Dobbelaere D, Mention K, Wemeau JL, Saudubray JM, Douillard C. Endocrine manifestations related to inherited metabolic diseases in adults. Orphanet J Rare Dis 2012;7:11.
309. Hempenius LM, Van Dam PS, Marx JJ, Koppeschaar HP. Mineralocorticoid status and endocrine dysfunction in severe hemochromatosis. J Endocrinol Invest 1999;22:369-76.

Minocycline and Crystals

310. Alexander CB, Herrara GA, Jaffe K, Yu H. Black thyroid: clinical manifestations, ultrastructural findings and possible mechanisms. Hum Pathol 1985;16:72-8.
311. Benitz KF, Roberts GK, Yusa A. Morphologic effects of minocycline in laboratory animals. Toxicol Appl Pharmacol 1967;11:150-70.
312. Billano RA, Ward WQ, Little WP. Minocycline and black thyroid. JAMA 1983;249:1887.
313. Gordon G, Sparano BM, Kramer AW, Kelley RG, Iatropoulos MJ. Thyroid gland pigmentation and minocycline therapy. Am J Pathol 1984;117:98-109.
314. Kurosumi M, Fujita H. Fine structural aspects of the fate of rat black thyroids induced by minocycline. Virchows Arch B Cell Pathol Incl Mol Pathol 1986;51;207-13.
315. LiVolsi VA. Surgical pathology of the thyroid: major problems in pathology; vol 22. Philadelphia: WB Saunders; 1990:125-8.
316. Matsubara F, Mizukami Y, Tanaka Y. Black thyroid. Morphological, biochemical and geriatric studies on the brown granules in the thyroid follicular cells. Acta Pathol Jpn 1982;32:13-22.
317. Yusim A, Ghofrani M, Ocal IT, Roman S. Black thyroid syndrome. Thyroid 2006;16:811-2.
318. Bell CD, Kovacs K, Horvath E, Rotondo F. Histologic, immunohistochemical, and ultrastructural findings in a case of minocycline-associated "black thyroid." Endocr Pathol 2001;12:443-51.
319. Oertel YC, Oertel JE, Dalal K, Mendoza MG, Fadeyi EA. Black thyroid revisited: cytologic diagnosis in fine-needle aspirates is unlikely. Diagn Cytopathol 2006;34:106-11.
320. Thompson AD, Pasieka JL, Kneafsey P, DiFrancesco LM. Hypopigmentation of a papillary carcinoma arising in a black thyroid. Mod Pathol 1999;12:1181-5.
321. Fayemi AO, Ali M, Braun EV. Oxalosis in hemodialysis patients: a pathologic study of 80 cases. Arch Pathol Lab Med 1979;103:58-62.

322. MacMahon HE, Lee HY, Rivelis CF. Birefringent crystals in human thyroid. Acta Endocrinol 1968;58:172-6.
323. Reid JD, Choi CH, Oldroyd NO. Calcium oxalate crystals in the thyroid. Their identification, prevalence, origin and possible significance. Am J Clin Pathol 1987;87:443-54.
324. Richter MN, McCarty KS. Anisotropic crystals in the human thyroid gland. Am J Pathol 1954;30:545-53.
325. Katoh R, Suzuki K, Hemmi A, et al. Nature and significance of calcium oxalate crystals in normal human thyroid gland. A clinicopathological and immunohistochemical study. Virchows Arch A Pathol Anat Histopathol 1993;422:301-6.
326. Gross S. Granulomatous thyroiditis with anisotropic crystalline material. AMA Arch Pathol 1955;59:412-8.
327. Scully RE, Young RH, Clement PB. Tumors of the ovary, maldeveloped gonads, fallopian tube, and broad ligament. AFIP Atlas of Tumor Pathology, 3rd Series, Fascicle 23. Washington, DC: American Registry of Pathology; 1998:290.
328. Isotalo PA, Lloyd RV. Presence of birefringent crystals is useful in distinguishing thyroid from parathyroid gland tissues. Am J Surg Pathol 2002;26:813-4.

Teflon in Thyroid

329. LiVolsi VA. Surgical pathology of the thyroid: major problems in pathology, vol 22. Philadelphia: WB Saunders; 1990:125-8.
330. Sanfilippo F, Shelburne J, Ingram P. Analysis of a polytef granuloma mimicking a cold thyroid nodule 17 months after laryngeal injection. Ultrastruct Pathol 1980;1:471-5.
331. Walsh FM, Castelli JB. Polytef granuloma clinically simulating carcinoma of the thyroid. Arch Otolaryngol 1975;101:262-3.
332. Rubin HJ. Misadventures with injectable polytef (Teflon). Arch Otolaryngol 1975;101:114-6.

Infectious and Granulomatous Diseases

333. Hazard JB. Thyroiditis: a review. Am J Clin Pathol 1955;25:289-98.
334. LiVolsi V, LoGerfo P. Thyroiditis. Boca Raton: CRC Press; 1981.
335. Volpe R. The pathology of thyroiditis. Hum Pathol 1978;9:429-38.
336. Paes JE, Burman KD, Cohen J, et al. Acute bacterial suppurative thyroiditis: a clinical review and expert opinion. Thyroid 2010;20:247-55.
337. Berger SA, Zonszein J, Villamena P, Mittman N. Infectious diseases of the thyroid gland. Rev Infect Dis 1983;5:108-22.
338. Womack NA. Thyroiditis. Surgery 1944;16:777-82.
339. Volpe R. Acute suppurative thyroiditis. In: Werner SC, Ingbar SH, eds. The thyroid. New York: Harper & Row;1971:852.
340. Weissel M, Wolf A, Linkesch W. Acute suppurative thyroiditis caused by a pseudomonas aeruginosa. Br Med J 1977;2:580.
341. Hagan AD, Goffinet J, Davis JW. Acute streptococcal thyroiditis. JAMA 1967;202:842-3.
342. Robinson MF, Forgan-Smith WR, Craswell PW. Candida thyroiditis—treated with 5-fluoro-cytosine. Aust NZ J Med 1975;5:472-4.
343. Saksouk F, Salti IS. Acute suppurative thyroiditis caused by Escherichia coli. Br Med J 1977;2:23-4.
344. VanHeerden JA, O'Connell P. Acute suppurative thyroiditis due to Salmonella enteritides. Vir Med Mon 1971;98:556-7.
345. Nieburg PI, Gardner LI. Thyroiditis and congenital rubella syndrome. J Pediatr 1976;89:156.
346. Frank TS, LiVolsi VA, Connor AM. Cytomegalovirus infection of the thyroid in immuno- compromised adults. Yale J Biol Med 1987;60:1-8.
347. Guttler R, Singer PA, Axline SG, et al. Pneumocystis carinii thyroiditis. Report of three cases and review of the literature. Arch Intern Med 1993;153:393-6.
348. McAninch EA, Xu C, Lagari VS, Kim BW. Coccidiomycosis thyroiditis in an immunocompromised host post-transplant: case report and literature review. J Clin Endocrinol Metab 2014;99:1537-42.
349. Raman L, Murray J, Banka R. Primary tuberculosis of the thyroid gland: an unexpected cause of thyrotoxicosis. BMJ Case Rep 2014;2014:bcr201320792.
350. Larsen PP, Daves TI, Hay ID. The thyroid gland. In: Wilson JD, Foster DW, Kronenberg HM, Larson PR, eds. Williams textbook of endocrinology, 9th ed. Philadelphia: WB Saunders; 1990:479.
351. Thomas CG. Acute supperative thyroiditis: surgical treatment. In: Werner SC, Ingbar SH, eds. The thyroid. New York: Harper & Row; 1971:852.

Granulomatous Thyroiditis

352. Furszyfer J, McConahey WM, Wahner HW, Kurland LT. Subacute (granulomatous) thyroiditis in Olmsted County, Minnesota. Mayo Clinic Proc 1970;45:396-404.
353. Lazarus JH. Silent thyroiditis and subacute thyroiditis. In: Braverman LE, Utiger RD, eds. Werner and Ingbar's the thyroid: a fundamental and clinical text, 7th ed. Philadelphia: Lippincott Williams & Wilkins; 1996:577.
354. Lindsay S, Dailey ME. Granulomatous or giant cell thyroiditis; a clinical and pathologic study of thirty-seven patients. Surg Gynecol Obstet 1954;98:197-212.

355. Meachim G, Young MH. DeQuervain's subacute granulomatous thyroiditis: histological identification and incidence. J Clin Pathol 1963;16:189-99.
356. de Pauw BE, de Rooy HA. DeQuervain's subacute thyroiditis. A report of 14 cases and a review of the literature. Neth J Med 1975;18:70-8.
357. Singer PA. Thyroiditis. Acute, subacute and chronic. Med Clin North Am 1991;75:61-77.
358. Swann NH. Acute thyroiditis. Five cases associated with adenovirus infection. Metabolism 1964;13:908-10.
359. Nyulassy S, Hnilica P, Buc M, Guman M, Hirschova V, Stefanovic J. Subacute (deQuervain's) thyroiditis: associated with HLA-Bw35 antigen and abnormalities of the complement system, immunoglobulins and other serum proteins. J Clin Endocrinol Metab 1977;45:270-4.
360. Ohsako N, Tamai H, Sudo T, et al. Clinical characteristics of subacute thyroiditis classified according to human leukocyte antigen typing. J Clin Endocrinol Metab 1995;80:3653-6.
361. Rotenberg Z, Weinberger I, Fuchs J, Maller S, Agmon J. Euthyroid atypical subacute thyroiditis simulating systemic or malignant disease. Arch Intern Med 1986;146:105-7.
362. Fatourechi V, Aniszewski JP, Fatourechi GZ, et al. Clinical features and outcome of subacute thyroiditis in an incidence cohort: Olmsted County, Minnesota, study. J Clin Endocrinol Metab 2003;88:2100-5.
363. Larsen PR. Serum triiodothyronine, thyroxine and thyrotropin during hyperthyroid, hypothyroid and recovery phases of subacute nonsuppurative thyroiditis. Metabolism 1974;23:467-71.
364. Volpe R. The management of subacute (DeQuervain's) thyroiditis. Thyroid 1993;3:253-5.
365. Volpe R. Subacute (deQuervain's) thyroiditis. Clin Endocrinol Metab 1979;8:81-95.
366. Volpe R. Thyroiditis: current review of pathogenesis. Med Clin North Am 1975;59:1163-75.
367. Benker G, Olbricht T, Windeck R, et al. The sonographic and functional sequelae of deQuervain's subacute thyroiditis: long-term follow-up. Acta Endocrinol (Copenh) 1988;117:435-41.
368. Hay ID. Thyroiditis: a clinical update. Mayo Clinic Proc 1985;60:836-43.
369. Woolner LB. Thyroiditis: classification and clinicopathologic correlation. In: Hazard JB, Smith DE, eds. The thyroid. Baltimore: Williams & Wilkins; 1964:23.
370. Woolner LB, McConahey WB, Beahrs OH. Granulomatous thyroiditis (De Quervain's thyroiditis). J Clin Endocrinol Metab 1957;17:1202-21.
371. Sasaki H, Harada T, Eimoto T, Matsuoka Y, Okumura M. Concomitant association of thyroid sarcoidosis and Hashimoto's thyroiditis. Am J Med Sci 1987;249:441-3.
372. Abdulsalam F, Abdulaziz S, Mallik AA. Primary tuberculosis of the thyroid gland. Kuwait Med J. 2005;37:116-8.
373. Mpikashe P, Sathekge MM, Mokgoro NP, Ogunbanjo GA. Tuberculosis of the thyroid gland: a case report: case study. South Afr Fam Pract 2004;46:19-20.
374. Zivaljevic V, Paunovic I, Diklic A. Tuberculosis of thyroid gland: a case report. Acta Chir Belg 2007;107:70-2.
375. Gupta KB, Gupta R, Varma M. Tuberculosis of the thyroid gland. Pulmon 2008;9:65-8.
376. Markowicz H, Shanon E. Tuberculosis of the thyroid gland: report of a case. Ann Otol Rhinol Laryngol 1958;67:223-6.
377. Sachs MK, Dickinson G, Amazon K. Tuberculous adenitis of the thyroid mimicking subacute thyroiditis. Am J Med 1988;85:573-5.
378. Goldani LZ, Zavascki AP, Maia AL. Fungal thyroiditis: an overview. Mycopathologia 2006; 161:129-39.
379. Kakudo K, Kanokogi M, Mitsunobu M, et al. Acute mycotic thyroiditis. Acta Pathol Jpn 1983;33:147-51.
380. Manson CM, Cross P, De Sousa B. Post-operative necrotizing granulomas of the thyroid. Histopathology 1992;21:392-3.
381. Wilson GE. Haboubi NY, McWilliam LJ, Hirsch PJ. Postoperative necrotizing granulomata in the cervix and ovary. J Clin Pathol 1990:43;1037-8.
382. Schmitz KJ, Baumgaertek MW, Schmidt C, Sheu SY, Betzler M, Schmid KW. Wegener's granulomatosis in the thyroid mimicking a tumour. Virchows Arch 2008;452:571-4.
383. Yermakov VM, Hitti IF, Sutton AL. Necrotizing vasculitis associated with diphenylhydantoin: two fatal cases. Hum Pathol 1983;14:182-4.
384. Karlish AJ, MacGregor GA. Sarcoidosis, thyroiditis and Addison's disease. Lancet 1970;2:330-3.
385. Mizukami Y, Nomomura A, Michigishi T, Ohmura K, Matsubara S, Noguchi M. Sarcoidosis of the thyroid gland manifested initially as thyroid tumor. Pathol Res Pract 1994;190:1201-5.
386. Warshawsky ME, Shanies HM, Rozo A. Sarcoidosis involving the thyroid and pleura. Sarcoidosis Vasc Diffuse Lung Dis 1997;14:165-8.
387. Winnacker JL, Becker KL, Katz S. Endocrine aspects of sarcoidosis. N Engl J Med 1968; 278:483-92.
388. Manchanda A, Patel S, Jiang JJ, Babu AR. Thyroid: an unusual hideout for sarcoidosis. Endocr Pract 2013;19:e40-3.
389. Vailati A, Marena C, Aristia L, et al. Sarcoidosis of the thyroid: report of a case and a review of the literature. Sarcoidosis 1993;10:66-8.

390. Salih AM, Fatih SM, Kakamad FH. Sarcoidosis mimicking metastatic papillary thyroid cancer. Int J Surg Case Rep 2015;16:71-2.

Palpation Thyroiditis

391. Baloch ZW, LiVolsi VA. Thyroid and parathyroid. In: Mills SE, Greenson JK, Hornick JL, Longacre TA, Reuter VE, eds. Sternberg's diagnostic pathology, 6th ed. Philadelphia: Wolters Kluwer Health; 2015:595-646.
392. Espiritu RP, Dean DS. Parathyroidectomy-induced thyroiditis. Endocr Pract 2010;16:656-9.
393. Kobayashi A, Kuma K, Matsuzuka F, Hirai K, Fukata S, Sugawara M. Thyrotoxicosis after needle aspiration of thyroid cyst. J Clin Endocrinol Metab 1992;75:21-4.
394. Leckie RG, Buckner AB, Bornemann M. Seat belt-related thyroiditis documented with thyroid Tc-99m pertechnetate scans. Clin Nucl Med 1992;17:859-60.
395. Mai VQ, Glister BC, Clyde PW, Shakir KM. Palpation thyroiditis causing new-onset atrial fibrillation. Thyroid 2008;18:571-3.
396. Stang MT, Yim JH, Challinor SM, et al. Hyperthyroidism after parathyroid exploration. Surgery 2005;138:1058-64.

Alterations Following Fine Needle Aspiration Biopsy

397. Baloch ZW, LiVolsi VA. Post fine-needle aspiration histologic alterations of thyroid revisited. Am J Clin Pathol 1999;112:311-6.
398. Layfield LJ, Lones MA. Necrosis in thyroid nodules after fine needle aspiration biopsy. Report of two cases. Acta Cytol 1991;35:427-30.
399. LiVolsi VA, Merino MJ. Worrisome histologic alterations following fine-needle aspiration of the thyroid (WHAFFT). Pathol Annu 1994;29:99-120.

Hashimoto Thyroiditis

400. Ban Y, Tozaki T, Tobe T, et al. The regulatory T cell gene FOXP3 and genetic susceptibility to thyroid autoimmunity: an association analysis in Caucasian and Japanese cohorts. J Autoimmun 2007;28:201-7.
401. Dittmar M, Kahaly GJ. Polyglandular autoimmune syndromes: immunogenetics and long-term follow-up. J Clin Endocrinol Metab 2003;88:2983-92.
402. Baloch ZW, Solomon AC, LiVolsi VA. Primary mucoepidermoid carcinoma and sclerosing mucoepidermoid carcinoma with eosinophilia of the thyroid gland: a report of nine cases. Mod Pathol 2000;13:802-7.
403. Tomer Y, Menconi F, Davies TF, et al. Dissecting genetic heterogeneity in autoimmune thyroid diseases by subset analysis. J Autoimmun 2007;29:69-77.
404. Tomer Y, Davies TF. Infection, thyroid disease, and autoimmunity. Endocr Rev 1993;14:107-20.
405. Hashimoto H. Zur Kenntnis der lymphomatöen Veränderung der Schilddrüse (Struma lymphomatosa). Arch F Klin Chir 1912;97:219-48.
406. Larsen PR, Davies TF, Hay ID. The thyroid gland. In: Wilson JD, Foster DW, Kronenberg AM, Larsen PR, eds. Williams textbook of endocrinology, 9th ed. Philadelphia: WB Saunders; 998:389-515.
407. Hollowell JG, Staehling NW, Flanders WD, et al. Serum TSH, T(4), and thyroid antibodies in the United States population (1988 to 1994): National Health and Nutrition Examination Survey (NHANES III). J Clin Endocrinol Metab 2002;87:489-99.
408. Del Prete GF, Maggi E, Mariotti S, et al. Cytolytic T lymphocytes with natural killer activity in thyroid infiltrates of patients with Hashimoto's thyroiditis: analysis at the clonal level. J Clin Endocrinol Metab 1986;62:52-7.
409. Tomer Y, Davies TF. Searching for the autoimmune thyroid disease susceptibility genes: from gene mapping to gene function. Endocr Rev 2003;24:694-717.
410. Tamai H, Kimura A, Dong RP, et al. Resistance to autoimmune thyroid disease is associated with HLA-DQ. J Clin Endocrinol Metab 1994;78:94-7.
411. Burman P, Totterman TH, Oberg K, Karlsson FA. Thyroid autoimmunity in patients on long-term therapy with leukocyte-derived interferon. J Clin Endocrinol Metab 1986;63:1086-90.
412. Chopra IJ, Solomon DH, Chopra V, Yoshihara E, Terasaki PI, Smith F. Abnormalities in thyroid function in relatives of patients with Graves' disease and Hashimoto's thyroiditis: lack of correlation with inheritance of HLA-B8. J Clin Endocrinol Metab 1977;45:45-54.
413. Menconi F, Monti MC, Greenberg DA, et al. Molecular amino acid signatures in the MHC class II peptide-binding pocket predispose to autoimmune thyroiditis in humans and in mice. Proc Natl Acad Sci U S A 2008;105:14034-9.
414. Honda K, Tamai H, Morita T, Kumar K, Nishimura Y, Sasazuki T. Hashimoto's thyroiditis and HLA in Japanese. J Clin Endocrinol Metab 1989;69:1268-73.
415. Doniach D, Boltazzo GF, Russell RC. Goitrous autoimmune thyroiditis (Hashimoto's disease). Clin Endocrinol Metab 1979;8:63-80.
416. Ewins DL, Rossor MN, Butler J, Roques PK, Mullan MJ, McGregor AM. Association between autoimmune thyroid disease and familial Alzheimer's disease. Clin Endocrinol 1991;35:93-6.
417. Kennedy RL, Jones TH, Cuckle HS. Down's syndrome and the thyroid. Clin Endocrinol 1992;37:471-6.

418. Roitt IM, Doniach D. A reassessment of studies on the aggregation of thyroid autoimmunity in families of thyroiditis patients. Clin Exp Immunol 1967;2:727-36.
419. Harach HR, Escalante DA, Onativia A, Lederer Outes J, Sarvia Duy E, Williams ED. Thyroid carcinoma and thyroiditis in endemic goiter region before and after iodine prophylaxis. Acta Endocrinol (Copenh) 1985;108:55-60.
420. Trip MD, Wiersinga W, Plomp TA. Incidence, predictability, and pathogenesis of amiodaroneinduced thyrotoxicosis and hypothyroidism. Am J Med 1991;91:507-11.
421. Lazarus JH, John R, Bennie EH, Crockett RJ, Crockelt G. Lithium therapy and thyroid function: a long-term study. Psychol Med 1981;11:85-92.
422. Sauter NP, Atkins MB, Mier JW, Lechan RM. Transient thyrotoxicosis and persistent hypothyroidism due to acute autoimmune thyroiditis after interleukin-2 and interferon-alpha therapy for metastatic carcinoma: a case report. Am J Med 1992;92:441-4.
423. Somerset DA, Zheng Y, Kilby MD, et al. Normal human pregnancy is associated with an elevation in the immune suppressive CD25+ CD4+ regulatory T-cell subset. Immunology 2004;112:38-43.
424. Raghupathy R. Th1-type immunity is incompatible with successful pregnancy. Immunol Today 1997;18:478-82.
425. Othman S, Phillips DI, Parkes AB, et al. A longterm follow-up of postpartum thyroiditis. Clin Endocrinol (Oxf) 1990;32:559-64.
426. Srinivasappa J, Garzelli C, Onodera T, Ray U, Notkins AL. Virus-induced thyroiditis. Endocrinology 1988;122:563-6.
427. Okayasu I, Hara Y, Nakamura K, Rose NR. Racial and age-related differences in incidence and severity of focal autoimmune thyroiditis. Am J Clin Pathol 1994;101:698-702.
428. Bagchi N, Brown TR, Parish RF. Thyroid dysfunction in adults over 55 years. A study in an urban U.S. community. Arch Intern Med 1990;150:785-7.
429. Parle JV, Franklyn JA, Cross KW, Jones SC, Sheppard MC. Prevalence and follow-up of abnormal thyrotropin (TSH) concentration in the elderly in the United Kingdom. Clin Endocrinol 1991;34:77-83.
430. Weetman AP, McGregor AM. Autoimmune thyroid disease: further developments in our understanding. Endocr Rev 1994;15:788-830.
431. Dayan CM, Daniels GH. Chronic autoimmune thyroiditis. N Engl J Med 1996;335:99-107.
432. Heufelder AE, Wenzel BE, Gorman CA, Bahn RS. Detection, cellular localization, and modulation of heat shock proteins in cultured fibroblasts from patients with extrathyroidal manifestations of Graves' disease. J Clin Endocrinol Metab 1991;73:739-45.
433. Louis DN, Vickery AL Jr, Rosai J, Wang CA. Multiple branchial cleft-like cysts in Hashimoto's thyroiditis. Am J Surg Pathol 1989;13:45-9.
434. Maran AG, Buchanan DR. Branchial cysts, sinuses, and fistulae. Clin Otolaryngol Allied Sci 1978;3:77-92.
435. Srinivasappa J, Saegusa J, Prabhakar BS, et al. Molecular mimicry: frequency of reactivity of monoclonal antiviral antibodies with normal tissues. J Virol 1986;57:397-401.
436. Valtonen VV, Ruutu P, Varis K, Ranki N, Malkamaki M, Makela PH. Serological evidence for the role of bacterial infections in the pathogenesis of thyroid diseases. Acta Med Scand 1986;219:105-11.
437. Arata N, Ando T, Unger P, Davies TF. By-stander activation in autoimmune thyroiditis: studies on experimental autoimmune thyroiditis in the GFP+ fluorescent mouse. Clin Immunol 2006;121:108-17.
438. Horwitz MS, Bradley LM, Harbertson J, T Krahl, Lee J, Sarvetnick N. Diabetes induced by Coxsackie virus: initiation by bystander damage and not molecular mimicry. Nat Med 1998;4:781-5.
439. Bottazzo GF, Todd I, Pujol-Borrell R. Hypotheses on genetic contributions to the aetiology of diabetes mellitus. Immunol Today 1984;5:230-1.
440. Londei M, Bottazzo GF, Feldmann M. Human T-cell clones from autoimmune thyroid gland: specific recognition of autologous thyroid cells. Science 1985;228:85-9.
441. Hanafusa T, Pujol-Borrell R, Chiovato L, Russell RC, Doniach D, Bottazzo GF. Aberrance expression of HLA-DR antigen on thyrocytes in Graves' disease: relevance for autoimmunity. Lancet 1983;2:1111-5.
442. Lloyd RV, Johnson TL, Blaivas M, Sisson JC, Wilson BS. Detection of HLA-DR antigens in paraffin-embedded thyroid epithelial cells with a monoclonal antibody. Am J Pathol 1985;120:106-11.
443. Eguchi K, Matsuoka N, Nagataki S. Cellular immunity in autoimmune thyroid disease. Balliere's Clin Endocrinol Metab 1995;9:71-94.
444. Konishi J, Iida Y, Kasagi K, et al. Primary myxedema with thyrotrophin-binding inhibitor immunoglobulins. Clinical and laboratory findings in 15 patients. Ann Intern Med 1985; 103:26-31.
445. Kraiem Z, Lahat N, Glaser B, et al. Thyrotrophin receptor blocking antibodies: incidence, characterization and in-vitro synthesis. Clin Endocrinol (Oxf) 1987;27:409-21.

446. Kahaly GJ, Diana T, Glang J, et al. Thyroid stimulating antibodies are highly prevalent in hashimoto's thyroiditis and associated orbitopathy. J Clin Endocrinol Metab 2016;101:1998-2004.
447. Giordano C, Stassi G, De Maria R, et al. Potential involvement of Fas and its ligand in the pathogenesis of Hashimoto's thyroiditis. Science 1997;275:960-3.
448. Rallison ML, Dobyns BM, Meikle AW, Bishop M, Lyon JL, Stevens W. Natural history of thyroid abnormalities: prevalence, incidence, and regression of thyroid disease in adolescents and young adults. Am J Med 1991;91:363-70.
449. Rotondi M, Capelli V, Locantore P, et al. Painful Hashimoto's thyroiditis: myth or reality? J Endocrinol Invest 2017;40:815-8.
450. Zimmerman RS, Brennan MD, McConahey WM, Goellner JR, Gharib H. Hashimoto's thyroiditis. An uncommon cause of painful thyroid unresponsive to corticosteroid therapy. Ann Intern Med 1986;104:355-7.
451. Fatourechi V, McConahey WM, Woolner LB. Hyperthyroidism associated with histologic Hashimoto's thyroiditis. Mayo Clin Proc 1971; 46:682-9.
452. Kraiem Z, Baron E, Kahana L, et al. Changes in stimulating and blocking TSH receptor antibodies in a patient undergoing three cycles of transition from hypo to hyper-thyroidism and back to hypothyroidism. Clin Endocrinol (Oxf) 1992;36:211-4.
453. Takasu N, Yamada T, Sato A, et al. Graves' disease following hypothyroidism due to Hashimoto's disease: studies of eight cases. Clin Endocrinol (Oxf) 1990;33:687-98.
454. Ramtoola S, Maisey MN, Clarke SE, Fogelman I. The thyroid scan in Hashimoto's thyroiditis: the great mimic. Nucl Med Commun 1988;9:639-45.
455. de Kerdanet M, Lucas J, Lemee F, Lecornu M. Turner's syndrome with X-isochromosome and Hashimoto's thyroiditis. Clin Endocrinol (Oxf) 1994;41:673-6.
456. Nordmeyer JP, Shafeh TA, Heckmann C. Thyroid sonography in autoimmune thyroiditis: a prospective study on 123 patients. Acta Endocrinol 1990;122:391-5.
457. Sostre S, Reyes MM. Sonographic diagnosis and grading of Hashimoto's thyroiditis. J Endocrinol Invest 1991;14:115-21.
458. Amino N, Hagen SR, Yamada N, Refetoff S. Measurement of circulating thyroid microsomal antibodies by the tanned red cell hemagglutination technique: its usefulness in the diagnosis of autoimmune thyroid diseases. Clin Endocrinol 1976;5:115-25.
459. Diana T, Krause J, Olivo PD, et al. Prevalence and clinical relevance of thyroid stimulating hormone receptor-blocking antibodies in autoimmune thyroid disease. Clin Exp Immunol 2017;189:304-9.
460. Hayashi Y, Tamai H, Fukata S, et al. A long term clinical, immunological, and histological follow-up study of patients with goitrous chronic lymphocytic thyroiditis. J Clin Endocrinol Metab 1985;61:1172-8.
461. Iwatani Y, Amino N, Mori H, et al. T lymphocyte subsets in autoimmune thyroid diseases and subacute thyroiditis detected with monoclonal antibodies. J Clin Endocrinol Metab 1982;56:251-4.
462. Reidbord HE, Fisher ER. Ultrastructural features of subacute granulomatous thyroiditis and Hashimoto's disease. Am J Clin Pathol 1973; 59:327-37.
463. Knecht H, Hedinger CE. Ultrastructural Hashimoto's thyroiditis and focal lymphocytic thyroiditis with reference to giant cell formation. Histopathology 1982;6:511-38.
464. Katz SM, Vickery AL Jr. The fibrous variant of Hashimoto's thyroiditis. Hum Pathol 1974;5: 161-70.
465. Davies TF, Martin A, Concepcion ES, Graves P, Cohen L, Ben-Nun A. Evidence of limited variability of antigen receptors on intrathyroidal T cells in autoimmune thyroid disease. N Engl J Med 1991;325:238-44.
466. Rallison ML, Dobyns BM, Keating FR, Rall JE, Tyler FH. Occurrence and natural history of chronic lymphocytic thyroiditis in childhood. J Pediatr 1975;86:675-82.
467. Woolner LB. Thyroiditis classification and clinicopathologic correlation. In: Hazad JB, Smith DE, eds. The thyroid. Baltimore: Williams & Wilkins; 1964:123-42.
468. Ha CS, Shadle KM, Medeiros LJ, et al. Localized non-Hodgkin lymphoma involving the thyroid gland. Cancer 2001;91:629-35.
469. Thieblemont C, Mayer A, Dumontet C, et al. Primary thyroid lymphoma is a heterogeneous disease. J Clin Endocrinol Metab 2002;87:105-11.
470. Kovacs CS, Mant MJ, Nguyen GK, Ginsberg J. Plasma cell lesions of the thyroid: report of a case of solitary plasmacytoma and a review of the literature. Thyroid 1994;41:65-71.
471. LiVolsi VA. The pathology of autoimmune thyroid disease: a review. Thyroid 1994;4:333-9.
472. Holm LE, Blomgren H, Lowhagen T. Cancer risks in patients with chronic lymphocytic thyroiditis. N Engl J Med 1985;312:601-4.
473. Kato I, Tajima K, Suchi T, et al. Chronic thyroiditis as a risk factor for B cell lymphoma in the thyroid gland. Jpn J Cancer Res 1985;76:1085-90.
474. Woolner LB, McConahey WM, Beahrs OH. Struma lymphomatosa (Hashimoto's thyroiditis) and related thyroidal disorders. J Clin Endocrinol Metab 1959;19:53-83.

475. Isaacson PG, Androulakis-Papachristou A, Diss TC, Pan L, Wright DH. Follicular colonization in thyroid lymphoma. Am J Pathol 1992; 141:43-52.
476. Zinzani PL, Magagnoli M, Galieni P, et al. Nongastrointestinal low-grade mucosa-associated lymphoid tissue lymphoma: analysis of 75 patients. J Clin Oncol 1999;17:1254-60.
477. Schubert MF, Kountz DS. Thyroiditis: a disease with many faces. Postgrad Med 1995;98:101-3.
478. Mariotti S, Caturegli P, Piccolo P, Barbesino G, Pinchera A. Antithyroid peroxidase autoantibodies in thyroid diseases. J Clin Endocrinol Metab 1990;71:661-9.
479. Takasu N, Yamada T, Takasu M, et al. Disappearance of thyrotropin-blocking antibodies and spontaneous recovery from hypothyroidism in autoimmune thyroiditis. N Engl J Med 1992;326:513-8.
480. Gonzalez-Aguilera B, Betea D, Lutteri L, et al. Conversion to Graves disease from Hashimoto thyroiditis: a study of 24 patients. Arch Endocrinol Metab 2018;62:609-14.
481. Matsuzuka F, Miyauchi A, Katayama S, et al. Clinical aspects of primary thyroid lymphoma: diagnosis and treatment based on our experience of 119 cases. Thyroid 1993;3:93-9.
482. Oertel JE, Hefferss CS. Lymphoma of the thyroid and related disorders. Semin Oncol 1987;14:333-42.
483. Galofré JC, Haber RS, Mitchell AA, et al. Increased postpartum thyroxine replacement in Hashimoto's thyroiditis. Thyroid 2010;20:901-8.

Graves Disease

484. Larsen PR, Davies TF, Hay ID. The thyroid gland. In: Wilson JD, Foster DW, Kronenberg AM, Larsen PR, eds. William's textbook of endocrinology, 9th ed. Philadelphia: WB Saunders; 1998:389-515.
485. Tunbridge WM, Evered DE, Hall R, et al. The spectrum of thyroid disease in a community: the Wickham Survey. Clin Endocrinol (Oxf) 1977;7:481-93.
486. Stenszky V, Kozma L, Balázs C, Rochlitz S, Bear JC, Farid NR. The genetics of Graves' disease: HLA and disease susceptibility. J Clin Endocrinol Metab 1985;61:735-40.
487. Tomer Y, Ban Y, Concepcion E, et al. Common and unique susceptibility loci in Graves and Hashimoto diseases: results of whole-genome screening in a data set of 102 multiplex families. Am J Hum Genet 2003;73:736-47.
488. Tomer Y, Davies TF. Searching for the autoimmune thyroid disease susceptibility genes: from gene mapping to gene function. Endocr Rev 2003;24:694-717.
489. Villanueva R, Greenberg DA, Davies TF, Tomer Y. Sibling recurrence risk in autoimmune thyroid disease. Thyroid 2003;13:761-4.
490. Chen QY, Huang W, She JX, Baxter F, Volpe R, Maclaren NK. HLA-DRB1*08, DRB1*03/DRB3*0101, and DRB3*0202 are susceptibility genes for Graves' disease in North American Caucasians, whereas DRB1*07 is protective. J Clin Endocrinol Metab 1999;84:3182-6.
491. Davies TF, Ando T, Lin RY, Tomer Y, Latif R. Thyrotropin receptor-associated diseases: from adenomata to Graves disease. J Clin Invest 2005;115:1972-83.
492. Latif R, Morshed SA, Zaidi M, Davies TF. The thyroid-stimulating hormone receptor: impact of thyroid-stimulating hormone and thyroid-stimulating hormone receptor antibodies on multimerization, cleavage, and signaling. Endocrinol Metab Clin North Am 2009;38:319-41.
493. Colobran R, Armengol Mdel P, Faner R, et al. Association of an SNP with intrathymic transcription of TSHR and Graves' disease: a role for defective thymic tolerance. Hum Mol Genet 2011;20:3415-23.
494. Stefan M, Wei C, Lombardi A, et al. Genetic-epigenetic dysregulation of thymic TSH receptor gene expression triggers thyroid autoimmunity. Proc Natl Acad Sci U S A 2014;111:12562-7.
495. Lytton SD, Kahaly GJ. Bioassays for TSH-receptor autoantibodies: an update. Autoimmun Rev 2010;10:116-22.
496. Massart C, Gibassier J, d'Herbomez M. Clinical value of M22-based assays for TSH-receptor antibody (TRAb) in the follow-up of antithyroid drug treated Graves' disease: comparison with the second generation human TRAb assay. Clin Chim Acta 2009;407:62-6.
497. Zöphel K, Roggenbuck D, von Landenberg P, et al. TSH receptor antibody (TRAb) assays based on the human monoclonal autoantibody M22 are more sensitive than bovine TSH based assays. Horm Metab Res 2010;42:65-9.
498. Saito T, Endo T, Kawaguchi A, et al. Increased expression of the Na+/I- symporter in cultured human thyroid cells exposed to thyrotropin and in Graves' thyroid tissue. J Clin Endocrinol Metab 1997;82:3331-6.
499. Rapoport B, McLachlan SM. Graves' hyperthyroidism is antibody-mediated but is predominantly a Th1-type cytokine disease. J Clin Endocrinol Metab 2014;99:4060-1.
500. Watson PF, Pickerill AP, Davies R, Weetman AP. Analysis of cytokine gene expression in Graves' disease and multinodular goiter. J Clin Endocrinol Metab 1994;79:355-60.
501. Brown RS. Immunoglobulins affecting thyroid growth: a continuing controversy. J Clin Endocrinol Metab 1995;80:1506-8.

502. Sridama V, Pacini F, DeGroot LJ. Decreased suppressor T lymphocytes in autoimmune thyroid diseases detected by monoclonal antibodies. J Clin Endocrinol Metab 1982;54:316-9.
503. Menconi F, Hasham A, Tomer Y. Environmental triggers of thyroiditis: hepatitis C and interferon-[a]. J Endocrinol Invest 2011;34:78-84.
504. Leclere J, Weryha G. Stress and auto-immune endocrine diseases. Horm Res 1989;31:90-3.
505. Winsa B, Adami HO, Bergstrom R, et al. Stressful life events and Graves' disease. Lancet 1991;338:1475-9.
506. Grubeck-Boebenstein B, Trieb K, Sztankay A, Holter W, Anderl H, Wick G. Retrobulbar T cells from patients with Graves' ophthalmology are CD8+ and specifically autologous fibroblasts. J Clin Invest 1994;93:2738-43.
507. Wall JR, Bernard N, Boucher A, et al. Pathogenesis of thyroid-associated ophthalmopathy: an autoimmune disorder of the eye muscle associated with Graves' hyperthyroidism and Hashimoto's thyroiditis. Clin Immunolmmunopathol 1993;68:1-8.
508. Wall JR, Salvi M, Bernard NF, Boucher A, Haegert D. Thyroid-associated ophthalmopathy—a model for the association of organ-specific autoimmune disorders. Immunol Today 1991;12:150-3.
509. Weetman AP, Cohen S, Gatter KC, Fells P, Shine B. Immunohistochemical analysis of the retrobulbar tissues in Graves' ophthalmopathy. Clin Exp Immunol 1989;75:222-7.
510. Eckstein AK, Plicht M, Lax H, et al. Thyrotropin receptor autoantibodies are independent risk factors for Graves' ophthalmopathy and help to predict severity and outcome of the disease. J Clin Endocrinol Metab 2006;91:3464-70.
511. Bahn RS. Graves' ophthalmopathy. N Engl J Med 2010;362:726-38.
512. Burch HB, Wartofsky L. Graves' ophthalmopathy: current concepts regarding pathogenesis and management. Endocr Rev 1993;14:747-93.
513. Doshi DN, Blyumin ML, Kimball AB. Cutaneous manifestations of thyroid disease. Clin Dermatol 2008;26:283.
514. Shishido M, Kuroda K, Tsukifuji R, Fujita, H Shinkai. A case of pretibial myxedema associated with Graves' disease: an immunohistochemical study of serum-derived hyaluronan-associated protein. J Dermatol 1995;22:948-52.
515. Ramtoola S, Maisey MN, Clarke SE, Fogelman I. The thyroid scan in Hashimoto's thyroiditis: the great mimic. Nucl Med Commun 1988;9:639-45.
516. Reinwein D, Benker G, Konig MP, Pinchera A, Schatz H, Schleusener A. The different types of hyperthyroidism in Europe. Results of a prospective survey of 924 patients. J Endocrinol Invest 1988;11:193-200.
517. LiVolsi V, LoGerfo P. Thyroiditis. Boca Raton: CRC Press; 1981.
518. Duh YC, Su IJ, Liaw KY, et al. Subpopulations of intrathyroidal lymphocytes in Graves' disease. J Formos Med Assoc 1990;89:121-5.
519. Eguchi K, Matsuoka N, Nagataki S. Cellular immunity in autoimmune thyroid disease. Balliere's Clin Endocinol Metab 1995;9:71-94.
520. Kabel PJ, Voorbij HA, De Haan M, van der Gaag RD, Drexhage HA. Intrathyroidal dendritic cells. J Clin Endocrinal Metab 1988;65:199-207.
521. Chang DC, Wheeler MH, Woodcock JP, et al. The effect of preoperative Lugol's iodine on thyroid blood flow in patients with Graves' hyperthyroidism. Surgery 1987;102:1055-61.
522. Curran RC, Eckert H, Wilson GM. The thyroid gland after treatment of hyperthyroidism by partial thyroidectomy or I-131. J Pathol Bacteriol 1958;76:541-60.
523. Hanson GA, Komorowski RA, Cerlety JM, Wilson SD. Thyroid gland morphology in young adults: normal subjects versus those with prior low-dose neck irradiation in childhood. Surgery 1983;94:984-8.
524. Lee KS, Kim K, Hur KB, Kim CK. The role of propranolol in the preoperative preparation of patients with Graves' disease. Surg Gynecol Obstet 1986;162:365-9.
525. Stout BD, Wiener L, Cox JW. Combined alpha and beta sympathetic blockade in hyperthyroidism. Clinical and metabolic effects. Ann Intern Med 1969;70:963-70.
526. Patchefsky AS, Hoch WS. Psammoma bodies in diffuse toxic goiter. Am J Clin Pathol 1972;57:551-6.
527. Shapiro SJ, Friedman NB, Perzik SL, Catz B. Incidence of thyroid carcinoma in Graves' disease. Cancer 1970;26:1261-70.
528. Feldberg NT, Sergott RC, Savino PJ, Blizzard JJ, Schatz NJ, Amsel J. Lymphocyte subpopulation in Graves' ophthalmology. Arch Ophthalmol 1985;103:656-9.
529. Wang PW, Hiromatsu Y, Laryea E, Wosu L, How J, Wall JR. Immunologically mediated cytotoxicity against human eye muscle cells in Graves' ophthalmology. J Clin Endocrinol Metab 1986;63:316-22.
530. Noppakun N, Bancheun K, Chandraprasert S. Unusual locations of localized myxedema in Graves' disease. Report of three cases. Arch Dermatol 1986;122:85-8.
531. Tachiwaki O, Wollman SH. Shedding of dense cell fragments into follicular lumen early in involution of the hyperplastic thyroid gland. Lab Invest 1982;47:91-8.

532. Hedberg CW, Fishbein DB, Janssen RS, et al. An outbreak of thyrotoxicosis caused by the consumption of bovine thyroid gland in ground beef. N Engl J Med 1987;316:993-8.
533. Lloyd RV, Johnson TL, Blairas M, Sisson JC, Wilson BS. Detection of HLA-DR antigens in paraffin-embedded thyroid epithelial cells with a monoclonal antibody. Am J Pathol 1985;120:106-11.
534. Okita N, How J, Topliss D, Lewis M, Row VV, Volpe R. Suppressor T lymphocyte dysfunction in Graves' disease: role of the H-2 histamine receptor-bearing suppressor T lymphocytes. J Clin Endocrinol Metab 1981;53:1002-7.
535. Madec AM, Allannic H, Genetet N, et al. T lymphocyte subsets at various stages of hyperthyroid Graves' disease: effect of carbimazole treatment and relationship with thyroid-stimulating antibody levels of HLA status. J Clin Endocrinol Metab 1986;62:117-21.
536. Mori H, Amino N, Iwatani Y, et al. Decrease of immunoglobulin G-Fc receptor-bearing T lymphocytes in Graves' disease. J Clin Endocrinol Metab 1982;55:399-402.
537. Hamburger JI. Solitary autonomously functioning thyroid lesions. Diagnosis, clinical features and pathogenic considerations. Am J Med 1975;58:740-8.
538. McKenzie JM. Hyperthyroidism caused by thyroid adenomata. J Clin Endocrinol Metab 1966;26:779-81.
539. Tomer Y, Davies TF. Infection, thyroid disease, and autoimmunity. Endocr Rev 1993;14:107-20.
540. Valenta L, Lemarchand-Beraud T, Nemec J, Griessen M, Bednar J. Metastatic thyroid carcinoma provoking hyperthyroidism with elevated circulating thyrostimulators. Am J Med 1970;40:72-6.
541. Johnson JR. Adenomatous goiters with and without hyperthyroidism; some aspects of relationship of microscopic appearance to hyperthyroidism. Arch Surg 1949;59:1088-99.
542. Miyai K, Tanizawa O, Yamamoto T, Azukizawa M, Kawai Y. Pituitary-thyroid function in trophoblastic disease. J Clin Endocrinol Metab 1976;42:254-9.
543. Fradkin JE, Wolff J. Iodide-induced thyrotoxicosis. Medicine (Baltimore) 1983;62:1-20.
444. Cooper DS, Wenig BM. Hyperthyroidism caused by an ectopic TSH-secreting pituitary tumor. Thyroid 1996;6:337-43.
445. Cooper DS, Ladenson PW, Nisula BC, Dunn JF, Chapman EM, Ridgway EC. Familial thyroid hormone resistance. Metabolism 1982;31:504-9.
546. Zellmann HE. Iatrogenic and factitious thyroidal disease. Med Clin N Am 1979;63:329-35.
547. Ross DS, Burch HB, Cooper DS, et al. 2016 American Thyroid Association guidelines for diagnosis and management of hyperthyroidism and other causes of thyrotoxicosis. Thyroid 2016;26:1343-421.
548. Franklyn JA. The management of hyperthyroidism. N Engl J Med 1994;330:1731-8.
549. Elfenbein DM, Schneider DF, Havlena J, Chen H, Sippel RS. Clinical and socioeconomic factors influence treatment decisions in Graves' disease. Ann Surg Oncol 2015;22:1196-9.
550. Burch HB, Burman KD, Cooper DS. A 2011 survey of clinical practice patterns in the management of Graves' disease. J Clin Endocrinol Metab 2012;97:4549-58.
551. Cantalamessa L, Baldini M, Orsatti A, Meroni L, Amodei V, Castagnone D. Thyroid nodules in Graves disease and the risk of thyroid carcinoma. Arch Intern Med 1999;159:1705-8.

Postpartum Thyroiditis

552. Alexander EK, Pearce EN, Brent GA, et al. 2017 Guidelines of the American Thyroid Association for the diagnosis and management of thyroid disease during pregnancy and the postpartum. Thyroid 2017;27:315-89.
553. De Groot L, Abalovich M, Alexander EK, et al. Management of thyroid dysfunction during pregnancy and postpartum: an Endocrine Society clinical practice guideline. J Clin Endocrinol Metab 2012;97:2543-65.
554. Nicholson WK, Robinson KA, Smallridge RC, Ladenson PW, Powe NR. Prevalence of postpartum thyroid dysfunction: a quantitative review. Thyroid 2006;16:573.
555. Stagnaro-Green A, Schwartz A, Gismondi R, Tinelli A, Mangieri T, Negro R. High rate of persistent hypothyroidism in a large-scale prospective study of postpartum thyroiditis in southern Italy. J Clin Endocrinol Metab 2011;96:652-7.
556. Kologlu M, Fung H, Darke C, Richards CJ, Hall R, McGregor AM. Postpartum thyroid dysfunction and HLA status. Eur J Clin Invest 1990; 20:56-60.
557. Gorman CA, Duick DS, Woolner LB, Wahner HW. Transient hyperthyroidism in patients with lymphocytic thyroiditis. Mayo Clin Proc 1978;53:359-65.
558. Lucas A, Pizarro E, Granada ML, Salinas I, Foz M, Sanmarti A. Postpartum thyroiditis: epidemiology and clinical evolution in a nonselected population. Thyroid 2000;10:71-7.
559. Lazarus JH, Hall R, Othman S, et al. The clinical spectrum of postpartum thyroid disease. QJM 1996;89:429-35.
560. Stagnaro-Green A. Approach to the patient with postpartum thyroiditis. J Clin Endocrinol Metab 2012;97:334-42.

561. Azizi F. The occurrence of permanent thyroid failure in patients with subclinical postpartum thyroiditis. Eur J Endocrinol 2005;153:367-71.
562. Tachi J, Amino N, Tamaki H, Aozasa M, Iwatani Y, Miyai K. Long term follow-up and HLA association in patients with postpartum hypothyroidism. J Clin Endocrinol Metab 1988;66:480-4.
563. Mizukami Y, Michigishi T, Nonomura A, et al. Postpartum thyroiditis. A clinical, histologic, and immunopathologic study of 15 cases. Am J Clin Pathol 1993;100:200-5.
564. Nikolai TF, Turney SL, Roberts RC. Postpartum lymphocytic thyroiditis. Prevalence, clinical course, and long-term follow-up. Arch Intern Med 1987;147:221-4.
565. Taylor HC, Sheeler LR. Recurrence and heterogeneity in painless thyrotoxic lymphocytic thyroiditis. Report of five cases. JAMA 1982;248:1085-8.
566. Othman S, Phillips DI, Parkes AB, et al. A long-term follow-up of postpartum thyroiditis. Clin Endocrinol (Oxf) 1990;32:559-64.

Painless/Silent Thyroiditis

567. Ross DS. Syndromes of thyrotoxicosis with low radioactive iodine uptake. Endocrinol Metab Clin North Am 1998;27:169-85.
568. Schwartz F, Bergmann N, Zerahn B, Faber J. Incidence rate of symptomatic painless thyroiditis presenting with thyrotoxicosis in Denmark as evaluated by consecutive thyroid scintigraphies. Scand J Clin Lab Invest 2013;73:240-4.
569. Pearce EN, Farwell AP, Braverman LE. Thyroiditis. N Engl J Med 2003;348:2646-55.
570. Nikolai TF, Brosseau J, Kettnernick MA, Roberts R, Beltaos E. Lymphocyte thyroiditis with spontaneously resolving hyperthyroidism (silent thyroiditis). Arch Intern Med 1980;140:478-82.
571. Nikolai TF, Coombs GJ, McKenzie AK. Lymphocytic thyroiditis with spontaneously resolving hyperthyroidism and subacute thyroiditis. Long-term follow-up. Arch Intern Med 1981;141:1455-8.
572. Farid NR, Hawe BS, Walfish PG. Increased frequency of HLA-DR3 and 5 in the syndromes of painless thyroiditis with transient thyrotoxicosis: evidence for an autoimmune aetiology. Clin Endocrinol (Oxf) 1983;19:699-704.
573. Iyer PC, Cabanillas ME, Waguespack SG, et al. Immune-related thyroiditis with immune checkpoint inhibitors. Thyroid 2018;28:1243-51.
574. LiVolsi VA. The pathology of autoimmune thyroid disease: a review. Thyroid 1994;4:333-9.
575. Mizukami Y, Michigishi T, Hashimoto T, et al. Silent thyroiditis: a histologic and immunohistochemical study. Hum Pathol 1988;19:423-31.
576. Woolf PD. Transient painless thyroiditis with hyperthyroidism: a variant of lymphocytic thyroiditis? Endocr Rev 1980;1:411-20.

Focal Lymphocytic Thyroiditis

577. Goudie RB, Anderson JR, Gray KG. Complement-fixing antithyroid antibodies in hospital patients with asymptomatic thyroid lesions. J Path Bact 1959;77:389-400.
578. Harach HR, Williams ED. Fibrous thyroiditis—an immunopathological study. Histopathology 1983;7:739-51.
579. Harland WA, Frantz VK. Clinicopathologic study of 261 surgical cases of so-called thyroiditis. J Clin Endocrinol Metab 1956;16:1433-7.
580. Williams ED, Doniach I. The post-mortem incidence of focal thyroiditis. J Pathol 1962;83:255-64.
581. Weaver D, Batsakis J, Nishiyama R. Relationship of iodine to "lymphocytic goiter." Arch Surg 1969;98:183-5.
582. Kurashima C, Hirokawa K. Focal lymphocytic infiltration in thyroids of elderly people. Histopathological and immunohistochemical studies. Surv Synth Pathol Res 1985;4:457-66.

Riedel Thyroiditis

583. Hay ID. Thyroiditis: a clinical update. Mayo Clin Proc 1985;60:836-43.
584. Beahrs OH, McConahey WM, Woolner LB. Invasive fibrous thyroiditis (Riedel's struma). J Clin Endocrinol Metab 1957;17:201-20.
585. Fatourechi MM, Hay ID, McIver B, Sebo TJ, Fatourechi V. Invasive fibrous thyroiditis (Riedel thyroiditis): the Mayo Clinic experience, 1976-2008. Thyroid 2011;21:765-72.
586. LiVolsi VA. Fibrosis in the thyroid. In: LiVolsi VA. Surgical pathology of the thyroid, volume 22: major problems in pathology. Philadelphia: WB Saunders; 1990:125-8.
587. Volpe R. Subacute and sclerosing thyroiditis. In: De Groot LJ, ed. Endocrinology, 3rd ed. Philadelphia: WB Saunders; 1995:745.
588. Zimmermann-Belsing T, Feldt-Rasmussen U. Riedel's thyroiditis: an autoimmune or primary fibrotic disease? J Intern Med 1994;235:271-4.
589. Erdogan MF, Anil C, Türkçapar N, Ozkaramanli D, Sak SD, Erdogan G. A case of Riedel's thyroiditis with pleural and pericardial effusions. Endocrine 2009;35:297-301.
590. Hennessey JV. Clinical review: Riedel's thyroiditis: a clinical review. J Clin Endocrinol Metab 2011;96:3031-41.
591. Stan MN, Sonawane V, Sebo TJ, et al. Riedel's thyroiditis association with IgG4-related disease. Clin Endocrinol (Oxf) 2017;86:425-30.
592. Dahlgren M, Khosroshahi A, Nielsen GP, Deshpande V, Stone JH. Riedel's thyroiditis and multifocal fibrosclerosis are part of the IgG4-related systemic disease spectrum. Arthritis Care Res (Hoboken) 2010;62:1312-8.

593. Chopra D, Wool MS, Crosson A, Sawin CT. Riedel's struma associated with subacute thyroiditis, hypothyroidism, and hypoparathyroidism. J Clin Endocrinol Metab 1978;46:869-71.
594. Lo JC, Loh KC, Rubin AL, Cha I, Greenspan FS. Riedel's thyroiditis presenting with hypothyroidism and hypoparathyroidism: dramatic response to glucocorticoid and thyroxine therapy. Clin Endocrinol (Oxf) 1998;48:815-8.
595. Schwaegerle SM, Bauer TW, Esselstyn CB Jr. Riedel's thyroiditis. Am J Clin Pathol 1988;90:715-22.
596. de Lange WE, Freling NJ, Molenaar WM, Doorenbos H. Invasive fibrous thyroiditis (Riedel's struma): a manifestation of multifocal fibrosclerosis? A case report with review of the literature. Q J Med 1989;72:709-17.
597. Cho MH, Kim CS, Park JS, et al. Riedel's thyroiditis in a patient with recurrent subacute thyroiditis: a case report and review of the literature. Endocr J 2007;54:559-62.
598. Papi G, Corrado S, Cesinaro AM, Novelli L, Smerieri A, Carapezzi C. Riedel's thyroiditis: clinical, pathological and imaging features. Int J Clin Pract 2002;56:65-7.
599. Meyer S, Hausman R. Occlusive phlebitis in multifocal fibrosclerosis. Am J Clin Pathol 1976;65:274-83.
600. Woolner L, McConahey W, Beahrs O. Invasive fibrous thyroiditis (Riedel's struma). J Clin Endocrinol Metab 1957;17:201-20.
601. Munro JM, Van der Walt JD, Cox EL. A comparison of cytoplasmic immunoglobulins in retroperitoneal fibrosis and abdominal aortic aneurysms. Histopathology 1986;10:1163-9.
602. Heufelder AE, Goellner JR, Bahn RS, Gleich GJ, Hay ID. Tissue eosinophilia and eosinophil degranulation in Riedel's invasive fibrous thyroiditis. J Clin Endocrinol Metab 1996;81:977-84.
603. Kurashima C, Hirokawa K. Focal lymphocytic infiltration of thyroids in elderly people. Histopathological and immunohistochemcial studies. Survey Synth Pathol Res 1985;4:457-66.
604. Wan SK, Chan JK, Tang SK. Paucicellular variant of anaplastic thyroid carcinoma. A mimic of Reidel's thyroiditis. Am J Clin Pathol 1996;105:388-93.
605. Larsen PR, Davies TF, Hay ID. The thyroid gland. In: Wilson JP, Foster DW, Kronenberg HM, Larsen PR, eds. Williams textbook of endocrinology, 9th ed. Philadelphia: WB Saunders; 1998:389-515.
606. Vaidya B, Harris PE, Barrett P, Kendall-Taylor P. Corticosteroid therapy in Riedel's thyroiditis. Postgrad Med J 1997;73:817-9.
607. Few J, Thompson NW, Angelos P, Simeone D, Giordano T, Reever T. Riedel's thyroiditis: treatment with tamoxifen. Surgery 1996;120:993-8.
608. Jung YJ, Schaub CR, Rhodes R, Rich FA, Muehlenbein SJ. A case of Riedel's thyroiditis treated with tamoxifen: another successful outcome. Endocr Pract 2004;10:483-6.
609. Soh SB, Pham A, O'Hehir RE, Cherk M, Topliss DJ. Novel use of rituximab in a case of Riedel's thyroiditis refractory to glucocorticoids and tamoxifen. J Clin Endocrinol Metab 2013;98:3543-9.
610. Levy JM, Hasney CP, Friedlander PL, Kandil E, Occhipinti EA, Kahn MJ. Combined mycophenolate mofetil and prednisone therapy in tamoxifen- and prednisone-resistant Reidel's thyroiditis. Thyroid 2010;20:105-7.

Simple Nontoxic or Multinodular Goiter

611. Al-Moussa M, Beck JS. Histometry of thyroids containing few and multiple nodules. J Clin Pathol 1986;39:483-8.
612. Hull OH. Critical analysis of two hundred twenty-one thyroid glands—study of thyroid glands obtained at necropsy in Colorado. AMA Arch Pathol 1955;59:291-311.
613. Maloof F, Wang CA, Vickery AL Jr. Nontoxic goiter-diffuse or nodular. Med Clin North Am 1975;59:1221-32.
614. Borowski GD, Garofaro CD, Rose LI, et al. Effect of long-term amiodarone therapy on thyroid hormone levels and thyroid function. Am J Med 1985;78:443-50.
615. Burke G, Silverstein GE, Sorkin AI. Effect of longterm sulfonylurea therapy on thyroid function in man. Metabolism 1967;16:651-7.
616. Mehbod H, Swartz CD, Brest AN. The effect of prolonged thiazide administration on thyroid function. Arch Intern Med 1967;119:283-6.
617. Bray GA. Increased sensitivity of the thyroid in iodine-depleted rats to the goitrogenic effects of thyrotropin. J Clin Invest 1968;47:1640-7.
618. Dige-Petersen H, Hummer L. Serum thyrotropin concentration under basal conditions and after stimulation with thyrotropin-releasing hormone in idiopathic nontoxic goiter. J Clin Endocrinol Metab 1977;44:1115-20.
619. Brown RS. Immunoglobulins affecting thyroid growth: a continuing controversy. J Clin Endocrinol Metab 1995;80:1506-8.
620. Drexhage HA, Bottazzo GF, Doniach D, Bitensky L, Chayen J. Evidence for thyroid growth stimulating immunoglobulins in some goitrous thyroid diseases. Lancet 1980;2:287-92.
621. Smyth PP, Neylan D, O'Donovan DK. The prevalence of thyroid-stimulating antibodies in goitrous disease assessed by cytochemical section bioassay. J Clin Endocrinol Metab 1982; 54:357-61.

622. Valente WA, Vitti P, Rotellar CM, et al. Antibodies that promote thyroid growth. A distinct population of thyroid-stimulating autoantibodies. N Engl J Med 1983;309:1028-34.
623. Marine D. Etiology and prevention of simple goiter. Medicine 1924;3:453-79.
624. Studer H, Peter HJ, Gerber H. Natural heterogeneity of thyroid cells: the basis for understanding thyroid function and nodular goiter growth. Endocr Rev 1989;10:125-35.
625. Lyon MF. Clones and X-chromosomes. J Pathol 1988;155:97-9.
626. Lyon MF. The William Allan memorial award address: X-chromosome inactivation and the location and expression of X-linked genes. Am J Hum Genet 1988;42:8-16.
627. Aeschimann S, Kopp PA, Kimura ET, et al. Morphological and functional polymorphism within clonal thyroid nodules. J Clin Endocrinol Metab 1993;77:846-51.
628. Hicks DG, LiVolsi VA, Neidich JA, Puck JM, Kant JA. Clonal analysis of solitary follicular nodules in the thyroid. Am J Pathol 1990;137:553-62.
629. Kopp P, Kimura ET, Aeschimann S, et al. Polyclonal and monoclonal thyroid nodules coexist within human multinodular goiters. J Clin Endocrinol Metab 1994;79:134-9.
630. Thomas GA, Williams D, Williams ED. The clonal origin of thyroid nodules and adenomas. Am J Pathol 1989;134:141-7.
631. Apel RL, Ezzat S, Bapat BV, Pan N, Livolsi VA, Asa SL. Clonality of thyroid nodules in sporadic goiter. Diagn Mol Pathol 1995;4:113-21.
632. Fujita M, Enomoto T, Wada H, Inoue M, Okudaira Y, Shroyer KR. Application of clonal analysis. Differential diagnosis for synchronous primary ovarian and endometrial cancers and metastatic cancer. Am J Clin Pathol 1996;105:350-9.
633. Greig WR, Boyle JA, Duncan A, et al. Genetic and non-genetic factors in simple goitre formation: evidence from a twin study. Q J Med 1967;36:175-88.
634. Murray IP, Thomson JA, McGirr EM, MacDonald EM, Kennedy JS, McLennan I. Unusual familial goiter associated with intrathyroidal calcification. J Clin Endocrinol Metab 1966;26:1039-49.
635. Chen AY, Bernet VJ, Carty SE, et al. American Thyroid Association statement on optimal surgical management of goiter. Thyroid 2014;24:181-9.
636. Huysmans DA, Hermus AR, Corstens FH, Barentsz JO, Kloppenborg PW. Large, compressive goiters treated with radioiodine. Ann Intern Med 1994;121:757-62.
637. Le Moli R, Wesche MF, Tiel-Van Buul MM, Wiersinga WM. Determinants of longterm outcome of radioiodine therapy of sporadic non-toxic goitre. Clin Endocrinol (Oxf) 1999;50:783-9.
638. Nygaard B, Faber J, Hegedüs L. Acute changes in thyroid volume and function following 131I therapy of multinodular goitre. Clin Endocrinol (Oxf) 1994;41:715-8.
639. Nygaard B, Hegedüs L, Gervil M, et al. Radioiodine treatment of multinodular non-toxic goitre. BMJ 1993;307:828-32.
640. Verelst J, Bonnyns M, Glinoer D. Radioiodine therapy in voluminous multinodular non-toxic goitre. Acta Endocrinol (Copenh) 1990;122:417-21.
641. Wesche MF, Tiel-v-Buul MM, Smits NJ, Wiersinga WM. Reduction in goiter size by 131I therapy in patients with non-toxic multinodular goiter. Eur J Endocrinol 1995;132:86-7.
642. Bonnema SJ, Bertelsen H, Mortensen J, et al. The feasibility of high dose iodine 131 treatment as an alternative to surgery in patients with a very large goiter: effect on thyroid function and size and pulmonary function. J Clin Endocrinol Metab 1999;84:3636-41.
643. Glinoer D, Lemone M. Goiter and pregnancy: a new insight into an old problem. Thyroid 1992;2:65-70.
644. Hegedüs L, Hansen BM, Knudsen N, Hansen JM. Reduction of size of thyroid with radioactive iodine in multinodular non-toxic goitre. BMJ 1988;297:661-2.
645. Izembart M, Heshmati HM, Dagousset F, de Cremoux P, Boutteville C, Vallee G. [Serum thyroglobulin is elevated in patients with heterogeneous goiter during radioiodine scintography but normal in those with homogeneous goiter.] Schweiz Med Wochenschr 1986;116:634-7. [French]
646. Rieu M, Bekka S, Sambor B, Berrod JL, Fombeur JP. Prevalence of subclinical hyperthyroidism and relationship between thyroid hormonal status and thyroid ultrasonographic parameters in patients with non-toxic nodular goitre. Clin Endocrinol 1993;39:67-71.
647. Ghossein RA, Rosai J, Heffess C. Dyshormonogenetic goiter: a clinicopathologic study of 56 cases. Endocr Pathol 1997;8:283-92.
648. Papotti M, Fara E, Ardeleanu C, Bussolati G. Occurrence and significance of vascular invasion in multinodular adenomatous goiter. Endocr Pathol 1994;5:35-9.
649. Vickery AL Jr. The diagnosis of malignancy in dyshormonogenic goitre. Clin Endocrinol Metab 1981;10:317-35.
650. Berghout A, Wiersinga WM, Drexhage HA, et al. The long-term outcome of thyroidectomy for sporadic non-toxic goitre. Clin Endocrinol 1989;31:193-9.

651. Berghout A, Wiersinga WM, Drexhage HA, et al. Comparison of placebo with L-thyroxine alone or with carbimazole for treatment of sporadic non-toxic goitre. Lancet 1990;336:193-7.
652. Cesareo R, Iozzino M, Isgrò MA, Annunziata F, Di Stasioet E. Short term effects of levothyroxine treatment in thyroid multinodular disease. Endocr J 2010;57:803-9.
653. Grussendorf M, Reiners C, Paschke R, Wegscheider K; LISA Investigators. Reduction of thyroid nodule volume by levothyroxine and iodine alone and in combination: a randomized, placebo-controlled trial. J Clin Endocrinol Metab 2011;96:2786-95.
654. Vanderpump MP, Tunbridge WM, French JM, et al. The incidence of thyroid disorders in the community: a twenty-year follow-up of the Whickham Survey. Clin Endocrinol (Oxf) 1995;43:55-68.
655. Hegedüs L, Bonnema SJ, Bennedbaek FN. Management of simple nodular goiter: current status and future perspectives. Endocr Rev 2003;24:102-32.

Dyshormonogenetic Goiter

656. Knobel M, Medeiros-Neto G. An outline of inherited disorders of the thyroid hormone generating system. Thyroid 2003;13:771-801.
657. Fisher DA, Klein AH. Thyroid development and disorders of thyroid function in the newborn. N Engl J Med 1981;304:702-12.
658. Szinnai G. Clinical genetics of congenital hypothyroidism. Endocr Dev 2014;26:60-78.
659. Thompson LD. Dyshormonogenetic goiter of the thyroid gland. Ear Nose Throat J 2005; 84:200.
660. Moreno JC, Bikker H, Kempers MJ, et al. Inactivating mutations in the gene for thyroid oxidase 2 (THOX2) and congenital hypothyroidism. N Engl J Med 2002;347:95-102.
661. Szinnai G, Kosugi S, Derrien C, et al. Extending the clinical heterogeneity of iodide transport defect (ITD): a novel mutation R124H of the sodium/iodide symporter gene and review of genotype-phenotype correlations in ITD. J Clin Endocrinol Metab 2006;91:1199-204.
662. Moreno JC, de Vijlder JJ, Vulsma T, Ris-Stalpers C. Genetic basis of hypothyroidism: recent advances, gaps and strategies for future research. Trends Endocrinol Metab 2003;14:318-26.
663. Park SM, Chatterjee VK. Genetics of congenital hypothyroidism. J Med Genet 2005;42:379-89.
664. Ramesh BG, Bhargav RP, Rajesh BG, Devi NV, Vijayaraghavan R, Varma BA. Genotype-phenotype correlations of dyshormonogenetic goiter in children and adolescents from South India. Indian J Endocrinol Metab 2016;20:816-24.
665. Muir A, Daneman D, Daneman A, Ehrlich R. Thyroid scanning, ultrasound, and serum thyroglobulin in determining the origin of congenital hypothyroidism. Am J Dis Child 1988;142:214-6.
666. Ghossein RA, Rosai J, Heffess C. Dyshormonogenetic goiter: a clinicopathologic study of 56 cases. Endocr Pathol 1997;8:283-92.
667. Batsakis JG, Nishiyama RH, Schmidt RW. "Sporadic goiter syndrome": a clinicopathologic analysis. Am J Clin Pathol 1963;39:241-51.
668. Moore GH. The thyroid in sporadic goitrous cretinism. A report of three new cases, description of the pathologic anatomy of the thyroid glands, and a review of the literature. Arch Pathol 1962;74:35-46.
669. Leger J, Olivieri A, Donaldson M, et al. European Society for Paediatric Endocrinology consensus guidelines on screening, diagnosis, and management of congenital hypothyroidism. J Clin Endocrinol Metab 2014;99:363-84.
670. Williams JL, Paul DL, Bisset G III. Thyroid disease in children: part 1: state-of-the-art imaging in pediatric hypothyroidism. Pediatr Radiol 2013;43:1244-53.
671. Kennedy JS. The pathology of dyshormonogenic goitre. J Pathol 1969;99:251-64.
672. Tong GX, Chang Q, Hamele-Bena D, et al. Targeted next-generation sequencing analysis of a pendred syndrome-associated thyroid carcinoma. Endocr Pathol 2016;27:70-5.

Toxic Multinodular Goiter

673. Larsen PR, Davies TF, Hay ID. The thyroid gland. In: Wilson JD, Fasler DW, Kronenberg HM, Larsen PR, eds. Williams textbook of endocrinology, 9th ed. Philadelphia: W.B. Saunders; 1998:389-515.
674. Huysmans DA, Hermus AR, Corstens FH, Barentsz JO, Kloppenborg PW. Large compressive goiters treated with radioiodine. Ann Intern Med 1994;121:757-62.
675. Studer H, Peter HJ, Gerber H. Toxic nodular goitre. Clin Endocrinol Metab 1985;14:351-72.
676. Duprez L, Hermans J, Van Sande J, Dumont JE, Vassart G, Parma J. Two autonomous nodules of a patient with multinodular goiter harbor different activating mutations of the thyrotropin receptor gene. J Clin Endocrinol Metab 1997;82:306-8.
677. Holzapfel HP, Führer D, Wonerow P, Weinland G, Scherbaum WA, Paschke R. Identification of constitutively activating somatic thyrotropin receptor mutations in a subset of toxic multinodular goiters. J Clin Endocrinol Metab 1997;82:4229-33.
678. Parma J, Duprez L, Van Sande J, et al. Diversity and prevalence of somatic mutations in the thyrotropin receptor and Gs alpha genes as a cause of toxic thyroid adenomas. J Clin Endocrinol Metab 1997;82:2695-701.

679. Pinducciu C, Borgonovo G, Arezzo A, et al. Toxic thyroid adenoma: absence of DNA mutations of the TSH receptor and Gs alpha. Eur J Endocrinol 1998;138:37.
680. Laurberg P, Pedersen KM, Vestergaard H, Sigurdsson G. High incidence of multinodular toxic goitre in the elderly population in a low iodine intake area vs. high incidence of Graves' disease in the young in a high iodine intake area: comparative surveys of thyrotoxicosis epidemiology in East-Jutland Denmark and Iceland. J Intern Med 1991;229:415.
681. Pinchera A, Aghini-Lombardi F, Antonangeli L, Vitti P. [Multinodular goiter. Epidemiology and prevention]. Ann Ital Chir 1996;67:317-25. [Italian]
682. Medeiros-Neto G. Multinodular goiter. In: Feingold KR, Anawalt B, Boyce A, et al., eds. South Dartmouth: MDText.com, Inc.; 2000-.
683. Van Sande J, Parma J, Tonacchera M, Swillens S, Dumont J, Vassart G. Somatic germline mutations of the TSH receptor gene in thyroid disease. J Clin Endocrinol Metab 1995;80:2577-85.
684. Singer PA, Cooper DS, Levy EG, et al. Treatment guidelines for patients with hyperthyroidism and hypothyroidism. Standards of Care Committee, American Thyroid Association. JAMA 1995;273:808-12.

Endemic Goiters

685. Scrimshaw NS. The geographic pathology of thyroid disease. In: Hazard JB, Smith DV, eds. The thyroid. International Academy of Pathology Monograph, Publication 5. Baltimore: Williams & Wilkins; 1964:100-22.
686. Versmiglio F, Benvenga S, Melluso R, et al. Increased serum thyroglobulin concentrations and impaired thyrotropin response to thyrotropin- releasing hormone in euthyroid subjects with endemic goiter in Sicily: their relation to goiter size and nodularity. J Endocrinol Invest 1986;9:389-96.
687. Day TK, Powell-Jackson PR. Fluoride, water hardness, and endemic goitre. Lancet 1972;1:1135-8.
688. Hetzel BS. Iodine deficiency disorders (IDD) and their eradication. Lancet 1983;2:1126-9.
689. McCarrison R. Observations on endemic cretinism in the Chitral and Gilgit valleys. Ind Med Gaz 1908;43:441-9.
690. Correa P, Castro S. Survey of pathology of thyroid glands from Cali, Columbia—a goiter area. Lab Invest 1961;10:39-50.
691. Weetman AP. Is endemic goiter an autoimmune disease? J Clin Endocrinol Metab 1994;78:1017-9.
692. Maberly GF, Corcoran JM, Eastman CJ. The effect of iodized oil on goitre size, thyroid function, and the development of the Jod Basedow phenomenon. Clin Endocrinol (Oxf) 1982;17:253-9.
693. Kopp P, Kimura ET, Aeschimann S, et al. Polyclonal and monoclonal thyroid nodules coexist within human multinodular goiters. J Clin Endocrinol Metab 1994;79:134-9.
694. Tonacchera M, Agretti P, Chiovato L, et al. Activating thyrotropin receptor mutations are present in nonadenomatous hyperfunctioning nodules of toxic or autonomous multinodular goiter. J Clin Endocrinol Metab 2000;85:2270-4.

Amyloid Goiter

695. Falk RH, Comenzo RL, Skinner M. The systemic amyloidoses. N Engl J Med 1997;337:898-909.
696. Hamed G, Heffess CS, Shmookler BM, Wenig BM. Amyloid goiter. A clinicopathologic study of 14 cases and review of the literature. Am J Clin Pathol 1995;104:306-12.
697. Ozdemir D, Dagdelen S, Erbas T. Endocrine involvement in systemic amyloidosis. Endocr Pract 2010;16:1056-63.
698. Hirota S, Miyamoto M, Kasugai T, Kitamura Y, Morimura Y. Crystalline light chain deposition of amyloidosis in the thyroid gland and kidneys of a patient with myeloma. Arch Pathol Lab Med 1990;114:429-31.
699. Kanoh T, Shimada H, Uchino H, Matsumura K. Amyloid goiter with hypothyroidism. Arch Pathol Lab Med 1989;113:542-4.
700. Sinha RN, Plehn JF, Kinlaw WB. Amyloid goiter due to primary systemic amyloidosis: a diagnostic challenge. Thyroid 1998;8:1051-4.
701. Cabrejas Gómez M del C, González Cabrera N, Gómez González C, Bergara Elorza S. Amyloid goiter as an initial manifestation of systemic amyloidosis. Reumatol Clin 2015;11:404-5.
702. Amado JA, Ondiviela R, Palacios S, Casanova D, Manzanos J, Freijanls J. Fast growing goitre as the first clinical manifestation of systemic amyloidosis. Postgrad Med J 1982;58:171-2.
703. Arean VM, Klein RE. Amyloid goiter. Review of literature and report of a case. Am J Clin Pathol 1961;36:341-55.
704. Himmetoglu C, Yamak S, Tezel GG. Diffuse fatty infiltration in amyloid goiter. Pathol Int 2007;57:449-53.
705. Jacques TA, Stearns MP. Diffuse lipomatosis of the thyroid with amyloid deposition. J Laryngol Otol 2013;127:426-8.
706. Law JH, Dean DS, Scheithauer B, Earnest F 4th, Sebo TJ, Fatourechi V. Symptomatic amyloid goiters: report of five cases. Thyroid 2013;23:1490-5.

C-Cell Hyperplasia

707. Albores-Saavedra J. C-cell hyperplasia. Am J Surg Pathol 1989;13:987-9.
708. Albores-Saavedra J, Monforte H, Nadji M, Morales AR. C-cell hyperplasia in thyroid tissue adjacent to follicular cell tumors. Hum Pathol 1988;19:795-9.

709. Asaadi AA. Ultrastructure in C-cell hyperplasia in asymptomatic patients with hypercalcitonemia and a family history of medullary thyroid carcinoma. Hum Pathol 1981;12:617-22.
710. Biddinger PW, Brennan MF, Rosen PP. Symptomatic C-cell hyperplasia associated with chronic lymphocytic thyroiditis. Am J Surg Pathol 1991;15:599-604.
711. De Lellis RA, Wolfe HJ. The pathobiology of the human calcitonin (C) cell: a review. Pathol Annu 1981;16:25-52.
712. Libbey NP, Nowakowski KJ, Tucci JR. C-cell hyperplasia of the thyroid in a patient with goitrous hypothyroidism and Hashimoto's thyroiditis. Am J Surg Pathol 1989;13:71-7.
713. Lips CJ, Leo JR, Berends MJ, et al. Thyroid C-cell hyperplasia and micronodules in close relatives of MEN-2A patients: pitfalls in early diagnosis and re-evaluation of criteria for surgery. Henry Ford Hosp Med J 1987;35:133-8.
714. LiVolsi V, DeLellis R, Komminoth P, et al. Multiple endocrine neoplasia type 2. In: Lloyd RV, Osamura RY, Kloppel G, Rosai J, eds. WHO classification of tumours of endocrine organs, 4th ed. Lyon: IARC Press; 2017:248-50.
715. LiVolsi VA, Feind CR, Lo Gerfo P, Tashjian AH Jr. Demonstration by immunoperoxidase staining of hyperplasia of parafollicular cells in the thyroid gland in hyperparathyroidism. J Clin Endocrinol Metab 1973;37:550-9.
716. Tashjian AH, Wolfe HJ, Voelkl EF. Human calcitonin: immunologic assay, cytologic localization and studies on medullary thyroid carcinoma. Am J Med 1974;56:840-9.
717. Williams ED, Ponder BJ, Craig RK. Immunohistochemical study of calcitonin gene-related peptide in human medullary carcinoma and C-cell hyperplasia. Clin Endocrinol (Oxf) 1987;27:107-14.
718. Ekblom M, Valimaki M, Pelkonen R, Jansson R, Sivula A, Franssila K. Familial and sporadic medullary thyroid carcinoma: clinical and immunohistochemical findings. Q J Med 1987;65:899-910.
719. Fuchs TL, Bell SE, Chou A, Gill AJ. Revisiting the significance of prominent C cells in the thyroid. Endocr Pathol 2019;30:113-7.
720. Gibson WC, Peng TC, Croker BP. C-cell nodules in adult human thyroid. A common autopsy finding. Am J Clin Pathol 1981;75:347-50.
721. Guyetant S, Rousselet MC, Durigon M, et al. Sex-related C-cell hyperplasia in the normal human thyroid: a quantitative autopsy study. J Clin Endocrinol Metab 1997;82:42-7.
722. O'Toole K, Fenoglio-Preiser C, Pushparaj N. Endocrine changes associated with the human aging process: III. Effect of age on the number of calcitonin immunoreactive cells in the thyroid gland. Hum Pathol 1985;16:991-1000.
723. Perry A, Molberg K, Albones-Saavedra J. Physiologic versus neoplastic C-cell hyperplasia of the thyroid: separation of distinct histologic and biologic entities. Cancer 1996;77:750-6.
724. Scopsi L, DiPalma S, Ferrari C, Holst JJ, Rehfeld JF, Rilke F. C-cell hyperplasia accompanying thyroid diseases other than medullary thyroid carcinoma: an immunocytochemical study by means of antibodies to calcitonin and somatostatin. Mod Pathol 1991;4:297-304.
725. DeLellis RA, Nunnemacher G, Wolfe HJ. C-cell hyperplasia: an ultrastructural analysis. Lab Invest 1977;36:237-48.
726. Melvin KE, Miller HH, Tashjian AH Jr. Early diagnosis of medullary carcinoma of the thyroid gland by means of calcitonin assay. N Engl J Med 1971;285:1115-20.
727. Rosai J, Carcangiu ML, DeLellis RA. Tumors of the thyroid gland. AFIP Atlas of Tumor Pathology, 3rd Series, Fascicle 5. Washington DC: American Registry of Pathology; 1992.
728. Saggiorato E, Rapa I, Garino F, et al. Absence of RET gene point mutations in sporadic thyroid C-cell hyperplasia. J Mol Diag 2007;9:214-9.
729. Wolfe HJ, DeLellis RA, Scott RT, Tashijian AH Jr. C-cell hyperplasia in chronic hypercalcemia in man [Abstract]. Am J Pathol 1975;78:20A.
730. Wolfe HL, Melvin KE, Cervi-Skinner SJ, et al. C-cell hyperplasia preceding medullary thyroid carcinoma. N Engl J Med 1973;289:437-41.
731. Eng C. The RET proto-oncogene in multiple endocrine neoplasia type 2 and Hirschsprung's disease. N Engl J Med 1996;335:943-51.
732. Eng C, Thomas GA, Neuberg DS, et al. Mutation of the RET proto-oncogene is correlated with RET immunostaining in subpopulations of cells in sporadic medullary thyroid carcinoma. J Clin Endocrinol Metab 1998;83:4310-3.
733. Komminoth P, Kunz EK, Matias-Guiu X, et al. Analysis of RET proto-oncogene point mutations distinguishes heritable from nonheritable medullary thyroid carcinomas. Cancer 1995;76:479-89.
734. Lloyd RV. RET proto-oncogene mutations and rearrangements in endocrine diseases. Am J Pathol 1995;147:1539-44.
735. McGregor LM, McCune BK, Graff JR, et al. Roles of the trk family neurotropin receptors in medullary thyroid carcinoma development and progression. Proc Natl Acad Sci U S A 1999; 96:4540-5.
736. Reagh J, Bullock M, Andrici J, et al. NRASQ61R mutation-specific immunohistochemistry also identifies the HRASQ61R mutation in medullary thyroid cancer and may have a role in triaging genetic testing for MEN2. Am J Surg Pathol 2017;41:75-81.

737. Schürch W, Babäi F, Boivin Y, Verdy M. Light-electron microscopic and cytochemical studies on the morphogenesis of familial medullary thyroid carcinoma. Virchows Arch A Pathol Anat Histol 1977;376:29-46.
738. Yamaoka Y. Solid cell nest (SCN) of the human thyroid gland. Acta Pathol Jpn 1973;23:493-506.
739. Vollenweider I, Hedinger C. Solid cell nests (SCN) in Hashimoto's thyroiditis. Virchows Arch A Pathol Anat Histopathol 1988;412:357-63.
740. Wolfe HJ, DeLellis RA. Familial medullary thyroid carcinoma and C cell hyperplasia. Clin Endocrinol Metab 1981;10:351-65.
741. Dralle H, Gimm O, Simon D, et al. Prophylactic thyroidectomy in 75 children and adolescents with hereditary medullary thyroid carcinoma: German and Austrian experience. World J Surg 1998;22:744-50.
742. Gagel RF, Tashjian AH Jr, Cummings T, et al. The clinical outcome of prospective screening for multiple endocrine neoplasia type 2a. An 18-year experience. N Engl J Med 1988;318:478-84.
743. Kaserer K, Scheuba C, Neuhold N, et al. C-cell hyperplasia and medullary thyroid carcinoma in patients routinely screened for serum calcitonin. Am J Surg Pathol 1998;22:722-8.
744. Lallier M, St-Vil D, Giroux M, et al. Prophylactic thyroidectomy for medullary thyroid carcinoma in gene carriers of MEN 2 syndrome. J Pediatr Surg 1998;33:846-8.
745. Fuchs TL, Nassour AJ, Glover A, et al. A proposed grading scheme for medullary thyroid carcinoma based on proliferative activity (Ki67 and mitotic count) and coagulative necrosis. Am J Surg Pathol 2020;44:1419-28.
746. Wells SA Jr, Ontjes DA, Cooper CW, et al. The early diagnosis of medullary carcinoma of the thyroid in patients with multiple endocrine neoplasia type II. Ann Surg 1975;182:362-70.

4 ADRENAL GLAND

NORMAL ADRENAL GLAND

Embryology

The adrenal gland consists of two embryologically and physiologically distinct parts: the cortex, which is responsible for the production of steroid hormones; and the medulla, which is responsible for synthesizing catecholamines. The adrenal cortex is derived from mesoderm and starts to develop from a condensation of the coelomic epithelium medial to the mesonephros and urogenital ridge, beginning around 3 to 4 weeks of gestation (1–3). Around the 30th day of gestation, cells from the mesonephric glomerulus migrate toward the adrenal primordium, and by the 35th day, the coelomic epithelial cells adjacent to the adrenal primordium penetrate into the primordium. Between days 35 and 45 of gestation, the adrenal glands enlarge significantly, and the primitive sympathetic cells, along with nerve tracts, migrate to form the medulla. By the 45th day of gestation, the gland is elliptical in shape, and weighs about 1 mg (4).

By 8 weeks of gestation the adrenal glands weigh 4 to 6 mg each and there is marked proliferation and differentiation into two distinct zones that persist until birth: the outer definitive zone and the inner fetal zone (3). The fetal zone is composed of large eosinophilic cells secreting large amounts of dehydroepiandrosterone sulphate (DHEAS), which is converted into estrogens in the placenta and play an important role in its function. The definitive zone is composed of a thin layer of small, tightly packed basophilic cells and is thought to function as a reservoir of progenitor cells that move centripetally to differentiate into fetal zone cells (3,5). Around the end of gestation, a third layer (transitional zone) develops between the definitive and fetal zones, with intermediate characteristics (6).

By the ninth week of gestation, the adrenal blastema is completely enclosed by the adrenal capsule. At the same time, an extensive network of sinusoidal capillaries develops between the cords of the fetal zone. This vasculature predominates in the central portion of the fetal zone and persists throughout fetal life. Consequently, the adrenal cortex is one of the most highly vascularized organs in the fetus. This abundant vascularization is likely required to facilitate access of hormonal products to the circulation.

At 4 months of gestation, the adrenal glands are slightly larger than the kidneys and are mostly composed of fetal cortex. They continue to grow until the third trimester, reaching about 4 g each at birth (4). There is no difference in the weight of the adrenal glands by sex during embryonic development. Postnatally, the fetal adrenal zone rapidly involutes to disappear by about 6 months of age, and the definitive zone expands to form the zona glomerulosa and zona fasciculata (4).

The adrenal medulla is derived from cells of the neural crest in association with the development of the rest of the sympathetic nervous system (7). Around day 37 of development, bundles of nerve fibers and sympathicoblasts in primitive small cells enter the adrenal glands. The medullary primordium is visible at about day 45 of gestation, and results from the assembly of sympathogonia from the sympathetic chain (of ectodermal origin) in the region near the developing mesodermal cortical primordium (8). While the fetal cortex is forming, the invading sympathogonia cells become arranged in clusters and cords of neuroblasts. These cells give rise to pheochromoblasts, which are dispersed along the course of the nerve fibers (1,7).

Beginning around day 56 of gestation and continuing throughout the fetal and into the neonatal period, the pheochromoblasts undergo maturation to give rise to chromaffin cells.

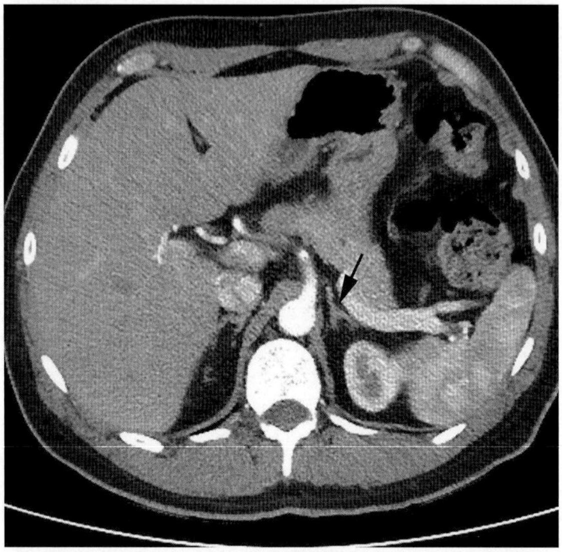

Figure 4-1

NORMAL ADULT ADRENAL GLAND

Contrast-enhanced computerized tomography (CT) of the upper abdomen demonstrates the normal left adrenal gland (arrow) as an inverted Y-shaped structure anterior to the upper pole of the left kidney.

However, some of these primitive cells may persist as neuroblastic nodules, which increase with age and peak at 17 to 20 weeks' gestation (7,9). They may then regress in older fetuses, or persist until birth or early infancy. Nodules range in diameter from 60 µm to 400 µm, and may be confused with small neuroblastomas (1,9).

Gross Anatomy

The adrenal glands are located in the retroperitoneum, superior and slightly anterior and medial to the kidney (fig. 4-1). At birth, the adrenal glands together weigh about 8 g and are nearly the size of adult adrenal glands (10,11). Therefore, by weight, newborn adrenal glands are 10 to 20 times proportionally larger than adult adrenal glands (12).

The fetal zone of the adrenal gland involutes markedly starting at about 2 weeks after birth, and by the end of the first month of life, about 50 percent of the weight of this zone has been lost (13,14). Recent studies have shown a role for apoptosis in this involution, especially during the first 2 to 4 weeks of life (15,16). By the end of the first year of life, only the stroma of the degenerated fetal cortex remains, and the adrenal gland approaches normal adult dimensions (approximately 5.0 x 3.0 x 0.6 cm) and weight (4 to 6 g each) (10). With chronic illness and elevated adrenocorticotropin hormone (ACTH), or corticotropin, levels, the weight can increase by 0.5 to 1.0 g per gland.

The adrenal glands of neonates have a smooth external surface. Cut sections show a dark red-brown color caused by regression of the fetal cortex and congestion. Medullary tissue is not visible in the neonatal gland, since it is less than 1 percent of the total gland volume at this time. In adults, the right adrenal gland tends to be more triangular in shape than the larger, more crescent-shaped left adrenal gland (33). The surface of the adrenal gland has a characteristic bright yellow-orange color due to lipid accumulation. The zona reticularis or inner zone is thinner and darker.

The medulla has an ellipsoid shape near the head of the gland and is more elongated at the body. The medulla is gray-tan and is best seen in the head and body of the gland. In a review of several studies, the calculated weight of the adrenal medulla varied from 0.37 to 0.48 g in individual glands and accounted for about 9 percent of the total adrenal volume (18,19). If the cortex becomes atrophic, the medulla appears more prominent. In some areas of the adrenal gland, the medulla is absent, and opposing parts of the cortex may be separated by a raphe (20).

The adrenal gland can be divided grossly into the head, which contains most of the medulla; the body, which has some medullary tissue; and the tail, which consists only of cortical tissue (21–23). The wings, or alae, are 2- to 3-mm–thick structures adjacent to the long axis of the body; they do not contain medulla, except when medullary hyperplasia develops.

The weight of the adrenal gland in adults who die suddenly without prior illness is 4.0 to 4.2 g ± 15 percent (22). Ninety percent of the gland weight is contributed by the cortex (19). The mean weight of adrenal glands in individuals of African descent dying suddenly is 3.8 ± 0.8 g, suggesting racial differences in weight (22). There is no correlation between absolute body weight and adrenal weight, and no difference between the weight of the left and right gland. Although historically it has been reported that the adrenal gland is heavier in males than in females (24), these observations are not supported by more recent studies (14,22,25). Pregnancy- and sex-related changes in adrenal weights are seen in some mammals (26), but not in humans (22). When death is preceded by a chronic illness, there is a significant increase in the mean adrenal weight (mean, 5.8 to 6.2 g ± 25 percent).

There are a number of ways to evaluate the gross anatomy of the non-neoplastic adrenal glands obtained surgically or at autopsy. In the absence of an overtly neoplastic process, the periadrenal fibroadipose tissue should be carefully dissected. Accurate weighing of the gland may provide valuable information in the evaluation of cortical or medullary abnormalities. The gland should be sectioned in the transverse plane at narrow intervals of 2 to 3 mm, and individual slices laid out sequentially. In this manner, the distribution and size of the medulla can be evaluated, and the cortex and appropriate representative areas can be submitted for histologic study. As always, the results should be correlated with endocrinologic and other clinical data to make a precise diagnosis.

Blood Supply

Adrenal capillaries drain into a central vein that exits the cortex at the inferomedial aspect of the gland. Each gland has a dominant vein, although smaller accessory veins are often found adjacent to the adrenal arteries (27). Anatomic variants have been reported in up to 50 percent of patients, most of which are minor (28). Significant variations probably occur in 3 to 5 percent of patients (28,29). Multiple adrenal veins draining via their usual pathway into the inferior vena cava on the right, and into the left renal vein on the left are the most common anomalies (30). In anatomically normal individuals, the right adrenal vein drains directly into the inferior vena cava and is less than 1 cm in length. The left adrenal vein is slightly longer and merges with the left inferior phrenic vein prior to draining into the left renal vein (27). It has been suggested that the shortness and decreased compliance of the right adrenal vein may be responsible for the higher incidence of perinatal adrenal hemorrhage in the right compared to the left gland (20). The major effluent vessels have eccentrically arrayed bundles of longitudinal smooth muscle. These muscles are absent in neonates and are not well developed in young children. They may also have important functions during hemorrhage (31).

Unlike the venous system, the arterial branches to the adrenal glands are not as distinct. There are 50 to 60 small branches that arise from three principal arteries: the inferior phrenic arteries superiorly, the abdominal aorta medially, and the renal arteries inferiorly (17,27,32). Some of these small branches partially supply the adrenal capsule.

The arteries enter on the medial aspect of the glands and give rise to a dense network of vessels that supply the three layers of the adrenal cortex and medulla (11). Most of the cortical tissue is perfused by blood that enters the parenchyma from the capsular arterial plexus. Thus, the blood supply to the cortex is derived from capillary sinusoids, which are lined by endothelial cells with fenestrae or pores closed by a single membrane. Because there is no direct arterial supply to the deeper cortical layers, these cells rely on the blood that has passed through the central zones and which contains secreted steroid hormones. Most of the blood entering the cortex in the head of the gland eventually reaches the medulla, where there is a corticomedullary portal system. The zona glomerulosa always lies nearest the source of arterial blood. The medulla is also supplied by the arterial medullae which pass directly to it from the capsular arterial plexus (22).

Lymphatic Supply

Lymphatic vessels are present only in the capsule of the gland. Cortical and medullary parenchymal tissues are devoid of lymphatics, which are present only in the adventitia of the central vein and its major tributaries (23). On the right, adrenal lymphatics drain into paracrural, para-aortic, and paracaval lymph nodes. On the left, adrenal lymphatics initially drain into para-aortic and left renal hilar lymph nodes. Adrenal lymphatics may also drain directly into the thoracic duct and posterior mediastinal nodes (27,32). Abnormalities of the adrenal lymphatic vessels may contribute to adrenal cysts.

Innervation

The adrenal gland is richly innervated by the autonomic nervous system (33). Its capsule contains nerve plexuses of fibers originating

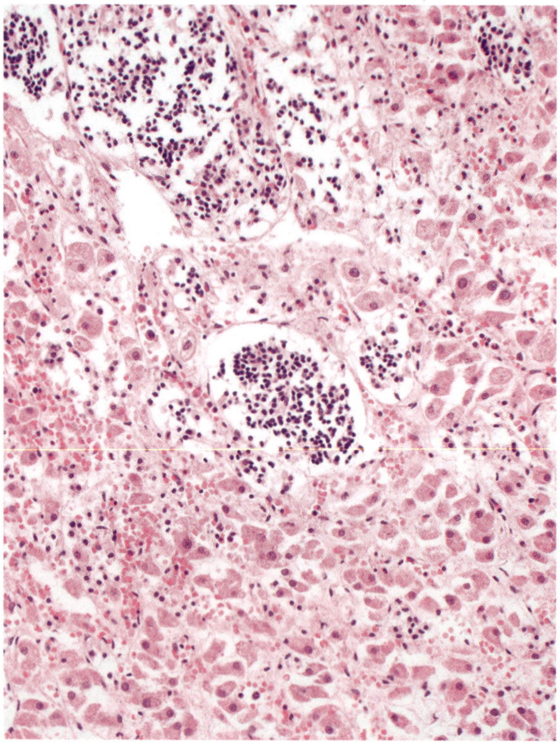

Figure 4-2

DEVELOPING FETAL ADRENAL GLAND

Provisional or fetal adrenal cortex at 3 months of development shows eosinophilic adrenal cortical cells and small nodules of primitive neuroblastic cells admixed with the fetal cortex.

from the celiac ganglion, celiac plexus, and greater splanchnic nerve (33,34). From these plexuses, nerve bundles traverse the cortex, mostly in association with blood vessels and connective tissue trabeculae. Most of these nerve bundles terminate in the medulla or in the smooth muscle of the major vessels (35). Unlike other visceral organs which receive stimuli from postganglionic nerve fibers, the adrenal medulla is unique in that it is supplied by preganglionic fibers that directly synapse to chromaffin cells which produce catecholamines. There are no postganglionic nerve fibers (36).

Histology

At birth, between 70 and 85 percent of the adrenal cortex is made up of the provisional cells of the fetal zone (fig. 4-2). This zone undergoes rapid involution during the first few weeks to months of life. As a consequence, the total weight of the gland decreases by approximately 50 percent (37–40). The atrophy of the fetal zone appears to occur by apoptosis, with large numbers of apoptotic nuclei visible histologically in the postnatal period (41). Anencephalic fetuses have a normal-appearing adrenal cortex at up to 20 weeks of gestation, but at birth the cortex is much thinner than that of the normal neonate (20).

Microscopic cystic changes have been reported in the adrenal glands of some neonates as well as premature and stillborn infants. These changes have been attributed to in utero stress (42). Other degenerative changes, including vacuolization of the cortical cells, have been reported in infants with erythroblastosis fetalis and thalassemia major, and these conditions may also be related to intrauterine stress caused by hypoxia (24).

The adrenal cortex is composed of two histologically distinct layers in fetal life. The most superficial layer of the cortex, located directly beneath the capsule, is known as the definitive zone, adult zone, or neocortex (43). The definitive zone is thin (10 to 20 cells thick), and in the fetal adrenal gland represents only a small fraction of the entire cortex. The neonatal definitive zone is 0.1 to 0.2 mm in thickness; by 9 days after birth, it is 0.5 mm thick, and by the 12th year of life, it is 1.0 mm thick. It is recognizable histologically by its basophilic appearance on hematoxylin and eosin (H&E)-stained sections. The cells are polygonal, with round nuclei and scant pale eosinophilic, sometimes vacuolated, cytoplasm.

During fetal life, the stratification that is present in the adult adrenal cortex into zona glomerulosa, fasciculata, and reticularis is not visible. However, near term, the beginning of this zonation is seen as the zona glomerulosa becomes organized into small clusters distinct from the cords of the zona fasciculata.

The adult adrenal cortex consists of three readily identifiable cell types, arranged in concentric zones or layers (fig. 4-3) (38). The zona glomerulosa, which is the outer zone, comprises 5 to 10 percent of the cortex. The cells here are smaller, with less cytoplasm, than the other cortical cells (fig. 4-4) and forms an incomplete layer. The middle zone, or fasciculata, makes up about 70 percent of the cortex. It is composed of

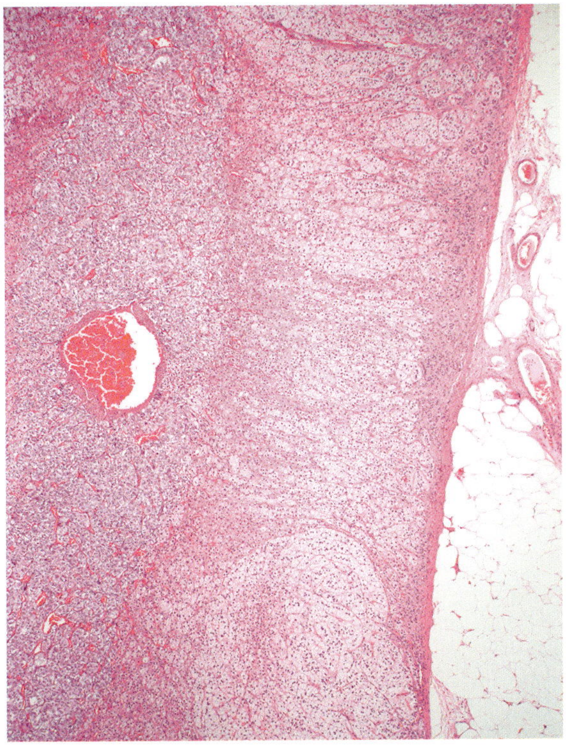

Figure 4-3

NORMAL ADULT ADRENAL GLAND

Normal adrenal cortex and medulla in an adult shows lipid-rich cortical cells of the zona fasciculata and medullary cells with basophilic cytoplasm adjacent to the central vein.

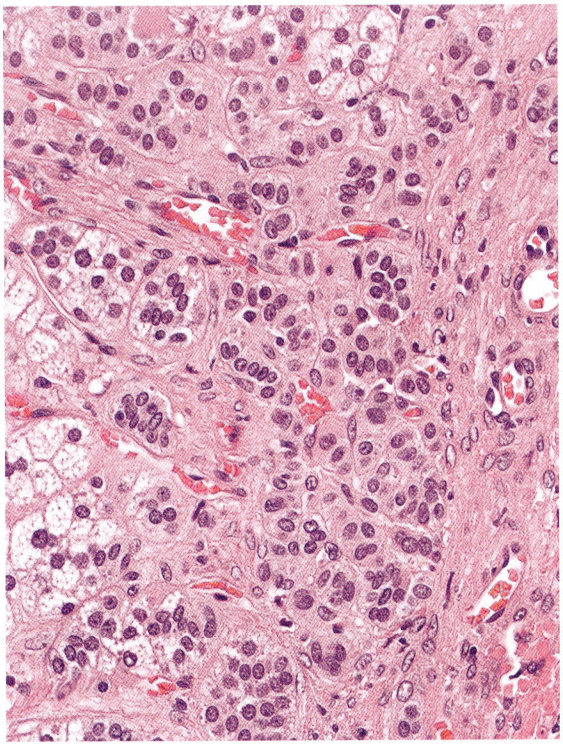

Figure 4-4

NORMAL ADULT ADRENAL GLAND

High magnification of the zona glomerulosa shows small cells with a high nuclear to cytoplasmic ratio and lipid-filled clear cytoplasm.

radial cords of cells with abundant pale-staining lipid-filled cytoplasm (fig. 4-5). The inner portion of the zona fasciculata merges with the inner cortical zone, or zona reticularis. This zone is composed of cells with compact, finely granular, eosinophilic cytoplasm (fig. 4-6). These cells have variable amounts of lipochrome pigment which contribute to the brown color; they are designated as compact cells. Adrenal cortical cells are also found around the central vein and its branches, and are termed the "cortical cuff." All three zones may be present in the cuff.

Cortical extrusions are found frequently in the adrenal glands of adults (44). They are characterized by nodular groups of cortical cells that extend into the periadrenal fat. Typically, they are attached to the adjacent cortex by a small pedicle and are surrounded by a fibrous capsule; however, they may be completely separated from the gland in some instances or may lack a capsule. In the assessment of adrenal cortical neoplasms for invasive growth, it is important to recognize that extrusions are a normal finding and do not indicate invasive growth. Focal aggregates of lymphocytes are an incidental finding in the adrenal cortices of normal adults and increase in frequency with age. Most of the lymphocytes are of T lineage (45).

The adrenal medulla is composed of chromaffin cells that are organized in nests and anastomosing cords separated by a rich capillary network (fig. 4-7). The cells have basophilic to amphophilic granular cytoplasm and indistinct cell borders (fig. 4-8). There is moderate variation in cell size, and mitotic figures are extremely uncommon. A few medullary cells, particularly those in the juxtacortical regions, may have enlarged, hyperchromatic nuclei, which increase in number with age (46). In addition, the cytoplasm of medullary cells frequently contains periodic acid–Schiff (PAS)-positive hyaline globules, which also tend to be

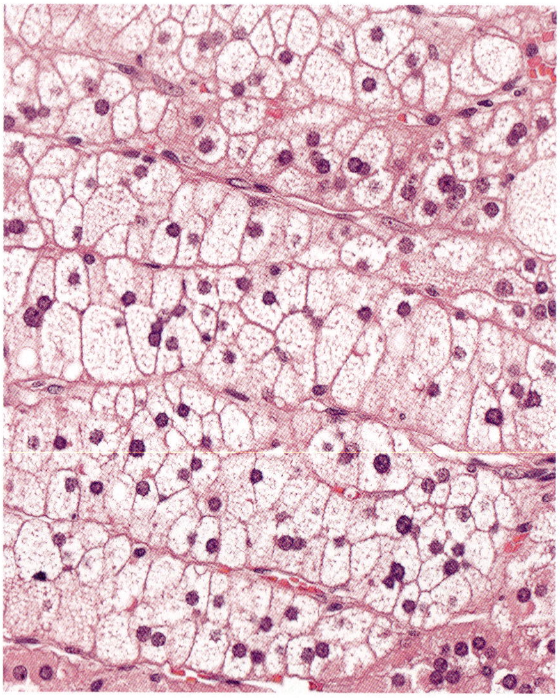

Figure 4-5

NORMAL ADULT ADRENAL GLAND

The zona fasciculata consists of large cells with predominantly clear cytoplasm organized in columns.

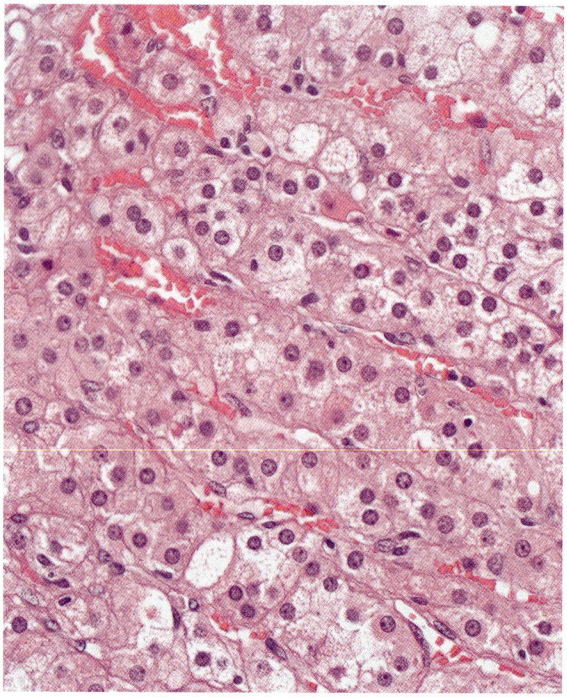

Figure 4-6

NORMAL ADULT ADRENAL GLAND

The zona reticularis consists of cells with less cytoplasmic lipid than the fasciculata. The cytoplasm is composed of eosinophilic compact cells organized in cords and containing brown lipofuscin pigment.

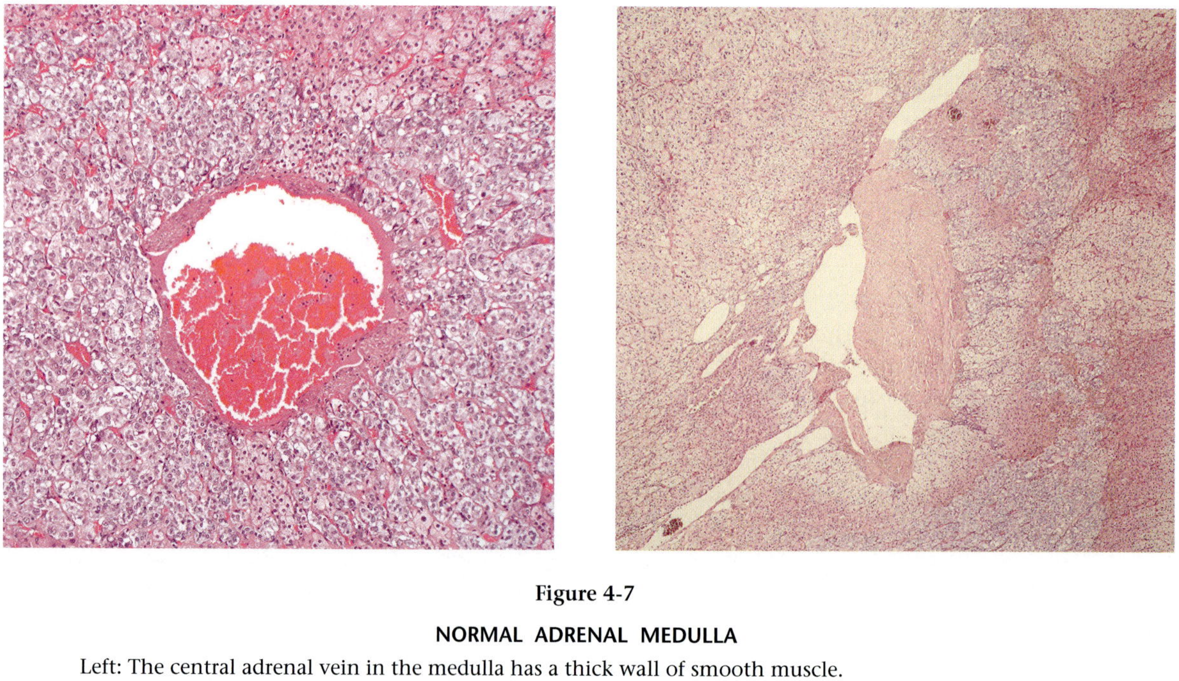

Figure 4-7

NORMAL ADRENAL MEDULLA

Left: The central adrenal vein in the medulla has a thick wall of smooth muscle.
Right: The smooth muscle wall surrounding the central adrenal vein is characteristically incomplete in areas.

more prominent in juxtacortical regions and are thought to represent dense core secretory granules (47). They may be present in up to 80 percent of adrenal medullae, and are more common in patients with chronic neurologic disorders such as Parkinson disease (20).

A few ganglion cells are also present within the medulla, either as single cells or in small cell clusters, and some may be associated with myelinated nerve bundles (38,48). S-100 protein-positive sustentacular cells are present at the peripheries of the medullary cords and nests, and are also evident around the ganglion cells (fig. 4-9). Another cell type that has been termed the small, intensely fluorescent or small, granule-containing cell is also seen in the medulla of most mammals. These cells may function as interneurons (49).

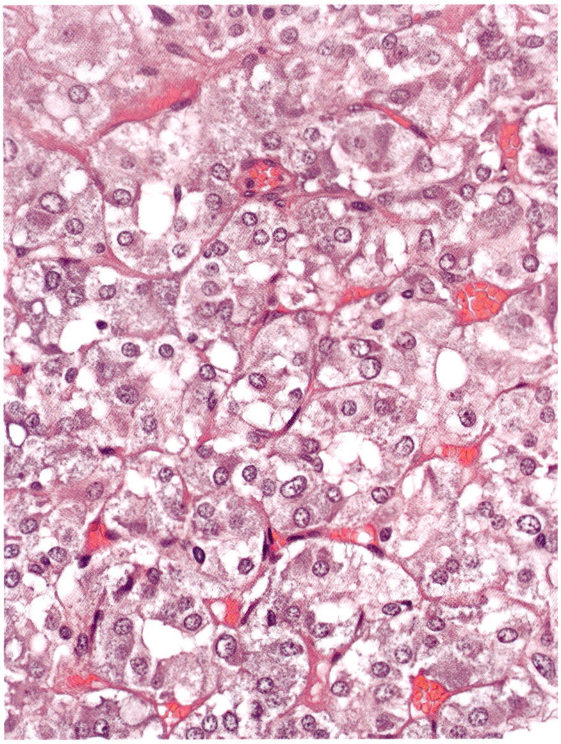

Figure 4-8

NORMAL ADRENAL MEDULLA

The adrenal medulla consists of round cells with basophilic cytoplasm arranged in nests and anastomosing cords.

Ultrastructure

Some ultrastructural features are shared by the three layers of the cortex, reflecting their common function: synthesis of steroid hormones. The cells feature voluminous endoplasmic reticulum, stacks of rough endoplasmic reticulum, a well-developed Golgi apparatus, lysosomes, and many mitochondria (50). Nevertheless, there are distinct ultrastructural features in each cortical zone, allowing for easy recognition and separation of optimally fixed, surgically excised tissue (51,52).

The cells of the zona glomerulosa have sparse intracellular lipid, elliptical mitochondria, and lamellar or plate-like cristae, resulting in a ladder-like internal structure similar to that found in many other tissues (50). Microvillous projections may be present on the cell surface. Lysosomes and lipofuscin granules are uncommon in this zone. Little smooth endoplasmic reticulum is present in the zona glomerulosa.

The cells of the zona fasciculata have distinct mitochondria that are round to oval, with short and long tubular cristae. Variably sized lipid droplets are present in the cytoplasm of these cells. Smooth endoplasmic reticulum is prominent, especially in the inner fasciculata. Rough endoplasmic reticulum is present in moderate amounts. Microvillous cytoplasmic projections are more prominent in the inner fasciculata, and lysosomes are increased in number compared to those in the glomerulosa.

The cells of the zona reticularis have spherical to ovoid mitochondria with short and long tubular invaginations of the inner membrane (51,52). The cells have abundant lipofuscin granules and prominent membrane-bound organelles, with a moderately dense matrix that contains dense granules and clear lipid globules. Glycogen is also present (50).

The fetal adrenal cortex contains spherical to ovoid mitochondria with tubular cristae. The cells have abundant smooth endoplasmic reticulum and short cytoplasmic microvilli (53).

Catecholamine-secreting (chromaffin) cells dominate the medulla. Two cell types, epinephrine- and norepinephrine-secreting, distinguished by granule type, are present (50,54). In tissue fixed in glutaraldehyde, cells that contain epinephrine feature granules measuring about 120 µm in diameter, with a moderately dense but not opaque, finely granular texture filling the enclosing membrane. Norepinephrine-secreting

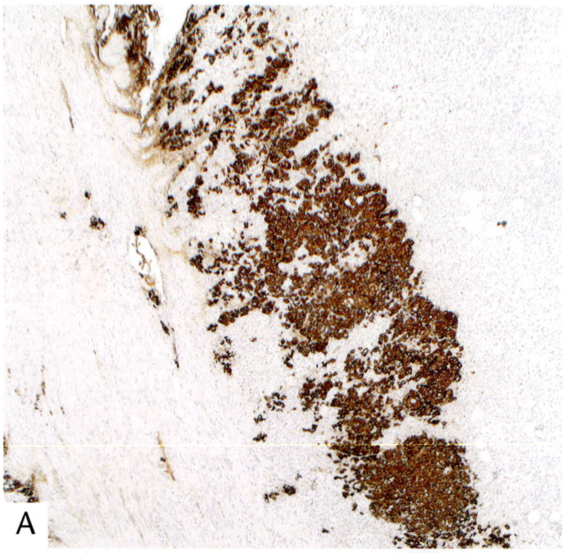

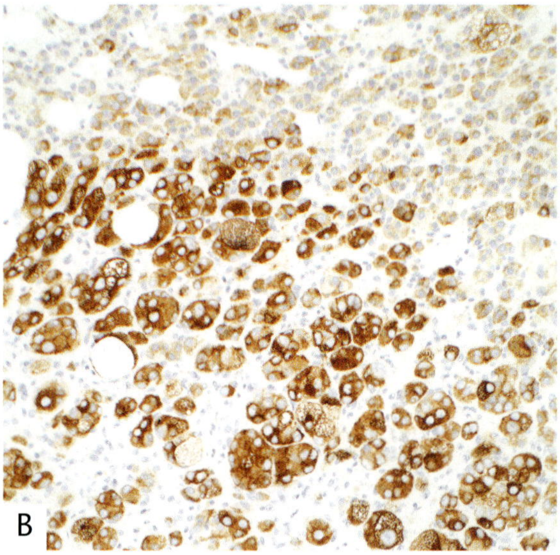

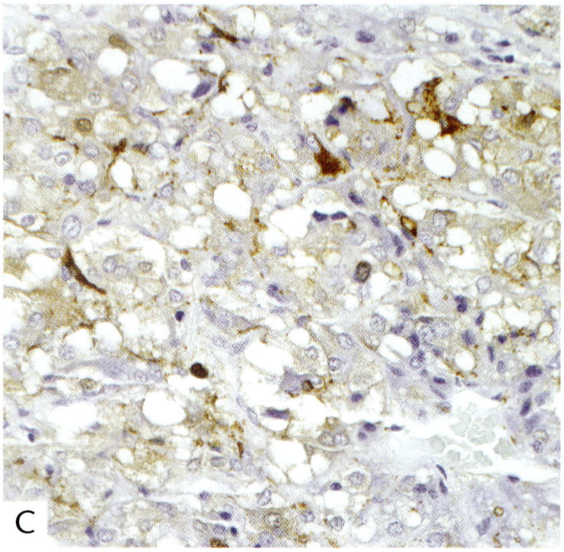

Figure 4-9

NORMAL ADRENAL MEDULLA

A: Immunostain for chromogranin A outlines the adrenal medulla in brown. The adrenal cortex is negative for chromogranin A.

B: Synaptophysin strongly stains the adrenal medulla. The adrenal cortex is weakly positive.

C: S-100 protein immunostain highlights the sustentacular cells, which are intimately associated with the pheochromocytes.

cells have granules that are electron opaque, often located eccentrically within a dilated sac, and measuring about 250 µm in diameter. Both cell types contain moderate amounts of rough endoplasmic reticulum and interdigitating blunt cytoplasmic processes with poorly developed cell junctions.

The sustentacular cells are in close proximity to the chromaffin cells. They have oval indented nuclei and thin elongated cytoplasmic processes (55). These cells partially surround the chromaffin cells without the interposition of a basal lamina. They have moderate amounts of rough endoplasmic reticulum and occasional lipid droplets, but no secretory granules.

Immunohistochemistry

Cells of the normal adrenal cortex are immunoreactive for intermediate filaments, including pancytokeratins (56–58). Alpha-inhibin protein is also expressed in adrenal cortical cells, mostly confined to the zona reticularis (59). This dimeric 32-kd peptide is composed of an alpha and beta subunit; higher concentrations of the alpha subunit are present in the adrenal cortex. Inhibin is not specific for adrenal cortical cells because other tissues, such as placenta, pituitary gland, and gonad, also produce this peptide (59,60).

The melanoma marker, melan A (MART1), has also been useful in differentiating adrenal cortical cells and neoplasms because of its selective immunoreactivity with steroid-producing cells (61). Other peptides detected in the adrenal cortex are: calretinin; CD99; steroidogenic factor-1 (SF-1) (62); growth factors such as transforming growth factor alpha, epidermal growth factor (63), and insulin-like growth factor II; and class II major histocompatibility complexes including human leukocyte antigen, D-related (HLA-DR) (64). Cells of all three cortical zones are also variably synaptophysin positive, but are negative for chromogranin and S-100 protein.

The antiapoptotic protein, BCL2, although initially thought to be positive in the adrenal cortex but negative in the medulla, has now been found to have variable results related to antigen retrieval methodologies. In a study by Zhang et al. (65), BCL2 immunoreactivity was detected focally in only 20 percent of adrenal cortical tumors but more diffusely in 86 percent of pheochromocytomas and in normal adrenal medullae.

The β-catenin target gene, known as *DAX1* (dosage-sensitive sex reversal, adrenal hypoplasia congenita critical region on the X chromosome, gene 1), regulates adrenocortical stem/progenitor cell pluripotency (66). The gene is mutated in X-linked cytomegalic adrenal hypoplasia congenita; it has been molecularly defined by its role as an inhibitor of SF-1–mediated gene transcription, particularly of steroidogenic enzymes (67–69). Immunohistochemical expression of the DAX1 protein is seen in both the nucleus and cytoplasm of adrenal cortical cells, and can be used to aid in the diagnosis of adrenocortical neoplasms (70,71).

Demonstration of steroidogenic enzymes, particularly those unique to the adrenal cortex such as P450c11 (11β-hydroxylase) (72), may help in the diagnosis of adrenocortical carcinoma. The patterns of expression of these enzymes in adrenocortical carcinoma, however, differ from adrenocortical adenoma, and some carcinoma cells do not express the steroidogenic enzymes involved in corticosteroidogenesis (73–75). Currently, there are no immunohistochemical markers of steroidogenic enzymes that are specific for adrenocortical carcinoma (70). Tartour et al. (76) reported that nuclear immunoreactivity recognized by the D11 monoclonal antibody was highly specific for adrenocortical carcinoma, but immunoreactivity was only observed in 44 percent of carcinomas and was restricted to "well-differentiated" tumors. Therefore, the practical value of the D11 monoclonal antibody in the diagnosis of adrenocortical carcinoma is limited. Nevertheless, the aforementioned antibodies allow functional classification of the three zones of the adrenal cortex based on historical and biochemical studies, and provide a direct method to study corticosteroidogenesis in situ.

Similar immunohistochemical studies have also shown that the immune mediatory protein, interleukin 6, which is involved in communication between the immune and endocrine systems, is also produced by cells in the inner zone (reticularis) of the adrenal cortex (77). The presence of HLA-DR and interleukin 6 in these cells suggests an important immunologic function for some adrenal cortical cells (78). In addition, alterations in immunohistochemical expression of interleukin 6 in neoplastic adrenocortical cells suggest that it may be an important step in the process of adrenal tumorigenesis (79).

Immunohistochemical studies of the adrenal medulla show that the cells (pheochromocytes) and tumors that arise from them are diffusely positive for neuroendocrine markers, including chromogranin (fig. 4-9A), synaptophysin (fig. 4-9B), PGP9.5, CD56, and INSM1, as well as IGF2 (80,81). Cells of the adrenal cortex also express synaptophysin (albeit less strongly than cells of the medulla) but are negative for other neuroendocrine markers. Pheochromocytes and tumors derived from them are usually negative for keratin and vimentin, but often express neurofilament protein (82). The sustentacular cells in the medulla, as well as some of the pheochromocytes, stain with S-100 protein (fig. 4-9C) (83,84). A subpopulation of sustentacular cells may be positive for glial fibrillary acid protein (GFAP) (85).

A number of peptides have been identified in the adrenal medulla, including methionine, enkephalin, and corticotropin. In addition, antibodies against catecholamine-synthesizing enzymes (e.g., tyrosine hydroxylase) as well as catecholamines can be detected by immunohistochemistry (59,84,86,87). These are not usually documented as part of the diagnostic procedure, with the exception of certain hormone expression (e.g., ACTH) that can aid in confirming a tumor as the source of hormone secretion in ectopic hormone syndromes (88).

Molecular Biology and Physiology

Adrenal Cortex. Adrenal cortical homeostasis is regulated by a feedback mechanism between the hypothalamus, pituitary gland, and adrenal cortex. Glucocorticoids exert a negative feedback influence on the hypothalamus and pituitary gland to maintain normal homeostasis.

The precursor of all steroid hormones is cholesterol, which is derived from circulating low density lipoproteins (89). After internalization into the cortical cells, the lipoproteins are hydrolyzed, which results in the production of cholesterol esters, which yield cholesterol and free fatty acids (90,91). There are four cytochrome P450 enzymes that are involved in the biosynthesis of adrenal steroids ($P450_{SCC}$, $P450_{c17}$, P450c21, P450c11) (89). The transcription factor AdBP/SF-1 regulates the expression of the *CYP* genes (92). The principal glucocorticoids are deoxycorticosteroid, 11-deoxycortisol, corticosteroid, and cortisol. Aldosterone is the principal mineralocorticoid and is derived from corticosteroid.

In the fetal adrenal gland, DHEAS production begins at 8 to 10 weeks of gestation and increases considerably during the second and third trimesters (93–95). De novo cortisol production likely occurs transiently early in gestation (around 7 to 10 weeks' gestation), although due to the lack of 3-beta-hydroxysteroid dehydrogenase (3β-HSD), it appears to be suppressed until late gestation when cortisol production in the definitive zone escalates (96–98). In the adult, different zones produce different glucocorticoids (92).

Mineralocorticoids, mainly aldosterone, are produced by the zona glomerulosa in response to angiotensin II and, to a lesser extent, ACTH. Glucocorticoids, mainly cortisol and corticosterone, are produced by the zona reticularis in response to ACTH and by the fasciculata after prolonged stimulation. Adrenal androgens, mainly dihydroepiandrosterone, as well as estrogens, are produced by the zona fasciculata and zona reticularis (92).

The pulsatile release of cortisol is under direct stimulation by ACTH released from the anterior pituitary. ACTH, or corticotropin, is synthesized in the anterior pituitary as a large precursor, pro-opiomelanocortin (POMC). ACTH stimulates cortisol release by binding to a Gαs protein-coupled plasma membrane melanocortin 2 receptor (MCR2) on adrenocortical cells, resulting in activation of adenylate cyclase, an increase in cycline adenosine monophosphate, and activation of protein kinase A. Protein kinase A phosphorylates the enzyme cholesteryl-ester hydrolase, increasing its enzymatic activity and leading to increased cholesterol availability for hormone synthesis (99).

The release of ACTH from the anterior pituitary is regulated by the hypothalamic peptide corticotropin-releasing hormone (CRH). Cortisol synthesized in the adrenal cortex enters the circulation, crosses the blood-brain barrier, and reaches the hypothalamus and anterior pituitary; here it binds to a glucocorticoid receptor and inhibits the biosynthesis and secretion of CRH and ACTH in a classic negative feedback loop. This closely regulated circuit is known as the hypothalamic-pituitary-adrenal (HPA) axis. This axis is exquisitely sensitive to environmental and internal factors such as light, sleep, stress, and disease (92,100).

Glucocorticoids, principally cortisol, exert multisystemic effects because virtually all cells express glucocorticoid receptors. Glucocorticoids, as their name implies, play an important role in regulation of glucose homeostasis. The glucocorticoid receptor is a 94-kd protein that is associated with heat shock protein 90 (Hsp90) as well as several others. These proteins assist in the folding and assembly of the glucocorticoid receptor protein complex (101). Glucocorticoids also regulate the transcriptional activity of many genes, including those for POMC, prolactin, and the glycoprotein hormones.

Aldosterone synthesis and release in the adrenal zona glomerulosa is predominantly regulated by angiotensin II and extracellular potassium and, to a lesser extent, by ACTH. Aldosterone is part of the renin-angiotensin-aldosterone system, which is responsible for preserving circulatory homeostasis in response to a loss of salt and water (99). Alterations in specific serum electrolytes, including potassium and sodium, stimulate aldosterone production via angiotensin II release (99).

Several compounds inhibit adrenal steroidogenesis by interfering with one or more of the enzymes in the steroidogenic pathway.

These compounds have been used to study the enzymes involved in steroid synthesis, to evaluate the HPA axis, and to treat some patients with endogenous excesses of cortisol, androgen, or mineralocorticoid secretion (102). Ketoconazole, metyrapone, etomidate, and mitotane block one or more of the enzymes in the steroid synthesis pathways (102). The subsequent decrease in cortisol secretion results in a compensatory rise in ACTH release, which tends to override the drug-induced cortisol blockade, but can also result in accumulation of precursor steroids (102). Mitotane is classified as an adrenolytic agent, since it has a cytotoxic effect on adrenal tissue. It can be used to produce a "medical adrenalectomy" in patients with Cushing disease, sometimes in association with pituitary radiotherapy (103). Aminoglutethimide, another antisteroidogenic drug that works via inhibition of CYPIIAI, is no longer available (102).

Adrenal Medulla. Catecholamines are amino acid–derived hormones, synthesized from the amino acid tyrosine. Tyrosine is actively transported into the catecholamine-producing cells, where it undergoes a number of enzymatic cytosolic reactions for its conversion to epinephrine (99). First, tyrosine is hydroxylated to 3,4-dihydroxyphenylalanine (L-dopa) by the enzyme tyrosine hydroxylase. L-dopa is then converted to dopamine, followed by norepinephrine, and finally epinephrine. The enzymes that regulate catecholamine synthesis are in turn regulated by some of the catecholamines themselves, such as norepinephrine. The phenylethanolamine N-methyltransferase enzyme, which converts norepinephrine to epinephrine, is in turn regulated by glucocorticoids.

Catecholamines are released in response to sympathetic nerve stimulation of the adrenal medulla and are central to the stress response to a physical or psychological insult such as severe blood loss, decrease in blood glucose concentration, traumatic injury, surgical intervention, or a fearful experience (99). Acetylcholine released from the preganglionic sympathetic nerve terminals binds to nicotinic cholinergic receptors (ligand-gated ion channels) in the plasma membrane of the chromaffin cells, leading to exocytosis of secretory granules, which release catecholamines into the interstitial space, from where they are transported in the circulation to their target organs (99). The chromaffin granules contain catecholamines and various peptides.

Chromogranins are the major soluble proteins within the chromaffin granule. They are part of a large family of acid proteins that are present in all neuroendocrine cells containing secretory granules (104). Other proteins in the secretory granule are dopamine beta-hydroxylase and cytochrome b-561. The latter is involved in transmembrane electron transport, linking the ascorbic acid cycle with dopamine beta-hydroxylase and other mono-oxygenases (105). Other peptides present in the chromaffin granules include: neuropeptide Y, a 36-amino acid peptide stored in the adrenal medulla that acts to decrease norepinephrine release; and galanin, a 29-amino acid peptide of the adrenal medulla and nerve tissue that has various functions including contraction of smooth muscle, inhibition of ion transport, and regulation of feeding behavior.

REACTIVE, HEREDITARY, AND DEVELOPMENTAL DISORDERS

Adrenal Cytomegaly

Definition. *Adrenal cytomegaly* is characterized by the presence of large polyhedral cells with eosinophilic granular cytoplasm and enlarged nuclei in the adrenal cortex.

General Features. Adrenal cytomegaly may occur as an isolated incidental finding, or in association with a number of other conditions. It is reported in 0.8 percent of pediatric autopsies (106), and 3 to 7 percent of premature infants (107). It is commonly seen in association with Rh incompatibility (108–112), and has been reported in up to 69 percent of cases of hydrops fetalis in some series (106,108,110,113). Adrenal cytomegaly is also a well-known finding in Beckwith-Wiedemann syndrome and is often seen in cases with associated adrenal hemorrhage (108,110,114,115).

Although the pathogenesis of adrenal cytomegaly is not well understood, it has been postulated to reflect adrenal cortical cell adaptation to physiologic stress (108). However, given that most cases of fetal death are not associated with adrenal cytomegaly, intrauterine stress cannot be the sole basis for the development of this finding. As further evidence for this, adrenal cytomegaly has also been reported in association with simple abnormalities, such as omphalocele and Meckel diverticulum (108,116), as well as in apparently normal fetuses.

Usually only present up to a few months of age, adrenal cytomegaly may persist in older children or adults, particularly in patients with Addison disease (106,117–122). On image analysis studies, adrenal cytomegalic cells contain increased amounts of DNA, with ploidy values ranging from 3 to 9 times normal (and more than 25 times in some cases), while adjacent normal-appearing nuclei are euploid (106). In Beckwith-Wiedemann syndrome, one of the associated tumors is adrenal cortical carcinoma, but whether this arises in adrenal cytomegaly is unknown. Adrenal cytomegaly is not considered to be a precursor to neoplasia (108,113,119,123).

Microscopic Findings. The cytomegalic cells are present in the adrenal cortex and can range in number from only a few affected cells to more widespread changes. The cells may be as large as 150 µm in diameter and have disproportionately enlarged pleomorphic and hyperchromatic nuclei (106). Nuclear pseudoinclusions or cytoplasmic invaginations are also seen. Mitotic figures are rare and no atypical forms are seen (106).

Differential Diagnosis. Adrenal cytomegaly may be confused with the virally infected cells of cytomegalic inclusion disease, most commonly caused by cytomegalovirus. The virally infected cells have basophilic rather than eosinophilic cytoplasm, and the large intranuclear inclusions are typically surrounded by a clear halo. In contrast, the pseudoinclusions in adrenal cytomegaly represent cytoplasmic invaginations into the nucleus. The distinction between these two entities is usually based on macroscopic findings (glands with adrenal cytomegaly are often hyperplastic), viral polymerase chain reaction (PCR), and other clinical findings such as features of Beckwith-Wiedemann syndrome or Rh incompatibility.

Focal Adrenalitis

Focal adrenalitis is characterized by focal aggregates of lymphocytes and plasma cells within the adrenal cortex (124). It is a common incidental finding, particularly in elderly patients, and may be present in up to half of autopsied patients (125). The lymphocytes are predominantly CD4-positive T cells, with fewer numbers of CD8-positive cells (126).

Focal adrenalitis is associated with chronic inflammatory disorders in the retroperitoneum, such as chronic pyelonephritis (127). However, it should be distinguished from autoimmune adrenalitis which typically shows more diffuse inflammation, sometimes with lymphoid follicles, and is often associated with marked cortical atrophy (127).

Ovarian Thecal Metaplasia

Ovarian thecal metaplasia of the adrenal gland is a rare tumor-like mesenchymal lesion characterized by a hyalinized, radial scar-like proliferation of fibroblastic/myofibroblastic spindle cells within or just beneath the adrenal capsule (128). The lesion usually contains entrapped adrenal parenchymal cells and dystrophic calcifications (123). True ovarian stromal elements are rare (123,129). The lesions range from 0.1 to 2.0 cm (127). In up to half of cases, the nodules are multifocal and some may be bilateral (130). A number of other terms have been suggested for this entity, including *nodular hyperplasia of adrenocortical blastema, thecal metaplasia,* or *radial scar-like spindle cell myofibroblastic nodule of the adrenal gland* (129).

Ovarian thecal metaplasia is present in up to 4 percent of women and is usually detected postmenopausally. The pathogenesis of this entity is not fully known. It has been suggested that the foci of theca-like stroma in the adrenal gland may represent metaplasia from undifferentiated but embryologically competent mesenchymal cells of the adrenal capsule as a response to elevated gonadotropin levels in postmenopausal women (127,131). On rare occasions, it also occurs in men. The clinical significance of these lesions is not fully understood, but they can be functional, and rare neoplastic transformation has been reported (127). Surgical excision remains the treatment of choice for these lesions, which should be distinguished from true adrenal neoplasms (128).

Adrenal Malformations

Adrenal union (fusion) is a developmental abnormality that results in a single butterfly- or horseshoe-shaped adrenal gland above the aorta. It is occasionally associated with midline congenital defects, including spinal dysraphism, indeterminate visceral situs, and the Cornelia de Lange syndrome (associated with mental and growth retardation, low-set ears, anteverted nostrils, and spade-like hands with short tapering fingers) (132). It can also occur in patients with bilateral renal agenesis (133). The histologic appearance of the glands is usually normal (134).

In contrast to adrenal union, where there is intermingling of different parenchymal cells, *adrenal adhesion (accreta)* results in two tissues adhering to each other with an intervening connective tissue capsule separating them. *Adrenorenal* and *adrenohepatic unions* occur in 0.4 to 3.0 percent of unselected autopsy cases (135). In these cases, the organs share a common capsule, but there is no intermingling of the parenchymal elements. The embryologic defect in adhesion and union is most likely related to failure of the periadrenal capsular mesenchyme to separate completely early in development.

Adrenal Rests and Accessory Adrenal Tissues

Congenital anomalies of the adrenal gland rarely come to the attention of the surgical pathologist. The most common is *accessory adrenal tissue* (136). Although the term "heterotopia" is commonly used for this condition, a more accurate descriptor is accessory adrenal tissue, because in most cases, orthotopic adrenal gland is also present (137–142).

Most accessory adrenal glands consist exclusively of cortical tissue, but a few examples, particularly those in the region of the celiac ganglion, also contain medulla (143). Accessory adrenal cortex is found most frequently in the retroperitoneal space along the course of the urogenital ridges. In addition, accessory adrenal tissue may be discovered incidentally just beneath the renal capsule in the upper pole, at the hilar regions of the ovaries and testes, along the course of the spermatic cord, and within inguinal hernia sacs (140,141,144). Rare sites of accessory adrenal tissue include pancreas, spleen, liver, mesentery, lung, and brain (140,145).

Adrenal rests within the spermatic cord are particularly common in children (fig. 4-10), reported in almost 4 percent of children at the time of inguinoscrotal surgery (142). They also have been reported in 9 percent of children at the time of surgery for undescended testes (146). They may also be found as incidental nodules in salpingo-oophorectomy specimens (fig. 4-11).

Accessory adrenal tissue may undergo hyperplasia in response to increased levels of ACTH and may serve as the site of origin of adrenocortical neoplasms (147–150). *Ectopic adrenal tissue* in the pancreas may be confused with metastatic renal cell tumors or with clear cell neuroendocrine tumors. Similarly, rare adrenal cortical tumors have been reported in the liver (145) and scrotum (151). True heterotopic adrenal glands may be fused with the liver or kidney, and are typically surrounded by a common connective tissue capsule (152,153).

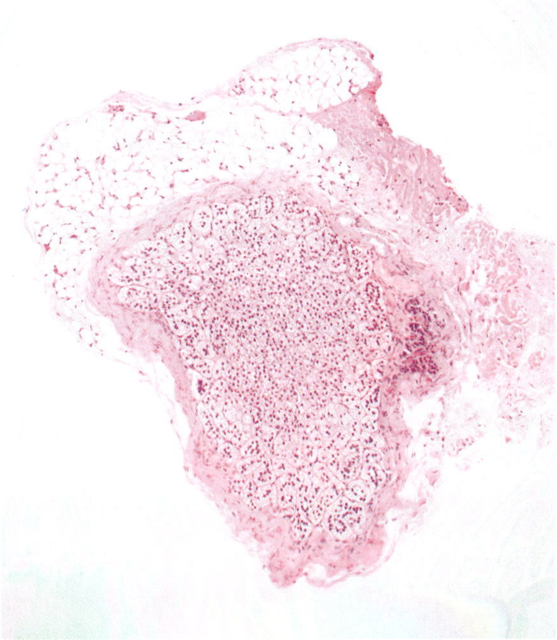

Figure 4-10

ECTOPIC ADRENAL TISSUE

Adrenal rests consisting of cortical but no medullary tissue were discovered incidentally in the spermatic cord.

Adrenal Hypoplasia

Congenital hypoplasia (or rarely, *aplasia*) of the adrenal gland is most commonly reported in association with anencephaly. Anencephaly occurs in up to 1 in 450 live births, with a female to male ratio of 4 to 1 (154). In anencephalic infants, the neurohypophysis is usually absent or very small, while the anterior pituitary gland is present but may lack some cell types. With the absence of an intact HPA axis, the sella turcica is frequently flat and filled with spongy vascular tissue. The pituitary gland has decreased numbers of ACTH cells while the other cells of the

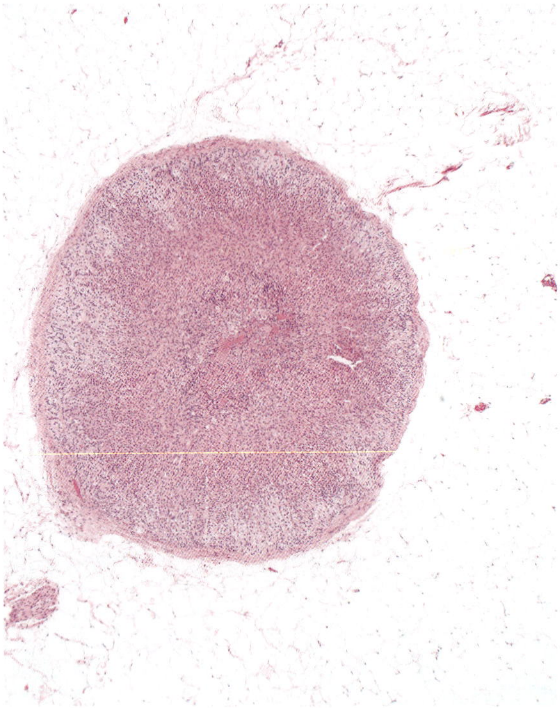

Figure 4-11

ECTOPIC ADRENAL TISSUE

Ectopic adrenal cortex in the periadnexal adipose tissue was discovered incidentally in a salpingo-oophorectomy specimen. These adrenal rest tissues can become hyperplastic in disorders associated with excessive adrenocorticotropic hormone (ACTH) production.

anterior pituitary are usually normal in number (155,156). Other central nervous system defects such as microcephalia, occipital cephalomeningocele, and hydrocephalus may be associated with adrenal hypoplasia. Rare cases of congenital absence or hypoplasia of the pituitary gland and neurohypophyseal aplasia may be associated with adrenal hypoplasia, resulting in perinatal adrenal failure (157,158).

Unilateral absence of the adrenal gland has been reported in up to 1 in 10,000 live births (159). In approximately 10 percent of patients with unilateral renal agenesis, the ipsilateral adrenal gland is also absent.

Over the past two decades, the genetic bases for several forms of familial adrenal insufficiency syndromes have been elucidated. *X-linked adrenal hypoplasia* is caused by a mutation or deletion of the *DAX1* gene (dosage-sensitive sex reversal adrenal hypoplasia congenita critical region of the X chromosome, also called the *NR0B1* gene) on the X chromosome (160–163). This form is usually associated with hypogonadotropic hypogonadism (164,165). It may be part of a contiguous chromosome deletion, which may include congenital adrenal hypoplasia, Duchenne muscular dystrophy, and glycerol kinase deficiency (166–168).

Autosomal recessive adrenal hypoplasia is due to a mutation or deletion of the gene that codes for SF-1 on chromosome 9q33 (169). This form, of adrenal hypoplasia is also associated with hypogonadotropic hypogonadism. Other forms of autosomal recessive adrenal hypoplasia have been identified, including rare syndromes such as Meckel-Gruber syndrome and Pena-Shokeir syndrome, as well as an autosomal recessive form of uncertain etiology (170).

Intrauterine growth retardation, metaphyseal dysplasia, adrenal hypoplasia congenita, genital anomalies (IMAGe) is a form of adrenal hypoplasia associated with intrauterine growth retardation, metaphyseal dysplasia, and genital abnormalities (171).

POMC deficiency leads to adrenal insufficiency as well. POMC (pro-opiomelanocortin) is a complex pro-peptide, encoded on chromosome 2p23.2. It is expressed in several tissues including the hypothalamus, pituitary gland, skin, and components of the immune system, and plays an important role in processing and release of ACTH (172). Mutations in the *POMC* gene result in adrenal insufficiency, obesity, ACTH deficiency, and alteration in pigmentation (173,174).

Infants with congenital adrenal hypoplasia usually present with weight loss, vomiting, dehydration, and severe electrolyte disturbances. The adrenal glands are small for age, with a decreased fetal zone. There are two histologic patterns seen in the adrenal cortices of patients with congenital adrenal hypoplasia: the miniature adult and cytomegalic forms (170). The miniature adult form is characterized by a small amount of residual adrenal cortex, which is composed primarily of permanent adult cortex with normal structural organization. This form is either sporadic or inherited in an autosomal recessive manner, and is frequently associated with abnormal central nervous system (CNS) development, including anencephaly or pituitary gland abnormalities (170).

In the cytomegalic form of congenital adrenal hypoplasia, the residual fetal adrenal cortex is structurally disorganized, with scattered irregular nodular formations of eosinophilic cells; the adult permanent zone is absent or nearly absent (175–177). Enlarged (cytomegalic) cells are present, some with abundant vacuolated cytoplasm. The cytomegalic form is generally considered to be X-linked, but there may be one or more autosomal genes associated with this phenotype (178–180).

ACTH deficiency due to any cause and defects in the ACTH receptor can result in *secondary hypoplasia* of the adrenal cortex at any stage of life. A common cause of adrenal atrophy is chronic treatment with exogenous glucocorticoids (fig. 4-12). These patients usually have to be weaned slowly from the exogenous therapy to allow their adrenal gland to return to normal function.

Hereditary Adrenal Cortical Unresponsiveness to Adrenocorticotropic Hormone (Familial Glucocorticoid Deficiency)

Familial glucocorticoid deficiency (FGD) is a rare autosomal recessive disorder that commonly presents in the first 2 years of life with hyperpigmentation, recurrent hypoglycemia, seizures, and muscle weakness (181). It is characterized by low levels of serum cortisol and high levels of plasma ACTH, with no cortisol response to exogenous ACTH (182–184). Plasma renin and aldosterone levels are normal.

The syndrome is caused by inactivating mutations in the genes encoding the ACTH receptor (185). Defects in the *MC2R* and *MRAP* genes account for 40 to 45 percent of cases, while the remaining 55 to 60 percent of patients have no identifiable genetic defect (186). Histologically, the adrenal glands show atrophy of zona fasciculata-reticularis, with a well-preserved but disorganized zona glomerulosa (183,187–190).

EXOGENOUS INJURY

Drugs that Inhibit Steroid Hormone Synthesis

Iatrogenic suppression of the HPA axis is an increasing challenge for clinicians, particularly in the era of widespread use of novel steroid medications to manage chronic diseases in an aging society (191,192). New drugs targeting common disorders such as hypertension and cancer can interfere with adrenal steroid synthesis or metabolism, and have the potential to induce adrenal suppression (193).

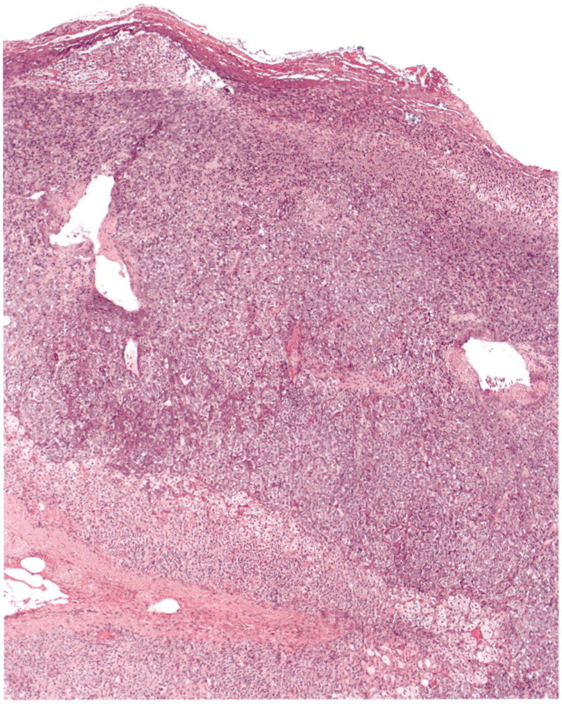

Figure 4-12

ADRENAL ATROPHY

The adrenal medulla appears more prominent with decreased thickness of the cortex.

Hormonal deprivation therapy is part of the routine management of a number of common malignancies, such as prostate and breast cancer (193). However, in tumors refractory to the standard treatments, novel selective steroid hormone synthesis inhibitors have shown promise in a number of clinical trials. For example, 17α-hydroxylase (CYP17A1) inhibitors (such as abiraterone for prostate cancer) can suppress androgen production but may also induce cortisol deficiency (194). Patients treated with these inhibitors require glucocorticoid replacement therapy to avoid adrenal crisis, to lower progesterone and aldosterone accumulation, and to decrease hypertension (193). However, such replacement therapy (up to 20 mg prednisone per day) exposes patients to excess glucocorticoids and the risk of acute adrenal crisis (194).

Aldosterone synthase (CYP11B2) inhibition has been used to prevent metabolic, cardiac, and renal dysfunction in patients with metabolic syndrome (195). At higher doses, however, these compounds also induce a general inhibition of 11β-hydroxylase activity, which lowers aldosterone and cortisol production (193).

In patients with excessive production of glucocorticoids, several pharmacologic agents have been used to inhibit adrenal steroidogenesis. Mitotane, or 2,2-bis (2-chlorophenyl-4-chlorophenyl)-1,1-dichloroethane, is a cytotoxic drug that alters cortisol metabolism by leading to formation of 6-hydroxycortisol rather than the 17-hydroxycorticosteroids. It is used to produce a "medical adrenalectomy" in patients with Cushing syndrome (196,197). Metyrapone is a pyridine derivative that inhibits CYP11B1 (11-beta-hydroxylase), thereby inhibiting synthesis of cortisol as well as aldosterone (198). It is used mainly for the diagnostic testing of the HPA axis and for treating patients with Cushing syndrome.

Immune Checkpoint Inhibitors

The development of immune checkpoint inhibitors (ICIs) has been a revolutionary milestone in the field of immuno-oncology. Many patients with a range of solid organ tumors, cutaneous neoplasms, and hematologic malignancies are now treated with these antibodies, which act by blocking key immune checkpoint proteins, such as CTLA4 or PD1, and promoting immune-mediated elimination of tumor cells (199). It has become apparent, however, that *ICI-induced endocrinopathies* are a frequent adverse effect of these medications (193). Over 50 percent of these endocrinopathies involve an impairment of HPA axis function (200).

The mechanisms of ICI-induced HPA axis dysregulation are not fully understood. However, it is hypothesized that all endocrinopathies associated with ICI therapy are autoimmune in etiology. Checkpoint regulator proteins diminish the immune response to antigen and hence act as a brake on the immune system (200). Blockage of these checkpoint regulator proteins releases this brake and allows the patient's immune system to attack cancer cells as well as damage certain healthy tissues by autoimmune mechanisms (200). For example, it has been postulated that blocking expression of CTLA4 on endocrine cells with ipilimumab treatment leads to site-specific deposition of complement components, pituitary infiltration and hypophysitis (200). This inflammation of the pituitary gland in turn blunts ACTH release. Direct effects on the adrenal gland are less common side effects of ICIs (191,200).

Hypophysitis is less common among patients treated with anti-PD1/PD-L1 drugs. This is thought to be because monoclonal antibodies to nivolumab and pembrolizumab are in the immunoglobulin (IgG) 4 class (201), as opposed to ipilimumab, which is a CTLA-4 antibody of the IgG1 class, which can activate the classic complement pathway (202).

There has also been an increase in reports of adrenal crisis in patients treated with ICIs (191,200). The first onset of symptoms can occur as early as 1 week or as late as 416 weeks after treatment (193). The diagnosis of ICI-induced adrenal insufficiency is usually made clinically, and rarely comes to the attention of the surgical pathologist unless it is an incidental finding in an organ resected for other reasons. Nonetheless, the microscopic findings are expected to resemble those seen in primary autoimmune adrenalitis, which is characterized by a nonspecific infiltrate of chronic inflammatory cells in the adrenal cortex.

Radiation

Radiation-induced adrenal injury has not been extensively described in the literature (203,204), and the adrenal glands are reported to be relatively radioresistant compared to other endocrine tissues (205,206). Adrenal cortical hypertrophy/hyperplasia has been described after radiotherapy, and is presumed to be a nonspecific response of the adrenal glands to stress (205). Radiation exposure to the HPA axis and decreased ACTH secretion can result in loss of adrenal cortical cells and atrophy of the adrenal gland (207).

Low-dose irradiation of the adrenal glands in mice is known to cause adrenal cortical adenomas (205,208); however, there are no known oncogenic effects of radiation therapy in human adrenal glands (205). High-dose irradiation to the abdominal, pelvic, and lumbar regions can lead to hyaline fibrosis (206). The fibrosis is usually present in the inner cortex (zona reticularis) and is accompanied by a reduction in the zona fasciculata; however, no change in cortical function has been reported in this setting (203,206).

Miscellaneous Drugs and Chemicals

Some medications other than glucocorticoids may suppress HPA axis function (209). Progestational agents such as medroxyprogesterone and megestrol have glucocorticoid activity (210). Enzyme inducers such as rifampicin and carbamazepine enhance clearance of some synthetic glucocorticoids (209,211). Inhibitors of cortisol synthesis include ketoconazole, aminoglutethimide, and etomidate (an anaesthetic agent that selectively inhibits adrenal 11-β-hydroxylase, the enzyme that converts 11-deoxycortisol to cortisol) (211,212).

A number of chemicals cause morphologic or functional lesions in the adrenal gland. Some of the lesions are localized to specific anatomic zones of the adrenal cortex, and the resulting functional deficits depend on the physiologic role of the affected zone. In addition, metabolic activation is an important factor contributing to the gland's vulnerability to chemical injury. For example, carbon tetrachloride (CCl4) causes adrenocortical necrosis, but only of the innermost zone of the gland, the zona reticularis. The apparent reason for the localized effect of CCl4 in the adrenal cortex is that only the cells of the zona reticularis have the enzymatic capacity to activate CCl4, resulting in lipid peroxidation and covalent binding to cellular macromolecules (213).

Deposition of iron in the adrenal glands has been reported histologically in an autopsy series of patients with iron overload (214). Additionally, functional alterations in the adrenal glands due to iron deposition have been described in the literature (215). Iron overload can also indirectly affect the adrenal glands via iron deposition in the pituitary gland, which in turn leads to destruction of anterior pituitary cells and decreased ACTH production (216,217).

INFECTIOUS DISEASES

The adrenal gland can be infected by a myriad of pathogens, including fungi, viruses, parasites, and bacteria. Although *Mycobacterium tuberculosis* is the most common infectious cause of primary adrenal insufficiency (Addison disease) (218), its incidence has greatly decreased since the introduction of antituberculosis drugs (219). Other infectious and noninfectious causes of adrenal insufficiency include autoimmune adrenalitis, other infections (including fungi and human immunodeficiency virus [HIV]), hemorrhagic infarction, metastatic malignant disease, infiltrative disease (including amyloidosis, sarcoidosis, and hemochromatosis), and drug reactions.

Tuberculous Adrenalitis

Definition. *Tuberculous adrenalitis* is a primary or secondary infection of the adrenal glands caused by the *Mycobacterium* species.

General Features. Although tuberculosis is now a rare cause of adrenal insufficiency in the United States and Western Europe, it remains the primary cause in parts of the world where tuberculosis is endemic (220,221). In order to induce Addison disease, tuberculosis must involve both glands, with complete or near complete destruction of the adrenal cortex. In a large retrospective study of 13,762 patients, active tuberculosis was found in 6.5 percent of all cases, with 6.0 percent of the patients with active *M. tuberculosis* disease demonstrating adrenal insufficiency (222). In 25 percent of the patients with adrenal involvement, the infection was restricted to the adrenal gland.

M. tuberculosis appears to cause direct adrenal dysfunction by inducing degeneration of cells within the adrenal cortex (223). The adrenal glands are also indirectly affected by disruptions to the HPA axis (224). Infection causes significant increases in the inflammatory cytokine tumor necrosis factor alpha (TNFα), which stimulates macrophage aggregation and granuloma formation, while simultaneously activating the HPA axis (225). The upregulation of the HPA axis and subsequent rise in cortisol levels promotes a predominantly Th2 response (224). Decreased levels of DHEAS are also seen in patients with tuberculosis, suggesting a role for decreased androgens as well as increased cortisol in the pathogenesis of the disease (226,227). The interplay between infection, endocrine function, and immunomodulatory cytokines dictates the overall response both systemically and locally at the adrenal level, thereby delineating the role of the HPA axis and adrenal hormones in general in causing disease both in the adrenal gland and at distant sites.

Gross Findings. The macroscopic appearance of the adrenal glands correlates with the length and activity of infection (224). Large glands occur in recent active infection, with mean combined weights of 25 g, but weights of up to 50 g have been reported (229). In contrast, small, atrophic, or calcified glands appear to represent inactive or remote infection (223). The glands are yellow-gray, with gray-red nodules, depending on the stage of the lesion. The cut surface shows destruction of both the cortex and medulla and replacement by caseous material (222). Focal areas of calcification may be seen, but extensive calcification is not usually present.

Microscopic Findings. Histologically, caseous necrosis is a common finding in adrenal tuberculosis, often with a rim of residual viable cortex. However, the typical granulomatous inflammation with Langhans giant cells is seen in less than half of cases (222). This lack of granulomatous inflammation is postulated to be related to the local immunosuppressive effect of steroids secreted in the adrenal cortex (222). The acid-fast bacilli are detected in about half of cases with special stains such as Ziehl-Neelsen, and in even more cases with fluorescent stains. Fibrosis is usually minimal, which is also thought to be related to the inhibitory effects of the glucocorticoids produced in the adrenal glands.

Medullary tissue is usually also obliterated. It is thought that the destruction begins in the medulla and extends to the cortex, often sparing only a thin rim of outer cortex, which may contain a few clusters of viable clear cells (229,230).

Differential Diagnosis. The histologic differential diagnosis of tuberculous adrenalitis includes other granulomatous diseases such as fungal infections. Special stains with methanamine silver and Ziehl-Neelsen can usually detect diagnostic organisms. Using a number of different stains is particularly important in immunosuppressed patients who may have multiple infections. Infection with *M. avium-intracellulare* is typically associated with the presence of confluent masses of histiocytes containing the acid-fast organisms. Other entities include noninfectious causes of granulomatous adrenalitis, such as sarcoidosis and xanthogranulomatous adrenalitis.

Treatment and Prognosis. Although the adrenal cortex has considerable capacity for regeneration, Addison disease due to tuberculosis is generally regarded as irreversible, with only a few reports of patients showing recovery of adrenal function in the literature (222,231). In these cases, antituberculous chemotherapy does not appear to restore function. In addition, rifampicin is a potent hepatic enzyme inducer that inhibits metabolism of glucocorticoids (232). Failure to increase the dose of steroid replacement therapy in patients treated with rifampicin may result in the development of adrenal crisis (233).

Fungal Infections

General Features. Many pathogenic fungi are known to affect the adrenal glands in both immunocompetent and immunocompromised individuals, with the highest incidence occurring in individuals with defects in cell-mediated immunity (224). Adrenal involvement most frequently occurs in the setting of disseminated fungal infection, particularly in endemic areas (224). For example, up to half of patients with severe, disseminated *Histoplasma capsulatum* infection have adrenal infection (234,235). It is postulated that the reason for the tropism of *H. capsulatum* for the adrenal gland is due to the local production and release of glucocorticoids and a relative lack of reticuloendothelial cells within the gland (236).

Adrenal insufficiency may occur as a result of a number of other fungal infections (117,237–244). *Histoplasma* and *Cryptococcus* are the most common etiologic fungi in nontropical regions, while *Paracoccidioides* is most common in tropical regions. *Blastomyces dermatitidis* has a high affinity for the adrenal gland, however, it rarely causes overt adrenal failure (224).

Gross Findings. The adrenal glands are enlarged, with variable degrees of necrosis and fibrosis. Calcifications may be seen. In patients with disseminated disease, the adrenal glands are affected bilaterally.

Microscopic Findings. The predilections of certain fungi for the adrenal glands and their pathologic effects give rise to a variety of histologic appearances. *Paracoccidioides brasiliensis* primarily causes destruction by embolic infection of small vessels leading to

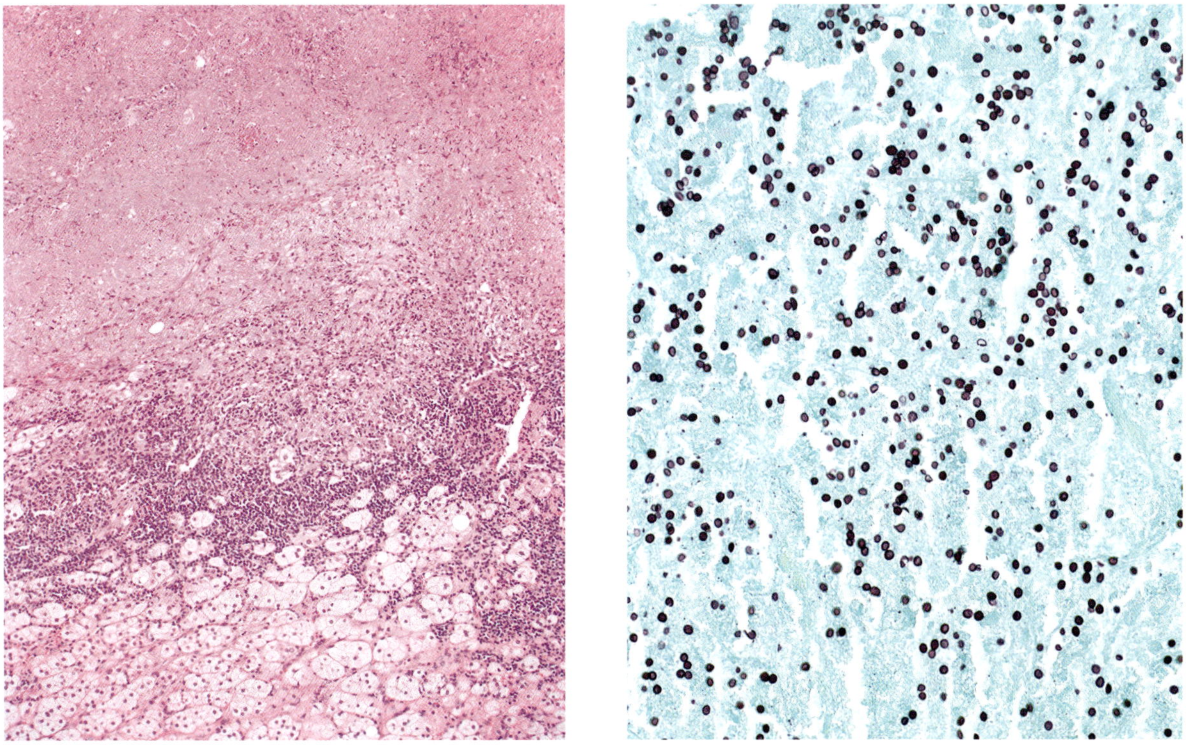

Figure 4-13

CRYPTOCCOCAL ADRENALITIS

Left: The adrenal gland is largely necrotic, with lymphocytic and granulomatous inflammation. A rim of residual adrenal cortex is present at the bottom of the image.

Right: Grocott methenamine silver (GMS) stain is positive for fungal organisms.

endovasculitis and granuloma formation (224). Caseating necrosis is responsible for the largest loss of glandular tissue and is more likely a consequence of local tissue ischemia secondary to the fungal emboli (241,245,246). *H. capsulatum* adrenal lesions are found most commonly in the zona reticularis where there are elevated levels of corticosteroids (235,238). For *P. brasiliensis,* the development of adrenal vasculitis can cause extensive glandular destruction, leading to caseating necrosis and massive glandular infarction (224). Special stains for fungal organisms help to establish the diagnosis (fig. 4-13).

Differential Diagnosis. Tuberculous adrenalitis often has a similar histologic appearance. The use of special stains, in addition to microbiologic studies, and in some cases PCR analysis, is important in making the correct diagnosis.

Treatment and Prognosis. Although isolated adrenal involvement has been reported in immunocompetent hosts with a variety of fungal infections (246–249), bilateral adrenal involvement in the setting of disseminated mycoses is much more common (250). In this setting, adrenal function remains unaffected unless more than 90 percent of the adrenal cortex is obliterated, and so patients with disseminated mycoses may die of their infection before such destruction occurs (251–254).

Addison disease is reported with a wide range of fungal infections (246,253–257). Prompt recognition and glucocorticoid supplementation are fundamental for survival in such patients. Most cases of adrenal insufficiency are irreversible and necessitate lifelong glucocorticoid supplementation. Nevertheless, recovery of adrenal function has been reported with antifungal treatment in some cases (237,241,258).

Antifungal agents (azoles) affect glucocorticoid and mineralocorticoid function by inhibiting CYP450-dependent enzymes involved in adrenal steroidogenesis (250). Specifically, ketoconazole is a dose-dependent reversible inhibitor of cholesterol desmolase, 17,20-lyase, 11β-hydroxylase, 17α-hydroxylase, and 18-hydroxylase, resulting in inhibition of adrenal steroidogenesis (257). Careful titration and monitoring of adrenal function are crucial in patients being treated with these agents.

Human Immunodeficiency Virus (HIV)-Associated Adrenal Dysfunction

Definition. Adrenal dysfunction may be related to HIV, either directly, or as a result of opportunistic infections or antiretroviral therapy.

General Features. Functional adrenal insufficiency is one of the most common endocrine manifestations in patients with HIV/acquired immunodeficiency syndrome (AIDS) (259). The changes in the adrenal gland are secondary to the effects of opportunistic infections (260,261); neoplasms such as Kaposi sarcoma and lymphoma (260); adrenal hemorrhage (especially Waterhouse-Friderichsen syndrome); drugs such as rifampicin, ketoconazole, and phenytoin; and autoimmune adrenalitis (262,263). The most common opportunistic infection in HIV patients with adrenal insufficiency is cytomegalovirus (CMV), with a histologic prevalence of up to 93 percent in those who are not on antiretroviral therapy (264). Other infectious etiologies include *Mycobacterium avium* complex, *Mycobacterium tuberculosis, Toxoplasma gondii, Pneumocystis jirovecii, Histoplasma capsulatum,* and *Cryptococcus neoformans* (259).

Patients with HIV/AIDS tend to have functional adrenal insufficiency, since they generally retain adequate or even high secretion of cortisol, albeit with relatively higher levels of ACTH (265). While most patients present with some chronic symptoms of adrenal insufficiency, acute symptoms (adrenal crisis) only tend to manifest in periods of stress when cortisol production is insufficient to meet the increased demands (259). Some studies have shown that during early HIV infection, adrenal stimulation with synthetic ACTH shows inadequate response in up to 14 percent of patients (265,266), while the proportion was up to 54 percent in patients with advanced HIV/AIDS. This implies that there is progressive glandular dysfunction with more advanced disease (267). A defect in the production of 17-deoxycorticosteroid by the zona fasciculata may explain these findings (266,267).

Glucocorticoid resistance is also seen in a subset of AIDS patients, and is characterized by a hyperfunctioning HPA axis, which results in hypercortisolemia and increased ACTH secretion (268). This may be explained in part by sustained chronic inflammation induced by HIV in the adaptive immune system of the gut, as this tends to play a major part in progression to advanced disease (269). Peripheral resistance to glucocorticoids is noted in some patients with AIDS, which may be related to a decreased affinity of type II glucocorticoid receptors despite increased receptor density (268).

Gross Findings. The gross appearance of the glands depends on the etiologic agent. As discussed previously, mycobacterial and fungal infections result in caseous necrosis within an enlarged adrenal gland. CMV infection is not associated with marked gland enlargement (260,261). Waterhouse-Friderichsen syndrome results in extensive hemorrhagic necrosis, which is usually bilateral (270). Autoimmune adrenalitis is associated with normal-sized glands initially, which become markedly shrunken in end-stage disease (270).

Microscopic Findings. The histologic features also depend on the etiologic agents. Mycobacterial and fungal infections result in necrotizing granulomatous inflammation. Viral infections are associated with intranuclear inclusions in the adrenocortical cells, while *Toxoplasma* infections are associated with intracytoplasmic organisms. Autoimmune adrenalitis is characterized by lymphoplasmacytic inflammation preferentially involving the adrenal cortex, with associated cortical cell necrosis (270).

Differential Diagnosis. A plethora of organisms may be present in the adrenal glands of patients with HIV/AIDS. Immunohistochemistry for CMV, adenovirus, and herpes simplex viruses 1 and 2 (HSV1/2) is helpful for detecting intracellular viral inclusions. Special stains for fungal and acid-fast organisms are also useful, particularly in the presence of caseating necrosis with multinucleated giant cells. Autoimmune adrenalitis shows preferential inflammation and destruction of the adrenal cortex, and can be distinguished from tuberculous adrenalitis which results in destruction of both cortex and medulla (270). Involvement by lymphoma or leukemia should also be considered in cases with a florid lymphocytic infiltrate, and immunohistochemistry for B- and T-cell markers can aid in excluding this differential diagnosis.

Miscellaneous Infectious Causes of Adrenalitis

A few other infections are rare causes of primary adrenal insufficiency. Bacterial infections such as secondary syphilis can lead to adrenal insufficiency in rare cases. The adrenal gland is sclerotic, with a lymphoplasmacytic infiltrate, and spirochetes are demonstrated with special stains such as Warthin-Starry (271). Parasitic infections of the adrenal gland are rare.

Case reports have demonstrated adrenal involvement with such diverse pathogens as *Microsporidia* (272,273), amebic species (274), *Trypanosoma* (275,276), *Leishmania* (277,278), and *Echinococcus* (279,280). The pathologic findings in the adrenal glands vary significantly depending on the microbe. *Echinococcus* causes hydatid disease, which presents with diffuse cystic involvement of visceral organs. The most commonly involved organs are the lungs and liver, with adrenal involvement occurring in 0.5 percent of cases (279,280). Visceral leishmaniasis can also cause cystic adrenal disease, both in immunocompetent (277) and immunocompromised individuals (278). African trypanosomiasis can also lead to adrenal insufficiency, and this is compounded by treatment with suramin, which can impair adrenal function when given at high doses (281).

ADRENAL CORTICAL HYPOFUNCTION

Adrenal insufficiency was first described by Thomas Addison in 1855; it is characterized by deficient production or action of glucocorticoids and/or mineralocorticoids and adrenal androgens (282). The disease may result from disorders affecting the adrenal cortex (primary adrenal insufficiency, or Addison disease), the anterior pituitary (secondary), or the hypothalamus (tertiary) (283,284). The prevalence of primary adrenal insufficiency is 93 to 144 cases per million population (285–289), and the currently estimated incidence is 4.4 to 6.0 new cases per million population per year (288). Secondary adrenal insufficiency occurs more frequently than primary adrenal insufficiency (283), with an estimated prevalence of 150 to 280 per million (287,290–293).

The clinical symptoms of adrenal insufficiency are nonspecific and include weakness, fatigue, anorexia, abdominal pain, weight loss, orthostatic hypotension, salt craving, and hyperpigmentation of the skin (294,295). Most cases can be easily treated with synthetic glucocorticoid substitution, however, there are still numerous challenges surrounding the diagnosis and treatment of patients with adrenal insufficiency.

Primary Adrenal Insufficiency

General Features. The etiology of *primary adrenal insufficiency* has changed over time. Prior to 1920, the most common cause of primary adrenal insufficiency was tuberculosis (283). However, since 1950, most cases (80 to 90 percent) have been attributed to autoimmune adrenalitis, which may either be confined to the adrenal gland (40 percent) or occur as part of an autoimmune polyendocrinopathy syndrome (60 percent) (283,296,297). Other causes of primary adrenal insufficiency include infectious disease caused by mycobacterial, fungal, or bacterial organisms; adrenal neoplasms; or uncommon hereditary disorders (298–305).

Autoimmune Adrenalitis. *Autoimmune adrenalitis,* previously termed idiopathic primary adrenal insufficiency, is caused by an autoimmune process that results in destruction of the adrenal cortex. The immunologic basis for the disease is evident in the activation of both cellular and humoral immune mechanisms, familial clustering, prevalence of specific HLA types, and involvement of other endocrine glands.

Autoantibodies that react with several steroidogenic enzymes (most often 21-hydroxylase) and all three zones of the adrenal cortex are present in the serum in up to 86 percent of patients with autoimmune adrenalitis (285). These antibodies are not found in normal subjects or in patients with other diseases. The detection of antibodies is more common in women than men. With longstanding disease, the antibody titers may decrease or even disappear (306).

Autoantibodies directed at all three zones of the adrenal cortex appear months to years before the onset of adrenocortical insufficiency (296,307,308). The zona glomerulosa is probably affected first (308), and may precede failure of the fasciculata by months to years (307). Up to 10 percent of first-degree relatives of patients with autoimmune adrenalitis also express these antibodies, and have an increased risk of developing adrenal insufficiency (296,308–314).

Many patients with autoimmune adrenal disease have antibodies against other endocrine glands; in contrast, these antibodies are uncommon in normal populations without adrenal or other endocrine diseases (315,316). About 60 percent of patients with autoimmune adrenal disease also have thyroid antimicrosomal antibodies (317). The presence of antibodies to gastric parietal cells and intrinsic factor usually correlates with atrophic gastritis and pernicious anemia (306). The detection of antigonadal antibodies is associated with gonadal failure; there is a higher incidence of ovarian failure than testicular failure. At the opposite end of the spectrum, patients who have other autoimmune endocrine diseases, but do not have autoimmune adrenalitis, usually do not have antiadrenal antibodies in their serum, with the exception of patients with hypoparathyroidism (315).

Immunologic studies also show evidence of cell-mediated immunity, including autoantibodies to CYP21A2 of the IgG1 or IgG2a subclass, implicating a type 1 T-helper cell (Th1) response (318,319). The Th1 response is characterized by the production of interferon gamma, an antigen-specific, T-helper cell–driven process with cytotoxic T lymphocytes and activated macrophages mediating the destruction of the adrenal cortex. The presence of lymphocytic infiltration of the adrenal glands further supports a possible role for cellular immunity. Decreased suppressor T-cell function and increased circulating Ia-positive T lymphocytes have also been described in these patients (320–322).

The genetics of autoimmune adrenal insufficiency have been studied extensively (311,323–325). The disease may be familial or nonfamilial. It is somewhat less likely to be familial when it occurs in isolation. Approximately one third of such patients have affected family members, compared with approximately half of patients who have adrenal insufficiency as part of the polyglandular autoimmune syndrome (324,326–328).

Genetic susceptibility to autoimmune adrenal insufficiency is strongly associated with HLA-B8, -DR3, and -DR4 alleles, except when it occurs as part of polyglandular autoimmune syndrome type 1, where no HLA association is seen (327,329,330). Several specific HLA antigens, including DQA1*0301, DQA1*0501, DQB1*0201, DQB1*0302, DRB1*0404, and DRB1*0301, are associated with sporadic autoimmune adrenal insufficiency and polyglandular autoimmune syndrome type 2 (325,331–333). Chronic autoimmune thyroiditis, pernicious anemia, and premature ovarian failure, all components of the polyglandular autoimmune syndrome type 2, are not closely linked to any HLA haplotypes (329,330). Normal adrenocortical cells express HLA-DR type II antigens, so inappropriate expression of these antigens is probably not linked directly to the pathogenesis of autoimmune adrenal insufficiency (306,323).

Polyglandular Autoimmune Syndrome. Adrenal insufficiency is associated with the *polyglandular autoimmune syndrome* (PGA) type II in 100 percent of cases and with PGA type I in 60 percent of cases (324,327). About 50 percent of patients with autoimmune adrenal insufficiency have one or more autoimmune endocrine disorders (306). In contrast, patients with the more common autoimmune endocrine disorders such as Hashimoto thyroiditis or insulin-dependent diabetes mellitus rarely develop adrenal insufficiency (306).

PGA type I, also known as *autoimmune polyendocrinopathy-candidiasis-ectodermal dystrophy* (APECED), is a rare autosomal recessive disorder caused by mutations in the so-called autoimmune regulator (*AIRE*) gene on chromosome 21q22.3 (334–336). Females are affected slightly more frequently than males (327,337,338). The first clinical manifestations are usually hypoparathyroidism or chronic mucocutaneous candidiasis, which usually appear during childhood or early adolescence, always by the early twenties (327,328,338,339). Adrenal insufficiency usually develops later, at age 10 to 15 years.

PGA type II is much more common than type I, and primary adrenal insufficiency is its principal manifestation (327,339,340). Autoimmune thyroid disease, usually chronic autoimmune thyroiditis but occasionally Graves disease, and type 1 diabetes mellitus are also common. Approximately half of cases are familial and several modes of inheritance (autosomal recessive, autosomal dominant, and polygenic) have been reported (327,328,339). Women are affected up to three times more often than men. The age of onset ranges from childhood to late adulthood, with most cases occurring between age 20 and 40 years (311,327,328,340).

Clinical and Radiologic Features. Patients with autoimmune adrenal insufficiency and with PGA are predominantly female (about 70 percent) (306). Patients with isolated autoimmune adrenalitis are predominantly male (71 percent) during the first two decades of life, are equally male and female in the third decade, and are predominantly female (81 percent) thereafter (328). Adrenal insufficiency (Addison disease) is the final result of autoimmune adrenalitis; the initial phase is subclinical, but symptoms of adrenal failure appear after at least 90 percent of the adrenal gland has been destroyed (307).

Acute adrenal insufficiency is a life-threatening disease that presents with severe hypotension or hypovolemia, acute abdominal pain, and nausea and vomiting (283). Other signs and symptoms include anorexia, fever, hypoglycemia, weakness, fatigue, lethargy, and confusion. Most patients have predisposing factors that acutely increase their need for corticosteroids, such as trauma, surgery, and infections. Other precipitating factors include mineralocorticoid deficiency, adrenal infarction, hemorrhage, sepsis, or thromboembolism (306). Shock that is unresponsive to volume and vasoconstrictor agents is a typical finding. Adrenal shock can progress to coma and death without appropriate diagnosis and therapy (306). Adrenal hemorrhage and death are associated with the Waterhouse-Friderichsen syndrome, which is secondary to sepsis from *Meningococcus* or *Pseudomonas aeruginosa* (306,307).

Chronic adrenal insufficiency is associated with signs and symptoms of glucocorticoid, mineralocorticoid, and androgen deficiencies. The most common presenting complaint is fatigue, and patients also present with weakness, lack of stamina, and weight loss. Gastrointestinal problems are also frequent and include nausea and abdominal pain, possibly related to a loss of gastrointestinal motility (341). Dizziness, irritability, and postural hypotension are also frequent complaints. Patients may survive without a diagnosis for many years, until a minor infection leads to cardiovascular collapse (342). Joint pain, calcification of articular cartilage, splenomegaly, and lymphadenopathy may also develop. Psychiatric manifestations include organic brain syndrome, depression, and psychosis (306). Hyperpigmentation, caused by stimulation of melanocytes by high ACTH levels, as well as salt craving are highly specific to primary adrenal insufficiency.

Abdominal CT usually reveals small noncalcified adrenal glands in autoimmune adrenalitis, in contrast to the larger volumes seen in tuberculosis and neoplasias (343). In tuberculous or fungal adrenalitis, hemorrhage may be associated with adrenal enlargement, but this is not seen in patients with autoimmune adrenal disease (344,345). Moreover, the atrophic adrenal glands seen in chronic tuberculous adrenalitis usually have associated calcification, which is not seen in autoimmune adrenal disease (344). In chronic primary adrenal insufficiency, the sella turcica may be enlarged on skull radiograph secondary to ACTH cell hyperplasia; this enlargement may be reduced by steroid treatment (326,346).

Laboratory Findings. The first evidence of autoimmune adrenal insufficiency is usually an increase in plasma renin activity in association with a normal or low serum aldosterone concentration, suggesting that the zona glomerulosa is involved initially (307,347). Several months to years later, zona fasciculata dysfunction becomes evident, first by a decreasing serum cortisol response to ACTH stimulation, later by increased basal serum ACTH concentrations, and finally by decreasing basal serum cortisol concentrations and symptoms (307,338). Serum testosterone levels in men are normal, but low in women (because the adrenal cortex is the source of approximately 50 percent of androgen in women). Serum thyroxine levels are normal or low, while TSH levels are often increased. Serum prolactin (PRL) may also be slightly elevated. Both TSH and PRL levels return to normal after treatment with glucocorticoids.

Electrolyte abnormalities, including hyponatremia and hyperkalemia, are present in most patients (341,348). The hyponatremia is related to aldosterone deficiency from the zona glomerulosa. Liver aspartate transaminase levels may be abnormal but usually return to normal after glucocorticoid therapy.

Gross Findings. The pathologic changes in autoimmune adrenal insufficiency vary with the stage of the disease. Initially, the glands may be enlarged; but with longstanding disease, the glands become atrophic and can be difficult to locate (300). Gland weights vary from 1.9 to 4.6 g, but smaller glands of 0.2 g have been reported (321). The capsule is thickened and fibrotic, and the cortex is completely destroyed, with relative sparing of the medulla.

Microscopic Findings. Histopathology reveals a widespread mononuclear cell infiltrate, composed of lymphocytes, plasma cells, and histiocytes, during the active phase of autoimmune adrenalitis (343). In some cases, there are lymphoid follicles with germinal centers. There is loss of the normal cortical cells, or islands of residual cortical cells in a necrotic background. The capsule is thickened, and is more prominent in smaller glands (349). The residual cells are compact, with eosinophilic cytoplasm and lipid depletion. Hypertrophy of these residual cells may be secondary to stimulation by the increased concentrations of serum ACTH. The medulla is usually spared and may extend to the inner aspect of the capsule because of the destruction of the overlying cortex. When accessory adrenal cortical tissue is present outside the cortex, similar histologic changes are usually present.

Differential Diagnosis. Certain conditions may mimic adrenal insufficiency, such as sepsis, hypovolemia, cardiogenic shock, exogenous intoxication, hyperkalemia from renal failure, and abdominal pain from porphyria. Malignancy is an important differential diagnosis in those patients presenting with gastrointestinal complaints and weight loss. Clinical conditions and compounds that can produce hyperpigmentation (antimalarial agents, antineoplastic agents, tetracyclines, phenothiazines, zidovudine, hemochromatosis, porphyria cutanea tarda, and heavy metals) should also be excluded.

Conditions leading to atrophic adrenal glands include chronic glucocorticoid therapy, treatment with adrenal steroid synthesis inhibitors such as mitotane, end-stage tuberculosis, and fungal infections. Chronic glucocorticoid treatment leads to diffuse atrophy of the glands, but there is no destruction of the adrenocortical cells. Lipid-laden cells are seen histologically. Treatment with mitotane can lead to atrophic adrenal glands with areas of fibrosis and residual islands of cortical cells. In atrophic adrenal glands due to chronic tuberculosis, multinucleated giant cells in a background of necrosis and fibrosis are seen (350).

Inflammatory infiltrates are seen in other conditions, such as lymphocytic adrenalitis, myelolipomatous changes in the adrenal cortex, and Carney complex with bilateral pigmented adrenocortical disease. Myelolipomatous changes are often associated with hypercortisolism. The presence of fat cells, lymphocytes, and bone marrow elements, including myeloid and erythroid precursors, help make the distinction. In bilateral pigmented adrenocortical disease (Carney complex), the glands may be of normal or slightly increased size. The presence of lymphocytes is usually associated with enlarged zona reticularis-type cells, which form nodules, mainly at the corticomedullary junction. Focal aggregates of lymphocytes are a normal finding with increasing frequency in the adrenal glands of older patients, and should not be misconstrued as evidence of adrenalitis.

Treatment and Prognosis. The standard initial therapy is corticosteroid replacement. During an acute adrenal crisis, therapy should not be delayed to perform diagnostic studies or to await laboratory results. Intravenous fluids and glucocorticoids may be life-saving in these emergencies. Correction of hemodynamic and metabolic disturbances with large volumes of intravenous saline and glucose is crucial (343). Attempting to identify precipitating factors, particularly infections, is recommended, and if infections are suspected, antibiotics should be commenced appropriately (343).

Lifelong glucocorticoid and mineralocorticoid treatment is needed for primary adrenal insufficiency. Patients undergoing surgery usually require high-dose glucocorticoid prophylaxis to combat the stress of surgery (306). During pregnancy, the usual maintenance dose of glucocorticoids is increased in the third trimester and/or during labor and delivery (306).

Historically, the prognosis of patients with chronic adrenal insufficiency was grim, and more than 80 percent died during the first 2 years after diagnosis (313). Today, the survival rate of patients who have been adequately diagnosed and treated is similar to that of the normal population (296,343). The prognosis depends largely on the underlying disease. Heart failure, essential hypertension, and osteopenia are some complications in patients with chronic adrenal insufficiency treated with replacement therapy (306). Although spontaneous recovery of adrenal function has been described (351,352), others report that this is a rare occurrence (353).

Secondary and Tertiary Adrenal Insufficiency

Definition. These adrenal insufficiencies are related to failure of the pituitary gland to secrete adequate amounts of ACTH (*secondary adrenal insufficiency*) or of the hypothalamus to secrete CRH (*tertiary adrenal insufficiency*).

General Features. Any process that involves the pituitary gland and interferes with ACTH secretion can cause secondary adrenal insufficiency. The ACTH deficiency may be isolated or occur in conjunction with other pituitary hormone deficiencies (panhypopituitarism) (354). Pituitary tissue can be destroyed and hormone secretion reduced by large pituitary tumors or craniopharyngiomas, infectious diseases such as tuberculosis or histoplasmosis, infiltrative diseases, lymphocytic hypophysitis, head trauma, and large intracranial artery aneurysms. Postpartum pituitary necrosis secondary to infarction (Sheehan syndrome), pituitary apoplexy, and metastasis to the pituitary can also disrupt ACTH secretion and lead to secondary adrenal insufficiency.

Combined pituitary hormone deficiency (including ACTH deficiency) due to genetic pituitary abnormalities is rare. ACTH and cortisol deficiency have been described in patients with multiple pituitary hormone deficiencies due to mutations in the *PROP1* (Prophet of Pit-1) gene, even though PROP1 is not expressed in corticotrophs. The onset of cortisol deficiency, which may be severe, occurs from childhood to late adulthood (355–358).

Isolated ACTH deficiency is a rare disorder (359). The defect is probably in the pituitary gland because there is no ACTH secretory response to CRH or vasopressin, as there usually is in hypothalamic disorders (360–362). Occasional patients have hypothyroxinemia and hyperprolactinemia, which are corrected with glucocorticoid replacement (363,364). Anti-ACTH antibodies are present in some patients, some of whom had repeated administration of cosyntropin (365). Although an uncommon adverse reaction, ICIs are associated with isolated ACTH deficiency (366).

The frequent association with other autoimmune endocrine disorders (364), the presence of lymphocytic hypophysitis with selective corticotroph absence in some patients (367), and the presence of antipituitary antibodies in the serum of some patients (368,369), suggest that most cases of isolated ACTH deficiency are caused by an autoimmune process. Rarely, isolated ACTH deficiency is caused by a genetic condition. Loss of function mutations in the *POMC* gene give rise to a rare syndrome of early-onset obesity and secondary adrenal insufficiency, usually with red hair (370). Other genetic causes of isolated ACTH deficiency include *TPIT* gene mutations (371) and defects in the PC1 cleavage enzyme (372).

Tertiary adrenal insufficiency results from hypothalamic abnormalities that reduce CRH secretion. Suppression of HPA axis function by chronic administration of high doses of glucocorticoids is the most common cause of tertiary adrenal insufficiency. High doses of glucocorticoids decrease hypothalamic CRH synthesis and secretion. They also block the trophic and ACTH secretagogue actions of CRH on the anterior pituitary. This results in decreased synthesis of POMC and decreased secretion of ACTH and other POMC-derived peptides by the pituitary corticotrophs. As a result, pituitary corticotrophs decrease in size, and eventually, the number of identifiable corticotrophs decreases (373). The clinical features of secondary or tertiary adrenal insufficiency that help distinguish them from primary adrenal insufficiency include: cortisol production that can be restored by prolonged ACTH administration, and mineralocorticoid secretion that is nearly normal because this function depends mostly on the renin-angiotensin system rather than on ACTH.

Clinical and Radiologic Features. A number of clinical features associated with glucocorticoid deficiency seen in primary adrenal insufficiency also occur with secondary and tertiary adrenal insufficiencies. These include weakness, fatigue, muscle and joint pain, and psychiatric symptoms. However, in secondary or tertiary adrenal insufficiency, other clinical features are either less prominent or absent, and some features may be present in central, but not primary adrenal insufficiency. Hyperpigmentation is not present, because ACTH secretion is not increased (374). Dehydration is also absent, and hypotension is less prominent (359,375,376). However, hyponatremia and volume expansion may be present, caused by an inappropriate increase in vasopressin secretion or action due to cortisol deficiency. The hyponatremia can occur early in the disease, and may be the initial manifestation. Hypoglycemia is more common in secondary adrenal insufficiency than in primary adrenal insufficiency. This difference may be explained by the absence of dehydration and hypotension, which allows patients to tolerate their illness longer and present with symptoms of chronic glucocorticoid deficiency rather than mineralocorticoid deficiency.

There may be clinical manifestations of a pituitary or hypothalamic tumor, such as symptoms and signs of deficiency of other anterior pituitary hormones, headache, or visual field defects. Evidence of metastatic or primary lesions involving the pituitary or hypothalamic region may be detected by computed tomography (CT) or magnetic resonance imaging (MRI). Adrenal atrophy, however, is often difficult to detect even with sensitive radiologic techniques.

Laboratory Findings. In secondary and tertiary insufficiencies, some laboratory findings are similar to those of primary adrenal insufficiency, however, there are some key differences. In patients with primary adrenal insufficiency, cortisol secretion is deficient despite their ability to secrete ACTH. Conversely, patients with secondary or tertiary adrenal insufficiency have intrinsically normal but atrophic adrenal glands that are capable of producing cortisol but fail to do so because ACTH secretion is deficient. Measurement of basal plasma ACTH concentration can usually distinguish between these disorders. Other laboratory findings that support a diagnosis of secondary or tertiary insufficiency include fairly normal serum potassium and mineralocorticoid levels, as well as hypoglycemia (306).

Secondary and tertiary adrenal insufficiencies are differentiated by the administration of CRH, although from a therapeutic standpoint, this distinction is seldom important. There is little or no ACTH response in patients with secondary adrenal insufficiency, whereas patients with tertiary disease usually have an exaggerated and prolonged ACTH response (377–380).

Men with secondary or tertiary adrenal insufficiency present with glucocorticoid deficiency, while women have both glucocorticoid and androgen deficiencies. In both secondary and tertiary insufficiency, mineralocorticoid secretion is usually normal, so adrenal crisis is less common than with primary adrenal failure.

Gross Findings. The adrenal glands are variably decreased in size, depending on the etiology and duration of the disease. Cut sections show a prominent medulla and a normal zona glomerulosa, while the fasciculata and reticularis are thinner than normal.

Microscopic Findings. Histologic examination shows atrophy of the zona fasciculata and reticularis, while the zona glomerulosa, which is largely ACTH-independent, and the medulla are relatively normal. If the patient has been treated with glucocorticoids before the adrenal glands are examined, the atrophic zone may respond to therapy by enlargement of individual cells. Patients with longstanding secondary or tertiary adrenal insufficiency may develop atrophy of the zona glomerulosa, but these cells respond to substitution therapy with glucocorticoids.

Differential Diagnosis. Clinical and pathologic correlation distinguishes primary from secondary or tertiary adrenal insufficiency. Since most cases of primary adrenal insufficiency are autoimmune in origin, the presence of lymphocytes and plasma cells, along with fibrosis and involvement of all three cortical zones, help distinguish it from secondary or tertiary insufficiency in which only the two inner zones are affected.

Treatment and Prognosis. Treatment is similar to that of primary chronic insufficiency, with the exception that mineralocorticoid replacement is usually not required (306). For secondary adrenal insufficiency, replacement of other pituitary hormones may be needed depending on their status.

The prognosis depends on the underlying cause of insufficiency. Metastatic tumors to the pituitary gland are associated with a poor prognosis while craniopharyngiomas are often treated successfully by surgery. For isolated ACTH deficiency, target hormone replacement therapy (hydrocortisone) is usually effective.

Isolated Mineralocorticoid Deficiency

Isolated mineralocorticoid deficiency is characterized by a deficiency of aldosterone production by the zona glomerulosa of the adrenal gland. The causes of hypoaldosteronism include both acquired and, less often, inherited disorders that affect adrenal aldosterone synthesis or renal (and perhaps adrenal) renin release (381). The most common acquired causes of hypoaldosteronism are hyporeninemic hypoaldosteronism, pharmacologic inhibition of angiotensin II, and heparin therapy. Primary adrenal insufficiency is an infrequent cause.

The syndrome of *hyporeninemic hypoaldosteronism* is characterized by both diminished renin release and an intra-adrenal defect, which together result in decreased systemic and intra-adrenal angiotensin II production. Reduced angiotensin II production contributes to the decline in aldosterone secretion (381,382). Causes of impaired release of renin from the kidney include diabetes mellitus, autoimmune diseases, amyloidosis, sickle cell anemia, chronic interstitial nephritis, and certain medications such as nonsteroidal anti-inflammatory drugs or calcineurin inhibitors (306,383,384).

Inherited causes of *primary hypoaldosteronism* include congenital isolated hypoaldosteronism and pseudohypoaldosteronism type 2 (385). *Congenital isolated hypoaldosteronism* is a rare autosomal recessive inherited disorder caused by a defect in the terminal enzyme in the aldosterone biosynthetic pathway, aldosterone synthase (CYP11B2), which converts the 18-hydrozyl group to an aldehyde (386,387). Infants with this condition present with recurrent hypovolemia, salt wasting, and failure to thrive (388,389).

Pseudohypoaldosteronism type 2 (Gordon syndrome) is caused by inherited mutations in two serine/threonine kinases, WNK1 and WNK4 (384). Defects in these proteins, which are localized to the distal nephron and affect the thiazide-sensitive Na-Cl cotransporter, result in increased chloride reabsorption and decreased potassium secretion (390–392). Patients with this condition present with hypertension, hyperkalemia, metabolic acidosis, normal renal function, and low or low-normal plasma renin activity and aldosterone concentrations (383,391,393–395).

Other causes of acquired hypoaldosteronism include heparin and low molecular weight heparin, which have a direct toxic effect on the adrenal zone glomerulosa cells; this may be mediated by a reduction in the number and affinity of adrenal angiotensin II receptors (381,396). In rare cases, metastatic carcinoma to the adrenal gland leads to primary hypoaldosteronism (303).

Adrenal Hemorrhage and Necrosis

Definition. *Hemorrhage* and *necrosis of the adrenal gland* are usually associated with bacterial infection.

General Features. Acute adrenal insufficiency may occur as a result of bilateral adrenal infarction caused by hemorrhage or adrenal vein thrombosis (397–399). Adrenal hemorrhage caused by bacterial infection (*Waterhouse-Friderichsen syndrome*) is commonly associated with meningococcemia (400), but other causative organisms have been reported, including *Pseudomonas aeruginosa, Streptococcus pneumoniae, Neisseria gonorrhoeae, Escherichia coli, Haemophilus influenzae,* and *Staphylococcus aureus* (401). The syndrome is most common during the first 2 years of life (402,403). Other causes of acute adrenal hemorrhage include burns, myocardial infarction, cardiac failure, hypertensive cardiac disease, hypothermia, and hemorrhage into other organs (403,404).

Bilateral adrenal hemorrhages are present in about 1 percent of routine autopsies (398,399) and 0.5 to 1.0 percent of neonatal autopsies (405). Unilateral hemorrhage, which is usually more common in the right adrenal gland, is also seen in newborn infants (406).

The major risk factors for adrenal hemorrhage include anticoagulant drug or heparin therapy (heparin-induced thrombocytopenia resulting in bilateral adrenal hemorrhage) (407), thromboembolic disease, hypercoagulable states such as antiphospholipid syndrome (408,409), physical trauma, the postoperative state, sepsis, and any cause of severe stress (398,410). However, in patients treated with anticoagulants, clotting test results are usually within the therapeutic range, and spontaneous bleeding elsewhere is not evident (397).

The pathogenesis of adrenal hemorrhage is unclear. Increased adrenal blood flow stimulated by ACTH secreted in response to stress may play a contributory role (398). Anticoagulation therapy is implicated in about one third of patients, but adrenal hemorrhage occurs rarely in patients who are anticoagulated; when it does, it is usually within the first 2 to 12 days of therapy (398,399,407).

Clinical and Radiologic Features. Symptoms and signs of acute adrenal hemorrhage include hypotension or shock (over 90 percent of patients); abdominal, back, flank, or lower chest pain (86 percent); fever (66 percent); anorexia, nausea, or vomiting (47 percent); confusion or disorientation (42 percent); and abdominal rigidity or rebound tenderness (22 percent) (397). In children with severe bacterial infection, typically with *Meningococcus* or *Pseudomonas*, petechial hemorrhages occur as part of the process of disseminated intravascular coagulation.

Mineralocorticoid levels are usually normal early on, but later, dehydration and hypotension develop as levels decrease. Bilateral adrenal hemorrhage may be seen in neonates from birth trauma. Hemorrhage also occurs during pregnancy, following idiopathic adrenal vein thrombosis, or as a complication of venography such as with infarction of an adenoma.

Radiologically, the hemorrhage of acute adrenal crisis is seen as enlargement of one or both adrenal glands. Before the development of CT and MRI, the diagnosis of adrenal hemorrhage was usually made at autopsy (399).

Laboratory Findings. Evidence of occult hemorrhage, such as a sudden fall in hemoglobin and hematocrit, and progressive hyperkalemia, hyponatremia, and volume depletion should suggest the diagnosis (401). However, the condition is difficult to recognize clinically and, despite dramatic symptoms, the diagnosis is often missed.

Gross Findings. The gross findings depend on the etiology of the hemorrhage. This may vary from massive hemorrhage, involving all of both adrenal glands, to small petechial hemorrhages in the adrenal cortex. In some cases, a localized hematoma is present. The adrenal glands are usually swollen but often retain their original shape. With acute hemorrhage, the changes are most striking in the zona reticularis and extend to the outer cortical layers. Necrosis, with sparing of the subcapsular cells, is often seen in the Waterhouse-Friderichsen syndrome. Neonatal hemorrhage and necrosis usually involve predominantly the fetal cortex, with extension to the adult cortex.

Hemorrhage and necrosis are usually seen when the capsular and emissary veins are occluded by adrenal vein thrombosis. In contrast to central vein thrombosis, the capsular and emissary veins may remain patent, without the development of hemorrhage and necrosis (404).

Microscopic Findings. The histologic findings support the gross evidence of hemorrhage and necrosis, with fibrin deposition and acute infiltration by neutrophils. In the Waterhouse-Friderichsen syndrome, cells in the outer zona glomerulosa may be spared, while the medullary cells are also involved with hemorrhage and necrosis.

Differential Diagnosis. The gross and microscopic findings distinguish different patterns of hemorrhage. With central adrenal vein thrombosis, most of the cortex may be spared early in the disorder. In the Waterhouse-Friderichsen syndrome, the outer cortex is spared while the rest of the adrenal gland shows central hemorrhagic necrosis.

Treatment and Prognosis. In the acute phase, treatment consists of replacement of circulating glucocorticoids, sodium, and water deficits. Vasoconstrictive agents such as dopamine are used in extreme cases to assist with volume replacement. After the acute adrenal hemorrhagic event, long-term glucocorticoid replacement, with or without mineralocorticoid replacement therapy, may be necessary, based on the results of the adrenal function tests (397,407).

Bilateral adrenal hemorrhage is associated with potentially life-threatening consequences related to the development of acute adrenal insufficiency when at least 90 percent of the glands are injured (411). Despite treatment with stress-dose glucocorticoids, some patients with adrenal hemorrhage die because of underlying disease or diseases associated with the adrenal hemorrhage itself (412). A mortality rate of 15 percent is reported; however, this varies according to the severity of the underlying predisposing illness and may be much higher if the adrenal insufficiency is not promptly diagnosed (407,410,413).

Metastatic Tumors Causing Adrenal Insufficiency

Definition. The massive replacement of adrenal cortical cells by metastatic tumors leads to adrenal insufficiency.

General Features. Infiltration of the adrenal glands by metastatic cancer is common, probably because of the rich sinusoidal blood supply. At autopsy, adrenal metastases are found in 40 to 60 percent of patients with disseminated lung or breast cancer, 30 percent of patients with melanoma, and 14 to 20 percent of patients with stomach or colon cancer, but clinically evident adrenal insufficiency is uncommon (414–416). Adrenal cortical carcinoma with metastasis to the contralateral gland is also implicated in adrenal cortical insufficiency (417). Massive involvement of the adrenal gland by malignant lymphoma may also lead to insufficiency (418–420). The low incidence of clinical adrenal insufficiency in patients with malignant disease is because 80 to 90 percent of the adrenal cortex must be destroyed before hypofunction becomes evident (415,416,419–424).

Clinical and Radiologic Features. The clinical findings of adrenal insufficiency secondary to massive bilateral metastases are similar to those of other causes of adrenal insufficiency: fatigue, weakness, weight loss, anorexia, nausea, vomiting, and electrolyte imbalances. These may be complicated by the underlying malignancy.

The use of MRI and CT scans has increased the sensitivity of detecting adrenal metastases. In one autopsy series, bilateral metastases were present in 23 percent of 91 patients who were studied by CT (425). Massive bilateral involvement of the adrenal glands on radiologic studies usually indicates metastatic disease. The glands are round to oval, with soft tissue density, unless there is extensive hemorrhage and necrosis. With hemorrhage, the MRI usually has a high signal intensity on T1- and T2-weighted images.

Gross Findings. Usually, there is bilateral involvement, with variable degrees of hemorrhage and necrosis. The tumors are usually tan-brown to black, which is different than the yellow of adrenal cortical tumors. Some metastases, such as from renal cell carcinomas which have abundant lipid, may simulate a primary adrenal cortical neoplasm. Renal cell carcinomas metastatic to the adrenal gland have been reported in 19 percent of autopsy cases (426), but the involvement is usually unilateral, since bilateral involvement from renal cell carcinoma is uncommon (427).

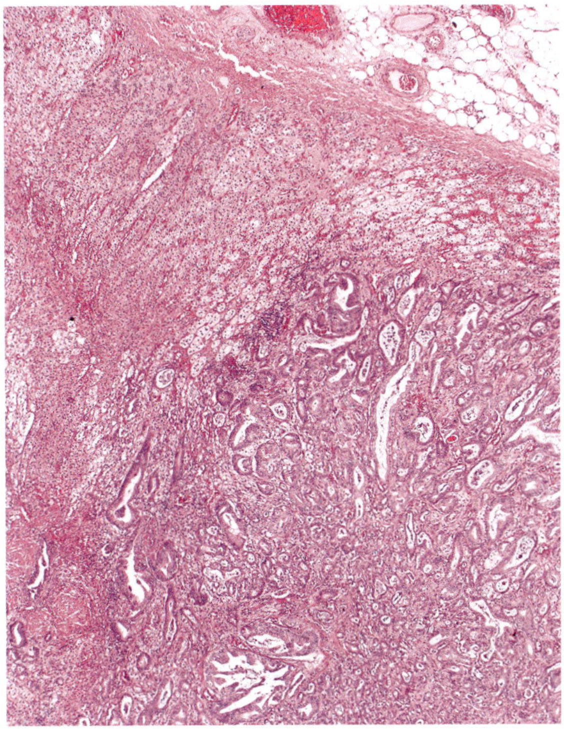

Figure 4-14

ADRENAL INSUFFICIENCY SECONDARY TO METASTATIC CARCINOMA

Histologic examination shows a deposit of metastatic adenocarcinoma with a thin rim of residual adrenal cortical tissue.

Microscopic Findings. The histologic diagnosis is usually uncomplicated, especially when the primary tumor is known and histologic slides are available for comparison (fig. 4-14). Immunohistochemical stains can help separate different types of poorly differentiated malignant tumors. This is especially true with malignant melanoma in which a primary may not be known. Immunohistochemical stains for S-100 protein, SOX10, HMB-45, melan A, and keratin are diagnostic in these cases, since all of these markers, except keratins, are positive in malignant melanoma. Massive involvement of the adrenal glands by lymphoma can be diagnosed readily by immunohistochemical characterization with CD45, along with specific B- and T-cell markers.

Differential Diagnosis. The differential diagnosis may be difficult when it is necessary to separate bilateral metastatic clear cell renal carcinoma from adrenocortical carcinoma metastatic to the contralateral gland. Immunostaining for keratin, CD10, carbonic anhydrase (CAIX), inhibin, melan A, SF-1, and synaptophysin help distinguish these lesions: the latter is usually positive for inhibin, synaptophysin, melan A, and SF-1 and weakly positive for keratin, whereas renal cell carcinomas are positive for keratin, CD10, and CAIX.

Treatment and Prognosis. In most cases of bilateral metastases, clinical recognition of adrenal insufficiency is uncommon since most of the adrenal cortex (more than 90 percent) has to be destroyed before hypofunction is manifested (414). Recent studies suggest that one fifth to one third of patients with bilateral adrenal metastases have partial adrenal insufficiency and benefit from glucocorticoid therapy (415,416).

Patients with adrenal metastases have a very poor prognosis. In one study, the 2-year survival rate was only 7 percent (428), considerably lower than the highly lethal primary adrenocortical carcinoma (429).

OTHER CONDITIONS ASSOCIATED WITH HYPOFUNCTION

Adrenal Amyloidosis

Definition. *Adrenal amyloidosis* is the deposition of amyloid in the adrenal glands, which may lead to adrenal insufficiency.

General Features. Amyloid deposition in the adrenal glands has been confirmed histologically in both systemic AL and AA amyloidoses (431–434); however, symptomatic primary adrenal insufficiency rarely occurs because extensive accumulation of amyloid and destruction of the adrenal cortex are required to produce clinical symptoms (435,436). In a study involving 374 patients with AA amyloidosis, adrenal amyloid deposits were found in 41 percent with the use of whole-body serum amyloid P component scintigraphy, but only 5 patients required long-term glucocorticoid replacement therapy (437).

Amyloid deposits in the adrenal gland may be the result of aging (438–440), with some studies reporting the presence of interstitial amyloid in the adrenal glands in up to 68 percent of autopsies (439). Multinodular amyloid deposits have also been seen in older patients (440).

Although clinically significant adrenal failure is rare, patients with systemic amyloidosis have a low cortisol response to stimulation with ACTH (441–445). In a study of the functional significance of amyloid deposits in the adrenal gland, Arik et al. (430) found that of 15 patients with renal amyloidosis without clinical evidence of adrenal insufficiency, 7 had abnormal cortisol responses, suggesting amyloid involvement of the adrenal cortex. In another study of 10 patients with renal amyloidosis, 2 had adrenal cortical dysfunction (446).

Gross Findings. The adrenal glands may be normal in size or slightly enlarged in some cases, but greatly enlarged glands weighing 30 to 34 g together have been reported (447). The cut surface is gray-yellow, and the glands appear firm and waxy when the amyloid deposition is more extensive.

Microscopic Findings. Histologic examination shows widespread infiltration of the cortex by amyloid, with destruction of the cortical cells. The zona fasciculata and zona reticularis are invariably extensively involved, while the zona glomerulosa tends to be less affected (448). The deposits are present between the adrenal cells and capillary endothelium. With increasing deposition of amyloid, the inner zone becomes a hyaline acellular mass, with obliteration of most adrenocortical cells resulting in adrenal cortical insufficiency. With senile and other secondary forms of amyloid deposition in the adrenal glands, the degree of involvement can be quite variable but is often focal.

Differential Diagnosis. Amyloid deposits should be distinguished from the hyaline sclerosis that may be seen in organizing infarction. Special stains such as Congo red, and immunohistochemistry for amyloid subtypes, as well as ultrastructural studies can readily distinguish these conditions.

Adrenoleukodystrophy

Definition. *Adrenoleukodystrophy* (ALD) is a genetic disorder characterized by primary adrenal insufficiency associated with progressive neurologic dysfunction. It is also known as *Addison-Schilder disease.*

General Features. ALD is a peroxisomal disorder of beta-oxidation that results in accumulation of very long-chain fatty acids in all tissues

(449). It is inherited as an X-linked recessive disorder and affects 1 in 20,000 Caucasian males (450). Abnormalities primarily affect the central nervous system, adrenal cortex, and Leydig cells in the testes (449).

Affected males have one of three main phenotypes, and can present from childhood through to adulthood (451). Female carriers often develop symptoms of myelopathy and peripheral neuropathy in adulthood. The clinical course in females is milder, and the onset is later (after age 35 years) than in affected males (449).

In the childhood forms of ALD, patients present between 3 and 10 years of age with progressive neurologic dysfunction that includes cognitive and behavioral abnormalities, blindness, and eventually the development of quadriparesis (452). Approximately 20 percent of affected boys have seizures, which may be the first manifestation in some cases (449). Most affected individuals have adrenal insufficiency. Some have hyperpigmented skin due to increased ACTH secretion (449).

Adrenomyeloneuropathy (AMN) is an adult form of disease that typically presents in males between 20 and 40 years of age and comprises approximately 40 to 45 percent of ALD cases (452,453). The primary manifestation is spinal cord dysfunction with progressive stiffness and weakness of the legs (spastic paraparesis), abnormal sphincter control, neurogenic bladder, polyneuropathy, and sexual dysfunction (449). The majority also have adrenal insufficiency. It is usually a slowly progressive disorder (454–456).

ALD/AMN is caused by mutations in the adenosine triphosphate (ATP)-binding cassette (ABC), subfamily D, member 1 gene (*ABCD1*), located at Xq28, that encodes an ABC transporter (457–461). *ABCD1* mutations may prevent normal transport of very long-chain fatty acids (VLCFAs) into peroxisomes, thereby preventing beta-oxidation and breakdown of VLCFAs (449). Accumulation of abnormal VLCFAs in affected organs (central nervous system, Leydig cells of the testes, and the adrenal cortex) is presumed to underlie the pathogenesis of ALD (454,462,463). In the adrenal gland, abnormal VLCFAs may directly alter cellular function by inhibiting the effects of ACTH on the adrenocortical cells, or indirectly by initiating an autoimmune response (449,464).

Gross Findings. The adrenal glands are smaller than usual, although the normal shape is retained. Each gland weighs less than 2 g, and may be less than 1 g.

Microscopic Findings. Histologically, the adrenal cortex shows changes mainly in the zona fasciculata and zona reticularis. These changes consist of variably enlarged cortical cells, with abundant cytoplasmic striations and macrovacuoles, often imparting a waxy appearance to the cytoplasm (466). These "balloon" cells are considered pathognomonic for ALD (467). Ultrastructural studies show that the ballooning and striations are a result of accumulation of smooth endoplasmic reticulum and lamellar-lipid profiles (467). The balloon cells often contain strictures and clefts, which are the result of lipid extraction during processing by xylene and alcohol. The ballooned cells may form nodules. The nuclei may be either vesicular or hyperchromatic. The zona glomerulosa and adrenal medulla are minimally affected in ALD. In extreme cases in which most of the cortical cells are destroyed, only medullary tissues remain.

Lymphocytic infiltrates are uncommon and found only in the most atrophic glands. Other tissues affected include the cerebral white matter, in which demyelinating changes and inflammation occur along with gliosis and macrophage infiltrates (456). The Schwann cells of the peripheral nerves and the Leydig cells of the testis are also abnormal. Spermatogenic arrest may be present in the testis in cases that develop in adulthood (468).

Differential Diagnosis. Distinction from autoimmune idiopathic Addison disease may be difficult in the end stage of ALD since both disorders are associated with cortical cells replaced by hyalinized connective tissue. However, the presence of ballooned cells, characteristic of ALD, is diagnostic. In addition, a lymphocytic infiltrate is characteristic of autoimmune Addison disease but is rare in ALD.

Treatment and Prognosis. Treatment approaches are targeted to specific phenotypes (469). Hematopoietic cell transplantation is the treatment of choice for boys with early stages of cerebral ALD (470). Adrenal insufficiency, if present, is treated with corticosteroid replacement. Treatment of patients with advanced cerebral ALD and those with pure AMN is supportive. Other interventions, such as dietary modifications (including Lorenzo's oil), statin medications, and other agents have not demonstrated clinical efficacy in limited observational studies and clinical trials (452,471–481).

Without treatment, childhood cerebral ALD has a rapid progression, with total disability in 6 months to 2 years, and death within 5 to 10 years of diagnosis. For boys who undergo successful hematopoietic cell transplant at an early stage of disease, the 5-year survival rate is greater than 90 percent (482). However, hematopoietic cell transplant is not curative and myelopathy symptoms may develop in adulthood (483).

Wolman Disease

Wolman disease (WD) is an autosomal recessive disorder resulting from complete deficiency of lysosomal acid lipase (484). The estimated incidence of WD worldwide is 1 per 350,000 newborns (485). The disease is characterized by massive lysosomal accumulation of cholesterol esters and triglycerides, particularly in the liver (485). Children with WD appear normal at birth, but after several weeks or months of life develop severe emesis, steatorrhea, abdominal distension, jaundice, and failure to thrive. Clinical signs include hepatosplenomegaly, anemia, and adrenal gland enlargement with calcification (485).

Adrenal histology shows cells with vacuolated cytoplasm in the zona fasciculata and zona reticularis. These cells were once filled with cholesterol esters, now represented by clefts resulting from losses during processing. There is usually necrosis, calcification, and fibrosis of the inner cortex.

Without treatment, patients with WD rapidly deteriorate and die during the first year of life (486,487). Even with successful hematopoietic cell transplantation, WD is fatal in most patients.

Adrenal Cysts

Definition. *Adrenal cysts* are tumefactive, fluid-filled cavities that primarily involve the adrenal cortex.

General Features. Adrenal cysts are uncommon lesions usually involving one adrenal gland (488,489,490), but occasionally they are bilateral (488). The frequency of adrenal cysts at autopsy is 0.06 percent (489). They account for 4 to 22 percent of adrenal incidentalomas (491). They occur more commonly in women, and are equally distributed bilaterally (492). They can occur at any age, but are rare in neonates and children (488,493,494).

Adrenal cysts have been subtyped into various groups by different investigators (488,489,495, 496). The most common types are endothelial, epithelial, parasitic, and pseudocysts (488,497).

Endothelial cysts are the most common type, constituting up to 45 percent of cases in the literature (488,489). They include lymphangiomatous (lymphangiectatic or serous) and angiomatous (hemangioma) cysts (492). These cysts are typically small and multiple.

Epithelial cysts represent less than 10 percent of adrenal gland cysts (488,495,498). They are also referred to as "glandular" cysts (492). Since the normal adrenal gland contains no glandular or ductal structures, it has been proposed that these epithelial cysts are derived from embryonic rests. This category includes cystic adenomas, glandular retention cysts, and embryonal cysts (492).

Adrenal *pseudocyst* is the second most common type, comprising up to 39 percent of cases in the literature (488). This is the most common clinically or surgically recognized type. Pseudocysts lack an epithelial lining and often occur as a result of hemorrhage or infarction. Calcification within the wall of a cyst is suspicious for either a pseudocyst or a parasitic cyst. Rarely, a benign or malignant adrenal tumor undergoes cystic degeneration and results in a pseudocyst (492). Hemorrhagic adrenal pseudocysts often contain thin-walled vascular channels (499). Immunohistochemical studies of these channels are consistent with a vascular origin, suggesting that many of these hemorrhagic cysts are endothelial cysts (490,495,500–502). It is likely that most adrenal pseudocysts arise from repeated episodes of hemorrhage, organization, and repair in association with endothelial cysts.

Parasitic cysts represent less than 10 percent of adrenal cysts. They are often discovered incidentally at autopsy. They are most commonly hydatid cysts, which are the result of an *Echinococcus* tapeworm infestation. *Echinococcus granulosus* is the most common parasite causing cystic echinococcosis (492); however, less than 0.5 percent of all patients with echinococcosis have adrenal involvement (495).

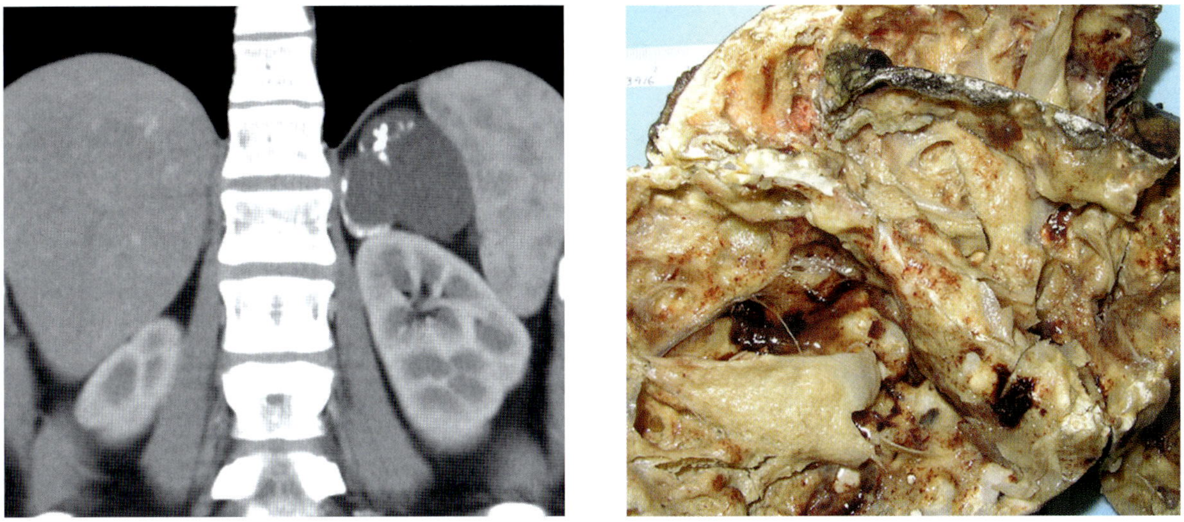

Figure 4-15

ADRENAL PSEUDOCYST

Left: Left adrenal pseudocyst presenting as a lobulated mass superior to the left kidney on contrast-enhanced CT. The cyst is nonenhancing, and contains thin central septations and peripheral calcifications.

Right: This adrenal pseudocyst measured 200 mm in diameter. It contained fibrin and altered blood. It is surrounded by a thick fibrous wall with readily identifiable dystrophic calcifications. Internal septations of variable thickness are noted.

Adrenal cysts occasionally simulate metastatic disease (493,495) or represent cystic degeneration of cortical or medullary neoplasms. The frequency with which adrenal pseudocysts are reported is increasing, which may reflect improved radiologic imaging methods. The possible etiology of these cysts includes cystic degeneration of an adrenal cortical or medullary neoplasm, a vascular neoplasm or malformation, or hemorrhagic degeneration in the adrenal gland parenchyma (501,503,504).

Clinical and Radiologic Features. Women have a 2- to 4-fold greater incidence of adrenal cysts than men. The most common presentation is as an incidentaloma. Other presentations include abdominal or flank pain, epigastric discomfort, or a palpable abdominal mass with larger cysts. Rare patients present with fever and leukocytosis. Chronic hemorrhage into large cysts can lead to anemia. Most adrenal cysts are not associated with endocrine symptoms. A few patients may have hypertension, which is cured by surgical removal of the cyst. The hypertension may be related to compression of the renal vein (505,506). In a series of eight adrenal pseudocysts reported by Medeiros et al. (502), four patients had symptoms of abdominal and/or flank pain, and three had hypertension which resolved in one patient after surgery. There was a remote history of trauma in two of the eight patients.

On imaging studies, an uncomplicated cyst appears as a well-defined lesion of water density that does not enhance on CT. MRI usually shows a low signal intensity on T1-weighted images and a high signal intensity on T2-weighted images with enhancement (507). These cysts have a thin wall and may have thin septa. Pseudocysts have thicker walls. Hemorrhage or debris within the cyst may cause increased internal attenuation. Their large size and variable density, with areas of necrosis and dystrophic calcification, may be concerning for malignancy on imaging (fig. 4-15, left). Both benign and malignant tumors may show cystic degeneration and necrosis.

Gross Findings. The cysts vary greatly in size, from microcysts that are 0.1 to 1.0 mm in diameter and are seen in the adrenal cortex of older fetuses and infants (494,506), to pseudocysts that are usually 4 to 10 cm in diameter but may be as large as 33 cm. Pseudocysts may contain dark brown fluid from hemorrhage (488). Cysts are unilocular or may contain many septations. Most are surrounded by an

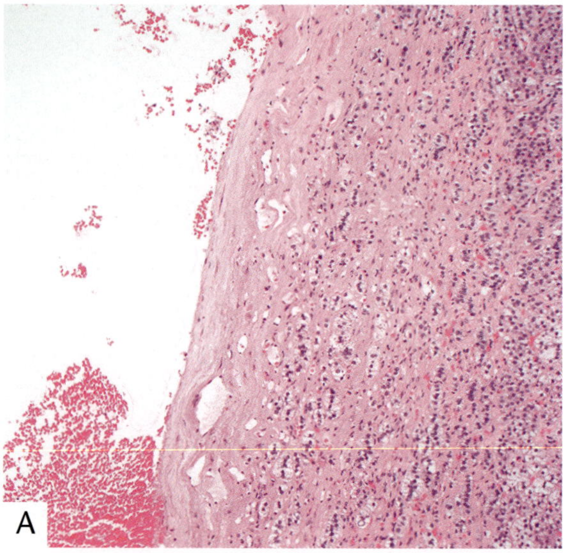

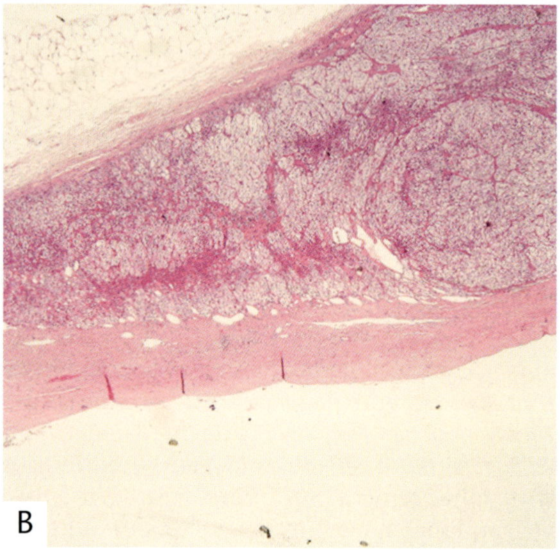

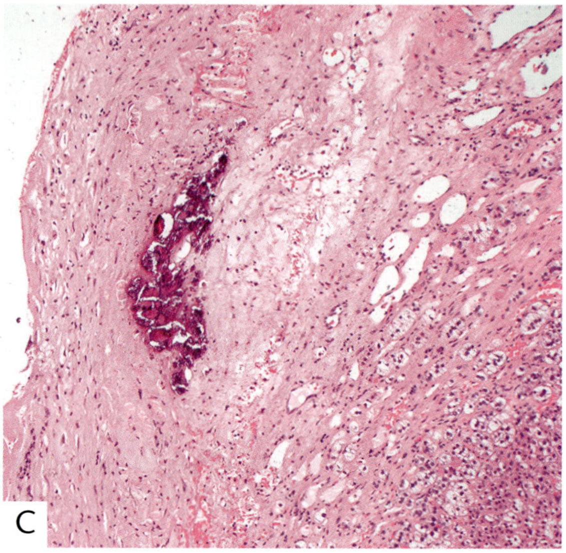

Figure 4-16

ADRENAL PSEUDOCYST

A: There are residual adrenal cortical cells in the wall of this adrenal pseudocyst.

B: Often, the adrenal cortical cells are separated from the cyst contents by a band of dense hyalinized connective tissue.

C: Dystrophic calcification in the areas of fibrosis is common.

irregular and variably thick fibrous wall with readily identifiable dystrophic calcifications (fig. 4-15, right). The focal yellowish areas present in the cyst wall usually represent residual adrenal cortical cells.

Microscopic Findings. The histologic features reflect the type of adrenal cyst present. Pseudocysts are among the most common and by definition lack an epithelial lining (fig. 4-16). They are thought to originate from hemorrhage into a previously normal adrenal gland. Liquefaction and organization of the necrotic material follows, with subsequent enlargement due to further hemorrhage and fluid accumulation. The cyst wall is composed of dense fibrous tissue, with areas of calcification and granulation tissue (fig. 4-16D). Although current thinking is that most adrenal pseudocysts arise in association with endothelial cysts after repeated episodes of hemorrhage and repair, an endothelial lining is either absent or identifiable only after a careful search, often with the assistance of immunohistochemistry (fig. 4-17).

True epithelial cysts include retention cysts, some of which are of mesothelial origin, embryonal cysts, and cystic neoplasms. They may represent a lymphangiomatous or angiomatous origin from the maldevelopment of lymphatics

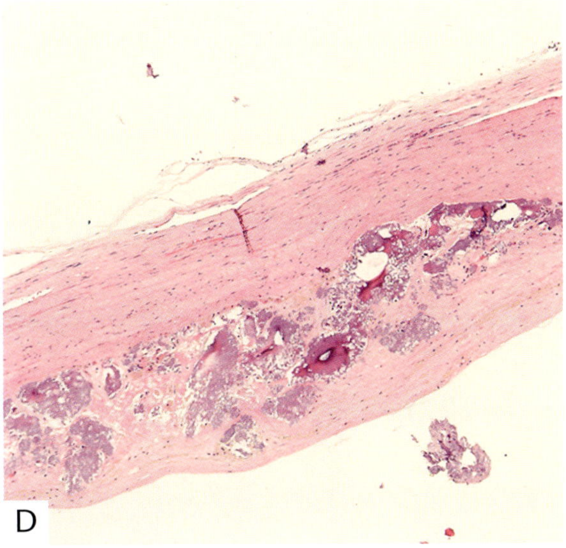

D

Figure 4-16, continued

D: The presence of dystrophic calcification in a cyst lining composed almost entirely of densely hyalinized connective tissue is typical.

E: Adrenal pseudocyst with recent and old hemorrhage in the wall.

F: Adrenal pseudocyst with an endothelial lining above the fibrotic cyst wall.

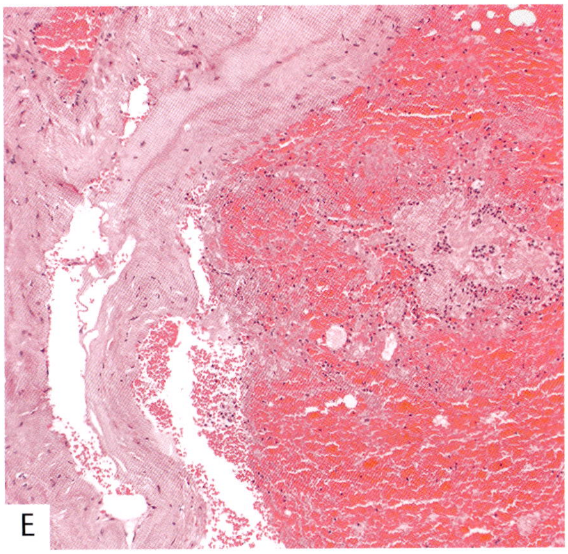

E

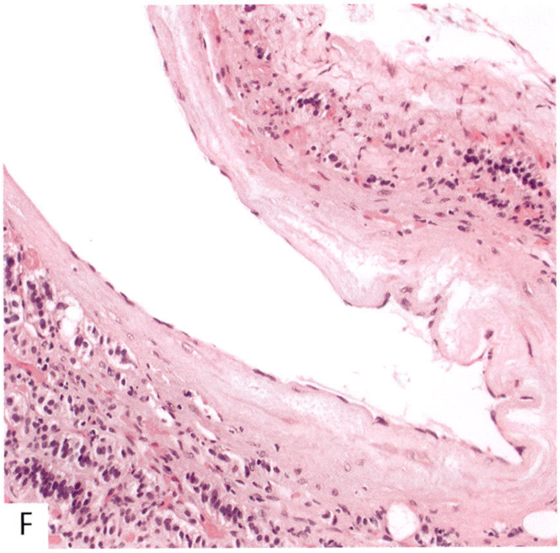

F

associated with blood vessels in the adrenal capsule and the capillary sinusoids. Alternatively, the endothelial lining may represent recanalization in a hemorrhagic pseudocyst. Parasitic cysts usually contain echinococcal or other organisms, and the wall has many eosinophils.

Immunohistochemical Findings. Analyses of pseudocysts have found factor VIII immunoreactivity in the inner wall, but no keratin (508). In one case report of an epithelial-lined cyst of the adrenal gland, the lining cells were positive for keratin but were negative for factor VIII, vimentin, and epithelial membrane antigen, and a mesothelial origin of the epithelial lining was proposed (508). Another study of a primary adrenal pseudocyst showed staining for laminin and type IV collagen, suggesting a vascular origin. Immunohistochemistry for D2-40 can be used to identify cystic lymphangiomas.

Differential Diagnosis. The major entity in the differential diagnosis is a cystic neoplasm. Cystic adrenal cortical or medullary neoplasms as well as cystic metastatic carcinomas can simulate adrenal cysts (495). Histologic identification of neoplastic adrenal cortical cells, pheochromocytoma cells, or metastatic carcinoma cells on H&E-stained sections usually provides the diagnosis. Occasionally, immunohistochemical

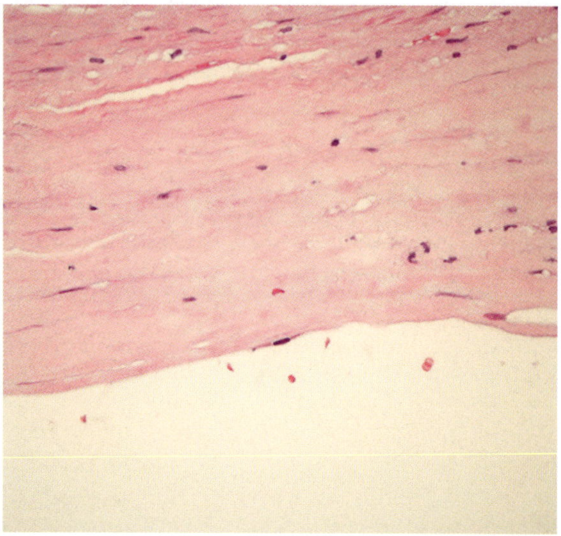

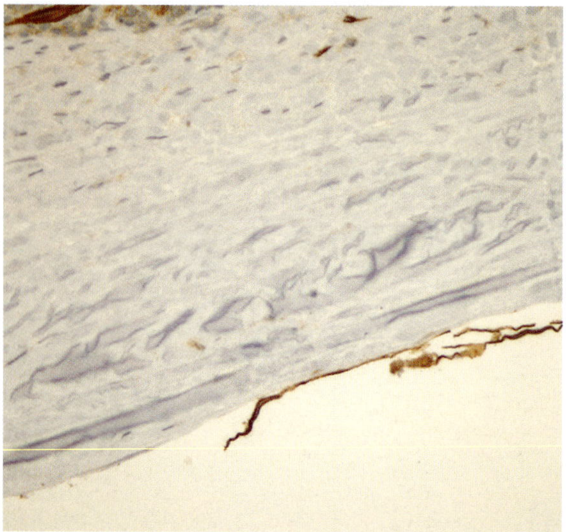

Figure 4-17

ADRENAL PSEUDOCSYST

Left: In the inner lining of this adrenal pseudocyst, the presence of a single endothelial cell is the only morphological clue that it has arisen after repeated episodes of hemorrhage and organization associated with an endothelial cyst.

Right: Immunohistochemistry for the endothelial marker CD31 highlights the partial endothelial lining of this adrenal pseudocyst, which would otherwise not be discerned.

stains are needed to distinguish between cystic adrenal neoplasms and metastatic tumors.

Treatment and Prognosis. Surgical excision is the treatment of choice. Laparoscopic surgery, which is used for benign adrenal disease, may be an option for smaller cysts (509,510,511). The prognosis for patients with benign adrenal cysts is excellent (488,502).

ADRENAL CORTICAL NODULES AND HYPERPLASIA

Hyperplasia of the adrenal cortex, is defined as an increase in cortical mass usually resulting from stimulation by ACTH, and is associated with a wide variety of clinical syndromes. These syndromes range from congenital hyperplasia due to inborn errors of metabolism to Cushing syndrome, and even rarer conditions such as primary pigmented adrenocortical disease. These conditions are usually associated with a hyperfunctioning adrenal cortex that produces excessive amounts of steroid hormones.

Adrenal cortical nodules, especially small nodules, discovered incidentally at autopsy or during surgical removal of the adrenal glands for other reasons, are nonfunctional. These nodules occur in asymptomatic patients and are often associated with age-related changes in the adrenal vasculature.

Adrenal cortical hyperplasia associated with hypercortisolism can be pathophysiologically categorized as either ACTH-dependent or ACTH-independent. Alternatively, a morphologic classification, which is probably more useful for pathologists, takes into account the nodule size, pattern of distribution, color, and other features that may be used to classify adrenal cortical hyperplasia. Other conditions associated with adrenal cortical hyperplasia include congenital adrenal hyperplasia and the adrenogenital syndrome secondary to deficiencies of the enzymes needed for steroid biosynthesis.

Incidental Nonfunctional Adrenal Cortical Nodules

Definition. These are incidentally discovered small nodules composed of microscopically normal adrenocortical tissue within the adrenal cortex or protruding through the capsule into the adjacent fat.

General Features. A great deal of confusion exists regarding the classification of *nonfunctional adrenal cortical nodules*. Many of these nodules are discovered incidentally, being reported in up to 54 percent of unselected autopsies (512); while others are diagnosed incidentally during radiologic studies (513–518). The designation of these nodules as neoplasms rather than hyperplastic nodules has often been made arbitrarily, without documentation of the monoclonal nature of the lesion as would be expected with true neoplasms. If the historic definition of an adenoma as a solitary nodule in the adrenal gland measuring at least 3 to 5 mm in diameter is used, then 1 to 3 percent of autopsied adrenal glands contain adenomas (514,516,519,520). These nodules are more common in older patients and in patients with hypertension. In one study, a solitary nodule was found in 29 percent of 100 consecutive autopsies in women of an average age of 81 years (521). Another study reports a prevalence of about 7 percent in people older than 70 years (522). Solitary nodules are found in up to 20 percent of autopsied patients with a history of hypertension (517). Patients with diabetes mellitus also have an increased incidence of solitary adrenal nodules at autopsy (523).

Most small adrenal nodules probably represent localized overgrowth of adrenocortical cells, which may be related to aging and/or response to adrenal vascular changes (513,521). Dobbie (521) has elegantly outlined the various types and stages of nodule development. The earliest lesions are entirely within the cortex, while further enlargement leads to compression of the surrounding tissues. Nodules continue to grow and form a mushroom-like mass that protrudes through the capsule or expands within the gland, attaining a size of up to 2 to 3 cm. The term cortical extrusion refers to the extension of cortical nodules into the periadrenal tissue. The cortex may show an "hour-glass" pattern of extrusion or a mushroom pattern, or the cortical cells may stream into the adjacent periadrenal adipose tissue (521).

These extrusions should not be mistaken for carcinomas. The capsular arteries of adrenal glands containing nodules often show hyalinization and intimal proliferation, changes that are sometimes associated with obliteration of the lumen (521). These changes may result from ischemia, leading to focal cortical atrophy.

Molecular biologic studies using the Lyon hypothesis have shown that many benign nodules are monoclonal (524). Similar studies have been done in the adrenal cortex (525). Although all adrenocortical carcinomas are monoclonal, adrenocortical nodules may be monoclonal (43 percent) or polyclonal (28.5 percent), with various intermediate phases (28.5 percent) (525). The significance of polyclonal nodules is that either they are not true adenomas or that some adenomas are truly polyclonal. Because some hyperplastic nodules, as in bilateral pigmented adrenocortical hyperplasia, have been shown to be monoclonal (526), clonality studies may not provide a direct answer when distinguishing hyperplastic nodules from adenomas.

Most small adrenocortical nodules are nonfunctional, or at the very most, minimally functional. Suzuki et al. (527) examined 15 small adrenocortical nodules that showed no clinical evidence of biologic activity of steroid production. They found immunohistochemical evidence of various steroidogenic enzymes, including 3-beta-hydroxysteroid dehydrogenase, C21-hydroxylase, 17-alpha-hydroxylase, and 11-beta-hydroxylase (in particular, CYP11B2), indicating that these nodules were capable of producing cortisol. However, since the enzymatic activity of these proteins was not investigated, the possibility that the immunoreactivity was not reflective of true enzymatic activity could not be excluded.

Some adrenal cortical nodules are pigmented when seen at autopsy or after surgical resection. In one series, there were focally pigmented nodules in 37 percent of autopsied adrenal glands (528). Multiple pigmented nodules, some of which are bilateral, have also been incidental findings. Neuromelanin has been detected in some of these pigmented nodules.

Some incidental nonfunctional nodules are discovered in patients undergoing radiologic studies of the upper abdomen; they have been reported in 0.6 to 1.3 percent of patients undergoing CT scans (515,529,530). High-resolution CT scans detect lesions in approximately 4 percent of people (531). In a Mayo Clinic series, the average age was 62 years, with females constituting more than half of the patients (532). The most common diagnosis of these nodules is a benign nonfunctioning adrenal cortical nodule, and less commonly, metastatic tumors from lung, breast, and other sites (524).

Clinical and Radiologic Features. Small nodules usually do not cause symptoms unless they are functional. Radiologic findings, including CT and MRI, often show one or more nodules in the adrenal gland. High-resolution CT scans can detect nodules smaller than 1 cm in diameter. The size and imaging characteristics may help determine whether an adrenal nodule is benign or malignant. In a study of 887 patients with adrenal incidentalomas from the National Italian Study Group on Adrenal Tumors (533), adrenocortical carcinomas were significantly associated with mass size, with 90 percent being more than 4 cm in diameter when discovered.

Gross Findings. Nonfunctional benign nodules of the adrenal gland are usually multiple. The studies of Dobbie (521) and Neville (513) showed that nodularity of the adrenal gland is common, being reported in 65 percent of more than 100 consecutive autopsies. Nodules are frequently bilateral, and may extrude into the adrenal capsule. They are usually small, but larger nodules are greater than 2 cm in diameter. The cut surface is yellow, with focal brown areas. Smaller nodules are usually unencapsulated, whereas larger single nodules may be surrounded by a fibrous capsule. In some areas, the nodule appears to infiltrate the periadrenal fat. Nodules may also completely detach from the adrenal capsule.

Microscopic Findings. The nodules are composed predominantly of fasciculata-type cells. Myxoid areas or myxoid-predominant tumors can be seen, as in adenomas associated with Conn syndrome or adrenocortical carcinomas (534,535). The cells of the nodules are arranged in various architectural patterns, including trabecular, pseudoglandular, solid, and alveolar. Metaplastic changes include myelolipomatous and osseous metaplasia. In contrast to functional adenomas associated with glucocorticoid production, the cortex adjacent to nonfunctional nodules is not atrophic. Degenerative or retrogressive changes, such as hyalinization, hemorrhage, and calcification, may be present in larger nodules.

Pigmented cortical nodules usually contain neuromelanin (528). These nodules are frequently located in the zona reticularis. They are unencapsulated and composed of eosinophilic cells with varying degrees of lipid depletion and lipofuscin pigment.

Differential Diagnosis. Incidental cortical nodules should be distinguished from adrenal cortical adenomas and carcinomas. The cytologic features of carcinomas, including increased mitotic activity with atypical mitoses and confluent necrosis in nodules that are larger than 5 cm, usually are diagnostic. Distinction of incidental nodules from adenomas or benign neoplasms may be more problematic. The presence of a solitary nodule that is often circumscribed usually indicates an adenoma. Nodular hyperplasia often consists of multiple nodules of varying sizes which may be bilateral.

Treatment and Prognosis. Nonfunctioning incidental nodules may be followed clinically if they are small (less than 4 cm). Nodules greater than 4 cm are often treated surgically. Small nonfunctional nodules that are surgically excised and shown to be benign are associated with an excellent prognosis.

Adrenal Cortical Hyperplasia

Definition. *Adrenal cortical hyperplasia* is characterized by increased adrenal cortical mass resulting from stimulation by ACTH, which can be derived from the pituitary gland or from a variety of extrapituitary sources.

General Features. Adrenocortical hyperplasia is categorized as either ACTH-dependent or ACTH-independent. The causes of *ACTH-dependent Cushing syndrome* include: Cushing disease (pituitary hypersecretion of ACTH; 65 to 70 percent); ectopic secretion of ACTH by nonpituitary tumors (10 to 15 percent); ectopic secretion of CRH by nonhypothalamic tumors causing pituitary hypersecretion of ACTH (less than 1 percent); and iatrogenic or factitious Cushing syndrome due to administration of exogenous ACTH (less than 1 percent) (536). The causes of *ACTH-independent Cushing syndrome* are: iatrogenic or factitious, which is by far the most common cause; adrenocortical adenomas and carcinomas (18 to 20 percent); primary pigmented nodular adrenocortical disease (less than 1 percent); and bilateral macronodular adrenal hyperplasia (less than 1 percent) (536).

Adrenocortical hyperplasia may also be classified on the basis of morphology as either diffuse or nodular. In adults, simple or diffuse hyperplasia is most common (62 percent), while nodular hyperplasia and hyperplasia with ectopic ACTH syndrome constitute 20 percent and 18 percent of cases, respectively (537,538). In children, simple hyperplasia is most common (62 percent), with nodular hyperplasia and hyperplasia with ectopic ACTH syndrome constituting 23 percent and 15 percent, respectively (538). An ACTH-secreting pituitary adenoma is the most frequent cause of diffuse hyperplasia (538).

Nodular hyperplasia involving both adrenal glands is seen in adults and children. The adrenal glands contain one or more prominent yellow nodules greater than 0.5 cm in diameter with associated background cortical hyperplasia (537). The nodules can be as large as 2.0 to 2.5 cm. In children, bilateral nodular hyperplasia is seen most frequently in the first year of life. In adults, some investigators have suggested that the hyperplasia develops secondary to vascular damage in capsular arteries, with continued increase in size of micronodules (539–541). In most cases of nodular hyperplasia, the adjacent adrenal cortex shows diffuse or simple hyperplasia and Cushing syndrome is cured only by bilateral adrenalectomy, suggesting that diffuse and nodular hyperplasia are part of the same morphologic spectrum (537,538,542). Some authors have observed unilateral nodular hyperplasia with low plasma ACTH levels (537,543). However, this may represent true adenoma rather than nodular hyperplasia (537). The existence of unilateral adrenal cortical hyperplasia associated with Cushing syndrome has been reported (544,545). These patients are usually cured by unilateral adrenalectomy with no recurrent disease up to several years after surgery.

Hyperplasia is seen in association with a variety of neoplasms producing ACTH or CRH (546,547). In most series, bronchial carcinoid tumors and small cell carcinomas account for the majority of cases. Other tumors associated with the *paraneoplastic ACTH syndrome* include pancreatic neuroendocrine neoplasms, medullary thyroid carcinoma, thymic carcinoid tumors, and pheochromocytomas (548–550). CRH is commonly produced by bronchial carcinoid tumors, although other tumors can also produce this peptide ectopically (551,552). Unusual cases of ectopic Cushing syndrome, such as a pituitary adenoma arising in a benign cystic teratoma of the ovary, have been reported (548).

Clinical and Radiologic Features. The symptoms and signs of Cushing syndrome result directly from chronic exposure to excess glucocorticoids. There is a large spectrum of manifestations, ranging from subclinical to overt syndrome, depending on the duration and intensity of excess steroid production (553). In adults, signs and symptoms most suggestive of hypercortisolism include proximal muscle weakness, facial plethora, wasting of the extremities with increased fat in the abdomen and face, wide purplish striae, bruising with no obvious trauma, and supraclavicular fat pads (553–555). Other signs and symptoms include cardiovascular complications, skin atrophy, glucose intolerance, hirsutism, and psychiatric complications (556,557).

Radiologic findings depend on the etiology of the Cushing syndrome. In patients with Cushing disease, a small ACTH-producing pituitary neuroendocrine tumor (PitNET) is detected by MRI in 50 percent of cases. Bilateral inferior petrosal sinus sampling for ACTH is required in most patients with ACTH-dependent Cushing syndrome (558). Scintigraphy with ^{131}I-labeled cholesterol and adrenal arteriograms are occasionally performed (557). In patients with ectopic sources of ACTH or CRH, a tumor may be detected by CT or MRI. The adrenal glands appear normal in size or show bilateral diffuse or macronodular enlargement. Nodules smaller than 1 cm are occasionally detected (537).

Laboratory Findings. The diagnosis of Cushing syndrome requires a stepwise approach. First, an endogenous and autonomous cause for the hypercortisolemia is confirmed, based on hypercortisoluria, abnormal response of the HPA axis to the administration of dexamethasone, and/or blunted diurnal cortisol variation (554). The cause of hypercortisolemia should then be sought, although this can be challenging due to some overlap between the biochemical profiles of the different etiologies (559). Low serum ACTH levels suggest primary adrenal Cushing syndrome, whereas high ACTH levels suggest ACTH-dependent disease (554). In patients with suspected ACTH-dependent Cushing syndrome, high dexamethasone suppression (560) and/or CRH stimulation test (561) may be used for identifying the cause, followed by the imaging modalities described above.

Gross Findings. Diffuse hyperplasia results in enlargement of both adrenal glands, although gland weights may be increased only slightly in the initial stages (562). In more advanced

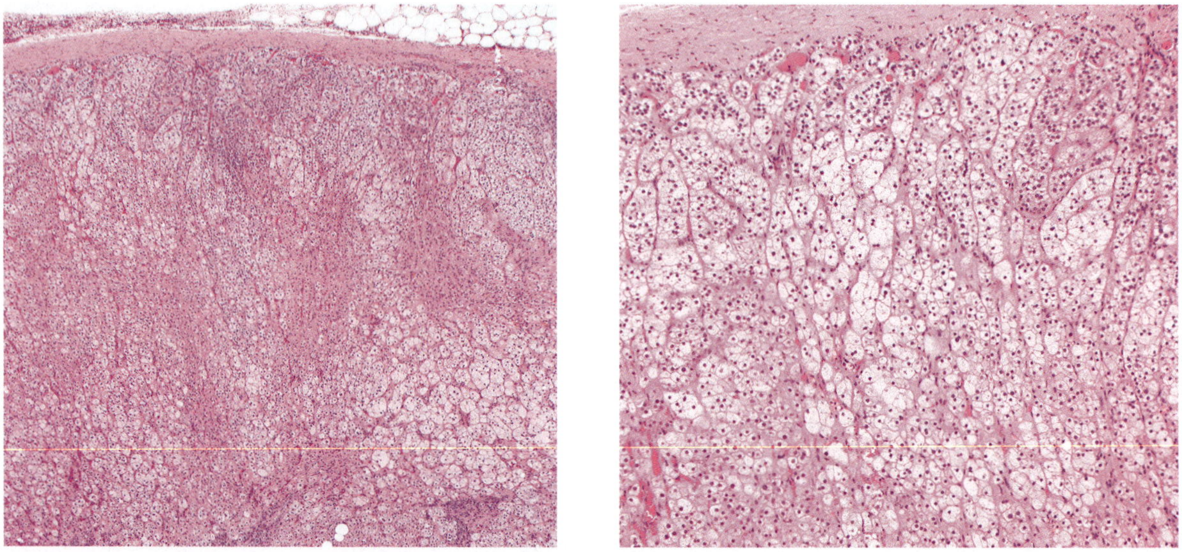

Figure 4-18

DIFFUSE ADRENOCORTICAL HYPERPLASIA

Left: The zona fasciculata is the predominant region that is expanded.
Right: A slight nodularity of lipid-rich and compact cells is noted in this diffusely enlarged adrenal gland.

disease, the combined average weight ranges from 12 to 24 g (563). The glands have rounded contours, rather than the sharp outlines typical of normal glands. The inner portion of the cortex is widened and often appears pale brown or tan. The outer layers of the cortex are typically yellow. Small nodules of less than 0.25 cm may be seen in diffuse hyperplasia; this is most prominent in prepubertal children.

Glands with bilateral nodular hyperplasia are somewhat heavier than those with diffuse hyperplasia. A difference of 2 g or more between the two glands is seen with nodular hyperplasia, but less commonly with diffuse hyperplasia. Nodules are present in both glands and are greater than 0.5 cm in diameter, and many are 2.0 to 2.5 cm. The nodules are multiple and project from one pole of the gland or compress the adjacent cortex. In some cases, multiple cortical nodules are superimposed on a diffusely hyperplastic cortex. This type of change has been referred to as *diffuse and micronodular hyperplasia* if the nodules are less than 1.0 cm in diameter, and *diffuse and macronodular hyperplasia* if the nodules are greater than 1.0 cm in diameter.

Microscopic Findings. Histologic examination of the adrenal glands in diffuse hyperplasia shows that the inner brown zone corresponds to lipid-depleted cells of the fasciculata, whereas the cells of the outer cortex are more characteristically vacuolated (fig. 4-18) (564). The glomerulosa in adults with Cushing disease is often difficult to identify, but in children, the glomerulosa may appear slightly hyperplastic (565). In adrenal glands in which the weights are normal or near normal, the thickening of the zona reticularis is a distinct sign of hyperplasia. The individual cells are typically normal in size and appearance, although slight hypertrophy may be present. The ultrastructural features are similar to those of the normal adrenal cortex (566,567).

At autopsy, the adrenal glands from patients with Cushing syndrome contain all compact or zona reticularis-type cells, which extend out to the capsule or to the zona glomerulosa. These additional changes are probably related to the stresses associated with dying, in this case from Cushing syndrome.

Another histologic change associated with Cushing syndrome is the presence of small micronodules, usually around the central vein, consisting of lipid-laden clear cells similar to those of the zona fasciculata. Aggregates of fat cells may be present in the zona reticularis. Hypertrophy of the cells in the zona fasciculata and zona reticularis, along with nuclear pleomorphism, is probably secondary to ACTH stimulation.

Occasional cases of spontaneous remission of Cushing syndrome with bilateral adrenocortical hyperplasia have been reported (562,568). Some of these have occurred after adrenal venography and administration of ACTH, which may be related to hemorrhage and necrosis of the glands (569).

Adrenal glands with nodular hyperplasia show nodules of clear cells and compression of the adjacent cortex (fig. 4-19). Clusters of compact cells, corresponding to the brown color of some nodules, may be present. The capsular arteries usually show hyalinization and sclerosis. Cellular hypertrophy and nuclear pleomorphism may be seen in some nodules. Although some authors have equated cellular pleomorphism with tumor development in these nodules (570,571), this is probably not accurate, since in the adrenal glands and in other endocrine organs, pleomorphism is not related to anaplasia. Examination of the adjacent cortex in nodular hyperplasia shows evidence of hyperplasia, while in nonfunctioning nodules it is normal and in functioning adenomas it is atrophic.

Ultrastructural examination shows normal-appearing adrenal cortical cells. In the areas adjacent to the nodules there is increased perivascular collagen and basement membrane material. The cells in the nodules may contain more prominent smooth endoplasmic reticulum (572).

The adrenal glands in ectopic ACTH or CRH syndrome weigh between 14 and 16 g each and may reach up to 20 g. Cut sections show a thickened cortex with a diffuse brown color. Histologic examination shows more prominent compact or zona reticularis–type cells which extend from the medulla up to or almost to the zona glomerulosa or capsule. Foci of clear cells are also present. The compact and clear cells are hypertrophic and show variable degrees of nuclear pleomorphism. Occasionally, metastatic tumors are detected in the hyperplastic adrenal cortex.

Differential Diagnosis. The principal differential diagnosis is between nodular hyperplasia and benign neoplasms, usually adenomas of the adrenal cortex. Nodular hyperplasia usually consists of multiple nodules, while adenomas consist of a single nodule. To distinguish functional from nonfunctional nodules, it is necessary to examine the uninvolved cortex. In nodular hyperplasia, the other areas of the adrenal cortex away from the nodules are also hyperplastic, while in functional nodules, there is atrophy of the uninvolved cortex.

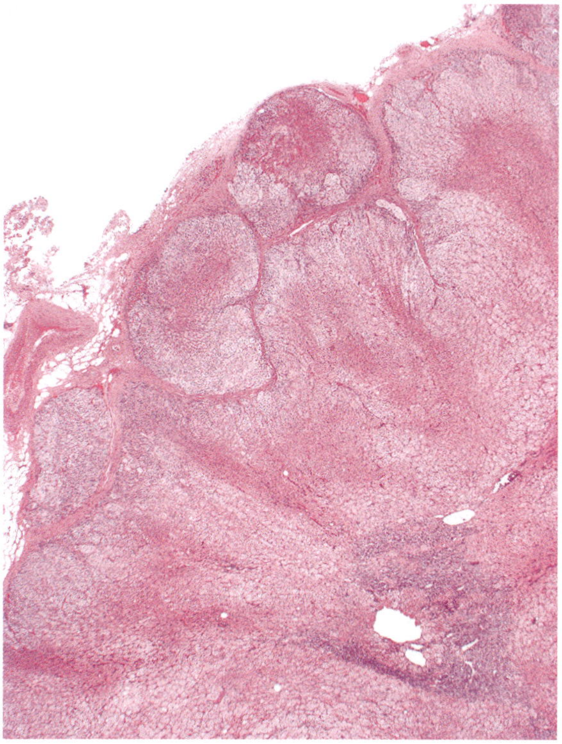

Figure 4-19

NODULAR HYPERPLASIA

Multiple nodules of predominantly lipid-rich cells are seen in a gland with nodular hyperplasia.

Treatment and Prognosis. The treatment of choice for Cushing disease (ACTH-producing PitNET) is transsphenoidal microadenomectomy when a clearly circumscribed microadenoma is identified at surgery (573). In the remaining patients, subtotal resection of the anterior pituitary may be indicated if future fertility is not desired.

In cases where surgery is delayed, contraindicated, or unsuccessful, medical therapy is often required. Adrenal enzyme inhibitors are the most commonly used drugs, but adrenolytic agents, drugs that target the pituitary, and glucocorticoid-receptor antagonists have also been used (573). Medical therapy targeting the corticotroph tumor, such as cabergoline or pasireotide, can result in normalization of 24-hour urinary free cortisol in 20 to 40 percent of cases, especially if the hypercortisolism is mild. The glucocorticoid (as well as progestin) antagonist mifepristone is sometimes used to treat glucose intolerance in patients with Cushing syndrome who are not surgical candidates (574).

For patients in whom fertility is an important concern and in whom a tumor is not found or is not cured by surgical resection, pituitary irradiation is one of the next treatment options. It may also be considered as primary therapy for children under 18 years of age (573).

Bilateral total adrenalectomy with lifelong daily glucocorticoid and mineralocorticoid replacement therapy is the final definitive cure, and may be preferred by some patients instead of radiation therapy. In one series, laparoscopic adrenalectomy was successful in 42 patients with Cushing disease who had not been cured by previous pituitary surgery, radiotherapy, or medical therapy (575).

The optimal therapy for ectopic ACTH or CRH syndrome is surgical excision of the tumor, thereby removing the source of ACTH or CRH and curing the metabolic disorder. In patients with metastases limited to the liver, after resection of the primary tumor, resection or cryoablation of the metastases or even liver transplantation may result in cure (576). Depending on the tumor type, chemotherapy and/or radiotherapy may be helpful. For those patients with nonresectable tumors, the hypercortisolism can be controlled with adrenal enzyme inhibitors, such as ketoconazole, metyrapone, and etomidate (573). In patients with indolent tumors that cannot be cured surgically, and with a long life expectancy, mitotane can be used to achieve a medical adrenalectomy. Bilateral surgical adrenalectomy or long-term treatment with steroidogenesis inhibitors may be used as an alternative to mitotane (577–579).

Untreated Cushing syndrome is often fatal, with most deaths being attributed to cardiovascular, thromboembolic, or hypertensive complications or bacterial or fungal infections. Untreated Cushing syndrome in pregnancy is associated with spontaneous abortion, premature delivery, and rarely, neonatal adrenal insufficiency. Maternal deaths occur in 4 percent of patients (580). Fortunately, Cushing syndrome is almost always curable, although rare patients die of perioperative or other complications (581).

Patients with ectopic ACTH and CRH secretion may have a poor prognosis associated with the underlying tumor. The prognosis is dictated by the nature of the tumor and the severity of the hypercortisolism. Most patients with overt metastases at the time of presentation die of malignant disease within 1 year, although patients with indolent tumors may survive for many years.

Adrenocortical Macronodular Hyperplasia

Definition. Primary *adrenocortical macronodular hyperplasia* is characterized by bilaterally enlarged adrenal glands containing multiple nonpigmented nodules that are greater than 10 mm in diameter. Synonyms include *primary bilateral macronodular adrenal hyperplasia, ACTH-independent macronodular adrenal hyperplasia, massive macronodular hyperplasia,* and *macronodular adrenal dysplasia.*

General Features. Primary bilateral macronodular adrenal hyperplasia is an uncommon cause (less than 2 percent) of endogenous Cushing syndrome. It is characterized by tumefactive enlargement of both adrenal glands with multiple nodules that are greater than 10 mm in diameter but can be as large as 30 to 40 mm (582–584). It is diagnosed in patients with Cushing syndrome but is more often seen in those with incidentally found bilateral macronodular adrenal glands and subclinical secretion of cortisol.

Pituitary studies, including petrosal sampling for ACTH, are normal (585,586). Endocrinologic studies reveal elevated plasma cortisol levels with low levels of plasma ACTH. Dexamethasone testing does not suppress adrenal cortisol secretion (582,583,587). The hypercortisolism associated with primary macronodular adrenal hyperplasia was previously considered to be ACTH-independent. However, it is now known that the cortisol secretion is at least partially regulated by intra-adrenal ACTH, which is produced by steroidogenic cells in the hyperplastic glands (588,589).

Abnormal adrenal expression of ectopic receptors or increased activity of ectopic peptide hormone receptors has been implicated in the pathogenesis of macronodular adrenal hyperplasia. Cortisol secretion becomes driven by a hormone that escapes cortisol-mediated feedback. The most common aberrantly expressed hormone receptors are vasopressin (590–594), serotonin, gastric inhibitory polypeptide (595), beta-adrenergic (596), and interleukin 1 (597).

Several molecular causes have been identified in the pathogenesis of primary adrenocortical macronodular hyperplasia, indicating that it is a heterogeneous disease (592). Although most cases were initially thought to be sporadic (598–601), there are now several reports of familial cases whose presentation suggests autosomal dominant transmission (602–610). Germline mutations of *ARMC5* account for approximately 25 percent of apparently sporadic cases (611–614). Activating mutations of Gs alpha are responsible for macronodular adrenal hyperplasia in the context of McCune-Albright syndrome (615,616), however, these mutations are also occasionally found in the nodules of pediatric and adult patients without any other features of McCune-Albright syndrome (583,617). Other syndromic associations include multiple endocrine neoplasia syndrome type 1 (MEN1), familial adenomatous polyposis (592,618), and hereditary leiomyomatosis and renal cell carcinoma (HLRCC) (619).

The diagnosis of bilateral adrenocortical nodular hyperplasia can be made in the following scenarios: overt Cushing syndrome and suppressed ACTH levels with bilaterally enlarged nodular adrenal glands found on abdominal CT, or incidentally found bilateral adrenal hyperplasia with several macronodules and subnormal suppression of cortisol following an overnight dexamethasone suppression test (584,620).

Immunohistochemical and in situ hybridization studies by Sasano et al. (621–623) showed immunoreactivity for steroidogenic enzymes in the nodular cortical cells but not in the atrophied adrenal cortex. Immunoreactivity for P450 cholesterol side chain cleavage and P450 21-alpha-hydroxylase was observed in both clear and compact cells, while 3-beta-hydroxysteroid dehydrogenase was present only in clear cortical cells (622,623). The conclusion is that ineffective corticosteroidogenesis may contribute to the relatively low production of cortisol (622).

Clinical and Radiologic Features. Patients with bilateral macronodular adrenal hyperplasia and Cushing syndrome typically present in the fifth and sixth decades (598–601), a later age of onset compared with unilateral adrenal cortisol-producing adenomas, primary pigmented nodular adrenocortical disease, or pituitary corticotroph adenomas (Cushing disease). The sex distribution is variable from one series to the next but is probably equal. The time to diagnosis is usually between 1 to 2 years.

Most patients have only mild cortisol production and present with bilateral adrenal incidentalomas (584). In these patients, no symptoms of Cushing syndrome are present, and the nodular adrenal glands are identified when abdominal imaging is performed for unrelated reasons. In other cases, abdominal CT shows massively enlarged adrenal glands (fig. 4-20). CT and MRI of the pituitary gland are within normal limits. In patients in whom cortisol secretion is sufficient to produce overt Cushing syndrome, the symptoms and signs are similar to those seen with other etiologies of primary adrenal Cushing syndrome.

Gross Findings. The adrenal glands are markedly enlarged, with weights ranging from 25 to 500 g (normal range, 4 to 6 g per gland). The glands contain multiple nonpigmented nodules that are greater than 10 mm in diameter (598,599,600). The nodules may vary in size, reaching over 40 mm in diameter in some cases. The glands are coarsely nodular, with discrete nodules admixed with loosely aggregated nodules. The cut surfaces of the nodules are golden yellow, with tan areas. Nodules are unencapsulated and there may be compression of the adjacent cortex as well as the medulla.

Microscopic Findings. The nodules are composed of clear cells, compact cells, or admixtures of these cell types with variable amounts of lipid (fig. 4-21). Some cells appear vacuolated; mitotic figures and cellular pleomorphism are uncommon. Myelolipomatous change may be present, which is probably a marker of cortical hyperactivity. Osseous metaplasia may also be present. The cortex between the nodules is either hyperplastic or atrophic (582,622,623).

Differential Diagnosis. The differential diagnosis includes ACTH-dependent macronodular adrenal hyperplasia secondary to Cushing disease

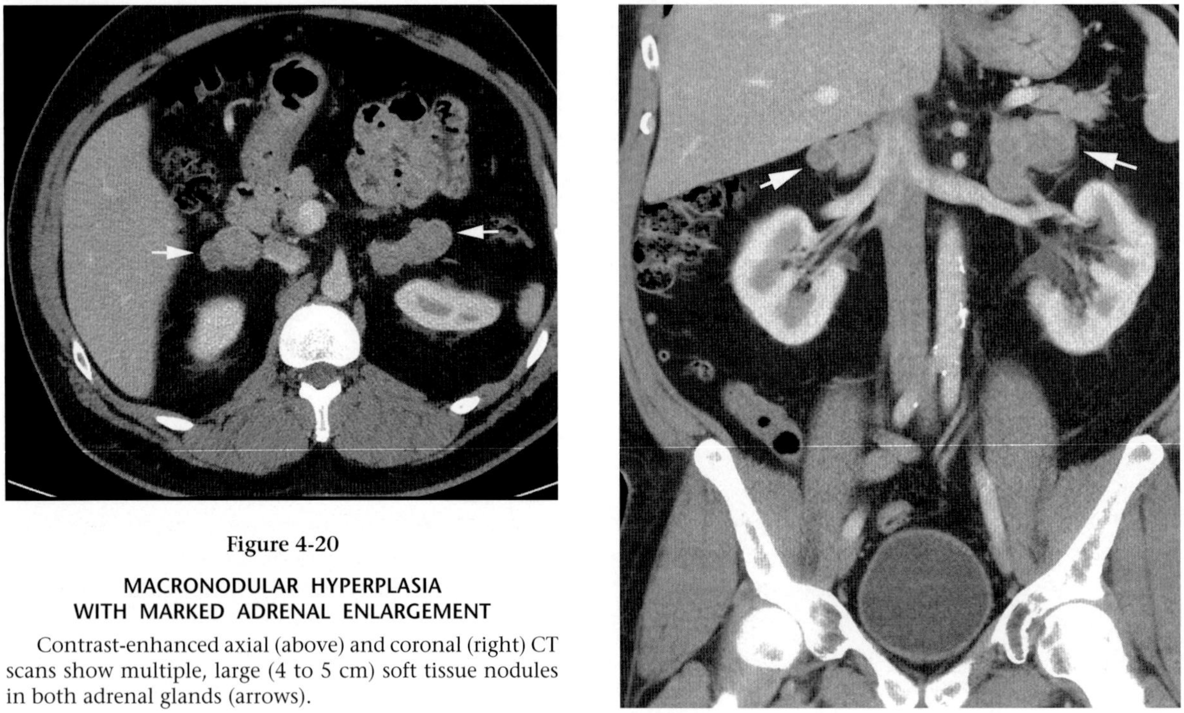

Figure 4-20

MACRONODULAR HYPERPLASIA WITH MARKED ADRENAL ENLARGEMENT

Contrast-enhanced axial (above) and coronal (right) CT scans show multiple, large (4 to 5 cm) soft tissue nodules in both adrenal glands (arrows).

or ectopic ACTH secretion. Clinicopathologic correlation helps make the distinction in difficult cases. Plasma ACTH is suppressed by excess cortisol in primary macronodular adrenal hyperplasia, whereas ACTH is inappropriately normal or slightly elevated in ACTH-dependent secondary hyperplasia. Nodules are present in both cases, but the glands are much smaller in ACTH-dependent nodular hyperplasia. However, in a few cases of macronodular hyperplasia with marked adrenal enlargement, one adrenal gland may weigh only 15 g or less (582). In ACTH-dependent nodular Cushing syndrome, the adrenal glands are seldom more than 15 g each, and the entire gland is hyperplastic, while atrophic areas may be present in ACTH-independent macronodular hyperplasia with marked adrenal enlargement.

Treatment and Prognosis. Bilateral surgical adrenalectomy is the treatment of choice for patients with overt Cushing syndrome and equally enlarged macronodular adrenal glands. This treatment is uniformly effective in these patients. In patients with moderately increased cortisol production (the majority of cases) but with clinical evidence of cortisol excess, unilateral adrenalectomy is recommended as it can restore cortisol levels to normal (624–627). However, as the cell mass subsequently increases in the contralateral adrenal gland, medical therapy with steroidogenesis inhibitors or a second adrenalectomy may become necessary (624–626, 628,629). In patients with mild cortisol excess, clinical follow-up and biochemical assessment every 6 to 12 months and yearly CT scans are sufficient, since this is a benign process that has not been shown to become malignant.

In some patients in whom aberrant hormone receptors have been identified, specific pharmacologic therapies blocking the aberrant receptors may be effective as alternatives to adrenalectomy. For example, beta-blocker therapy is used in patients with aberrant beta-adrenergic receptors (590,596,630), and gonadotropin-releasing hormone agonists are used for patients with luteinizing hormone or human chorionic gonadotropin receptors (631,632). Despite normalization of cortisol secretion with these therapies, no tumor regression has been observed.

Primary Pigmented Nodular Adrenocortical Disease

Definition. *Primary pigmented nodular adrenocortical disease* (PPNAD) is a form of ACTH-independent Cushing syndrome characterized by

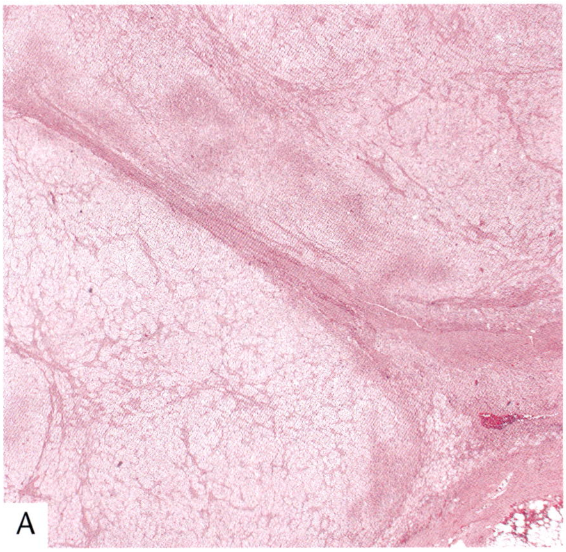

Figure 4-21

MACRONODULAR HYPERPLASIA WITH MARKED ADRENAL ENLARGEMENT

A: This histologic section accentuates the nodules and the distortion of the gland.

B: The nodules are composed of lipid-rich fasciculata-type cells and some lipid-depleted cells.

C: There is atrophy between the lipid-rich nodules. Fatty metaplasia is also present focally, which is usually associated with hyperactivity.

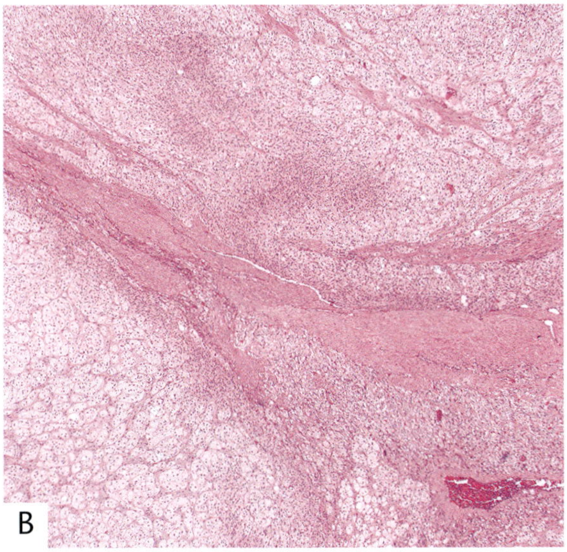

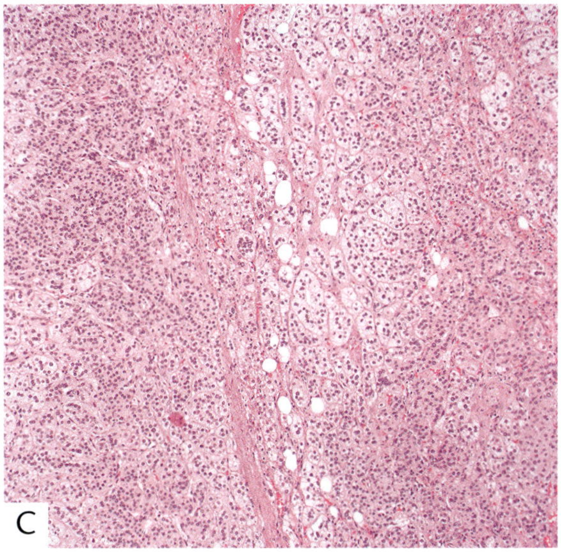

bilateral micronodular hyperplasia of pigmented nodules in the adrenal cortex. Synonyms include *adrenocortical dysplasia* and *bilateral micronodular hyperplasia*.

General Features. PPNAD is a rare disorder, accounting for fewer than 1 percent of cases of Cushing syndrome. It occurs as a sporadic (approximately 33 percent) or familial disorder (approximately 66 percent), either as part of Carney complex or as isolated PPNAD. Patients with the nonfamilial sporadic form of PPNAD are young (always less than 30 years of age) and may be infants (633,634).

Carney Complex. Carney complex or syndrome is an autosomal dominantly inherited multiple neoplasia syndrome characterized by spotty skin pigmentation (pigmented lentigines and blue nevi on the face, neck, and trunk); endocrine tumors, including PPNAD, testicular large cell calcifying Sertoli cells, growth-hormone-secreting PitNETs, thyroid adenomas and carcinomas, and ovarian cysts; and nonendocrine tumors, including cutaneous myxomas, breast ductal adenomas, atrial myxomas, psammomatous melanotic schwannomas, and osteochondromyxomas (633,635–643).

In one series of 88 patients with PPNAD, 40 patients had Carney complex (633). Of these, 32 (80 percent) had pigmented skin lesions or cutaneous myxomas, 29 (72 percent) had one or more cardiac myxomas, and 18 (45 percent) had Cushing syndrome. Breast masses or testicular tumors were seen in 10 women (42 percent of the female patients) and 9 men (56 percent of the male patients). Overt Cushing syndrome caused by PPNAD occurs in 25 to 45 percent of all patients with the Carney complex; subclinical, atypical, or periodic Cushing syndrome occurs in others; and histologic changes in the adrenal cortex are found at autopsy in almost all patients (633,642).

Linkage analysis has isolated three genetic loci for the Carney complex, at 2p16 (CNC2), 17q22-24 (CNC1), and at 17p12-13.1 (644,645). To date, four responsible disease genes have been identified: protein kinase A regulatory 1-alpha subunit (*PRKAR1A*), phosphodiesterase 11A isoform 4 gene (*PDE11A*), phosphodiesterase 8B gene (*PDE8B*), and myosin heavy chain gene (*MYH8*) (646–658).

Clinical and Radiologic Features. Clinically evident Cushing syndrome is present in 84 percent of patients with PPNAD, while 6 percent have only biochemical evidence of adrenocortical autonomy *(subclinical PPNAD)*, and 10 percent have *latent PPNAD* without a firm diagnosis. Patients with Cushing syndrome present with the typical signs and symptoms of hypercortisolism, including weight gain, obesity, hypertension, and menstrual cycle disorders. Presentation occurs before the age of 30 years; 50 percent present before 15 years of age (633, 639,643). Short stature due to stunted growth occurs in about one third of patients due to the early onset of hypercortisolism.

In many patients with PPNAD, the signs and symptoms of hypercortisolism are subtle and develop slowly over years (633). In others, the hypercortisolism is irregular (659) or cyclic (660,661), with either progressive or rapidly appearing Cushing habitus, followed by periods of remission. In these patients, establishing the diagnosis of Cushing syndrome may be challenging, because plasma ACTH levels may be incompletely suppressed. In one study of patients with the sporadic form of PPNAD, the median time from manifestation of symptoms until diagnosis was 1 year, but intervals of up to 18 years were also reported (662).

The appearance of the adrenal glands on imaging in patients with PPNAD is often initially interpreted as normal, which is different from other ACTH-independent disorders where relatively large tumors are seen (633,663,664). In PPNAD, the adrenal gland is often not enlarged, but instead is occupied by several small nodules spread in an otherwise atrophic cortex; this can be seen as a "string of beads" on thin-section high-resolution CT scan (665).

Laboratory Findings. Plasma cortisol is usually moderately elevated but without a diurnal rhythm. Plasma ACTH is low or undetectable, consistent with ACTH-independent disease. In addition, patients with PPNAD may have cyclic or episodic hypercortisolism. The hypercortisolism is resistant to dexamethasone suppression, metyrapone stimulation, and CRH stimulation.

Gross Findings. The adrenal glands are normal or slightly enlarged, weighing between 4 and 17 g (633). The surface has scattered pigmented nodules ranging from 1 to 4 mm in diameter. A wide range in the size and distribution of the pigmented nodules is noted. Most nodules are less than 4 mm, unencapsulated, but sharply demarcated from the adjacent atrophic cortex. They are commonly black, but some may be tan or yellow. A few macronodules up to 3 cm in diameter are seen, and some of these may not be pigmented.

Microscopic Findings. The nodules are round to oval or irregularly shaped, and unencapsulated but well demarcated from the adjacent cortex. They are typically located at the corticomedullary junction, but can extend into the periadrenal fat or involve the entire thickness of the cortex. The cells in the nodules are large and globular, with eosinophilic or clear cytoplasm; many of the cells contain coarsely granular brown pigment, identified as lipofuscin (633). Occasional binucleated or multinucleated cells are seen, but mitoses are uncommon. The adjacent cortex is atrophic and lacks the usual adrenal zonation (633,636,639,642,663).

Frequently, lipomatous or myelolipomatous change is associated with increased activity of the glands in Cushing syndrome. Lymphocytic infiltrates of mixed, cytotoxic, and T-helper cells (662) are often present.

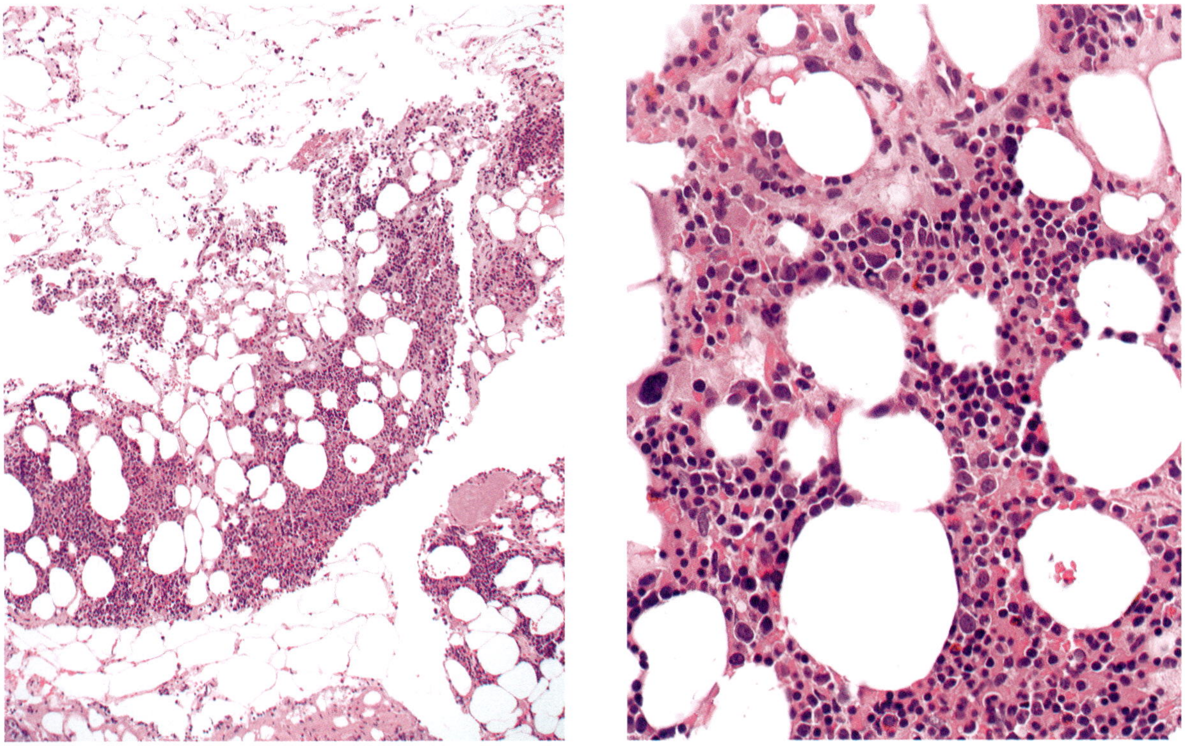

Figure 4-22

MYELOLIPOMA

Left: Myelolipoma is a tumor-like condition developing commonly in the adrenal gland. It is composed of mature adipose tissue and hematopoietic elements.

Right: Myelolipoma with hematopoietic elements, including megakaryocytes and erythroid precursors.

Immunohistochemical Findings. Diffuse synaptophysin expression in PPNAD nodules suggests a neuroendocrine phenotype (633,642, 666). Studies by Sasano et al. (667,668) with antibodies to steroid metabolizing enzymes show a marked increase in immunoreactivity in the nodules, suggesting increased glucocorticoid synthesis. This may explain, in part, the Cushing syndrome observed with these relatively small adrenal glands and the adjacent areas of atrophy.

Ultrastructural Findings. Adrenal cortical cells with ultrastructural features of zona reticularis–type cells are admixed with zona fasciculata–type cells. Both types contain abundant lipofuscin-type bodies, which would explain the dark pigment noted grossly and microscopically (669).

Differential Diagnosis. The findings in PPNAD are pathognomonic. One source of potential confusion is a metastatic malignant melanoma, especially if both adrenal glands are involved. If necessary, immunohistochemical staining for S-100 protein and HMB-45 can readily separate these two diseases.

Myelolipomatous metaplasia with metaplastic fat and hematopoietic precursors can be seen in the adrenal glands of patients with Carney complex, but are also seen in cases of ACTH-dependent and -independent Cushing syndrome, suggesting that these changes may be dependent on increased glucocorticoid activity. Myelolipomatous change should be distinguished from myelolipomas, which are tumefactive masses of myeloid and fat cells occurring in varying proportions (fig. 4-22). Myelolipomas are not associated with increased glucocorticoid production.

Treatment and Prognosis. Bilateral adrenalectomy is the treatment of choice for patients with either sporadic or familial PPNAD (633,636,665). Subtotal or unilateral adrenalectomy should not be performed, except in the rare occurrence of a large unilateral functional adrenal nodule with a contralateral normal-sized gland. Unilateral adrenalectomy can lead to recurrence of Cushing syndrome (633) even many decades after treatment (670).

Aldosterone Excess Due to Adrenocortical Hyperplasia

Definition. *Primary hyperaldosteronism* is the excessive secretion of aldosterone from the adrenal glands and is associated with suppression of plasma renin activity.

General Features. At least six subtypes of primary hyperaldosteronism have been recognized, including aldosterone-producing adenoma, idiopathic hyperaldosteronism, primary adrenal hyperplasia, aldosterone-producing adrenocortical carcinoma, aldosterone-producing ovarian tumor, and familial hyperaldosteronism (671). Familial hyperaldosteronism (FH) is subdivided into three groups: FH-I (glucocorticoid-remediable hyperaldosteronism), FH-II (aldosterone-producing adenoma and idiopathic hyperaldosteronism), and FH-III (hypertension and massive adrenocortical hyperplasia) (672).

Primary hyperaldosteronism is most often associated with adrenocortical adenomas; however, in approximately 40 percent of cases, the only apparent adrenal abnormality is hyperplasia of the zona glomerulosa, with or without the formation of micronodules (673–676). Although the pathogenesis of zona glomerulosa hyperplasia is uncertain, possible causes include other products from the *POMC* gene including α-melanocyte-stimulating hormone (MSH), β-MSH, and β-endorphin (677,678). The distinction between diffuse and nodular hyperplasia is important, because the latter has recently been shown to represent a spectrum of monoclonal lesions with somatic mutations, thus representing a neoplastic rather than a hyperplastic process.

The biochemical abnormalities and clinical manifestations in patients with hyperplasia are less severe than in those with adenomas (679,680). The hypertension caused by bilateral hyperplasia is usually not cured by total bilateral adrenalectomy and should be treated medically (680). One type of bilateral adrenal hyperplasia has been described as primary hyperplasia, because unilateral or subtotal (75 percent) adrenalectomy results in permanent cure (677). Approximately 60 percent of cases of primary hyperaldosteronism are idiopathic and result in bilateral hyperplasia of the zona glomerulosa (681).

Secondary hyperaldosteronism results from activation of the renin-angiotensin system and increased secretion of aldosterone in response to increased levels of circulating renin and angiotensin, which are secondary to extra-adrenal disease. Etiologies include hyperthyroidism, renal artery stenosis, and malignant hypertension. Patients have a low plasma sodium level, hyperplasia of the renal juxtaglomerular apparatus, and an increase in the width of the zona glomerulosa (682).

Tertiary aldosteronism (Bartter syndrome) is presumed to result from chronic elevations in plasma renin levels with accompanying elevated angiotensin II and aldosterone. Patients may present in infancy or during adult life. Some cases are inherited as part of an autosomal recessive condition with characteristic sets of metabolic abnormalities, including hypokalemia, metabolic alkalosis, hyperreninemia, hyperplasia of the juxtaglomerular apparatus, and hyperaldosteronism (683–687). It is not associated with hypertension (688). The adrenal changes in tertiary aldosteronism are not well characterized. Cases in which bilateral adrenalectomy were performed to control the hypokalemia show grossly normal adrenal glands, and biopsy specimens reportedly show hyperplasia of the zona glomerulosa (689).

Clinical and Radiologic Features. The classic presenting signs of primary hyperaldosteronism are hypertension and hypokalemia, but potassium levels are frequently normal in modern-day cases. When compared with patients with idiopathic hyperplasia, patients with aldosterone-producing adenomas tend to have more severe hypertension and are more frequently recognized to be hypokalemic. Other presenting signs and symptoms include frontal headache, muscle weakness, and flaccid paralysis caused by hypokalemia. The peak incidence is between 30 and 50 years of age.

The evaluation of a patient with possible primary hyperaldosteronism begins with measurement of the plasma renin activity (or plasma renin concentration) and plasma aldosterone concentration. Primary hyperaldosteronism is confirmed by a lack of suppression of aldosterone with volume expansion. The plasma renin activity is typically very low. Adrenal CT should be the initial study to determine the cause of hyperaldosteronism (adenoma versus hyperplasia). Bilateral adrenal gland thickening or micronodular changes suggest adrenal hyperplasia; however, patients with hyperplasia may also have normal-appearing adrenal glands on CT (690). High-resolution CT detects nodules as small as 5 mm in diameter (691). These nodules may be difficult to distinguish from small adenomas (fig. 4-23) (692). Similarly, the presence of a unilateral adrenal mass on CT does not exclude bilateral hyperplasia (693–697).

Gross Findings. The gross features of the adrenal glands are variable. They may be of normal size, weight, and appearance; less than normal weight; or increased size and weight. Macronodular glands have nodules between 0.25 and 1.0 cm in diameter but may be up to 3.0 cm. Larger nodules may be present at one pole or within the gland. Although the macronodules are usually multiple, a few may be single and simulate an adenoma. Cut sections show yellow nodules intraglandularly and intracortically, including adjacent to the central vein.

Microscopic Findings. Hyperplasia of the glomerulosa is characterized by thickening of this cell layer, with tongue-like projections of the glomerulosa extending toward the fasciculata. Micronodules, when present, are usually composed of clear fasciculata-type cells (676,698), and are thought to be a consequence of the associated hypertension (fig. 4-24). The hyperplastic changes may be focal, and many sections are often needed to make the correct diagnosis (699,700).

In patients treated with spironolactone, spironolactone bodies may be seen in the cells of the zona glomerulosa as well as in the zona fasciculata (fig. 4-25). Intermediate-type cells with eosinophilic and lipid-containing cytoplasm may be present below the hyperplastic zona glomerulosa. Columns of these cells extend through the fasciculata to the reticularis.

Differential Diagnosis. In approximately 10 percent of cases, it is not possible to distinguish a micronodule from a true adenoma associated with aldosterone production. The presence of one nodule usually indicates an adenoma, which is often unilateral. Nodular hyperplasia is usually a bilateral disorder, although it is now known to reflect a neoplastic process rather than true physiologic hyperplasia.

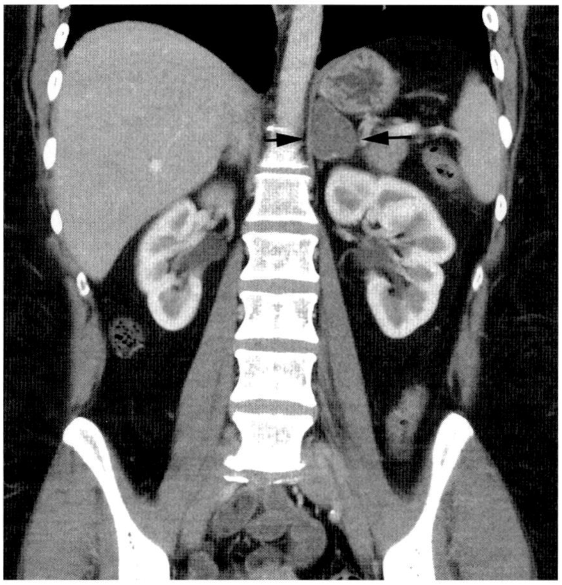

Figure 4-23

ADRENAL CORTICAL ADENOMA

Contrast-enhanced CT scan of the abdomen shows a left adrenal cortical adenoma as a low density, moderately enhancing mass. It may be difficult to distinguish adrenal adenomas from nodular hyperplasia with a dominant nodule.

In FH-III, the resected adrenal glands are massively enlarged (combined weights up to 82 g). There is hyperplasia of a single cortical compartment, with features of fasciculata or transitional cells but with atrophy of the glomerulosa (672).

Treatment and Prognosis. Surgery is the preferred treatment for patients with aldosterone-producing adenomas or unilateral hyperplasia. Unilateral adrenalectomy in these patients induces a marked reduction in aldosterone secretion and correction of hypokalemia in almost all patients (701–703). Hypertension is improved in all and is cured in 35 to 60 percent of patients (701–706). Some patients with primary bilateral hyperplasia may be treated with subtotal adrenalectomy. Laparoscopic adrenalectomy should be considered in patients with a unilateral hypodense macroadenoma detected on CT. In many cases, however, adrenal venous sampling is required to distinguish between unilateral and bilateral disease (707).

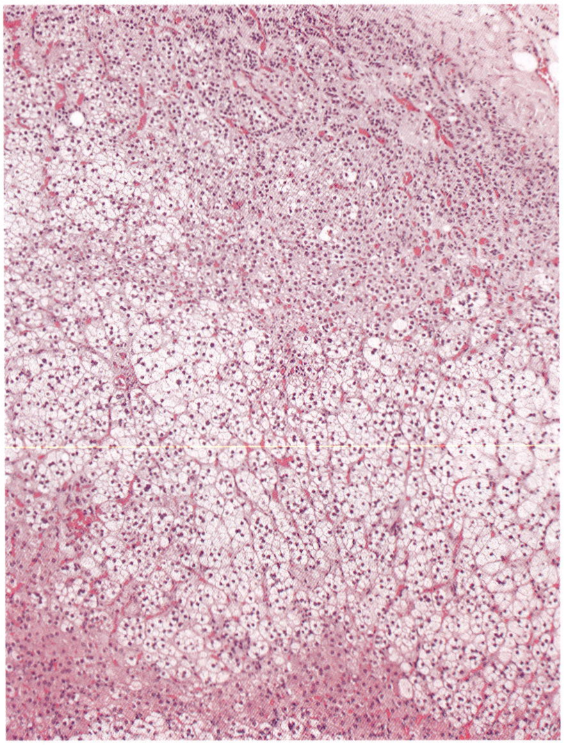

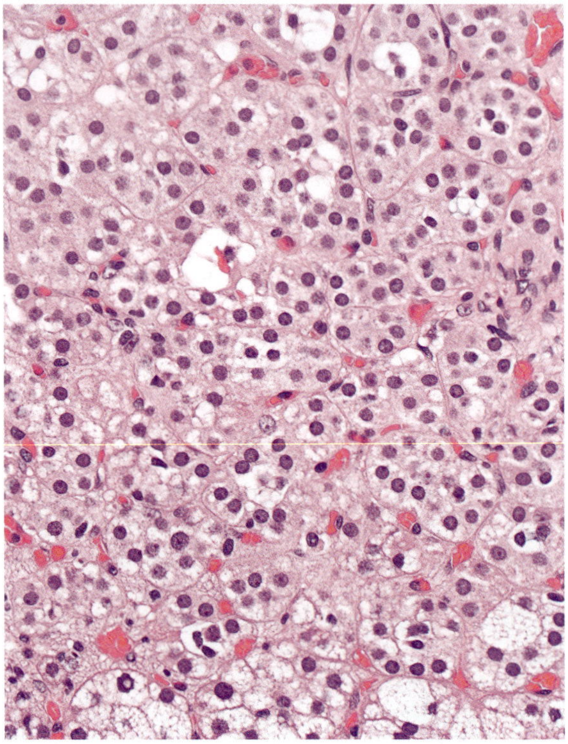

Figure 4-24

ZONA GLOMERULOSA NODULAR HYPERPLASIA

Left: The histologic view shows diffusely hyperplastic glomerulosa- and fasciculata-type cells.

Right: The hyperplastic cells are arranged in nests and cords, with a transition between glomerulosa- and fasciculata- or "hybrid-" type cells.

Figure 4-25

ZONA GLOMERULOSA HYPERPLASIA

Multiple spironolactone bodies are seen as eosinophilic inclusions in a patient treated with spironolactone before surgery.

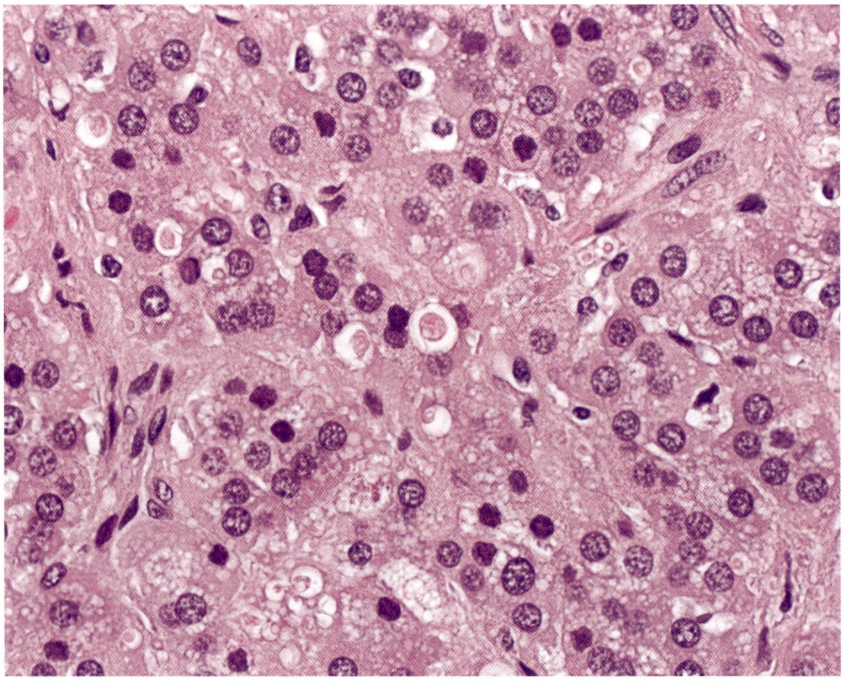

Medical therapy is the preferred treatment for patients with bilateral adrenal hyperplasia. Optimal treatment consists of mineralocorticoid receptor blockade with spironolactone. In patients with persistent hypertension, an additional antihypertensive agent, such as hydrochlorothiazide or an angiotensin-converting enzyme (ACE) inhibitor is used. Amiloride, a potassium-sparing diuretic, can be used in patients who are intolerant to spironolactone (708). Amiloride works by blocking the aldosterone-sensitive sodium channel in the collecting tubules, thereby lowering the blood pressure and raising the serum potassium concentration (708,709). However, amiloride is not recommended for first-line therapy because of persistence of the effect of the hyperaldosteronism at the mineralocorticoid receptor, with its possible deleterious cardiovascular effects.

The morbidity and mortality associated with primary hyperaldosteronism are primarily related to hypokalemia and hypertension (710,711). Hypokalemia, especially if severe, causes cardiac arrhythmias, which can be fatal. Complications from chronic hypertension include myocardial infarction, cerebrovascular disease, and congestive heart failure. Treatment can also lead to drug reactions and complications from surgery. The prognosis of patients with adenomas treated by unilateral adrenalectomy is usually much better than that of patients with hyperplasia.

Unilateral Adrenal Cortical Hyperplasia

There are a few reported cases of *unilateral adrenal cortical hyperplasia* in patients with Cushing syndrome or with hyperaldosteronism (712–717). This disorder is extremely uncommon, and the possibility of bilateral disease, with only a few microscopic nodules involving the other adrenal gland in an ACTH-independent manner must always be excluded. When unilateral nodular hyperplasia is diagnosed in the presence of low plasma ACTH levels, the most likely diagnosis is an adenoma (714,715). Careful examination of the attached adrenal tissue and contralateral adrenal gland for evidence of atrophy can usually help to determine whether there is a functional adenoma. The periodic production of ACTH by some pituitary and ectopic tumors may cause a cyclically functional tumor to be missed. An exception to the existence of unilateral hyperplasia occurs when ectopic ACTH production and inhibition of pituitary ACTH production, due to a metastasis to one adrenal gland, results in atrophy of the contralateral adrenal gland (717).

Congenital Adrenal Hyperplasia

Definition. *Congenital adrenal hyperplasia* (CAH) is a group of disorders associated with adrenocortical hyperplasia caused by inherited defects in cortisol biosynthesis. The syndromes of CAH (*congenital adrenogenital syndromes*) result from a series of autosomal recessive enzymatic defects in the biosynthesis of adrenal steroids (718,719). The resultant decrease in cortisol production negatively feeds back on pituitary ACTH secretion, leading to compensatory adrenal cortical hyperplasia.

General Features. Although first described more than 135 years ago (720), the pathophysiology of CAH was only discovered much later (721). The most common cause of CAH, which accounts for approximately 95 percent of cases, is 21-hydroxylase (CYP21A2) deficiency (718,722,723). The worldwide incidence of classic 21-hydroxylase deficiency is 1 in 14,500 births, with a heterozygote frequency of approximately 1 in 60 (724–727). The microsomal enzyme 21-hydroxylase is responsible for conversion of 17-hydroxyprogesterone to 11-deoxycortisol. With defects in enzymatic activity, there is decreased cortisol biosynthesis and accumulation of cortisol precursors, which are then converted to adrenal androgens. The gene for CYP21A2 is located on chromosome 6 within the major histocompatibility locus. Genetic alterations include deletions and other gene defects (728).

Deficiency of 11-beta-hydroxylase (CYP11B1) accounts for about 5 percent of all cases of CAH and is associated with increased production of androgens and deoxycorticosterone (729–731). It is seen in approximately 1 in 100,000 live births. This deficiency inhibits the conversion of 11-deoxycortisol and deoxycorticosterone to cortisol and corticosteroid, resulting in excessive androgen production and virilization of female fetuses. The disorder is caused by a mutation in the gene on chromosome 8 (718).

Other less common deficiencies include: 17-alpha-hydroxylase (CYP17) deficiency, in which many mutations are present in the *CYP17* genes; 3-beta-hydroxysteroid dehydrogenase deficiency, in which synthesis of all steroid hormone classes are impaired due to a mutation in the hydroxysteroid dehydrogenase 32 gene; and a deficiency of all adrenal and gonadal steroid hormones causing congenital lipoid adrenal hyperplasia, the rarest form of congenital hyperplasia. These patients have low cortisol and aldosterone secretion rates and increased activity of ACTH, follicle-stimulating hormone (FSH), luteinizing hormone (LH), and plasma renin. Multiple genetic mutations are present in a gene on chromosome 8.

Clinical and Radiologic Features. On clinical examination, affected patients have abnormalities of sexual development, salt wasting, hypertension, or acute adrenal insufficiency, depending on the specific defect. In females, common abnormalities associated with these defects include masculinization in utero, postpubertal masculinization, and primary amenorrhea (727). Males show signs of precocious puberty or pseudohermaphrodism. Patients with CAH are also prone to developing adrenal myelolipomas.

With 21-hydroxylase deficiency, the clinical spectrum ranges from mild to severe, depending on the degree of the deficiency. Three main clinical phenotypes have been described: classic salt-losing, classic non-salt-losing (simple virilizing), and nonclassic (late-onset) (732). Females with the classic form (salt-losing and non-salt-losing) present with genital atypia. Males with the salt-losing form who are not identified by neonatal screening present with failure to thrive, dehydration, hyponatremia, and hyperkalemia typically at 7 to 14 days of life. Males with the classic non-salt-losing form who are not identified by neonatal screening typically present at 2 to 4 years of age with early virilization (pubic hair, growth spurt, adult body odor). The nonclassic, or late-onset, form presents as early pubarche or sexual precocity in school-age children, hirsutism and menstrual irregularity in young women, or without symptoms (732).

Patients with 11-beta-hydroxylase deficiency present with androgen excess in the neonatal period, leading to ambiguous genitalia and later to hypertension. In contrast, 17-alpha-hydroxylase deficiency is diagnosed at the time of puberty because of hypertension, hypokalemia, and hypogonadism. Females usually have primary amenorrhea and absent secondary sexual characteristics, while males have pseudohermaphrodism with female external genitalia and intra-abdominal testes (733). Patients with 17-alpha-hydroxylase deficiency have increased levels of corticosteroid and deoxycorticosteroid. Patients with 3-beta-hydroxysteroid dehydrogenase deficiency present in early infancy with adrenal insufficiency, varying degrees of virilization in females, and problems with genital development. With congenital lipoid adrenal hyperplasia, patients have severe adrenal insufficiency during the neonatal period, with hyponatremia, hypokalemia, vomiting, and diarrhea.

Radiologic studies show enlarged adrenal glands and may also reveal incidental myelolipomas, which are overrepresented in patients with CAH. High-resolution CT scans and MRI studies detect even slightly enlarged glands in patients with the milder forms of CAH (734,735). Adrenal ultrasonography (showing adrenal limb width greater than 4 mm, lobulated surface, or abnormal echogenicity) has 92 percent sensitivity and 100 percent specificity in identifying neonates and children with untreated CAH (736).

Laboratory Findings. In many countries, including the United States, neonatal screening for 21-hydroxylase deficiency is an approved part of the neonatal screening program. The characteristic biochemical abnormality for diagnosis at any age is an elevated serum concentration of 17-alpha-hydroxylase, the normal substrate for 21-hydroxylase.

Patients with 21-hydroxylase deficiency also have elevated plasma ACTH and adrenal androgens including dihydroepiandrosterone (DHEA) and androstenedione. Patients with 11-beta-hydroxylase deficiency also have increased adrenal androgens and deoxycorticosterone. With 17-alpha-hydroxylase deficiency, there are increased plasma levels of corticosteroid and deoxycorticosteroid and decreased levels of cortisol, adrenal androgens, and gonadal steroids (737).

Gross Findings. The adrenal glands become markedly enlarged, with a mean weight at autopsy in untreated children of 15 g compared to the normal 1 to 4 g seen between the neonatal period and age 12 years. Single gland weight may reach as much as 30 g. The glands have a characteristic convoluted cerebriform appearance and a tan to dusky brown color. In lipoid CAH, however, the glands are pale yellow (727). Myelolipomas occur in up to 4 percent of patients with CAH and appear as unencapsulated, bright yellow adipose-like tissue with hemorrhagic foci.

Microscopic Findings. Histologically, there is diffuse or diffuse and nodular expansion of the zona fasciculata by populations of clear, lipid-containing cells and compact, lipid-depleted cells. A thin layer of clear cells from the outer zona fasciculata is usually seen next to the zona glomerulosa. In lipoid CAH, the cortical cells are vacuolated, with occasional formation of cholesterol clefts and an accompanying giant cell reaction.

In children, the zona glomerulosa is often hyperplastic, especially when the enzyme defect leads to decreased aldosterone secretion and the renin-angiotensin system is activated (738). Hyperplasia of the zona glomerulosa is not prominent in cases of 11-hydroxylase deficiency, although the remainder of the gland may be enlarged.

Glands from neonates and infants with CAH usually have a hyperplastic definitive zone consisting of compact cells with lipid-poor eosinophilic cytoplasm. In addition, these glands show the predictable degeneration and involution of the fetal zone.

Testicular Involvement with 21-Hydroxylase Deficiency. Patients with CAH may have multiple masses, histologically similar to Leydig cell tumors within the testes and peritesticular areas. They are commonly bilateral and are most typically located in the hilar regions of the testes. The cell of origin of these nodules is uncertain, but the component lesional cells contain abundant eosinophilic cytoplasm. These tumors have been shown to be dependent on ACTH for growth (739); malignant tumors with metastases have not been reported (726), suggesting that they represent hyperplastic foci. A striking difference between these lesions and true Leydig cell tumors is that the Reinke crystalloids seen in up to 35 percent of Leydig cell tumors (740) are not present in testicular tumors arising in patients with CAH.

Differential Diagnosis. The diagnosis of CAH is usually made clinically. If the clinical history is not available, the histopathologic appearance of the glands may be confused with bilateral hyperplasia secondary to ectopic ACTH production. The cerebriform appearance of the glands is more in keeping with CAH. Both disorders have prominent compact cells extending from the medulla to the zona glomerulosa. However, foci of clear cells are more commonly seen in ectopic ACTH hyperplasia. The presence of microscopic deposits of metastatic carcinoma in ectopic ACTH hyperplasia helps confirm that diagnosis.

Treatment and Prognosis. The goals for treatment of classic 21-hydroxylase deficiency are to prevent adrenal crisis and optimize growth, sexual maturation, and reproductive function. This is accomplished by replacing glucocorticoid and mineralocorticoid insufficiency to reduce the associated excessive CRH and ACTH secretion and hyperandrogenemia. Mineralocorticoid replacement is especially important in patients with the salt-losing form of the disorder to maintain normal serum electrolyte concentrations and extracellular fluid volume, but also may be beneficial for patients with classic non-salt-losing CAH.

Prenatal therapy with glucocorticoids, via maternal administration of dexamethasone, is an experimental approach aimed at reducing the virilization of an affected 46,XX fetus (723,741–743). If prenatal treatment is initiated sufficiently early (before the ninth postmenarchal week), approximately 85 percent of affected 46,XX infants are born with normal genitalia and the remainder are only slightly virilized (743). However, this intervention does not change the need for lifelong hormonal therapy. Reproductive function in women is impaired in untreated cases. Strict adherence to glucocorticoid therapy can correct the dysregulation of cyclic ovulatory function.

Infants with atypical genitalia require urgent medical attention. An appropriate therapeutic plan, usually including reconstructive surgery in females, should be developed after complete evaluation by an experienced interdisciplinary team of endocrinologists, surgeons, and mental health professionals with expertise in managing the psychosocial aspects of disorders of sex development.

The prognosis for patients with 21-hydroxylase deficiency is generally good with treatment

(737). The outcomes of therapy can be measured by evaluating growth and development, gonadal function and fertility, bone density, and quality of life. Mortality is increased threefold in patients aged 1 to 4 years, often because of an adrenal crisis after an infection (744). In adults, mortality rates are elevated compared with a healthy population; the main causes of death are adrenal crisis and cardiovascular disease.

Treatment of 11-beta-hydroxylase deficiency is by replacement of glucocorticoids. Genital malformation in females must be surgically corrected. 17-beta-hydroxylase and 3-beta-hydroxysteroid dehydrogenase deficiencies are also treated by replacement therapy.

Congenital lipoid adrenal hyperplasia is fatal during infancy in two thirds of reported patients (745), but some patients survive (746,747). Glucocorticoid and mineralocorticoid deficiency is managed by replacement therapy as in any form of adrenal insufficiency (745). At the time of puberty, estrogen therapy is provided to complete the development of secondary sexual characteristics, as all patients are phenotypically female.

Beckwith-Wiedemann Syndrome

Definition. *Beckwith-Wiedemann syndrome* (BWS) is a pediatric overgrowth disorder involving a predisposition to tumor development (748). Patients typically present with visceromegaly, gigantism, macroglossia, abdominal wall defects, craniofacial abnormalities, midfacial hypoplasia, and adrenal cortical hyperplasia. However, the clinical presentation is highly variable, and some cases lack the characteristic features originally described by Beckwith and Wiedemann (749,750). The syndrome is also referred to as the *exophthalmos, macroglossia, gigantism syndrome.*

General Features. BWS is a panethnic disorder with an estimated population prevalence of 1 in 10,300 to 13,700 (751,752). This figure most likely represents an underestimate because milder phenotypes may not be recognized. The prevalence is equal in males and females, with the notable exception of an increased frequency of female monozygotic twins versus male monozygotic twins (753). BWS usually occurs sporadically (85 percent), but familial transmission is seen in approximately 15 percent of cases. Subfertility and/or use of assisted reproductive technology (ART) is associated with an increased risk of imprinting disorders (754); one study found a 10-fold increased risk of BWS in live births after ART (755).

Dysregulation of imprinted gene expression in the chromosome 11p15.5 region can result in the BWS phenotype (756–758). The critical BWS genes in that region include insulin-like growth factor 2 (*IGF2*), *H19*, cyclin-dependent kinase inhibitor 1C (*CDKN1C*), potassium channel voltage-gated KQT-like subfamily member 1 (*KCNQ1*), and KCNQ1-overlapping transcript 1 (*KCNQ1OT1*, or long QT intronic transcript 1). A chromosome 11p15 molecular alteration is identified in only 80 percent of individuals with BWS. This is due, in part, to somatic mosaicism for some of the molecular alterations. Genomic loci outside of the chromosome 11p15.5 region have also been implicated in the etiology of BWS (759–761).

Clinical Features. The hallmark features of BWS are omphalocele (exomphalos), macroglossia, and macrosomia (gigantism) (749,750); however, there is significant clinical heterogeneity. Neonatal hypoglycemia can lead to brain damage, with mental retardation or even death in some cases (764). Children with BWS have an increased risk of neoplasia, usually Wilms tumor and hepatoblastoma, but also neuroblastoma, adrenocortical carcinoma, and rhabdomyosarcoma, as well as many other tumors, both malignant and benign (762). This increased risk for neoplasia is concentrated in the first 8 years of life; tumor development is uncommon after this age.

Individuals with uniparental disomy of 11p15.5 or gain of methylation at the *H19* imprinting center (IC1) carry the highest risk of tumor development, primarily for Wilms tumor and hepatoblastoma (16 and 30 percent, respectively), but also some benign tumors (756,759,761,763–767). Individuals with mutations in *CDKN1C* (p57Kip2) have a low but increased risk of tumor formation, with primarily neuroblastoma reported (768). Children clinically suspected to have, or diagnosed with, BWS who have no detectable molecular alteration have a significant risk for tumor development. This is probably due to somatic mosaicism for chromosome 11p15.5 uniparental disomy or other molecular anomalies. Tumor surveillance will probably become universally stratified by molecular etiology once reliable molecular methods are available, along with clinical correlation data from large studies of children with BWS (769).

Gross Findings. The adrenal glands are enlarged, with a combined weight of up to 16 g. They are often nodular and may have a cerebriform appearance (770).

Microscopic Findings. Adrenal cortical cytomegaly involving both adrenal glands, usually in the fetal cortex, is a common finding. The nuclei are large and pleomorphic, and cytoplasmic nuclear pseudoinclusions are often present. Hemorrhagic adrenal cortical macrocysts may be found (762,771). The chromaffin cells are also hyperplastic in the gland and at extra-adrenal sites (770,772).

ADRENAL MEDULLARY HYPERPLASIA

Definition. *Adrenal medullary hyperplasia* is characterized by an increase in the mass of the adrenal medullary cells and expansion of these cells into areas of the gland where they are not normally present, such as the tail.

General Features. Diffuse or nodular hyperplasia of the adrenal medulla is a distinct clinical and pathologic entity (773–775). The adrenal medulla normally comprises about 10 percent of the adrenal volume (774). This percentage is increased in sporadic and familial medullary hyperplasia. *Sporadic medullary hyperplasia* has a number of clinical associations, including: cystic fibrosis (776); sudden infant death syndrome (SIDS) (777); Cushing syndrome with adrenal cortical adenoma (777); hypertension in young patients (778); and the sporadic form of BWS (779,780). Other causes include drugs such as nicotine, hormones such as estrogen and growth hormone, and physical agents such as radiation (781–784).

Familial medullary hyperplasia is most commonly associated with MEN types 2a and 2b. MEN2 is an autosomal dominant disorder with an estimated prevalence of 1 per 30,000 in the general population (785). Both MEN2A and 2B are inherited with a very high degree of penetrance (786–790). The genetic defect in these disorders involves the *RET* proto-oncogene located on chromosome 10q, resulting in an activating mutation of the tyrosine kinase receptor (791–793). In both MEN2A and MEN2B syndromes, there is multicentric tumor formation in all organs where the *RET* proto-oncogene is expressed. The thyroid, parathyroid, and adrenal glands are at risk for developing tumors that may reduce life expectancy and quality of life. While medullary thyroid carcinoma and pheochromocytoma are seen in both MEN2A and MEN2B, parathyroid hyperplasia is usually associated with MEN2A while mucosal neuromas and ganglioneuromas are associated only with MEN2B (794,795).

Analysis of the specific mutations of the *RET* proto-oncogene in medullary carcinoma of the thyroid gland in patients with MEN2A, MEN2B familial medullary thyroid carcinoma (C-cell tumors not associated with the other stigmata of MEN2A or 2B), and sporadic medullary thyroid carcinoma has provided insight into the relationship of these conditions with respect to the location of the mutations. The most commonly mutated exons in MEN2A are exons 10 and 11, while mutations in exons 15 and 16 are most common in MEN2B (785,796).

Some patients with MEN2 syndrome present with adrenal medullary hyperplasia, which is considered a benign, generally noncatecholamine-overproducing lesion (797). The diagnosis of adrenal medullary hyperplasia should be made with considerable care. In the context of cortical atrophy, for example, the medulla may appear more prominent than usual. The distinction between adrenal medullary hyperplasia and pheochromocytoma is another controversial area (798–800). Molecular studies of microdissected nodules in patients with MEN2A-associated nodular adrenal medullary hyperplasia have shown that this disorder is a multifocal monoclonal proliferation. Interestingly, the same X chromosome is inactivated in individual nodules from the same patient. This observation has suggested an early clonal expansion of adrenal medullary precursors in these patients (801,802).

Clinical and Radiologic Features. The clinical features of patients with pheochromocytoma are well defined, while those of adrenal medullary hyperplasia are less clear, except in patients with familial diseases such as MEN2A and 2B. The most common findings in patients with pheochromocytoma are headaches (80 percent), excessive perspiration (71 percent), palpitations with or without tachycardia (64 percent), pallor (42 percent), followed by nausea, tremor, weakness, and exhaustion (803). Some of these signs and symptoms are present in patients with familial adrenal medullary hyperplasia, especially those with larger nodules approaching the size of true pheochromocytomas; however, it is often difficult to make the distinction between small pheochromocytomas and extensive nodular medullary hyperplasia with familial disease (773,804,805).

Radiologic studies of both adrenal medullary hyperplasia and pheochromocytoma have been enhanced by radionuclide scanning with iodinated meta-iodobenzylguanidine (MIBG) (806,807). Nodular hyperplasia can be readily detected by CT and MRI, especially by high resolution MRI. The localization technique should be done after biochemical diagnosis of the hyperplasia or pheochromocytoma (808). When evaluating a patient with previous surgery, MIBG scintigraphy is the preferred technique, since it is more sensitive than CT and is as sensitive as MRI but more specific with recurrent disease (806). The diagnosis of medullary hyperplasia may be suggested by the increased concentration of ^{131}I-MIBG in both glands when the abdominal CT is normal (807).

Laboratory Findings. The diagnosis of pheochromocytoma is usually made by demonstrating increased urinary excretion of catecholamines or catecholamine metabolites, including epinephrine, norepinephrine, vanillyl mandelic acid (VMA), and metanephrines. An elevated ratio of epinephrine to norepinephrine in the urine (774,776) and increased levels of tissue epinephrine are associated with medullary hyperplasia.

Gross Findings. The use of morphometric analysis may be required to diagnose adrenal medullary hyperplasia in subtle cases. Occasionally, the diagnosis is easily made on gross examination of the adrenal glands. Patients with MEN2A and 2B usually have diffuse or nodular hyperplasia, with multiple nodules involving both adrenal glands. In a study of 19 patients with MEN2A and 2B, most had bilateral gland involvement with diffuse or nodular hyperplasia, but normal glands or unilateral changes were present in some (773,809–811). The nodules and hyperplastic medulla often extend to both alae and the tail of the adrenal gland, a finding that is usually not present in the normal gland. The nodules are gray to tan, and they may compress the adjacent cortex. They do not have a capsule.

Microscopic Findings. The hyperplastic medullary cells may show various growth patterns: alveolar, trabecular, or solid. There may be progression from diffuse to nodular hyperplasia, with nodules of varying size. Some cases of diffuse hyperplasia are subtle. The presence of medullary tissue in both alae (normally present in only one) or in the tail of the gland (normally present only in the head and body), and a medullary volume greater than the normal 10 percent, point to medullary hyperplasia (774,812). Mixed cell types are common. The nodules may show nuclear atypia, but this is not associated with neoplasia or malignancy. In MEN2-associated medullary hyperplasia, hyaline globules may be present.

Differential Diagnosis. The distinction between nodular medullary hyperplasia and pheochromocytoma can be difficult. The use of a 1-cm cutoff size to separate nodular hyperplasia from small pheochromocytoma is arbitrary (773), since some pheochromocytomas are smaller than 1 cm. This is confounded by recent findings that benign adrenal nodules in patients with MEN2B can be monoclonal (801). It is probably best to consider adrenal medullary hyperplasia as a precursor lesion for pheochromocytoma and not as non-neoplastic hyperplasia (797).

The presence of unilateral versus bilateral disease has been suggested as a helpful distinguishing feature. However, unilateral medullary hyperplasia has been reported in sporadic as well as familial cases (810,811).

Treatment and Prognosis. Surgical resection is the recommended treatment for adrenal medullary hyperplasia (773,797). Such an approach may have the advantage of allowing cortex-sparing surgery, with less chance of recurrence compared to cortex-sparing surgery in pheochromocytoma, and may also lead to a lower risk of developing bilateral pheochromocytoma than thus far reported (813). Another advantage of early surgery for adrenal medullary hyperplasia is that lesions are often asymptomatic, which gives a lower chance for preoperative and intraoperative hypertensive crises (797).

Although the prognosis differs for patients with pheochromocytoma with MEN2A or 2B and sporadic tumors (809,814), there are no data to evaluate these differences. Some patients with von Recklinghausen disease may have adrenal medullary hyperplasia (784), and these patients are at increased risk for malignant tumors such as malignant nerve sheath tumors.

REFERENCES

Embryology

1. Beckwith JB, Perrin EV. In situ neuroblastomas: a contribution to the natural history of neural crest tumors. Am J Pathol 1963;43:1089-104.
2. Crowder RE. The development of the adrenal gland in man with special reference to origin and ultimate location of cell types and evidence in favor of the "cell migration" theory. Contributions to embryology, vol. 36, publication no. 251. Carnegie Institute of Washington; 1957:193-210.
3. Lalli E. Adrenal cortex ontogenesis. Best Pract Res Clin Endocrinol Metab 2010;24:853-64.
4. Kempna P, Fluck CE. Adrenal gland development and defects. Best Pract Res Clin Endocrinol Metab 2008;22:77-93.
5. Zajicek G, Ariel I, Arber N. The streaming adrenal cortex: direct evidence of centripetal migration of adrenocytes by estimation of cell turnover rate. J Endocr 1986;111:477-82.
6. Mesiano S, Jaffe RB. Developmental and functional biology of the primate fetal adrenal cortex. Endocr Rev 1997;18:378-403.
7. Turkel SB, Itabashi HH. The natural history of neuroblastic cells in the fetal adrenal glands. Am J Pathol 1974;76:225-44.
8. Pansky B. Review of medical embryology. Macmillan USA: Prentice Hall; 1982:180.
9. Neville AM. The adrenal medulla. In: Symington T, ed. Functional pathology of the adrenal gland. Baltimore: Williams & Wilkins; 1969:219-324.

Anatomy and Histology

10. Langlois D, Li JY, Saez JM. Development and function of the human fetal adrenal cortex. J Pediatr Endocrinol Metab 2002;15(Suppl 5):1311-22.
11. Vasudevan S, Brandt ML. Adrenal gland embryology, anatomy, and physiology. In: Ledbetter DJ, Johnson PR, eds. Endocrine surgery in children. Berlin: Springer; 2018:77-106.
12. Barwick TD, Malhotra A, Webb JA, Savage MO, Reznek RH. Embryology of the adrenal glands and its relevance to diagnostic imaging. Clin Radiol 2005;60:953-9.
13. Bartman J, Driscoll SG. Fetal adrenal cortex in erythroblastosis fetalis. Arch Pathol 1969;87:343-6.
14. Tahka H. On the weight and structure of the adrenal glands and the factors affecting them in children 0-2 years. Acta Paediatr Suppl 1951;40:1-95.
15. Bocian-Sobkowska J, Wozniak W, Malendowicz LK. Postnatal involution of the human adrenal fetal zone: stereologic description and apoptosis. Endocrine Res 1998;24:969-73.
16. Spencer SJ, Mesiano S, Lee JY, Jaffe RB. Proliferation and apoptosis in the human adrenal cortex during the fetal and perinatal periods: implications for growth and remodeling. J Clin Endocrinol Metab 1999;84:1110-5.
17. Linos D, Heerden JV. Adrenal glands: diagnostic aspects and surgical therapy. New York: Springer; 2005.
18. Kreiner E. Weight and shape of the human adrenal medulla in various age groups. Virchows Arch A Pathol Anat Histol 1982;397:7-15.
19. Quinan C, Berger AA. Observations on human adrenals with especial reference to the relative weight of the normal medulla. Ann Int Med 1933;6:1180-92.
20. Lack EE, Kozakewich HP. Embryology, developmental anatomy and selected aspects of non-neoplastic pathology. In: Lack EE, ed. Pathology of the adrenal gland. New York: Churchill Livingstone; 1990:1-74.
21. Carney JA. Adrenal gland. In: Sternberg SS, ed. Histology for pathologist. New York: Raven Press; 1992:321-46.
22. Neville AM, O'Hare MJ. The human adrenal cortex. Pathology and biology—an integrated approach. New York: Springer-Verlag; 1982.
23. Symington T. Functional pathology of the human adrenal gland. Edinburgh: Livingstone; 1969.

24. Studzinski GP, Hay DC, Symington T. Observations on the weight of the human adrenal gland and the effect of preparations of corticotropin of different purity on the weight and morphology of the human adrenal gland. J Clin Endocrinol Metab 1963;23:248-54.
25. Chester-Jones I. The adrenal cortex. Cambridge: University Press; 1957.
26. Avisse C, Marcus C, Patey M, Ladam-Marcus V, Delattre JF, Flament JB. Surgical anatomy and embryology of the adrenal glands. Surg Clin North Am 2000;80:403-15.
27. Parnaby CN, Galbraith N, O'Dwyer PJ. Experience in identifying the venous drainage of the adrenal gland during laparoscopic adrenalectomy. Clin Anat 2008;21:660-5.
28. Sebe P, Peyromaure M, Raynaud A, Delmas V. Anatomical variations in the drainage of the principal adrenal veins: the results of 88 venograms. Surg Radiol Anat 2002;24:222-5.
29. Brunt L, Cohen M. Adrenalectomy—open and minimally invasive. In: Fisher J, Bland K, eds. Mastery of surgery, 5th ed. Philadelphia: Lippincott Williams & Wilkins; 2007.
30. Black J, Williams DI. Natural history of adrenal haemorrhage in the newborn. Arch Dis Child 1973;48:183-90.
31. Clark O, Duh Q, Kebehew E. Textbook of endocrine surgery, 2nd ed. Philadelphia: WB Saunders; 2005.
32. Li Q, Johansson H, Grimelius L. Innervation of the human adrenal gland and adrenal cortical lesions. Virchows Arch 1999;435:580-9.
33. Mikhail Y. Innervation of the different zones of the adrenal cortex. J Comp Neurol 1961;117:365-9.
34. Mikhail Y, Amin F. Intrinsic innervation of the human adrenal gland. Acta Anat (Basal) 1969;72:25-32.
35. Tsapatsaris N, Breslin D. Physiology of the adrenal medulla. Urol Clin North Am 1989;16:439-45.
36. Bech K, Tygstrup I, Nerup J. The involution of the fetal adrenal cortex. A light microscopic study. Acta Pathol Microbiol Scand 1969;76:391-400.
37. Carney JA. Adrenal. In: Mills SE, ed. Histology for pathologists. Philadelphia: Lippincott Williams & Wilkins; 2012:1231-54.
38. Conran RM, Chung E, Dehner LP, et al. The pineal, pituitary, parathyroid, thyroid and adrenal glands. In: Stocker JT, Dehner LP, Husain AN, eds. Stocker and Dehner's pediatric pathology, 3rd ed. Philadelphia: Lippincott Williams & Wilkins; 2011:941-74.
39. Sadler TW. Langman's medical embryology. Philadelphia: Lippincott Williams & Wilkins; 2006:315-6.
40. Ishimoto H, Jaffe RB. Development and function of the human fetal adrenal cortex: a key component in the feto-placental unit. Endocr Rev 2011;32:317-55.
41. Oppenheimer EH. Cyst formation in the outer adrenal cortex. Studies in the human fetus and newborn. Arch Pathol 1969;87:653-9.
42. Holmes RO, Moon HD, Rinehart JF. A morphologic study of adrenal glands with correlation of body size and heart size. Am J Pathol 1951; 27:724-6.
43. Ernst LM. Adrenal gland. In: Ernst LM, Ruchelli EG, Huff DS, eds. Color atlas of fetal and neonatal histology. New York: Springer; 2011.
44. Mangray S, DeLellis RA. Adrenal glands. In: Mills SE, Greenson JK, Hornick JL, Longacre TA, Reuter VE, eds. Sternberg's diagnostic pathology, 6th ed. Philadelphia: Wolters Kluwer Health; 2015:595-646.
45. Hayashi Y, Hiyoshi T, Takemura T, Kurashima C, Hirokawa K. Focal lymphocytic infiltrating in the adrenal cortex of the elderly: immunohistological analysis of infiltrating lymphocytes. Clin Exp Immunol 1998;77:101-5.
46. Page DL, DeLellis RA, Hough AJ. Tumors of the adrenal. Atlas of Tumor Pathology, 2nd series, Fascicle 23. Washington DC: Armed Forces Institute of Pathology; 1985.
47. Dekker A, Oehrle JS. Hyaline globules of the adrenal medulla of man. Arch Pathol 1971;91:353-64.
48. Lack EE. Tumors of the adrenal gland and extra-adrenal paraganglia. AFIP Atlas of Tumor Pathology, 4th Series, Fascicle 8. Washington DC: American Registry of Pathology; 2007.
49. Kobayashi S, Coupland RF. Two populations of microvesicles in the SGC (small granule chromaffin) cells of the mouse adrenal medulla. Arch Histol Jpn 1977;40:251-9.

Ultrastructure

50. Carney JA. Adrenal. In: Mills SE, ed. Histology for pathologists. Philadelphia: Lippincott Williams & Wilkins; 2012:1231-54.
51. Long JA, Jones AL. Observations on the fine structure of the adrenal cortex of man. Lab Invest 1967;17:355-70.
52. Tannebaum M. Ultrastructural pathology of the adrenal cortex. Pathol Ann 1973;8:109-56.
53. McNutt NS, Jones AL. Observations on the ultrastructure of cytodifferentiation in the human fetal adrenal cortex. Lab Invest 1970;22:513-27.
54. Brown WJ, Barajas L, Latta H. The ultrastructure of the human adrenal medulla with comparative studies of white rat. Anat Rec 1971;169:173-83.
55. Lauriola L, Maggiano N, Sentinelli S, Michetti F, Cocchia D. Satellite cells in the normal human adrenal gland and in pheochromocytomas. An immunohistochemical study. Virchows Arch B Cell Pathol Incl Mol Pathol 1985;49:13-21.

Immunohistochemistry and In Situ Hybridization

56. Cote RJ, Cordon-Cardo C, Reuter VE, Rosen PP. Immunopathology of adrenal and renal cortical tumors. Coordinated change in antigen expression is associated with neoplastic conversion in the adrenal cortex. Am J Pathol 1990;136:1077-84.
57. Gaffey MJ, Traweek ST, Mills SE, et al. Cytokeratin expression in adrenocortical neoplasia: an immunohistochemical and biochemical study with implications for the differential diagnosis of adrenocortical, hepatocellular, and renal cell carcinoma. Hum Pathol 1992;23:144-53.
58. Miettinen M, Lehto VP, Virtanen I. Immunofluorescence microscopic evaluation of the intermediate filament expression of the adrenal cortex and medulla and their tumors. Am J Pathol 1985; 118:360-6.
59. McCluggage WG, Burton J, Maxwell P, Sloan JM. Immunohistochemical staining of normal, hyperplastic and neoplastic adrenal cortex with a monoclonal antibody against alpha inhibin. J Clin Pathol 1998;51:114-6.
60. Pelkey TJ, Frierson HF Jr, Mills SE, Stoler MH. The alpha subunit of inhibin in adrenal cortical neoplasia. Mod Pathol 1998;11:516-24.
61. Jungbluth AA, Bunsan KJ, Gerald WL, et al. A103: an anti-melan-A monoclonal antibody for the detection of malignant melanoma in paraffin-embedded tissues. Am J Surg Pathol 1998;22:595-602.
62. Enriquez ML, Lal P, Ziober A, Wang L, Tomaszewski JE, Bing Z. The use of immunohistochemical expression of SF-1 and EMA in distinguishing adrenocortical tumors from renal neoplasms. Appl Immunohistochem Mol Morphol 2012;20:141-5.
63. Sasano H, Suzuki T, Shizawa S, Kato K, Nagura H. Transforming growth factor alpha, epidermal growth factor and epidermal growth factor receptor expression in normal and diseased human adrenal cortex by immunohistochemistry and in situ hybridization. Mod Pathol 1994;7:741-6.
64. Khoury EL, Greenspan JS, Greenspan FS. Adrenocortical cells of the zona reticularis normally express HLA-DR antigenic determinants. Am J Pathol 1987;127:580-91.
65. Zhang PJ, Genega EM, Tomaszewski JE, Pasha TL, LiVolsi VA. The role of calretinin, inhibin, melan-A BCL-2 and c-kit in differentiated adrenal cortical and medullary tumors: an immunohistochemical study. Mod Pathol 2003;16:591-7.
66. Simon DP, Hammer GD. Adrenocortical stem and progenitor cells: implications for adrenocortical carcinoma. Mol Cell Endocrinol 2012;351:2-11.
67. Crawford PA, Dorn C, Sadovsky Y, Milbrandt J. Nuclear receptor DAX-1 recruits nuclear receptor corepressor N-CoR to steroidogenic factor 1. Mol Cell Biol 1998;18:2949-56.
68. Ito M, Yu R, Jameson JL. DAX-1 inhibits SF-1-mediate transactivation via a carboxy-terminal domain that is deleted in adrenal hypoplasia congenita. Mol Cell Biol 1997;17:1476-83.
69. Yu RN, Ito M, Jameson JL. The murine Dax-1 promoter is stimulated by SF-1 (steroidogenic factor-1) and inhibited by COUP-TF (chicken ovalbumin upstream promoter-transcription factor) via a composite nuclear receptor-regulatory element. Mol Endocrinol 1998;12:1010-22.
70. Sasano H, Suzuki T, Moriya T. Recent advances in histopathology and immunohistochemistry of adrenocortical carcinoma. Endocr Pathol 2006; 17:345-54.
71. Shibata H, Ikeda Y, Mukai T, et al. Expression profiles of COUP-TF, DAX-1, and SF-1 in the human adrenal gland and adrenocortical tumors: possible implications in steroidogenesis. Mol Genet Metab 2001;74:206-16.
72. Sasano H, Okamoto M, Sasano N. Immunohistochemical study of cytochrome P-450 11 beta-hydroxylase in human adrenal cortex with moneralo- and glucocorticoid excess. Virchows Arch A Pathol Anat Histopathol 1988;413:313-8.
73. Sasano H. Localization of steroidogenic enzymes in adrenal cortex and its disorders. Endocrine J 1994;41:471-82.
74. Sasano H. New approaches in human adrenocortical pathology: assessment of adrenocortical function in surgical specimen of human adrenal glands. Endocr Pathol 1992;3:4-13.
75. Sasano H, Miyazaki S, Sawai T, et al. Primary pigmented nodular adrenocortical disease (PPNAD): immunohistochemical and in situ hybridization analysis of steroidogenic enzymes in eight cases. Mod Pathol 1992;5:23-9.
76. Tartour E, Caillou B, Tenenbaum F, et al. Immunohistochemical study of adrenocortical carcinoma: predictive value of the D11 monoclonal antibody. Cancer 1993;72:3296-303.
77. Gonzalez-Hernandez JA, Bornstein SR, Ehrhart-Bornstein M, Späth-Schwalbe E, Jirikowski G, Scherbaum WA. Interleukin-6 messenger ribonucleic acid expression in human adrenal gland in vivo: new clue to a paracrine or autocrine regulation of adrenal function. J Clin Endocrinol Metab 1994;79:1492-7.
78. Strickland J, McIlmoil S, Williams BJ, Seager DC, Porter JP, Judd AM. Interleukin-6 increases the expression of key proteins associated with steroidogenesis in human NCI-H295R adrenocortical cells. Steroids 2017;119:1-17.
79. Willenberg HS, Path G, Vögeli TA, Scherbaum WA, Bornstein SR. Role of interleukin-6 in stress response in normal and tumorous adrenal cells and during chronic inflammation. Ann N Y Acad Sci 2002;966:304-14.

80. Khorram-Manesh A, Ahlman H, Jansson S, Nilsson O. N-cadherin expression in adrenal tumors: upregulation in malignant pheochromocytoma and downregulation in adrenocortical carcinoma. Endocr Pathol 2002;13:99-110.
81. Rosenbaum JN, Guo Z, Baus RM, Wener H, Rehrauer WM, Lloyd RV. INSM1: a novel immunohistochemical and molecular marker for neuroendocrine and neuroepithelial neoplasms. Am J Clin Pathol 2015;144:579-91.
82. Lehto VP, Virtanen I, Miettinen M, Dahl D, Kahri A. Neurofilaments in adrenal and extra-adrenal pheochromocytoma. Demonstration using immunofluorescence microscopy. Arch Lab Med Pathol 1983;107:492-4.
83. Cocchia D, Michetti F. S-100 antigen in satellite cells of the adrenal medulla and the superior cervical ganglion of the rat. An immunochemical and immunocytochemical study. Cell Tissue Res 1981;215:103-12.
84. Lloyd RV, Blaivas M, Wilson BS. Distribution of chrom ogranin and S-100 protein in normal and abnormal adrenal medullary tissues. Arch Pathol Lab Med 1985;109:633-5.
85. McNicol AM. Histopathology and immunohistochemistry of adrenal medullary tumors and paragangliomas. Endocr Pathol 2006;17:329-36.
86. Lloyd RV, Shapiro B, Sisson JC, Kalff V, Thompson NW, Beierwaltes WA. An immunohistochemical study of pheochromocytomas. Arch Pathol Lab Med 1984;108:541-4.
87. Lloyd RV, Sisson JC, Shapiro B, Verhofstad AA. Immunohistochemical localization of epinephrine, norepinephrine, catecholamine-synthesizing enzymes and chromogranin in neuroendocrine cells and tumors. Am J Pathol 1986;125:45-54.
88. Alvarez P, Isidro L, Gonzalez-Martin M, Loidi L, Arnal F, Cordido F. Ectopic adrenocorticotropic hormone production by a noncatecholamine secreting pheochromocytoma. J Urol 2002;167:2514-5.

Molecular Biology and Physiology

89. Mangray S, DeLellis RA. Adrenal glands. In: Mills SE, Greenson JK, Hornick JL, Longacre TA, Reuter VE, eds. Sternberg's diagnostic pathology, 6th ed. Philadelphia: Wolters Kluwer Health; 2015:595-646.
90. Boggaram V, Funkenstein B, Waterman MR, Simpson ER. Lipoproteins and the regulation of adrenal steroidogenesis. Endocr Res 1985;10:387-409.
91. Bondy P. Disorders of the adrenal cortex. In: Wilson JD, Foster DW, eds. Williams' textbook of endocrinology. Philadelphia: WB Saunders; 1985:816-91.
92. Orth DN, Kovacs WJ. The adrenal cortex. In: Wilson JD, Foster DW, Kronenberg HM, Larsen PR, eds. Williams textbook of endocrinology, 9th ed. Philadelphia: WB Saunders; 1998:517-664.
93. Ishimoto H, Jaffe RB. Development and function of the human fetal adrenal cortex: a key component in the feto-placental unit. Endocr Rev 2011;32:317-55.
94. Nieman LK. Causes of primary adrenal insufficiency. UpToDate 2016. Available at https://www.uptodate.com.acs.hcn.com.au/contents/causes-of-primary-adrenal-insufficiency-addisons-disease?search=adrenal%20infections&source=search_result&selectedTitle=1~150&usage_type=default&display_rank=1.
95. Nieman LK. Causes of secondary and tertiary adrenal insufficiency in adults. UpToDate, 2016. Available at https://www.uptodate.com.acs.hcn.com.au/contents/causes-of-secondary-and-tertiary-adrenal-insufficiency-in-adults?search=secondary%20adrenal%20insufficiency&source=search_result&selectedTitle=1~141&usage_type=default&display_rank=1.
96. Lockwood CJ, Radunovic N, Nastic D, Petkovic S, Aigner R, Berkowitz GS. Corticotropin-relearing hormone and related pituitary-adrenal axis hormones in fetal and maternal blood during the second half of pregnancy. J Perinat Med 1996;24:243-51.
97. Murphy BE. Human fetal serum cortisol levels at delivery: a review. Endocr Rev 1983;4:150-4.
98. Murphy BE. Human fetal serum cortisol levels related to gestational age: evidence of a midgestational fall and a steep late gestational rise, independent of sex or mode of delivery. Am J Obstet Gynecol 1982;144:276-82.
99. Molina PE. Adrenal gland. In: Molina PE, ed. Endocrine physiology, 5th ed. New York: McGraw-Hill; 2008.
100. Vaughan GM, Becker RA, Allen JP, et al. Cortisol and corticotrophin in burned patients. J Trauma 1982;22:263-72.
101. Bohen SP, Kralli A, Yamamoto KR. Hold 'em and fold em: chaperones and signal transduction. Science 1995;268:1303-4.
102. Lacroix A. Pharmacology and toxicity of adrenal enzyme inhibitors and adrenolytic agents. In: Nieman LK, Martin KA, eds. UpToDate. From https://www.uptodate.com/contents/pharmacology-and-toxicity-of-adrenal-enzyme-inhibitors-and-adrenolytic-agents.
103. Baudry C, Coste J, Bou Khalil R, et al. Efficiency and tolerance of mitotane in Cushing's disease in 76 patients from a single center. Eur J Endocrinol 2012;167:473-81.

104. Lloyd RV, Jin L, Kulig E, Fields K. Molecular approaches for the analysis of chromogranins and secretogranins. Diagn Mol Pathol 1992;1:2-15.
105. Young JB, Landsberg L. Catecholamines and the adrenal medulla. In: Wilson JD, Foster DW, Kronenberg HM, Larsen PR, eds. Williams text¬book of endocrinology, 9th ed. Philadelphia: WB Saunders; 1998:665-728.

Reactive, Hereditary, and Developmental Disorders

106. Favara BE, Steele A, Grant JH, Steele P. Adrenal cytomegaly: quantitative assessment by image analysis. Pediatr Pathol 1991;11:521-36.
107. Craig JM, Landing BH. Anaplastic cells of fetal adrenal cortex. Am J Clin Pathol 1951;21:940-9.
108. Aterman K, Kerenyi N, Lee M. Adrenal cytomegaly. Virchows Arch A Pathol Anat 1972;355:105-22.
109. Gau GS, Bennett MJ. Fetal adrenal cytomegaly. J Clin Pathol 1979;32:305-6.
110. Lacson A, deSa D. Endocrine gland. In: Gilbert-Barness E, ed. Potter's pathology of the fetus, infant and child, 2nd ed. Philadelphia: Mosby Elsevier; 2007:1595-656.
111. Noguchi S, Masumoto K, Taguchi T, Takahashi Y, Tsuneyoshi M, Suita S. Adrenal cytomegaly: two cases detected by prenatal diagnosis. Asian J Surg 2003;26:234-6.
112. Ong BB, Wong KT. Adrenal cytomegaly associated with diaphragmatic hernia: report of a case. Malays J Pathol 1996;18:121-3.
113. Lloyd R, Douglas B, Young W. Adrenal gland. In: Lloyd R, Douglas B, Young W, eds. Endocrine diseases. AFIP Atlas of Nontumor Pathology, 1st Series, Fascicle 1. Washington, DC: American Registry of Pathology; 2002:171-237.
114. Akata D, Haliloglu M, Ozmen MN, Akhan O. Bilateral cystic adrenal masses in the neonate associated with the incomplete form of Beckwith-Wiedemann syndrome. Pediatr Radiol 1997;27:1-2.
115. Anoop P, Anjay MA. Bilateral benign haemorrhagic adrenal cysts in Beckwith-Wiedemann syndrome: case report. East Afr Med J 2004;81:59-60.
116. Camuto PM, Wolman SR, Perle MA, Greco MA. Flow cytometry of fetal adrenal glands with adrenocortical cytomegaly. Pediatr Pathol 1989; 9:551-8.
117. Guttman PH. Addison's disease. A statistical analysis of 566 cases and a study of the pathology. Arch Path 1930;10:742-85.
118. Kissane JM, Smith MG. Pathology of infancy and childhood. St Louis: CV Mosby; 1967.
119. Neville AM, O'Hare MJ. The human adrenal cortex: pathology and biology—an integrated approach. New York: Springer-Verlag; 1982.
120. Russfield AB. The endocrine glands after bilateral adrenalectomy compared with those in spontaneous adrenal insufficiency. Cancer 1955;8:523-37.
121. Wells HG. Addison's disease with selective destruction of suprarenal cortex (suprarenal cortex atrophy). Arch Pathol 1930;10:499-523.
122. Yamashina M. Focal adrenocortical cytomegaly observed in two adult cases. Arch Pathol Lab Med 1986;110:1072-5.
123. Lack EE. Tumors of the adrenal gland and extra-adrenal paraganglia. AFIP Atlas of Tumor Pathology, 3rd Series, Fascicle 19. Washington DC: American Registry of Pathology; 1997.
124. Orth DN, Kovacs WJ. The adrenal cortex. In: Wilson JD, Foster DM, Kronenberg HM, Larsen PR, eds. Williams textbook of endocrinology, 9th ed. Philadelphia: WB Saunders; 1998:517-664.
125 Griffel B. Focal adrenalitis. Its frequency and correlation with similar lesions in the thyroid and kidney. Virchows Archiv A Pathol Anat Histol 1974;364:191-8.
126. Hayashi Y, Hiyoshi T, Takemura T, Kurashima C, Hirokawa K. Focal lymphocytic infiltration in the adrenal cortex of the elderly: immunohistological analysis of infiltrating lymphocytes. Clin Exp Immunol 1989;77:101-5.
127. Fidler WJ. Ovarian thecal metaplasia in adrenal glands. Am J Clin Pathol 1977;67:318-23.
128. Wassal EY, Habra MA, Vicens R, Rao P, Elsayes KM. Ovarian thecal metaplasia of the adrenal gland in association with Beckwith-Wiedemann syndrome. World J Radiol 2014;6:919-23.
129. Mete O, Raphael S, Pirzada A, Asa SL. Is adrenal ovarian thecal metaplasia a misnomer? Report of three cases of radial scar-like spindle cell myofibroblastic nodule of the adrenal gland. Endocr Pathol 2011;22:222-5.
130. Reed RJ, Pabic JT. Nodular hyperplasia of the adrenal cortical blastema. Bull Tulane Univ Med Fac 1967;26:151-7.
131. Wont TW, Warner NE. Ovarian thecal metaplasia in the adrenal gland. Arch Pathol 1971;92:319-28.

Adrenal Malformations

132. Lack EE. Tumors of the adrenal gland and extra adrenal paraganglia. AFIP Atlas of Tumor Pathology, 4th Series, Fascicle 8. Washington DC: American Registry of Pathology; 2007.
133. Mangray S, DeLellis RA. Adrenal glands. In: Mills SE, Greenson JK, Hornick JL, Longacre TA, Reuter VE, eds. Sternberg's diagnostic pathology, 6th ed. Philadelphia: Wolters Kluwer Health; 2015:595-646.
134. Bell JE. Fused suprarenal glands in association with central nervous system defects in the first half of fetal life. J Pathol 1979;127:191-4.
135. Dolan MF, Jonovski NA. Adreno-hepatic union (adreno dystopia). Arch Pathol 1968;86:22-4.

Adrenal Rests and Accessory Adrenal Tissues

136. Gutowski WT 3rd, Gray GF Jr. Ectopic adrenal in inguinal hernia sacs. J Urol 1979;121:353-4.
137. Armin A, Castelli M. Congenital adrenal tissue in the lung with adrenal cytomegaly. Case report and review of the literature. Am J Clin Pathol 1984;82:225-8.
138. Graham LS. Celiac accessory adrenal glands. Cancer 1953;6:149-52.
139. Honoré LH. Intra-adrenal hepatic heterotopia. J Urol 1985;133:652-4.
140. Mangray S, DeLellis RA. Adrenal glands. In: Mills SE, Greenson JK, Hornick JL, Longacre TA, Reuter VE, eds. Sternberg's diagnostic pathology, 6th ed. Philadelphia: Wolters Kluwer Health; 2015:595-646.
141. Mares AJ, Shkolnik A, Sacks M, Feuchtwanger MM. Aberrant (ectopic) adrenocortical tissue along the spermatic cord. J Pediatr Surg 1980; 15:289-92.
142. Mitchell N, Angrist A. Adrenal rests in the kidney. Arch Pathol 1943;35:46-52.
143. Lack EE, Kozakewich HP. Embryology, developmental anatomy and selected aspects of non-neoplastic pathology. In: Lack EE, ed. Pathology of the adrenal gland. New York: Churchill Livingstone; 1990:1-74.
144. MacLennan A. On the presence of adrenal rests in hernial sac walls. Surg Gynecol Obstet 1919;29:387.
145. Nelson AA. Accessory adrenal cortical tissue. Arch Pathol 1939;27:955-65.
146. Symonds DA, Driscoll SG. An adrenal cortical rest within the fetal ovary: report of a case. Am J Clin Pathol 1973;60:562-4.
147. Burke EF, Gilbert E, Uehling DT. Adrenal rest tumors of the testes. J Urol 1973;109:649-52.
148. Johnson RE, Scheithauer B. Massive hyperplasia of testicular adrenal rests in a patient with Nelson's syndrome. Am J Clin Pathol 1982;77:501-7.
149. Page DL, DeLellis RA, Hough AJ. Tumors of the adrenal. Atlas of Tumor Pathology, 2nd series, Fascicle 23. Washington DC: Armed Forces Institute of Pathology; 1985.
150. Wallace EZ, Leonidas JR, Stanek AE, Avramides A. Endocrine studies in a patient with functioning adrenal rest tumor of the liver. Am J Med 1981;70:1122-5.
151. Morimoto Y, Hiwada K, Nanahoshi M, et al. Cushing's syndrome caused by malignant tumor in the scrotum: clinical, pathologic and biochemical studies. J Clin Endocr Metab 1971;32:201-10.
152. Dolan MF, Jonovski NA. Adreno-hepatic union. (adreno dystopia). Arch Pathol 1968;86:22-4.
153. O'Crowley CR, Martland HS. Arenal heterotopia, rests and the so-called Grawitz tumor. J Urol 1943;50:576.

Adrenal Hypoplasia

154. Rimoin DL, Schimke RN. Genetic disorders of the endocrine glands. St. Louis: Mosby; 1971:281.
155. Osamura RY. Functional prenatal development of anencephalic and normal anterior pituitary glands. In human and experimental animals studied by peroxidase-labeled antibody method. Acta Pathol Jpn 1977;27:495-509.
156. Salazar H, MacAulay MA, Charles D, Pardo M. The human hypophysis in anencephaly. I. Ultrastructure of the pars distalis. Arch Pathol 1969;87:201-11.
157. Aimone V, Campagnoli C. Severe adrenal hypoplasia in a live-born normocephalic infant with neurohypophyseal aplasia. Am J Obstet Gynecol 1970;107:327.
158. Moncrieff MW, Hill DS, Archer J, Arthur LJ. Congenital absence of pituitary gland and adrenal hypoplasia. Arch Dis Child 1972;47:136-7.
159. Wont TW, Warner NE. Ovarian thecal metaplasia in the adrenal gland. Arch Pathol 1971;92: 319-28.
160. Engiz O, Ozon A, Riepe F, Alikasifoglu A, Gonc N, Kendemir N. Growth hormone deficiency due to traumatic brain injury in a patient with X-linked congenital adrenal hypoplasia. Turk J Pediatr 2010;52:312-6.
161. Li N, Liu R, Zhang H, et al. Seven novel DAX1 mutations with loss of function identified in Chinese patients with congenital adrenal hypoplasia. J Clin Endocrinol Metab 2010;95:E104-11.
162. Pereira BD, Pereira I, Portugal JR, Goncalves J, Raimundo L. X-linked adrena hypoplasia congenita: clinical and follow-up findings of two kindreds, one with a novel NR0B1 mutation. Arch Endocrinol Metab 2015;59:181-5.
163. Zhang Z, Feng Y, Ye D, Li CJ, Dong FQ, Tong Y. Clinical and molecular genetic analylsis of a Chinese family with congenital X-linked adrenal hypoplasia caused by novel mutation 1268delA in the DAX-1 gene. J Zhejiang Univ Sci B 2015;16:963-8.
164. Habiby RL, Boepple P, Nachtigall L, Sluss PM, Crowley WF Jr, Jameson JL. Adrenal hypoplasia congenita with hypogonadotropic hypogonadism: evidence that DAX-1 mutations lead to combined hypothalamic and pituitary defects in gonadotropin production. J Clin Invest 1996; 98:1055-62.
165. Loureiro M, Reis F, Robalo B, Pereira C, Sampaio L. Adrenal hypoplasia congenita: a rare cause of primary adrenal insufficiency and hypogonadotropic hypogonadism. Pediatr Rep 2015;7:5936.

166. Hay ID, Smail PJ, Forsyth CC. Familial cytomegalic adrenocortical hypoplasia: an X-linked syndrome of pubertal failure. Arch Dis Child 1981;56:715-21.
167. Mortin MM, Mortin AL. The syndrome of congenital hereditary adrenal hypoplasia and hypogonadotropic hypo gonadism. Int J Adolesc Med Health 1985;1:119-37.
168. Prader A, Zachmann M, Illig R. Luteinizing hormone deficiency in hereditary congenital adrenal hypoplasia. J Pediatr 1975;86:421-2.
169. Achermann JC, Ito M, Ito M, et al. A mutation in the gene encoding steroidogenic factor-1 causes XY sex reversal and adrenal failure in humans. Nat Genet 1999;22:125-6.
170. Kyritsi EM, Sertedaki A, Charmandari E, Chrousos GP. Familial or sporadic adrenal hypoplasia syndromes. In: Feingold KR, Anawalt A, Boyce A, et al., eds. Endotext [Internet]. South Dartmouth: MDText.com, Inc.; 2018. Available at https://www.ncbi.nlm.nih.gov/books/NBK279132/.
171. Bergada I, Del Rey G, Lapunzina P, et al. Familial occurrence of the IMAGe association: additional clinical variants and a proposed mode of inheritance. J Clin Endocrinol Metab 2005;90:3186-90.
172. Yeo GS, Farooqi IS, Challis BG, Jackson RS, O'Rahilly S. The role of melanocortin signalling in the control of body weight: evidence from human and murine genetic models. QJM 2000;93:7-14.
173. Krude H, Biebermann H, Luck W, Horn R, Brabant G, Grüters A. Severe early-onset obesity, adrenal insufficiency and red hair pigmentation caused by POMC mutations in humans. Nat Gen 1998;19:155-7.
174. Mendiratta MS, Yang Y, Balazs AE, et al. Early onset obesity and adrenal insufficiency associated with a homozygous POMC mutation. Int J Pediatr Endocrinol 2011;2011:5.
175. Burris TP, Guo W, McCabe ER. The gene responsible for congenital adrenal hypoplasia DAX-1, encodes anuclear hormone receptor that defines a new class within the superfamily. Recent Prog Horm Res 1996;54:241-59.
176. Lack EE, Kozakewich HP. Embryology, developmental anatomy and selected aspects of non-neoplastic pathology. In: Lack EE, ed. Pathology of the adrenal gland. New York: Churchill Livingstone; 1990:1-74.
177. Wise JE, Matalon R, Morgan AM, et al. Phenotypic features of patients the congenital adrenal hypoplasia and glycerol kinase deficiency. Am J Dis Child 1987;141:744-7.
178. McCabe ER. Adrenal hypoplasias and aplasias. In: Valle D, Beaudet AL, Vogelstein B, Kinzler KW, Antonarakis SE, Mitchell G, eds. The online metabolic and molecular bases of inherited disease. New York: McGraw-Hill; 2019.
179. Phillips K, Arroyo MR, Duckworth LV. IMAGe association: report of two cases in siblings with adrenal hypoplasia and review of the literature. Pediatr Dev Pathol 2014;17:204-8.
180. Tan TY, Jameson JL, Campbell PE, Ekert PG, Zacharin M, Savarirayan R. Two sisters with IMAGe syndrome: cytomegalic adrenal histopathology, support for autosomal recessive inheritance and literature review. Am J Med Genet A 2006;140:1778-84.

Hereditary Adrenal Cortical Unresponsiveness to Adrenocorticotropic Hormone (Familial Glucocorticoid Deficiency)

181. Berberoglu M, Aycan Z, Ocal G, et al. Syndrome of congenital adrenocortical unresponsiveness to ACTH. Report of six patients. J Pediatr Endocrinol Metab 2001;14:1113-8.
182. Davidai G, Kahana L, Hochberg Z. Glomerulosa failure in congenital adrenocortical unresponsiveness to ACTH. Clin Endocrinol (Oxf) 1984;20:515-20.
183. Migeon CJ, Kenny EM, Kowarski A, et al. The syndrome of congenital adrenocortical unresponsiveness to ACTH. Report of six cases. Pediatr Res 1968;2:501-13.
184. Naville D, Barjhoux L, Jaillard C, et al. Demonstration by transfection studies that mutations in the adrenocorticotropin receptor gene are one cause of the hereditary syndrome of glucocorticoid deficiency. J Clin Endocrinol Metab 1996;81:1442-8.
185. Clark AJ, McLoughlin L, Grossman A. Familial glucocorticoid deficiency associated with point mutation in the adrenocortocotropin receptor. Lancet 1993;341:461-2.
186. Chan LF, Clark AJ, Metherell LA. Familial glucocorticoid deficiency: advances in the molecular understanding of ACTH action. Horm Res 2008;69:75-82.
187. Clark AJ, Weber A. Adrenocorticotropin insensitivity syndromes. Endocr Rev 1998;19:828-43.
188. Kelch RP, Kaplan SL, Biglieri EG, Daniels GH, Epstein CJ, Grumbach MM. Hereditary adrenocortical unresponsiveness to adrenocorticotropic hormone. J Pediatr 1972;81:726-36.
189. Shepard TH, Landing BH, Mason DG. Familial Addison's disease. Case reports of two sisters with corticoid deficiency unassociated with hypoaldosternism. Am J Dis Child 1959;97:154-62.
190. Yamaoka T, Kudo T, Takuwa Y, Kawakami Y, Itakura M, Yamashita K. Hereditary adrenocortical unresponsiveness to adrenocorticotropin with a postreceptor defect. J Clin Endocrinol Metab 1992;75:270-4.

Exogenous Injury

191. Bornstein SR. Predisposing factors for adrenal insufficiency. N Engl J Med 2009;360:2328-39.
192. Mebrahtu TF, Morgan AW, Keeley A, Baxter PD, Stewart PM, Pujades-Rodriguez M. Dose dependency of iatrogenic glucocorticoid excess and adrenal insufficiency and mortality: a cohort study in England. J Clin Endocrinol Metab 2019;104:3757-67.
193. Bornstein SR, Bornstein TD, Andoniadou CL. Novel medications inducing adrenal insufficiency. Nat Rev Endocrinol 2019;15:561-2.
194. Ferraldeschi R, Sharifi N, Auchus RJ, Attard G. Molecular pathways: inhibiting steroid biosynthesis in prostate cancer. Clin Cancer Res 2013;19:3353-9.
195. Hofmann A, Brunssen C, Peitzsch M, et al. Aldosterone synthase inhibition improves glucose tolerance in zucker diabetic fatty (ZDF) rats. Endocrinology 2016;157:3844-55.
196. Bergenstal DM, Hertz R, Lipsett MB, et al. Chemotherapy of adrenocortical cancer with o,p'-DDD. Ann Intern Med 1966;53:672-82.
197. Vilar O, Tullner WW. Effects of o,p'DDD on histology and 17-hydroxycorticosteroid output of the dog adrenal cortex. Endocrinology 1959;65:80-6.
198. Liddle GW, Island D, Lance EM, Harris AP. Alterations of adrenal steroid patterns in man resulting from treatment with a chemical inhibitor of 11-hydroxylation. J Clin Endocrinol Metab 1958;18:906-12.
199. Darvin P, Toor SM, Nair VS, Elkord E. Immune checkpoint inhibitors: recent progress and potential biomarkers. Exp Mol Med 2018;50:1-11.
200. Tan MH, Iyengar R, Mizokami-Stout K, et al. Spectrum of immune checkpoint inhibitors-induced endocrinopathies in cancer patients: a scoping review of case reports. Clin Diabetes Endocrinol 2019;5:1.
201. Byun D, Wolchok JD, Rosenberg LM, Girotra M. Cancer immunotherapy—immune checkpoint blockade and associated endocrinopathies. Nat Rev Endocrinol 2017;13:195-207.
202. Iwama S, de Remigis A, Callahan MK, Slovin S, Wolchok JD, Caturegli P. Pituitary expression of CTLA-4 mediates hypophysitis secondary to administration of CTLA-4 blocking antibody. Sci Transl Med 2014;6:230ra45.
203. Schieda N, Ramchandani P, Siegelman ES. Computed tomographic findings of radiation-induced acute adrenal injury with associated radiation nephropathy: a case report. Acta Radiologica Short Reports 2013;2:1-4.
204. Schieda N, Ramchandani P, Siegelman ES. Imaging manifestations of radiation therapy in the genitourinary tract: organ-based approach to expected findings, complications, treatment response and recurrent tumor [Abstract]. In: EDS. 98th Scientific Assembly and Annual Meeting of the Radiologic Society of North America. Chicago; 2012:25-30.
205. Fajardo LF, Berthrong M, Anderson RE. Radiation pathology. New York: Oxford University Press; 2001.
206. Sommers SC, Carter ME. Adrenocortical postirradiation fibrosis. Arch Pathol 1975;99:421-3.
207. Arlt W, Hove U, Muller B, et al. Frequent and frequently overlooked: treatment induced endocrine dysfunction in adult long term survivors of primary brain tumors. Neurology 1997;49:498-506.
208. Rosol TJ, Yarrington JT, Latendresse J, Capen CC. Adrenal gland: structure, function and mechanisms of toxicity. Toxicol Pathol 2001;29:41-8.
209. Jung C, Inder WJ. Management of adrenal insufficiency during the stress of medical illness and surgery. Med J Aust 2008;188:409-13.
210. Arafah BM. Hypothalamic pituitary adrenal function during critical illness: limitations of current assessment methods. J Clin Endocrinol Metab 2006;91:3725-45.
211. James VH, Few JD. Adrenocorticosteroids: chemistry, synthesis and disturbances in disease. Clin Endocrinol Metab 1985;14:867-92.
212. de Jong FH, Mallios C, Jansen C, Scheck PA, Lamberts SW. Etomidate suppresses adrenocortical function by inhibition of 11 beta-hydroxylation. J Clin Endocrinol Metab 1984;59:1143-7.
213. Colby HD. Adrenal gland toxicity: chemically induced dysfunction. J Coll Toxicol 1988;7:45-69.
214. Drakonaki E, Papakonstantinou O, Maris T, Vasiliadou A, Papadakis A, Gourtsoyiannis N. Adrenal glands in beta-thalassemia major: magnetic resonance (MR) imaging features and correlation with iron stores. Eur Radiol 2005;15:2462-8.
215. Sklar CA, Lew LQ, Yoon DJ, David R. Adrenal function in thalassemia major following long-term treatment with multiple transfusions and chelation therapy. Evidence for dissociation of cortisol and adrenal androgen secretion. Am J Dis Child 1987;141:327-30.
216. Baldini M, Mancarella M, Cassinerio E, Marcon A, Giacinto Ambrogio A, Motta I. Adrenal insufficiency: an emerging challenge in thalassemia? Am J Hematol 2017;92:E119-21.
217. Costin G, Kogut MD, Hyman CB, Ortega JA. Endocrine abnormalities in thalassemia major. Am J Dis Child 1979;133:497-502.

Infectious Diseases and Miscellaneous Conditions

218. Arlt W, Allolio B. Adrenal insufficiency. Lancet 2003;361:1881-93.

219. Huang YC, Tang YL, Zhang XM, Zeng NL, Li R, Chen TW. Evaluation of primary adrenal insufficiency secondary to tuberculous adrenalitis with computed tomography and magnetic resonance imaging: current status. World J Radiol 2015;7:336-42.
220. Kinjo T, Higuchi D, Oshiro Y, et al. Addison's disease due to tuberculosis that required differentiation from SIADH. J Infect Chemother 2009;15:239-42.
221. Mangray S, DeLellis RA. Adrenal glands. In: Mills SE, Greenson JK, Hornick JL, Longacre TA, Reuter VE, eds. Sternberg's diagnostic pathology, 6th ed. Philadelphia: Wolters Kluwer Health; 2015:595-646.
222. Lam K, Lo C. A critical examination of adrenal tuberculosis and a 28-year autopsy experience of active tuberculosis. Clin Endocrinol 2001;54:633-9.
223. Kelestimur F, Unlu Y, Ozesmi M, Tolu I. A hormonal and radiological evaluation of adrenal gland in patients with acute or chronic pulmonary tuberculosis. Clin Endocrinol (Oxf) 1994;41:53-6.
224. Paolo WF, Nosanchuk JD. Adrenal infections. Int J Infect Dis 2006;5:343-53.
225. Barnes PF, Abrams JS, Lu S, Sieling PA, Rea TH, Modlin RL. Patterns of cytokine production by mycobacterium reactive human T-cell clones. Infect Immun 1993;61:197-203.
226. Kelestimur F. The endocrinology of adrenal tuberculosis: the effect of tuberculosis on the hypothalamic-pituitary-adrenal axis and adrenocortical function. J Endocrinol Invest 2004;27:380-6.
227. Keven K, Uysal AR, Erdogan G. Adrenal function during tuberculous infection and effects of antituberculosis treatment on endogenous and exogenous steroids. Int J Tuberc Lung Dis 1998;2:419-24.
228. Sloper JC. The pathology of the adrenals, thymus and certain other endocrine glands in Addison's disease: an analysis of 37 necropsies. Proc R Soc Med 1955;48:625-8.
229. Guttman PH. Addison's disease: statistical analysis of 566 cases and a study of the pathology. Arch Pathol 1930;10:742-895.
230. Barker NW. The pathologic anatomy in twenty-eight cases of Addison's disease. Arch Pathol 1929;8:432-50.
231. Penrice J, Nussey SS. Recovery of adrenocortical function following treatment of tuberculous Addison's disease. Postgrad Med J 1992;68:204-5.
232. Upadhyay J, Sudhindra P, Abraham G, Trivedi N. Tuberculosis of the adrenal gland: a case report and review of the literature of infections of the adrenal gland. Int J Endocrinol 2014;2014:876037.
233. Bhatia E, Jain SK, Gupta RK, Pandey R. Tuberculous Addison's disease: lack of normalization of adrenocortical function after anti-tuberculous chemotherapy. Clin Endocrinol 1998;48:355-9.
234. Kumar N, Singh S, Govil S. Adrenal histoplasmosis: clinical presentation and imaging features in nine cases. Abdom Imaging 2003;28:703-8.
235. Roubsanthisuk W, Sriussadaporn S, Vawesorn N, et al. Primary adrenal insufficiency caused by disseminated histoplasmosis: report of two cases. Endocr Pract 2002;8:237-41.
236. Swartz MA, Scofield RH, Dickey WD, et al. Unilateral adrenal enlargement due to histoplasma capsulatum. Clin Infect Dis 1996;23:813-5.
237. Abad A, Gomez I, Velez P, Restrepo A. Adrenal function in paracoccidioidomycosis: a prospective study in patients before and after ketoconazole therapy. Infection 1986;14:22-6.
238. Frenkel JK. Pathogenesis of infections of the adrenal gland leading to Addison's disease in man: the role of corticoids in adrenal and generalized infection. Ann NY Acad Sci 1960;84:391-439.
239. Frenkel JK. Role of corticosteroids as predisposing factors in fungal diseases. Lab Invest 1962;11:1192-208.
240. Marsigli I, Pinto J. Adrenal cortical insufficiency associated with paracoccidioidomycosis (South American blastomycosis). Report of four patients. J Clin Endocrinol Metab 1966;26:1109-15.
241. Osa SR, Peterson RE, Roberts RB. Recovery of adrenal reserve following treatment of disseminated South American blastomycosis. Am J Med 1981;71:298-301.
242. Salyer WR, Moravec CL, Salyer DC, Guerin PF. Adrenal involvement in cryptococcosis. Am J Clin Pathol 1973;60:559-61.
243. Sarosi GA, Voth DW, Dahl BA, Doto IL, Tosh FE. Disseminated histoplasmosis: results of long-term follow-up. A center for disease control cooperation mycosis study. Ann Intern Med 1971;75:511-6.
244. Walker BF, Gunthel CJ, Bryan JA, Watts NB, Clark RV. Disseminated cryptococcosis in an apparently normal host presenting as primary adrenal insufficiency: diagnosis by fine needle aspiration. Am J Med 1989;86:715-7.
245. Do Valle A, Guimaraes M, Cuba J, Wanke B, Tendrich M. Recovery of adrenal function after treatment of paracoccidioidomycosis. Am J Trop Med Hyg 1993;48:626-9.
246. Torres C, Duarte E, Guimares JP, Moreira LF. Destructive lesion of the adrenal gland in South American blastomycosis (Letz' disease). Am J Pathol 1952;28:145-55.

247. Agarwal J, Agarwal G, Ayyagari A, Kar DK, Mishra SK, Bhatia E. Isolated pneumocystis carinii infection of adrenal glands causing Addison's disease in a non-immunocompromised adult. Endocr Pathol 2001;12:87-91.
248. Lio S, Cibin M, Marcello R, Viviani MA, Ajello L. Adrenal bilateral incidentaloma by reactivated histoplasmosis. J Endocrinol Invest 2000;23:476-9.
249. Papadopoulos KI, Castor B, Klingspor L, Dejmek A, Lorén I, Bramnert M. Bilateral isolated adrenal coccidioidomycosis. J Intern Med 1996;239:275-8.
250. Washburn RG, Bennett JE. Reversal of adrenal glucocorticoid dysfunction in a patient with disseminated histoplasmosis. Ann Intern Med 1989;110:86-7.
251. Lionakis MS, Samonis G, Kontoyiannis DP. Endocrine and metabolis manifestations of invasive fungal infections and systemic antifungal treatment. Mayo Clinic Proc 2008;83:1046-60.
252. Del Negro G, Melo EH, Rodbard D, Melo MR, Layton J, Wachslicht-Rodband H. Limited adrenal reserve in paracoccidiomycosis: cortical and aldosterone responses to 1-24 ACTH. Clin Endocrinol 1980;13:553-9.
253. Kent DC, Collier TM. Addison's disease associated with North American blastomycosis: a case report. J Clin Endocrinol Metab 1965;25:164-9.
254. Maloney PJ. Addison's disease due to chronic disseminated coccidioidomycosis. AMA Arch Intern Med 1952;90:869-78.
255. Shah B, Taylor HC, Pillay I, Chung-Park M, Dobrinich R. Adrenal insufficiency due to cryptococcosis. JAMA 1986;256:3247-9.
256. Alteras I, Cojocaru I, Balanescu A. Generalized candidiasis associated with Addison's disease. Mykosen 1969;12:575-7.
257. Karim M, Sheikh H, Alam M, Sheikh Y. Disseminated bipolaris infection in an asthmatic patient: case report. Clin Infect Dis 1993;17:248-53.
258. Lamberts SW, Bons EG, Bruining HA, de Jong FH. Differential effects of the imidazole derivatives etomidate, ketoconazole and miconazole and of metyrapone on the secretion of cortisol and its precursors by human adrenocortical cells. J Pharmacol Exp Ther 1987;240:259-64.

Human Immunodeficiency Virus (HIV)-Associated Adrenal Dysfunction

259. Nassoro DD, Mkhoi ML, Sabi I, Meremo AJ, Lawala PS, Habakuk Mwakyula I. Adrenal insufficiency: a forgotten diagnosis in HIV/AIS patients in developing countries. Int J Endocrinol 2019;20191:2342857.
260. Glasgow BJ, Steinsapir KD, Anders K, Layfield LJ. Adrenal pathology in the acquired immune deficiency syndrome. Am J Clin Pathol 1985;84:594-7.
261. Rotterdam H, Dembitzer F. The adrenal gland in AIDS. Endocr Pathol 1993;4:4-14.
262. Tripathy SK, Agrawala K, Baliarsinha A. Endocrine alterations in HIV-infected patients. Indian J Endocrinol Metab 2015;19:143-7.
263. Zapanti E, Terzidis K, Chrousos G. Dysfunction of the hypothalamic-pituitary-adrenal axis in HIV infection and disease. Hormones (Athens) 2008;7:205-16.
264. Hoshino Y, Yamashita N, Nakamura T, Iwamoto A. Prospective examination of adrenocortical function in advanced AIDS patients. Endocr J 2002;49:641-7.
265. Villette JM, Bourin P, Doinel C, et al. Circadian variations in plasma levels of hypophyseal, adrenocortical and testicular hormones in men infected with human immunodeficiency virus. J Clin Endocrinol Metab 1990;70:572-7.
266. Dobs AS, Dempsey MA, Landenson PW, Polk BF. Endocrine disorders in men infected with human immunodeficiency virus. Am J Med 1988;84:611-6.
267. Membreno L, Irony I, Dere W, Klein R, Biglieri EG, Cobb E. Adrenocortical function in acquired immunodeficiency syndrome. J Clin Endocrinol Metab 1987;65:482-7.
268. Norbiato G, Bevilacqua M, Vago T, et al. Cortisol resistance in acquired immunodeficiency syndrome. J Clin Endocrinol Metab 1992;74:608-13.
269. Brenchley JM, Douek DC. HIV infection and the gastrointestinal immune system. Mucosal Immunol 2008;1:23-30.
270. Nose V. Diagnostic pathology: endocrine, 2nd ed. Philadelphia: Elsevier; 2018.

Miscellaneous Infections

271. Guttman PH. Addison's disease: statistical analysis of 566 cases and study of the pathology. Arch Pathol 1930;10:742-895.
272. Mertens RB, Didier ES, Fishbein MC, Bertucci DC, Rogers LB, Orenstein JM. Encephalitozoon cuniculi microsporidiosis: infection of the brain, heart, kidneys, trachea, adrenal glands, and urinary bladder in a patient with AIDS. Mod Pathol 1997;10:68-77.
273. Tosoni A, Nebuloni M, Ferri A, et al. Disseminated microsporidiosis caused by encephalitozoon cuniculi III (dog type) in an Italian AIDS patient: a retrospective study. Mod Pathol 2002;15:577-83.
274. Riestra-Castaneda JM, Riestra-Castaneda R, Gonzalez-Garrido AA, et al. Granulomatous amebic encephalitis due to balamuthia mandrillaris (leptomyxiidae): report of four cases from Mexico. Am J Trop Med Hyg 1997;56:603-7.

275. Teixeira Vde P, Araujo MB, dos Reis MA, et al. Possible role of an adrenal parasite reservoir in the pathogenesis of chronic trypanosoma cruzi myocarditis. Trans R Soc Trop Med Hyg 1993;87:552-4.
276. Teixeira Vde P, Hial V, Gomes RA, et al. Correlation between adrenal central vein parasitism and heart fibrosis in chronic chagasic myocarditis. Am J Trop Med Hyg 1997;56:177-80.
277. Brenner DS, Jacobs SC, Drachenberg CB, Papadimitriou JC. Isolated visceral leishmaniasis presenting as an adrenal cystic mass. Arch Pathol Lab Med 2000;124:1553-6.
278. Mondain-Miton V, Toussaint-Gari M, Hofman P, et al. Atypical leishmaniasis in a patient infected with human immunodeficiency virus. Clin Infect Dis 1995;21:663-5.
279. Akcay M, Akcay G, Balik A, Boyuk A. Hydatid cysts of the adrenal gland: review of nine patients. World J Surg 2004;28:97-9.
280. Bastounis E, Pikoulis E, Leppaniemi A, Cyrochristos D. Hydatid disease: a rare cause of adrenal cyst. Am J Surg 1996;62:383-5.
281. Heppner C, Petzke F, Arlt W, et al. Adrenocortical insufficiency in Rhodesian sleeping sickness is not attributable to suramin. Trans R Soc Trop Med Hyg 1995;89:65-8.

Adrenal Cortical Hypofunction

282. Nicolaides NC, Chrousos GP, Charmandari E. Adrenal Insufficiency. Endotext [Internet]. South Dartmouth: MDText.com, Inc.; 2017. Available at https://www.ncbi.nlm.nih.gov/books/NBK279083/.
283. Arlt W, Allolio B. Adrenal insufficiency. Lancet 2003;361:1881-93.
284. Neary N, Nieman L. Adrenal insufficiency: etiology, diagnosis and treatment. Curr Opin Endocrinol Diabetes Obes 2010;17:217-23.
285 Erichsen MM, Lovas K, Skinningsrud B, et al. Clinical, immunological, and genetic features of autoimmune primary adrenal insufficiency: observation from a Norwegian registry. J Clin Endocrinol Metab 2009;94:4882-90.
286. Kong MF, Jeffocoate W. Eighty-six cases of Addison's disease. Clin Endocrinol (Oxf) 1994;41: 757-61.
287. Laureti S, Vecchi L, Santeusanio F, Falorni A. Is the prevalence of Addison's disease underestimated? J Clin Endocrinol Metab 1999;84:1762.
288. Løvås K, Husebye ES. High prevalence and increasing incidence of Addison's disease in western Norway. Clin Endocrinol (Oxf) 2002;56:787-91.
289. Willis AC, Vince FP. The prevalence of Addison's disease in Coventry, UK. Postgrad Med J 1997;73:286-8.
290. Bates AS, Van't Hoff W, Jones PJ, Clayton RN. The effect of hypopituitarism on life expectancy. J Clin Endocrinol Metab 1996;81:1169-72.
291. Nilsson B, Gustavasson-Kadaka E, Bengtsson BA, Jonsson B. Pituitary adenomas in Sweden between 1958 and 1991: incidence, survival, and mortality. J Clin Endocrinol Metab 2000; 85:1420-5.
292. Regal M, Páramo C, Sierra SM, Garcia-Mayor RV. Prevalence and incidence of hypopituitarism in an adult Caucasian population in northwestern Spain. Clin Endocrinol 2001;55:735-40.
293. Tomlinson JW, Holden N, Hills RK, et al. Association between premature mortality and hypopituitarism. West Midlands Prospective Hypopituitary Study Group. Lancet 2001;357:425-31.
294. Addison T. On the constitutional and local effects of disease of the supra-renal capsules. London: Samuel Highley; 1855.
295. Løvås K, Husebye ES. Addison's disease. Lancet 2005;365:2058-61.
296. Betterle C, Morlin L. Autoimmune Addison's disease. Endocr Dev 2011;20:161-72.
297. Mitchell AL, Pearce SH. Autoimmune Addison disease: pathophysiology and genetic complexity. Nat Rev Endocrinol 2012;8:306-16.
298. Dunlop D. Eighty-six cases of Addison's disease. Br Med J 1963;2:887-91.
299. Florkowski CM, Holmes SJ, Elliot JR, Donald RA, Espiner EA. Bone mineral density is reduced in female but not male subjects with Addison's disease. N Z Med J 1994;107:52-3.
300. Irvine WJ, Barnes EW. Addison's disease, ovarian failure and hypoparathyroidism. Clin Endocrinol Metab 1975;4:379.
301. Knowlton AI, Baer L. Cardiac failure in Addison's disease. Am J Med 1983;74:829-36.
302. Margaretten W, Nakai H, Landing BH. Septicemic adrenal hemorrhage. Am J Dis Child 1963;105:346-51.
303. Otabe S, Muto S, Asano Y, et al. Hyperreninemic hypo aldosteronism due to hepatocellular carcinoma metastatic to the adrenal gland. Clin Nephrol 1991;35:66-71.
304. Sloper JC. The pathology of the adrenals, thymus and certain other endocrine glands in Addison's disease: an analysis of 37 necropsies. Proc R Soc Med 1955;48:625-8.
305. Vita JA, Silverberg SJ, Goland RS, Austin JH, Knowlton AI. Clinical clues to the cause of Addison's disease. Am J Med 1985;78:461-6.
306. Orth DN, Kovacs WJ. The adrenal cortex. In: Wilson JD, Foster DW, Kronenberg HM, Larsen PR, eds. Williams textbook of endocrinology, 9th ed. Philadelphia: WB Saunders; 1998:517-664.

307. Betterle C, Scalici C, Presotto F, et al. The natural history of adrenal function in autoimmune patients with adrenal autoantibodies. J Endocrinol 1988;117:467-75.
308. Boscaro M, Betterle C, Sonino N, et al. Early adrenal hypofunction in patients with organ-specific autoantibodies and no clinical adrenal insufficiency. J Clin Endocrinol Metab 1994;79:452-5.
309. Husebye ES, Lovas K. Immunology of Addison's disease and premature ovarian failure. Endocrinol Metab Clin North Am 2009;38:389-405.
310. Irvine WJ, Barnes EW. Adrenocortical insufficiency. Clin Endocrinol Metab 1972;1:1549-94.
311. Nerup J. Addison's disease—clinical studies. A report of 108 cases. Acta Endocrinol 1974;76:127-41.
312. Peterson P, Krohn KJ. Mapping of B cell epitopes on steroid 17-alpha-hydroxylase, an autoantigen in autoimmune polyglandular syndrome type I. Clin Exp Immunol 1994;98:104-9.
313. Song YH, Connor EL, Muir A, et al. Autoantibody epitope mapping of the 21-hydroxylase antigen in autoimmune Addison's disease. J Clin Endocrinol Metab 1994;78:1108-12.
314. Wedlock N, Asawa T, Baumann-Antczak A, Smith BR, Furmaniak J. Autoimmune Addison's disease. Analysis of autoantibody binding sites on human steroid 21-hydroxylase. FEBS Lett 1993;332:123-6.
315. Blizzard RM, Chee D, Davis W. The incidence of parathyroid and other antibodies in the sera of patients with idiopathic hypoparathyroidism. Clin Exp Immunol 1966;1:119-28.
316. McHardy-Young S, Lessof MH, Maisey MN. Serum TSH and thyroid antibody studies in Addison's disease. Clin Endocrinol (Oxf) 1972; 1:45-56.
317. Zelissen PM, Bast EJ, Croughs RJ. Associated autoimmunity in Addison's disease. J Autoimmun 1995;8:121-30.
318. Brozzetti A, Marzotti S, La Torre D, et al. Autoantibody responses in autoimmune ovarian insufficiency and in Addison's disease are IgG1 dominated and suggest a predominant, but not exclusive, Th1 type of response. Eur J Endocrinol 2010;163:309-17.
319. Chen S, Sawicka J, Prentice L, et al. Analysis of autoantibody epitopes on steroid 21-hydroxylase using a panel of monoclonal antibodies. J Clin Endocrinol Metab 1998;83:2977-86.
320. Fairchild RS, Schimke RN, Abdou NI. Immunoregulation abnormalities in familial Addison's disease. J Clin Endocrinol Metab 1980:51:1074-7.
321. Rabinowe SL, Jackson RA, Dluhy RG, Williams GH. Ia-positive T lymphocytes in recently diagnosed idiopathic Addison's disease. Am J Med 1984;77:597-601.
322. Weetman AP. Autoimmunity to steroid-producing cells and familial polyendocrine autoimmunity. Baillieres Clin Endocrinol Metab 1995;9:157-74.
323. Jackson R, McNicol AM, Farquharson M, Foulis AK. Class II MHC expression in normal adrenal cortex and cortical cells in autoimmune Addison's disease. J Pathol 1988;155:113-20.
324. Leshin M. Polyglandular autoimmune syndromes. Am J Med Sci 1985;290:77-88.
325. Partanen J, Peterson P, Westman P, Aranko S, Krohn K. Major histocompatibility complex class II and III in Addison's disease. MHC alleles do not predict autoantibody specificity and 21-hydroxylase gene polymorphism has no independent role in disease susceptibility. Hum Immunol 1994;41:135-40.
326. Nerup J. Addison's disease—serological studies. Acta Endocrinol 1974;76:142-58.
327. Neufeld M, Maclaren NK, Blizzard RM. Two types of autoimmune Addison's disease associated with different polyglandular autoimmune (PGA) syndromes. Medicine 1981;60:355-62.
328. Spinner MW, Blizzard RM, Childs B. Clinical and genetic heterogeneity in idiopathic Addison's disease and hypoparathyroidism. J Clin Endocrinol Metab 1968;28:795-804.
329. Farid NR, Bear JC. The human major histocompatibility complex and endocrine disease. Endocr Rev 1981;2:50-86.
330. Maclaren NK, Riley WJ. Thyroid, gastric, and adrenal autoimmunities associated with insulin-dependent diabetes mellitus. Diabetes Care 1985;8(Suppl 1):34-8.
331. Badenhoop K, Walfish PG, Rau H, et al. Susceptibility and resistance alleles of human leukocyte antigen (HLA) DQA1 and HLA DQB1 are shared in endocrine autoimmune disease. J Clin Endocrinol Metab 1995;80:2112-7.
332. Ghaderi M, Gambelunghe G, Tortoioli C, et al. MHC2TA single nucleotide polymorphism and genetic risk for autoimmune adrenal insufficiency. J Clin Endocrinol Metab 2006;91:4107-11.
333. Yu L, Brewer KW, Gates S, et al. DRB1*04 and DQ alleles: expression of 21-hydroxylase autoantibodies and risk of progression to Addison's disease. J Clin Endocrinol Metab 1999;84:328-35.
334. Heino M, Scott HS, Chen Q, et al. Mutation analyses of North American APS-1 patients. Hum Mutat 1999;13:69-74.
335. Nagamine K, Peterson P, Scott HS, et al. Positional cloning of the APECED gene. Nat Genet 1997;17:393-8.
336. Scott HS, Heino M, Peterson P, et al. Common mutations in autoimmune polyendocrinopathy-candidiasis-ectodermal dystrophy patients of different origins. Mol Endocrinol 1998;12:1112-9.

337. Ahonen P, Myllärniemi S, Sipilä I, Perheentupa J. Clinical variation of autoimmune polyendocrinopathy-candidiasis-ectodermal dystrophy (APECED) in a series of 68 patients. N Engl J Med 1990;322:1829-36.
338. Trence DL, Morley JE, Handwerger BS. Polyglandular autoimmune syndromes. Am J Med 1984;77:107-16.
339. Dittmar M, Kahaly GJ. Polyglandular autoimmune syndromes: immunogenetics and long-term follow-up. J Clin Endocrinol Metab 2003;88:2983-92.
340. Valenzuela GA, Smalley WE, Schain DC, Vance ML, McCallum RW. Reversibility of gastric dysmotility in cortisol deficiency. Am J Gastroenterol 1987;82:1066-8.
341. Stewart PM. The adrenal cortex. In: Kronenberg HM, Melmed S, Polonsky KS, Larsen PR, eds. William's textbook of endocrinology, 12th ed. Philadelphia: Saunders; 2008:445-503.
342. Neto RA, de Carvalho JF. Diagnosis and classification of Addison's disease (autoimmune adrenalitis). Autoimmun Rev 2014;13:408-11.
343. Jarvis JL, Jenkins D, Sosman MC, et al. Roentgenologic observations in Addison's disease; a review of 120 cases. Radiology 1954;62:16-29.
344. Rao RH, Vagnucci AH, Amico JA. Bilateral massive adrenal hemorrhage: early recognition and treatment. Ann Intern Med 1989;110:227-35.
345. Kubota T, Hayashi M, Kabuto M, et al. Corticotroph cell hyperplasia in a patient with Addison's disease: case report. Surg Neurol 1992; 37:441-7.
346. Saenger P, Levine LS, Irvine WJ, et al. Progressive adrenal failure in polyglandular autoimmune disease. J Clin Endocrinol Metab 1982;54:863-7.
347. Ahonen P, Miettinen A, Perheentupa J. Adrenal and steroidal cell antibodies in patients with autoimmune polyglandular disease type I and risk of adrenocortical and ovarian failure. J Clin Endocrinol Metab 1987;64:494-500.
348. Oelkers W. Adrenal insufficiency. N Engl J Med 1996;335:1206-12.
349. Symington T. Functional pathology of the human adrenal gland. Edinburgh: Livingstone; 1969.
350. Buxi TB, Vohra RB, Sujatha S, et al. CT in adrenal enlargement due to tuberculosis: a review of the literature with five new cases. Clin Imaging 1992;16:102-8.
351. De Bellis AA, Falorni A, Laureti S, et al. Time course of 21-hydroxylase antibodies and long-term remission of subclinical autoimmune adrenalitis after corticosteroid therapy: case report. J Clin Endocrinol Metab 2001;86:675-8.
352. Smans LC, Zellisen PM. Partial recovery of adrenal function in a patient with autoimmune Addison's disease. J Endocrinol Invest 2008;31:672-4.
353. Smans LC, Zellisen PM. Does recovery of adrenal function occur in patients with autoimmune Addison's disease? Clin Endocrinol (Oxf) Apr 2011;74:434-7.
354. Grossman AB. Clinical review: the diagnosis and management of central hypoadrenalism. J Clin Endocrinol Metab 2010;95:4855-63.
355. Agarwal G, Bhatia V, Cook S, Thomas PQ. Adrenocorticotropin deficiency in combined pituitary hormone deficiency patients homozygous for a novel PROP1 deletion. J Clin Endocrinol Metab 2000;85:4556-61.
356. Böttner A, Keller E, Kratzsch J, et al. PROP1 mutations cause progressive deterioration of anterior pituitary function including adrenal insufficiency: a longitudinal analysis. J Clin Endocrinol Metab 2004;89:5256-65.
357. Parks JS, Brown MR, Hurley DL, et al. Heritable disorders of pituitary development. J Clin Endocrinol Metab 1999;84:4362-70.
358. Pernasetti F, Toledo SP, Vasilyev VV, et al. Impaired adrenocorticotropin-adrenal axis in combined pituitary hormone deficiency caused by a two-base pair deletion (301-302delAG) in the prophet of Pit-1 gene. J Clin Endocrinol Metab 2000;85:390-7.
359. Stacpoole PW, Interlandi JW, Nicholson WE, Rabin D. Isolated ACTH deficiency: a heterogeneous disorder. Critical review and report of four new cases. Medicine (Baltimore) 1982; 61:13-24.
360. Cantalamessa L, Catania A, Baldini M, Orsatti A, Motta P, Peracchi G. CRH and lysine-vasopressin stimulation tests in the diagnosis of hypoadrenalism secondary to hypothalamic or pituitary disorders. Horm Metab Res 1990;22:389-93.
361. Koide Y, Kimura S, Inoue S, et al. Responsiveness of hypophyseal-adrenocortical axis to repetitive administration of synthetic ovine corticotropin-releasing hormone in patients with isolated adrenocorticotropin deficiency. J Clin Endocrinol Metab 1986;63:329-35.
362. Yoshida T, Arai T, Sugano J, Yarita H, Yanagisawa H. Isolated ACTH deficiency accompanied by "primary hypothyroidism" and hyperprolactinemia. Acta Endocrinol 1983;104:397-401.
363. Kanemaru Y, Noguchi T, Onaya T. Isolated ACTH deficiency associated with transient thyrotoxicosis and hyperprolactinemia. Endocrinol Jpn 1989;36:459-64.
364. Shigemasa C, Kouchi T, Ueta Y, Mitani Y, Yoshida A, Mashiba H. Evaluation of thyroid function in patients with isolated adrenocorticotropin deficiency. Am J Med Sci 1992;304:279-84.

365. Gan EH, MacArthur K, Mitchell AL, Joshi A, Crock P, Pearce SH. Spontaneous and tetracosactide-induced anti-ACTH antibodies in man. Clin Endocrinol (Oxf) 2016;84:489-95.
366. Chang LS, Barroso-Sousa R, Tolaney SM, Hodi FS, Kaiser UB, Min L. Endocrine toxicity of cancer immunotherapy targeting immune checkpoints. Endocr Rev 2019;40:17-65.
367. Jensen MD, Handwerger BS, Scheithauer BW, Carpenter PC, Mirakian R, Banks PM. Lymphocytic hypophysitis with isolated corticotropin deficiency. Ann Intern Med 1986;105:200-3.
368. Sugiura M, Hashimoto A, Shizawa M, et al. Heterogeneity of anterior pituitary cell antibodies detected in insulin-dependent diabetes mellitus and adrenocorticotropic hormone deficiency. Diabetes Res 1986;3:111-4.
369. Sauter NP, Toni R, McLaughlin CD, Dyess EM, Kritzman J, Lechan RM. Isolated adrenocorticotropin deficiency associated with an autoantibody to a corticotroph antigen that is not adrenocorticotropin or other proopiomelanocortin-derived peptides. J Clin Endocrinol Metab 1990;70:1391-7.
370. Clark AJ. 60 years of POMC: the proopiomelanocortin gene: discovery, deletion and disease. J Mol Endocrinol 2016;56:T27-37.
371. Vallette-Kasic S, Brue T, Pulichino AM, et al. Congenital isolated adrenocorticotropin deficiency: an underestimated cause of neonatal death, explained by TPIT gene mutations. J Clin Endocrinol Metab 2005;90:1323-31.
372. Nussey SS, Soo SC, Gibson S, et al. Isolated congenital ACTH deficiency: a cleavage enzyme defect? Clin Endocrinol (Oxf) 1993;39:381-5.
373. Phifer RF, Spicer SS, Orth DN. Specific demonstration of the human hypophyseal cells which produce adrenocorticotropic hormone. J Clin Endocrinol Metab 1970;31:347-61.
374. Bleicken B, Hahner S, Ventz M, Quinkler M. Delayed diagnosis of adrenal insufficiency is common: a cross-sectional study in 216 patients. Am J Med Sci 2010;339:525-31.
375. Burke CW. Adrenocortical insufficiency. Clin Endocrinol Metab 1985;14:947-76.
376. Charmandari E, Nicolaides NC, Chrousos GP. Adrenal insufficiency. Lancet 2014;383:2152-67.
377. Mineura K, Goto T, Yoneya M, Kowada M, Tamakawa Y, Kagaya H. Pituitary enlargement associated with Addison's disease. Clin Radiol 1987;38:435-7.
378. Nakahara M, Shibasaki T, Shizume K, et al. Corticotropin-releasing factor test in normal subjects and patients with hypothalamic-pituitary-adrenal disorders. J Clin Endocrinol Metab 1983;57:963-8.
379. Schulte HM, Chrousos GP, Avgerinos P, et al. The corticotropin-releasing hormone stimulation test: a possible aid in the evaluation of patients with adrenal insufficiency. J Clin Endocrinol Metab 1984;58:1064-7.
380. Taylor AL, Fishman LM. Corticotropin-releasing hormone. N Engl J Med 1988;319:213-22.
381. Rose BD, Post TW. Clinical physiology of acid-base and electrolyte disorders, 5th ed. New York: McGraw-Hill; 2001:900.
382. DeFronzo RA. Hyperkalemia and hyporeninemic hypoaldosteronism. Kidney Int 1980;17:118-34.
383. Schambelan M, Sebastian A, Rector FC Jr. Mineralocorticoid-resistant renal hyperkalemia without salt wasting (type II pseudohypoaldosteronism): role of increased renal chloride reabsorption. Kidney Int 1981;19:716-27.
384. Young WF. Etiology, diagnosis, and treatment of hypoaldosteronism (type 4 RTA). In: Sterns RH, Lacroix A, Forman JP, eds. UpToDate; 2013. Available at: https://www.uptodate.com.acs.hcn.com.au/contents/etiology-diagnosis-and-treatment-of-hypoaldosteronism-type-4-rta.
385. Schambelan M, Stockigt JR, Biglieri EG. Isolated hypoaldosteronism in adults. A renin-deficiency syndrome. N Engl J Med 1972;287:573-8.
386. White PC. Aldosterone synthase deficiency and related disorders. Mol Cell Endocrinol 2004;217:81.
387. Ulick S. Diagnosis and nomenclature of the disorders of the terminal portion of the aldosterone biosynthetic pathway. J Clin Endocrinol Metab 1976;43:92-6.
388. White PC. Disorders of aldosterone biosynthesis and action. N Engl J Med 1994;331:250-8.
389. Shizuta Y, Kawamoto T, Mitsuuchi Y, et al. Inborn errors of aldosterone biosynthesis in humans. Steroids 1995;60:15-21.
390. Lalioti MD, Zhang J, Volkman HM, et al. Wnk4 controls blood pressure and potassium homeostasis via regulation of mass and activity of the distal convoluted tubule. Nat Genet 2006;38:1124-32.
391. O'Shaughnessy KM. Gordon syndrome: a continuing story. Pediatr Nephrol 2015;30:1903-8.
392. Rossier BC. Negative regulators of sodium transport in the kidney: key factors in understanding salt-sensitive hypertension? J Clin Invest 2003;111:947-50.
393. Bonny O, Rossier BC. Disturbances of Na/K balance: pseudohypoaldosteronism revisited. J Am Soc Nephrol 2002;13:2399-414.
394. Gordon RD. Syndrome of hypertension and hyperkalemia with normal glomerular filtration rate. Hypertension 1986;8:93-102.
395. Take C, Ikeda K, Kurasawa T, Kurokawa K. Increased chloride reabsorption as an inherited renal tubular defect in familial type II pseudohypoaldosteronism. N Engl J Med 1991;324:472-6.

396. Oster JR, Singer I, Fishman LM. Heparin-induced aldosterone suppression and hyperkalemia. Am J Med 1995;98:575-86.

Adrenal Hemorrhage and Necrosis

397. Rao RH, Vagnucci AH, Amico JA. Bilateral massive adrenal hemorrhage: early recognition and treatment. Ann Intern Med 1989;110:227-35.
398. Streeten DH. Adrenal hemorrhage. Endocrinologist 1996;6:277.
399. Xarli VP, Steele AA, Davis PJ, Buescher ES, Rios CN, Garcia-Bunuel R. Adrenal hemorrhage in the adult. Medicine 1978;57:211-21.
400. Migeon CJ, Kenny FM, Hung W, Voorhess ML. Study of adrenal function in children with meningitis. Pediatrics 1967;40:163-83.
401. Lacroix A. Causes of primary adrenal insufficiency Addison's disease). In: Lacroix A, Martin KA, eds. UpToDate; 2021. Available at https://www.uptodate.com/contents/causes-of-primary-adrenal-insufficiency-addisons-disease.
402. Friderichsen C. Waterhouse-Friderichsen syndrome. Acta Endocrinol (Copenh) 1955;18:482-92.
403. Waterhouse R. A case of supraneural apoplexy. Lancet 1911;I:577-8.
404. Fox B. Venous infarction of the adrenal glands. J Pathol 1976;119:65-89.
405. Brown BS, Dunbar JS, Macewan DW. The radiologic features of acute massive adrenal hemorrhage of the newborn. J Can Assoc Radiol 1962;13:100-7.
406. Black J, Williams DI. Natural history of adrenal haemorrhage in the newborn. Arch Dis Child 1973;48:183-90.
407. Rosenberger LH, Smith PW, Sawyer RG, Hanks JB, Adams RB, Hedrick TL. Bilateral adrenal hemorrhage: the unrecognized cause of hemodynamic collapse associated with heparin-induced thrombocytopenia. Crit Care Med 2011;39:833-8.
408. Caron P, Chabannier MH, Cambus JP, Fortenfant F, Otal P, Suc JM. Definitive adrenal insufficiency due to bilateral adrenal hemorrhage and primary antiphospholipid syndrome. J Clin Endocrinol Metab 1998;83:1437-9.
409. Espinosa G, Santos E, Cervera R, et al. Adrenal involvement in the antiphospholipid syndrome: clinical and immunologic characteristics of 86 patients. Medicine (Baltimore) 2003;82:106-18.
410. Kovacs K, Lam Y, Pater J. Bilateral massive adrenal hemorrhage: assessment of putative risk factors by the case-control method. Medicine (Baltimore) 2001;80:45-53.
411. Kawashima A, Sandler CM, Ernst RD, et al. Imaging of nontraumatic hemorrhage of the adrenal gland. Radiographics 1999;19:949-63.
412. Di Serafino M, Severino R, Coppola V, et al. Nontraumatic adrenal hemorrhage: the adrenal stress. Radiol Case Rep 2017;12:483-7.
413. Sacerdote MG, Johnson PT, Fishman EK. CT of the adrenal gland: the many faces of adrenal hemorrhage. Emerg Radiol 2012;19:53-60.

Metastatic Tumors Causing Adrenal Insufficiency

414. Cedermark BJ, Blumenson LE, Pickren JW, Holyoke DE, Elias EG. The significance of metastases to the adrenal glands in adenocarcinoma of the colon and rectum. Surg Gynecol Obstet 1977; 144:537-46.
415. Redman BG, Pazdur R, Zingas AP, Loredo R. Prospective evaluation of adrenal insufficiency in patients with adrenal metastasis. Cancer 1987;60:103-7.
416. Seidenwurm DJ, Elmer EB, Kaplan LM, Williams EK, Morris DG, Hoffman AR. Metastases to the adrenal glands and the development of Addison's disease. Cancer 1984;54:552-7.
417. Sheeler LR, Myers JH, Eversman JJ, Taylor HC. Adrenal insufficiency secondary to carcinoma metastatic to the adrenal gland. Cancer 1983;52:1312-6.
418. Huminer D, Garty M, Lapidot M, Leiba S, Borohov H, Rosenfeld JB. Lymphoma presenting with adrenal insufficiency. Adrenal enlargement on computed tomographic scanning as a clue to diagnosis. Am J Med 1988;84:169-72.
419. Prayson RA, Segal GH, Stoler MH, Licata AA, Tubbs RR. Angiotropic large-cell lymphoma in a patient with adrenal insufficiency. Arch Pathol Lab Med 1991;115:1039-41.
420. Serrano S, Tejedor L, Garcia B, Hallal H, Polo JA, Algucil G. Addisonian crisis as the presenting feature of bilateral primary adrenal lymphoma. Cancer 1993;71:4030-3.
421. Barker NW. The pathologic anatomy in twenty-eight cases of Addison's disease. Arch Pathol 1929;8:432.
422. Cedermark BJ, Sjöberg HE. The clinical significance of metastases to the adrenal glands. Surg Gynecol Obstet 1981;152:607-10.
423. Kung AW, Punn KK, Lam K, Wang C, Leung CY. Addisonian crisis as presenting feature in malignancies. Cancer 1990;65:177-9.
424. Omoigui NA, Cave WT Jr, Chang AY. Adrenal insufficiency. A rare initial sign of metastatic colon carcinoma. J Clin Gastroenterol 1987; 9:470-4.
425. Allard P, Yankaskas BC, Fletcher RH, Parker LA, Halvorsen RA Jr. Sensitivity and specificity of computed tomography for the detection of adrenal metastatic lesions among 91 autopsied lung cancer patients. Cancer 1990;66:457-62.
426. Campbell CM, Middleton RG, Rigby OF. Adrenal metastasis in renal cell carcinoma. Urology 1983;21:403-5.
427. Selli C, Corini M, Barbanti G, Barbadgli G, Turin D. Simultaneous bilateral adrenal involvement by renal cell carcinoma: experience with 3 cases. J Urol 1987;137:480-2.

428. Patrova J, Jarocka I, Wahrenberg H, Falhammar H. Clinical outcomes in adrenal incidentaloma: experience from one center. Endocr Pract 2015; 21:870-7.
429. Calissendorff J, Calissendorff F, Falhammar H. Adrenocortical cancer: mortality, hormone secretion, proliferation and urine steroids—experience from a single centre spanning three decades. BMC Endocr Disord 2016:16:15.

Adrenal Amyloid

430. Guttman PH. Addison's disease: a statistical analysis of 566 cases and study of the pathology. Arch Pathol 1930;10:742-895.
431. Case Records of the Massachusetts General Hospital. Weekly clinicopathological exercises. Case 3-2000. A 66-year-old woman with diabetes, coronary disease, orthostatic hypotension and the nephrotic syndrome. N Engl J Med 2000;342:264-73.
432. Erdkamp FL, Gans RO, Hoorntje SJ. Endocrine organ failure due to systemic AA-amyloidosis. Neth J Med 1991;38:24-8.
433. Muro K, Kobayashi M, Shimizu Y, et al. [A case of systemic AA amyloidosis complicating Crohn's disease.] Nippon Jinzo Gakkai Shi 1998;40:284-9. [Japanese]
434. Tomita T. Amyloidosis of pancreatic islets in primary amyloidosis (AL type). Pathol Int 2005; 55:223-7.
435. Emeksiz H, Bakkaloglu S, Camurdan O, et al. Acute adrenal crisis mimicking familial Mediterranean fever attack in a renal transplant FMF patient with amyloid goiter. Rheumatol Int 2010;30:1647-9.
436. Keven K, Oztas E, Aksoy H, Duman N, Erbay B, Ertürk S. Polyglandular endocrine failure in a patient with amyloidosis secondary to familial Mediterranean fever. Am J Kidney Dis 2001; 38:E39.
437. Lachmann HJ, Goodman HJ, Gilbertson JA, et al. Natural history and outcome in systemic AA amyloidosis. N Engl J Med 2007;356:2361-71.
438. Ericksson L, Westermark P. Age-related accumulation of amyloid inclusions in adrenal cortical cells. Am J Pathol 1990;136:461-6.
439. Rocken C, Eick B, Saeger W. Senile amyloidoses of the pituitary and adrenal glands. Morphological and statistical investigations. Virchows Arch 1996;429:293-9.
440. Sasaki M, Kono M, Nakamura Y, Ishiura Y. Multinodular deposition of AA-type amyloid localized in the adrenal glands of an old man. Acta Pathol Jap 1992;42:893-6.
441. Arik N, Tasdemir I, Karaaslan Y, Yasavul U, Turgan C, Caglar S. Subclinical adrenocortical insufficiency in renal amyloidosis. Nephron 1990;56:246-8.
442. Danby P, Harris KP, Williams B, Feehally J, Walls J. Adrenal dysfunction in patients with renal amyloid. Q J Med 1990;76:915-22.
443. Gündüz Z, Kelestimur F, Durak AC, et al. The hormonal and radiological evaluation of adrenal glands, and the determination of the usefulness of low dose ACTH test in patients with renal amyloidosis. Ren Fail 2001;23:239-49.
444. Olofsson BO, Grankvist K, Boman K, Forsberg K, Lafvas I, Lithner F. Assessment of thyroid and adrenal function in patients with familial amyloidotic polyneuropathy. J Intern Med 1989;225:337-41.
445. Ozdemir D, Dagdelen S, Erbas T. Endocrine involvement in systemic amyloidosis. Endocr Pract 2010;16:1056-63.
446. el-Reshaid KA, Hakim AA, Hourani HA, Seshadri MS. Endocrine abnormalities in patients with amyloidosis. Ren Fail 1994;16:725-30.
447. Neville AM, O'Hare MJ. The human adrenal cortex: pathology biology, an integrated approach. New York: Springer-Verlag; 1982:263-4.
448. Cope S, Woodrow JC. Adrenal amyloidosis as a cause of Addison's disease. Postgrad Med J 1953;29:558-63.

Adrenoleukodystrophy

449. Wanders RJ. Adrenoleukodystrophy. In: Patterson MC, Hahn S, eds. UpToDate; 2020. Available at https://www.uptodate.com.acs.hcn.com.au/contents/ enoleukodystrophy&source=search_result&selectedTitle=1~42&usage_type=default&display_rank=1.
450. Pantanowitz L, Abu-Jawdeh G, Joseph J. Adrenoleukodystrophy. Fertil Steril 2000;74:162.
451. Moser HW, Loes DJ, Melhem ER, et al. X-linked adrenoleukodystrophy: overview and prognosis as a function of age and brain magnetic resonance imaging abnormality. A study involving 372 patients. Neuropediatrics 2000;31:227-39.
452. Percy AK, Rutledge SL. Adrenoleukodystrophy and related disorders. Ment Retard Dev Disabil Res Rev 2001;7:179-89.
453. Raymond GV, Moser AB, Fatemi A. X-linked adrenoleukodystrophy. In: In: Adam MP, Ardinger HH, Pagon RA, et al., eds. GeneReviews® [internet]. Seattle: University of Washington; 2018. Available at: https://www.ncbi.nlm.nih.gov/books/NBK1315/.
454. Griffin JW, Goren E, Schaumburg H, et al. Adrenomyeloneuropathy: a probable variant of adrenoleukodystrophy. I. Clinical and endocrinologic aspects. Neurology 1977;27:1107-13.
455. Moser HW, Moser AE, Singh I, et al. Adrenoleukodystrophy: survey of 303 cases: biochemistry, diagnosis, and therapy. Ann Neurol 1984;16:628-41.
456. Sarde CO, Mosser J, Kioschis P, et al. Genomic organization of the adrenoleukodystrophy gene. Genomics 1994;22:13-20.

457. Holzinger A, Kammerer S, Berger J, Roscher AA. cDNA cloning and mRNA expression of the human adrenoleukodystrophy related protein (ALDRP), a peroxisomal ABC transporter. Biochem Biophys Res Commun 1997;239:261-4.
458. McGuinness MC, Lu JF, Zhang HP, et al. Role of ALDP (ABCD1) and mitochondria in X-linked adrenoleukodystrophy. Mol Cell Biol 2003;23:744-53.
459. Migeon BR, Moser HW, Moser AB, Axelman J, Sillence D, Norum RA. Adrenoleukodystrophy: evidence for X linkage, inactivation, and selection favoring the mutant allele in heterozygous cells. Proc Natl Acad Sci U S A 1981;78:5066-70.
460. Mosser J, Douar AM, Sarde CO, et al. Putative X-linked adrenoleukodystrophy gene shares unexpected homology with ABC transporters. Nature 1993;361:726-30.
461. Mosser J, Lutz Y, Stoeckel ME, et al. The gene responsible for adrenoleukodystrophy encodes a paroxysomal membrane protein. Hum Mol Genet 1994;3:265-71.
462. Kemp S, Wanders R. Biochemical aspects of X-linked adrenoleukodystrophy. Brain Pathol 2010;20:831-7.
463. Kemp S, Huffnagel IC, Linthorst GE, et al. Adrenoleukodystrophy—neuroendocrine pathogenesis and redefinition of natural history. Nat Rev Endocrinol 2016;12:606-15.
464. Powers JM, Schaumberg HH. The adrenal cortex in adreno-leukodystrophy. Arch Pathol 1973; 96:305-10.
465. Chen X, DeLellis RA, Hoda SA. Adrenoleukodystrophy. Arch Pathol Lab Med 2003;127:119-20.
466. Powers JM, Schaumberg HH. Adreno-leukodystrophy. Similar ultrastructural changes in adrenal cortical cells and Schwann cells. Arch Neurol 1974;30:406-8.
467. Powers JM, Schaumberg HH. The testis in adreno-leukodystrophy. Am J Pathol 1981;102:90-8.
468. Kemp S, Wanders RJ. X-linked adrenoleukodystrophy: very long-chain fatty acid metabolism, ABC half-transporters and the complicated route to treatment. Mol Genet Metab 2007; 90:268-76.
469. Mahmood A, Dubey P, Moser HW, Moser A. X-linked adrenoleukodystrophy: therapeutic approaches to distinct phenotypes. Pediatr Transplant 2005;9(Suppl 7):55-62.
470. Turk BR, Theda C, Fatemi A, Moser AB. X-linked adrenoleukodystrophy: pathology, pathophysiology, diagnostic testing, newborn screening and therapies. Int J Dev Neurosci 2020;80:52-72.
471. Aubourg P, Adamsbaum C, Lavallard-Rousseau MC, et al. A two-year trial of oleic and erucic acids ("Lorenzo's oil") as treatment for adrenomyeloneuropathy. N Engl J Med 1993;329:745-52.
472. Bezman L, Moser HW. Incidence of X-linked adrenoleukodystrophy and the relative frequency of its phenotypes. Am J Med Genet 1998;76:415-9.
473. National Library of Medicine. A phase III trial of Lorenzo's oil in adrenomyeloneuropathy. Clinical Trials.gov; 2007. Available at: https://www.clinicaltrials.gov/ct2/show/NCT00545597.
474. Deon M, Wajner M, Sirtori LR, et al. The effect of Lorenzo's oil on oxidative stress in X-linked adrenoleukodystrophy. J Neurol Sci 2006;247:157-64.
475. Engelen M, Ofman R, Dijkgraaf MG, et al. Lovastatin in X-linked adrenoleukodystrophy. N Engl J Med 2010;362:276-7.
476. Gondcaille C, Depreter M, Fourcade S, et al. Phenylbutyrate upregulates the adrenoleukodystrophy-related gene as a nonclassical peroxisome proliferator. J Cell Biol 2005;169:93-104.
477. Moser HW, Moser AB, Hollandsworth K, et al. "Lorenzo's oil" therapy for X-linked adrenoleukodystrophy: rationale and current assessment of efficacy. J Mol Neurosci 2007;33:105-13.
478. Moser HW, Raymond GV, Lu SE, et al. Follow-up of 89 asymptomatic patients with adrenoleukodystrophy treated with Lorenzo's oil. Arch Neurol 2005;62:1073-80.
479. Pai GS, Khan M, Barbosa E, et al. Lovastatin therapy for X-linked adrenoleukodystrophy: clinical and biochemical observations on 12 patients. Mol Genet Metab 2000;69:312-22.
480. Singh I, Pahan K, Khan M. Lovastatin and sodium phenylacetate normalize the levels of very long chain fatty acids in skin fibroblasts of X- adrenoleukodystrophy. FEBS Lett 1998;426:342-6.
481. van Geel BM, Assies J, Haverkort EB, et al. Progression of abnormalities in adrenomyeloneuropathy and neurologically asymptomatic X-linked adrenoleukodystrophy despite treatment with "Lorenzo's oil." J Neurol Neurosurg Psychiatry 1999;67:290-9.
482. Peters C, Charnas LR, Tan Y, et al. Cerebral X-linked adrenoleukodystrophy: the international hematopoietic cell transplantation experience from 1982 to 1999. Blood 2004;104:881-8.
483. van Geel BM, Poll-The BT, Verrips A, Boelens JJ, Kemp S, Engelen M. Hematopoietic cell transplantation does not prevent myelopathy in X-linked adrenoleukodystrophy: a retrospective study. J Inherit Metab Dis 2015;38:359-61.

Wolman Disease

484. Reynolds T. Cholesteryl ester storage disease: a rare and possibly treatable cause of premature vascular disease and cirrhosis. J Clin Pathol 2013;66:918-23.

485. Marin-Valencia I, Pascual JM. Wolman disease. In: Rosenberg RN, Pascual JM, eds. Rosenberg's molecular and genetic basis of neurological and psychiatric disease, 5th ed. London: Elsevier; 2014:403-9.
486. Ben-Haroush A, Yogev Y, Levit O, Hod M, Kaplan B. Isolated fetal ascites caused by Wolman disease. Ultrasound Obstet Gynecol 2003;21:297-8.
487. Ozmen MN, Aygun N, Kilic I, Kuran L, Yalcin B, Besim A. Wolman's disease: ultrasonographic and computed tomographic findings. Pediatr Radiol 1992;22:541-2.

Adrenal Cysts

488. Abeshouse GA, Goldstein RB, Abeshouse BS. Adrenal cysts: review of the literature and report of three cases. J Urol 1959;81:711-9.
489. Barron SH, Emanuel B. Adrenal cyst. A case report and review of the pediatric literature. J Pediatr 1961;59:592-9.
490. Ghandur-Mnaymneh L, Slim M, Muakassa K. Adrenal cysts: pathogenesis and histological identification with a report of 6 cases. J Urol 1979;122:87-91.
491. Kloos RT, Gross MD, Francis IR, Korobkin M, Shapiro B. Incidentally discovered adrenal masses. Endocr Rev 1995;16:460-84.
492. Richards ML. Miscellaneous adrenal neoplasms (cysts, myelolipoma, hemangioma, lymphangioma). In: Linos D, van heerden JA, eds. Adrenal glands: diagnostic aspects and surgical therapy. Heidelberg: Springer; 2005:223-9.
493 Levin SE, Collins DL, Kaplan GW, Weller MH. Neonatal adrenal pseudocysts mimicking metastatic disease. Ann Surg 1974;179:186-9.
494. Oppenheimer EH. Cyst formation in the outer adrenal cortex. Studies in the human fetus and newborn. Arch Pathol 1969;87:653-9.
495. Foster DG. Adrenal cysts. Review of the literature and report of a case. Arch Surg 1966;92: 131-43.
496. Neri LM, Nance FC. Management of adrenal cysts. Am Surg 1999;2:151-63.
497. Lack EE. Tumors of the adrenal gland and extra adrenal paraganglia. AFIP Atlas of Tumor Pathology, 4th Series, Fascicle 8. Washington, DC: American Registry of Pathology; 2007.
498. Kearney GP, Mahoney EM, Maher E, Harrison JH. Functioning and nonfunctioning cysts of the adrenal cortex and medulla. Am J Surg 1977;134:363-8.
499. Torres C, Ro JY, Batt MA, Park YW, Ordonez NG, Ayala AG. Vascular adrenal cysts: a clinicopathologic and immunohistochemical study of six cases and a review of the literature. Mod Pathol 1997;10:530-6
500. Groben PA, Roberson JB, Anger SR, Askin FB, Price WG, Siegal GP. Immunohistochemical evidence for the vascular origin of primary adrenal pseudocysts. Arch Pathol Lab Med 1986;110:121-3.
501. Incze JS, Lui PS, Merriam JC, Austen G, Widrich WC, Gerzof SG. Morphology and pathogenesis of adrenal cysts. Am J Pathol 1979;95:423-32.
502. Medeiros LJ, Lewandrowski KB, Vickery AL Jr. Adrenal pseudocyst: a clinical and pathologic study of eight cases. Human Pathol 1989; 20:660-5.
503. Hodges FV, Ellis FR. Cystic lesions of the adrenal glands. Arch Pathol Lab Med 1958;66:53-8.
504. Symington T. The adrenal cortex. In: Bloodworth JM, ed. Endocrine pathology; gneral and surgical, 2nd ed. Baltimore: William & Wilkins; 1982:437.
505. Lynn RB. Cystic lymphangioma of the adrenal associated with arterial hypertension. Can J Surg 1965;8:92-5.
506. Rodin AE, Hsu FL, Whorton EB. Microcysts of the permanent adrenal cortex in perinates and infants. Arch Pathol 1976;100:499-502.
507. Lack EE. Tumors of the adrenal gland and extra-adrenal paraganglia. Atlas of Tumor Pathology, 3rd Series, Fascicle 19. Washington, DC: Armed Forces Institute of Pathology; 1997:172-4.
508. Medeiros LJ, Weiss LM, Vickery AL Jr. Epithelial-lined (true) cyst of the adrenal gland: a case report. Hum Pathol 1989;20:491-2.
509. Thompson GB, Grant CS, van Heerden JA, et al. Laparoscopic versus open posterior adrenalectomy: a case-control study of 100 patients. Surgery 1997;122:1132-6.
510. Wahl HR. Adrenal cysts. Am J Pathol 1951;27:758.
511. Wells SA, Merke DP, Cutler GB Jr, Norton JA, Lacroix A. Therapeutic controversy. The role of laparoscopic surgery in adrenal disease. J Clin Endocrinol Metab 1998;83:3041-9.

Incidental, Nonfunctional Adrenal Cortical Nodules

512. Reinhard C, Saeger W, Schubert B. Adrenocortical nodules in post-mortem series: development, functional significance, and differentiation from adenomas. Gen Diagn Pathol 1996;141:203-8.
513. Neville AM. The nodular adrenal. Invest Cell Pathol 1978;1:99-111.
514. O'Leary TJ, Ooi TC. The adrenal incidentaloma. Can J Surg 1986;29:6-8.
515 Prinz RA, Brooks MH, Churchill R, et al. Incidental asymptomatic adrenal masses detected by computed tomographic scanning. Is operation required? JAMA 1982;248:701-4.
516. Russi S, Blumenthol HT, Gray SH. Small adenomas of the adrenal cortex in hypertension and diabetes. Arch Intern Med 1945;76:284-91.

517. Shamnan AH, Goddard JW, Sommen SC. A study of the adrenal states in hypertension. J Chronic Dis 1958;8:587-95.
518. Spain DM, Weinsaft P. Solitary adrenal cortical adenoma in an elderly female; frequency. Arch Pathol 1964;78:231-3.
519. Commons RR, Callaway CP. Adenomas of the adrenal cortex. Arch Intern Med 1984;81:37-41.
520. Kokko JP, Brown TC, Berman MM. Adrenal adenoma and hypertension. Lancet 1967;1:468-70.
521. Dobbie JW. Adrenocortical nodular hyperplasia: the ageing adrenal. J Pathol 1969;99:1-18.
522. Kloos RT, Gross MD, Francis IR, Korobkin M, Shapiro B. Incidentally discovered adrenal masses. Endocr Rev 1995;16:460-83.
523. Hedeland H, Ostberg G, Hokfelt B. On the prevalence of adrenocortical adenomas in an autopsy material in relation to hypertension and diabetes. Acta Med Scan 1968;184:211-4.
524. Woodruff MF. Tumor clonality and its biological significance. Adv Cancer Res 1988;50:197-229.
525. Gicquel C, Leblond-Francillard M, Bertagna X, et al. Clonal analysis of human adrenocortical carcinomas and secreting adenomas. Clin Endocrinol 1994;40:465-77.
526. Witkiewicz AK, Blaszyk H, Kulig E, Carnes JA, Lloyd RV. Laser captured microdissection in the clonality evaluation of benign multicentric adrenal cortical proliferations. Mod Pathol 1999;12:71A-401.
527. Suzuki T, Sasano H, Sawai T, et al. Small adrenocortical tumors without apparent clinical endocrine abnormalities: immunolocalization of steroidogenic enzymes. Path Res Pract 1992;188:883-9.
528. Damron TA, Schelper RL, Sorensen L. Cytochemical demonstration of neuromelanin in black pigmented adrenal nodules. Am J Clin Pathol 1987;87:334-41.
529. Abecassis M, McLoughlin MJ, Langer B, Kudlow JE. Serendipitous adrenal masses: prevalence, significance and management. Am J Surg 1985;149:783-8.
530. Glazer GM, Woolsey EJ, Borrello J, et al. Adrenal tissue characterization using MR imaging. Radiology 1986;158:73-9.
531. Bovio S, Cataldi A, Reimondo G, et al. Prevalence of adrenal incidentaloma in a contemporary computerized tomography series. J Endocrinol Invest 2006;29:298-302.
532. Herrera MF, Grant CS, van Heerden JA, Sheedy PF II, Ilstrup DM. Incidentally discovered adrenal tumors: an institutional perspective. Surgery 1991;110:1014-21.
533. Angeli A, Osella G, Alì A, Terzolo M. Adrenal incidentaloma: an overview of clinical and epidemiological data from the National Italian Study Group. Horm Res 1997;47:279-83.
534. Brown FM, Gaffey TA, Wold LE, et al. Myxoid neoplasms of the adrenal cortex: a rare histological variant. Am J Surg Pathol 2000;24:396-401.
535. Zhang J, Sun J, Liang Z, et al. Myxoid adrenocortical neoplasms: a study of the clinicopathologic features and EGFR gene status of ten Chinese cases. Am J Clin Pathol 2011;136:783-92.

Adrenal Cortical Hyperplasia

536. Nieman LK. Causes and pathophysiology of Cushing's syndrome. In: Lacroix A, Martin KA, eds. Adrenal diseas. UpToDate; 2020. Available at https://www.uptodate.com.acs.hcn.com.au/ contents/causes-and-pathophysiology-of-cushings.
537. Doppman JL, Miller DL, Dwyer AJ, et al. Macronodular adrenal hyperplasia in Cushing disease. Radiology 1988;166:347-52.
538. Dupont AG, Somers G, van Steistegheim AC, Warson F, Vanhaelot L. Ectopic adrenocorticotropin production: disappearance after removal of inflammatory tissue. J Clin Endocrinol Metab 1984;58:654-8.
539. Neville AM, O'Hare MJ. The human adrenal cortex: pathology and biology, an integrated approach. New York: Springer Verlag; 1982.
540. Neville AM, O'Hare MJ. The human adrenal gland: aspects of structure functional pathology. In: James VH, ed. The adrenal cortex. New York: Raven Press; 1979.
541. Neville AM, Symington T. The pathology of the adrenal gland in Cushing's syndrome. J Pathol Bacteriol 1967;93:19-35.
542. Drasin GF, Lynch T, Temes GP. Ectopic ACTH production and mediastinal lipomatosis. Radiology 1978;127:610.
543. Lam KY, Lo CY. The clinicopathologic significance of unilateral adrenal cortical hyperplasia: report of an unusual case and a review of the literature. Endocr Pathol 1999;10:243-9.
544. Aron DC, Schnall AM, Sheeler LR. Cushing's syndrome and pregnancy. Am J Obstet Gynecol 1990;162:244-52.
545. Josse RG, Bear R, Kovacs K, Higgins HP. Cushing's syndrome due to unilateral nodular adrenal hyperplasia. A new pathophysiological entity. Acta Endocrinol (Copenh) 1980;93:495-504.
546. Carey RM, Varma SK, Drake CR, et al. Ectopic secretion of corticotropin releasing factor as a cause of Cushing's syndrome: a clinical, morphological and biochemical study. N Engl J Med 1984;311:13-20.
547. Upton GV, Amatruda TT Jr. Evidence for the presence of tumor peptide with corticotropin-releasing factor like activity in the ectopic ACTH syndrome. N Engl J Med 1971;285:419-24.

548. Axiotis CA, Lippes HA, Merino MJ, de Lanerolle NC, Stewart AF, Kinder B. Corticotroph cell pituitary adenoma within an ovarian teratoma. A new cause of Cushing's syndrome. Am J Surg Pathol 1987;11:218-24.
549. Gagner M, Lacroix A, Prinz RA, et al. Early experience with laparoscopic approach for adrenalectomy. Surgery 1993;114:1120-4.
550. Grua JR, Nelson DH. ACTH-producing tumors. Endocrinol Metab Clin N Am 1991;20:319-62.
551. Schteingart DE, Lloyd RV, Akil H, et al. Cushing's syndrome secondary to ectopic corticotropin-releasing hormone—adrenocorticotropin secretion. J Clin Endocrinol Metab 1986;63:770-5.
552. Zárate A, Kovacs K, Flores M, Moran C, Felix I. ACTH and CRF producing bronchial carcinoid associated with Cushing's syndrome. Clin Endocrinol 1986;24:523-9.
553. Nieman LK. Cushing's syndrome: update on signs, symptoms and biochemical screening. Eur J Endocrinol 2015;173:M33-8.
554. Nieman L, Biller B, Findling J, et al. The diagnosis of Cushing's syndrome: an Endocrine Society Clinical Practice Guideline. Clin Endocrinol Metab 2008;95:1526-40.
555. Ross EJ, Linch DC. Cushing's syndrome—killing disease: discriminatory value of signs and symptoms aiding early diagnosis. Lancet 1982;2:646.
556. McArthur RG, Cloutier MD, Hayles AB, Sprague RG. Cushing's disease in children. Findings in 13 cases. Mayo Clin Proc 1972;47:318-26.
557. Orth DN, Kovacs WJ. The adrenal cortex. In: Wilson JD, Foster DW, Kronenberg HM, Larsen PR, eds. Williams textbook of endocrinology, 9th ed. Philadelphia: WB Saunders; 1998:517-664.
558. Oldfield EH, Doppman JL, Nieman LK, et al. Petrosal sinus sampling with and without corticotropin-releasing hormone for the differential diagnosis of Cushing's syndrome. N Engl J Med 1991;325:897-905.
559. Boscaro M, Arnaldi G. Approach to the patient with possible Cushing's syndrome. J Clin Endocrinol Metab 2009;94:3121-31.
560. Bruno OD, Rossi MA, Contreras LN, et al. Nocturnal high dose dexamethasone suppression test in the aetiological diagnosis of Cushing's syndrome. Acta Endocrinol 1985;109:158-62.
561. Nieman LK, Oldfield EH, Wesley R, Chrousos GP, Loriaux DL, Cutler GB Jr. A simplified morning ovine corticotropin-releasing hormone stimulation test for the differential diagnosis of adrenocorticotropin-dependent Cushing's syndrome. J Clin Endocrinol Metab 1993;77:1308-12.
562. Symington T. Functional pathology of the human adrenal gland. Edinburgh: Livingstone; 1969.
563. Symington T. The adrenal cortex. In: Bloodworth JM Jr, ed. Endocrine pathology general and surgical. Baltimore: Williams & Wilkins; 1982:419-72.
564. Reibord H, Fisher ER. Electron microscopic study of adrenal cortical hyperplasia in Cushing's syndrome. Arch Pathol 1968;86:419-26.
565. Neville AM, Symington T. Bilatral adrenal cortical hyperplasia in children with Cushing's syndrome. J Pathol 1972;107:95-106.
566. MacKay A. Atlas of human adrenal cortex ultrastructure. In: Symington T, ed. Functional pathology of the human adrenal gland. Edinburgh: Livingstone; 1964:345-489.
567. Neville AM, MacKay AM. The structure of the human adrenal cortex in health and disease. Clin Endocrinol Metab 1972;1:361.
568. Pasqualini RQ, Gurevich N. Spontaneous remission in a case of Cushing's syndrome. J Clin Endocrinol Metab 1956;16:406-11.
569. Pratt JH, Sawin CT, Melby JC. Remission of Cushing's disease after administration of adreno corticotropin. Am J Med 1974;57:949-52.
570. Anderson DC, Child DF, Sutcliffe CH, Buckley CH, Davies D, Longson D. Cushing's syndrome, nodular adrenal hyperplasia and virilizing carcinoma. Clin Endocrinol 1978;9:1-14.
571. Kay S. Hyperplasia and neoplasia of the adrenal gland. Pathol Ann 1976;11:103-39.
572. Tannenbaum M. Ultrastructural pathology of the adrenal cortex. Pathol Ann 1973;8:109-56.
573. Nieman LK, Biller BM, Findling JW, et al. Treatment of Cushing's syndrome: an Endocrine Society Clinical Practice Guideline. J Clin Endocrinol Metab 2015;100:2807-31.
574. Fleseriu M, Biller BM, Findling JW, et al. Mifepristone, a glucocorticoid receptor antagonist, produces clinical and metabolic benefits in patients with Cushing's syndrome. J Clin Endocrinol Metab 2012;97:2039-49.
575. Chow JT, Thompson GB, Grant CS, et al. Bilateral laparoscopic adrenalectomy for corticotrophin-dependent Cushing's syndrome: a review of the Mayo Clinic experience. Clin Endocrinol (Oxf) 2008;68:513-9.
576. Miller CA, Ellison EC. Therapeutic alternatives in metastatic neuroendocrine tumors. Surg Oncol Clin N Am 1998;7:863-79.
577. Aniszewski JP, Young WF Jr, Thompson GB, Grant CS, van Heerden JA. Cushing syndrome due to ectopic adrenocorticotropic hormone secretion. World J Surg 2001;25:934-40.
578. Ilias I, Torpy DJ, Pacak K, Mullen N, Wesley RA, Nieman LK. Cushing's syndrome due to ectopic corticotropin secretion: twenty years' experience at the National Institutes of Health. J Clin Endocrinol Metab 2005;90:4955-62.

579. Isidori AM, Kaltsas GA, Pozza C, et al. The ectopic adrenocorticotropin syndrome: clinical features, diagnosis, management, and long-term follow-up. J Clin Endocrinol Metab 2006;91:371-7.
580. Thompson GB, Grant CS, van Heerden JA, et al. Laparoscopic versus open posterior adrenalectomy: a case-control study of 100 patients. Surgery 1997;122:1132-6.
581. Jäger F, Heintz A, Junginger T. Synchronous bilateral endoscopic adrenalectomy: experiences after 18 operations. Surg Endosc 2004;18:314-8.

Adrenocortical Macronodular Hyperplasia

582. Aiba M, Hirayama A, Iri H, et al. Adrenocorticotropic hormone-independent bilateral adrenocortical macronodular hyperplasia as a distinct subtype of Cushing's syndrome. Enzyme histochemical and ultrastructural study of four cases with a review of the literature. Am J Clin Pathol 1991;96:334-40.
583. Boston BA, Mandel S, LaFranchi S, Bliziotes M. Activating mutation in the stimulatory guanine nucleotide-binding protein in an infant with Cushing's syndrome and nodular adrenal hyperplasia. J Clin Endocrinol Metab 1994;79:890-3.
584. Lacroix A. ACTH-independent macronodular adrenal hyperplasia. Best Pract Res Clin Endocrinol Metab 2009;23:245.
585. Doppman JL, Nieman LK, Travis WD, et al. CT and MR imaging of massive macronodular adrenocortical disease: a rare cause of autonomous primary adrenal hyper cortisolism. J Computr Assist Tomogr 1991;15:773-9.
586. Hidai H, Fujii H, Otsuka K, Abe K, Shimizu N. Cushing's syndrome due to huge adrenocortical multinodular hyperplasia. Endocrinol Jpn 1975;22:555-60.
587. Hashimoto K, Kawada Y, Murakami K, et al. Cortisol responsiveness to insulin-induced hypoglycemia in Cushing's syndrome with huge nodular adrenocortical hyperplasia. Endocrinol Jpn 1986;33:479-87.
588. Lacroix A. Heredity and cortisol regulation in bilateral macronodular adrenal hyperplasia. N Engl J Med 2013;369:2147-9.
589. Louiset E, Duparc C, Young J, et al. Intraadrenal corticotropin in bilateral macronodular adrenal hyperplasia. N Engl J Med 2013;369:2115-25.
590. El Ghorayeb N, Bourdeau I, Lacroix A. Multiple aberrant hormone receptors in Cushing's syndrome. Eur J Endocrinol 2015;173:M45-60.
591. Hofland J, Hofland LJ, van Koetsveld PM, et al. ACTH-independent macronodular adrenocortical hyperplasia reveals prevalent aberrant in vivo and in vitro responses to hormonal stimuli and coupling of arginine-vasopressin type 1a receptor to 11β-hydroxylase. Orphanet J Rare Dis 2013;8:142.
592. Hsiao HP, Kirschner LS, Bourdeau I, et al. Clinical and genetic heterogeneity, overlap with other tumor syndromes, and atypical glucocorticoid hormone secretion in adrenocorticotropin-independent macronodular adrenal hyperplasia compared with other adrenocortical tumors. J Clin Endocrinol Metab 2009;94:2930-7.
593. Libé R, Coste J, Guignat L, et al. Aberrant cortisol regulations in bilateral macronodular adrenal hyperplasia: a frequent finding in a prospective study of 32 patients with overt or subclinical Cushing's syndrome. Eur J Endocrinol 2010;163:129-38.
594. Mircescu H, Jilwan J, N'Diaye N, et al. Are ectopic or abnormal membrane hormone receptors frequently present in adrenal Cushing's syndrome? J Clin Endocrinol Metab 2000;85:3531-6.
595. Lacroix A, Bolte E, Tremblay J, et al. Gastric inhibitory polypeptide-dependent cortisol hypersecretion—a new cause of Cushing's syndrome. N Engl J Med 1992;327:974-80.
596. Lacroix A, Tremblay J, Rousseau G, Bouvier M, Hamet P. Propranolol therapy for ectopic beta-adrenergic receptors in adrenal Cushing's syndrome. N Engl J Med 1997;337:1429-34.
597. Willenberg HS, Stratakis CA, Marx C, Ehrhart-Bornstein M, Chrousos GP, Bornstein SR. Aberrant interleukin-1 receptors in a cortisol-secreting adrenal adenoma causing Cushing's syndrome. N Engl J Med 1998;339:27-31.
598. Cheitlin RA, Westphal M, Cabrera CM, et al. Cushing's syndrome due to bilateral adrenal macronodular hyperplasia with undetectable ACTH: cell culture of adenoma cells on extracellular matrix. Horm Res 1988;29:162-7.
599. Kirschner MA, Powell RD, Lipsett MB. Cushing's syndrome: nodular cortical hyperplasia of adrenal glands with clinical and pathological features suggesting adrenocortical tumor. J Clin Endocr Metab 1964;24:947-55.
600. Malchoff CD, Rosa J, DeBold CR, et al. Adrenocorticotropin-independent bilateral macronodular adrenal hyperplasia: an unusual cause of Cushing's syndrome. J Clin Endocrinol Metab 1989;68:855-60.
601. Swain JM, Grant CS, Schlinkert RT, et al. Corticotropin-independent macronodular adrenal hyperplasia: a clinicopathologic correlation. Arch Surg 1998;133:541-5.
602. Findlay JC, Sheeler LR, Engeland WC, Aron DC. Familial adrenocorticotropin-independent Cushing's syndrome with bilateral macronodular adrenal hyperplasia. J Clin Endocrinol Metab 1993;76:189-91.

603. Imöhl M, Köditz R, Stachon A, et al. [Catecholamine-dependent hereditary Cushing's syndrome—follow-up after unilateral adrenalectomy.] Med Klin (Munich) 2002;97:747-53. [German]
604. Someya T, Koyano H, Osawa Y. ACTH-independent macronodular adrenocortical hyperplasia (AIMAH) in two brothers and a sister. Folia Endocrinol 1996;72:762.
605. Lee S, Hwang R, Lee J, et al. Ectopic expression of vasopressin V1b and V2 receptors in the adrenal glands of familial ACTH-independent macronodular adrenal hyperplasia. Clin Endocrinol (Oxf) 2005;63:625-30.
606. Minami S, Sugihara H, Sato J, et al. ACTH independent Cushing's syndrome occurring in siblings. Clin Endocrinol (Oxf) 1996;44:483-8.
607. Miyamura N, Taguchi T, Murata Y, et al. Inherited adrenocorticotropin-independent macronodular adrenal hyperplasia with abnormal cortisol secretion by vasopressin and catecholamines: detection of the aberrant hormone receptors on adrenal gland. Endocrine 2002;19:319-26.
608. Nies C, Bartsch DK, Ehlenz K, et al. Familial ACTH-independent Cushing's syndrome with bilateral macronodular adrenal hyperplasia clinically affecting only female family members. Exp Clin Endocrinol Diabetes 2002;110:277-83.
609. Vezzosi D, Cartier D, Régnier C, et al. Familial adrenocorticotropin-independent macronodular adrenal hyperplasia with aberrant serotonin and vasopressin adrenal receptors. Eur J Endocrinol 2007;156:21-31.
610. Watson TD, Patel SJ, Nardi PM. Case 121: familial adrenocorticotropin-independent macronodular adrenal hyperplasia causing Cushing syndrome. Radiology 2007;244:923-6.
611. Alencar GA, Lerario AM, Nishi MY, et al. ARMC5 mutations are a frequent cause of primary macronodular adrenal hyperplasia. J Clin Endocrinol Metab 2014;99:E1501-9.
612. Assié G, Libé R, Espiard S, et al. ARMC5 mutations in macronodular adrenal hyperplasia with Cushing's syndrome. N Engl J Med 2013; 369:2105-14.
613. Espiard S, Drougat L, Libé R, et al. ARMC5 mutations in a large cohort of primary macronodular adrenal hyperplasia: clinical and functional consequences. J Clin Endocrinol Metab 2015;100:E926-35.
614. Faucz FR, Zilbermint M, Lodish MB, et al. Macronodular adrenal hyperplasia due to mutations in an armadillo repeat containing 5 (ARMC5) gene: a clinical and genetic investigation. J Clin Endocrinol Metab 2014;99:E1113-9.
615. Ringel MD, Schwindinger WF, Levine MA. Clinical implications of genetic defects in G proteins. The molecular basis of McCune-Albright syndrome and Albright hereditary osteodystrophy. Medicine (Baltimore) 1996;75:171-84.
616. Shenker A, Weinstein LS, Moran A, et al. Severe endocrine and nonendocrine manifestations of the McCune-Albright syndrome associated with activating mutations of stimulatory G protein GS. J Pediatr 1993;123:509-18.
617. Fragoso MC, Domenice S, Latronico AC, et al. Cushing's syndrome secondary to adrenocorticotropin-independent macronodular adrenocortical hyperplasia due to activating mutations of GNAS1 gene. J Clin Endocrinol Metab 2003;88:2147-51.
618. Kartheuser A, Walon C, West S, et al. Familial adenomatous polyposis associated with multiple adrenal adenomas in a patient with a rare 3' APC mutation. J Med Genet 1999;36:65-7.
619. Matyakhina L, Freedman RJ, Bourdeau I, et al. Hereditary leiomyomatosis associated with bilateral, massive, macronodular adrenocortical disease and atypical cushing syndrome: a clinical and molecular genetic investigation. J Clin Endocrinol Metab 2005;90:3773-9.
620. Someya T, Koyano H, Ozana V. ACTH-independent macronodular adrenocortical hyperplasia (AIMAH) in two brothers and a sister. Folia Endorinol 1996;72:762.
621. Aiba M, Hirayama A, Iri H, et al. Primary adrenocortical micronodular dysplasia: enzyme histochemical and ultrastructural studies of two cases with a review of the literature. Hum Pathol 1990;21:503-11.
622. Sasano H, Miyazaki S, Sawai T, et al. Primary pigmented nodular adrenocortical disease (PPNAD): immunohistochemical and in situ hybridization analysis of steroidogenic enzymes in eight cases. Mod Pathol 1992;5:23-9.
623. Sasano H, Suzuki T, Nagura H. ACTH-independent macronodular adrenocortical hyperplasia: immunohistochemical and in situ hybridization studies of steroidogenic enzymes. Mod Pathol 1994;7:215-9.
624. Boronat M, Lucas T, Barceló B, et al. Cushing's syndrome due to autonomous macronodular adrenal hyperplasia: long-term follow-up after unilateral adrenalectomy. Postgrad Med J 1996;72:614-6.
625. Iacobone M, Albiger N, Scaroni C, et al. The role of unilateral adrenalectomy in ACTH-independent macronodular adrenal hyperplasia (AIMAH). World J Surg 2008;32:882-9.
626. Lamas C, Alfaro JJ, Lucas T, Lecumberri B, Barceló B, Estrada J. Is unilateral adrenalectomy an alternative treatment for ACTH-independent macronodular adrenal hyperplasia?: Long-term follow-up of four cases. Eur J Endocrinol 2002;146:237-40.

627. Xu Y, Rui W, Qi Y, et al. The role of unilateral adrenalectomy in corticotropin-independent bilateral adrenocortical hyperplasias. World J Surg 2013;37:1626-32.
628. Debillon E, Velayoudom-Cephise FL, Salenave S, et al. Unilateral adrenalectomy as a first-line treatment of cushing's syndrome in patients with primary bilateral macronodular adrenal hyperplasia. J Clin Endocrinol Metab 2015; 100:4417-24.
629. Osswald A, Quinkler M, Di Dalmazi G, et al. Long-term outcome of primary bilateral macronodular adrenocortical hyperplasia after unilateral adrenalectomy. J Clin Endocrinol Metab 2019;104:2985-93.
630. Mazzuco TL, Thomas M, Martinie M, et al. Cellular and molecular abnormalities of a macronodular adrenal hyperplasia causing beta-blocker-sensitive Cushing's syndrome. Arq Bras Endocrinol Metabol 2007;51:1452.
631. Goodarzi MO, Dawson DW, Li X, et al. Virilization in bilateral macronodular adrenal hyperplasia controlled by luteinizing hormone. J Clin Endocrinol Metab 2003;88:73-7.
632. Lacroix A, Hamet P, Boutin JM. Leuprolide acetate therapy in luteinizing hormone-dependent Cushing's syndrome. N Engl J Med 1999;341:1577-81.

Primary Pigmented Nodular Adrenocortical Disease

633. Carney JA, Young WF Jr. Primary pigmented nodular adrenocortical disease and its associated conditions. Endocrinologist 1992;2:6-21.
634. Meador CK, Bowdoin B, Owen WC, et al. Primary adrenocortical nodular dysplasia: a rare cause of Cushing's syndrome. J Clin Endocrinol Metab 1967;27:1255-63.
635. Böhm N, Lippmann-Grob B, von Petrykowski W. Familial Cushing's syndrome due to pigmented multi nodular adrenocortical dysplasia. Acta Endocrinol 1983;102:428-35.
636. Carney JA, Gordon H, Carpenter PC, et al. The complex of myxomas, spotty pigmentation, and endocrine overactivity. Medicine (Baltimore) 1985;64:270-83.
637. Hedinger C. [Combination of heart myxoma with primary nodular adrenal cortex dysplasia, spot-shaped skin pigmentation and myxoma-like tumors in other locations—a rare familial symptom complex ("Swiss syndrome").] Schweiz Med Wochenschr 1987;117:591-4. [German]
638. Lack EE, Travis WD, Oertel JE. Adrenal cortical nodules, hyperplasia, and hyperfunction. In: Lack EE, ed. Pathology of the adrenal glands. New York: Churchill Livingstone; 1990:75-113.
639. Larsen JL, Cathey WJ, Odell WD. Primary adrenocortical nodular dysplasia, a distinct subtype of Cushing's syndrome. Case report and review of the literature. Am J Med 1986;80:976-84.
640. Stratakis CA, Carney JA, Lin JP, et al. Carney complex, a familial multiple neoplasia and lentiginosis syndrome. Analysis of 11 kindreds and linkage to the short arm of chromosome 2. J Clin Invest 1996;97:699-705.
641. Stratakis CA, Courcoutsakis NA, Abati A, et al. Thyroid gland abnormalities in patients with the syndrome of spotty skin pigmentation, myxomas, endocrine overactivity, and schwannomas (Carney complex). J Clin Endocrinol Metab 1997;82:2037-43.
642. Stratakis CA, Kirschner LS, Carney JA. Clinical and molecular features of the Carney complex: diagnostic criteria and recommendations for patient evaluation. J Clin Endocrinol Metab 2001;86:4041-6.
643. Young WF Jr, Carney JA, Musa BU, et al. Familial Cushing's syndrome due to primary pigmented nodular adrenocortical disease. Reinvestigation 50 years later. N Engl J Med 1989;321:1659-64.
644. Casey M, Mah C, Merliss AD, et al. Identification of a novel genetic locus for familial cardiac myxomas and Carney complex. Circulation 1998; 98:2560-6.
645. Stratakis CA, Jenkins RB, Pras E, et al. Cytogenetic and microsatellite alterations in tumors from patients with the syndrome of myxomas, spotty ski n pigmentation, and endocrine overactivity (Carney complex). J Clin Endocrinol Metab 1996;81:3607-14.
646. Boikos SA, Horvath A, Heyerdahl S, et al. Phosphodiesterase 11A expression in the adrenal cortex, primary pigmented nodular adrenocortical disease, and other corticotropin-independent lesions. Horm Metab Res 2008;40:347-53.
647. Casey M, Vaughan CJ, He J, et al. Mutations in the protein kinase A R1alpha regulatory subunit cause familial cardiac myxomas and Carney complex. J Clin Invest 2000;106:R31-8.
648. D'Andrea MR, Qiu Y, Haynes-Johnson D, Bhattacharjee S, Kraft P, Lundeen S. Expression of PDE11A in normal and malignant human tissues. J Histochem Cytochem 2005;53:895-903.
649. Groussin L, Horvath A, Jullian E, et al. A PRKAR1A mutation associated with primary pigmented nodular adrenocortical disease in 12 kindreds. J Clin Endocrinol Metab 2006;91:1943-9.
650. Groussin L, Jullian E, Perlemoine K, et al. Mutations of the PRKAR1A gene in Cushing's syndrome due to sporadic primary pigmented nodular adrenocortical disease. J Clin Endocrinol Metab 2002;87:4324-9.

651. Groussin L, Kirschner LS, Vincent-Dejean C, et al. Molecular analysis of the cyclic AMP-dependent protein kinase A (PKA) regulatory subunit 1A (PRKAR1A) gene in patients with Carney complex and primary pigmented nodular adrenocortical disease (PPNAD) reveals novel mutations and clues for pathophysiology: augmented PKA signaling is associated with adrenal tumorigenesis in PPNAD. Am J Hum Genet 2002;71:1433-42.
652. Horvath A, Giatzakis C, Robinson-White A, et al. Adrenal hyperplasia and adenomas are associated with inhibition of phosphodiesterase 11A in carriers of PDE11A sequence variants that are frequent in the population. Cancer Res 2006;66:11571-5.
653. Kirschner LS, Carney JA, Pack SD, et al. Mutations of the gene encoding the protein kinase A type I-alpha regulatory subunit in patients with the Carney complex. Nat Genet 2000;26:89-92.
654. Libé R, Fratticci A, Coste J, et al. Phosphodiesterase 11A (PDE11A) and genetic predisposition to adrenocortical tumors. Clin Cancer Res 2008;14:4016-24.
655. Nadella KS, Kirschner LS. Disruption of protein kinase a regulation causes immortalization and dysregulation of D-type cyclins. Cancer Res 2005;65:10307-15.
656. Pereira AM, Hes FJ, Horvath A, et al. Association of the M1V PRKAR1A mutation with primary pigmented nodular adrenocortical disease in two large families. J Clin Endocrinol Metab 2010;95:338-42.
657. Veugelers M, Bressan M, McDermott DA, et al. Mutation of perinatal myosin heavy chain associated with a Carney complex variant. N Engl J Med 2004;351:460-9.
658. Veugelers M, Wilkes D, Burton K, et al. Comparative PRKAR1A genotype-phenotype analyses in humans with Carney complex and prkar1a haploinsufficient mice. Proc Natl Acad Sci U S A 2004;101:14222-7.
659. Ruder HJ, Loriaux DL, Lipsett MB. Severe osteopenia in young adults associated with Cushing's syndrome due to micronodular adrenal disease. J Clin Endocrinol Metab 1974;39:1138-47.
660. Carson DJ, Sloan JM, Cleland J, Russell CF, Atkinson AB, Sheridan B. Cyclical Cushing's syndrome presenting as short stature in a boy with recurrent atrial myxomas and freckled skin pigmentation. Clin Endocrinol (Oxf) 1988;28:173-80.
661. Gunther DF, Bourdeau I, Matyakhina L, et al. Cyclical Cushing syndrome presenting in infancy: an early form of primary pigmented nodular adrenocortical disease, or a new entity? J Clin Endocrinol Metab 2004;89:3173-82.
662. Teding van Berkhout F, Croughs RJ, Kater L, et al. Familial Cushing's syndrome due to nodular adrenocortical dysplasia. A putative receptor-antibody disease? Clin Endocrinol 1986;24:299-310.
663. Doppman JL, Travis WD, Nieman L, et al. Cushing syndrome due to primary pigmented nodular adrenocortical disease: findings at CT and MR imaging. Radiology 1989;172:415-20.
664. Stratakis CA, Sarlis N, Kirschner LS, et al. Paradoxical response to dexamethasone in the diagnosis of primary pigmented nodular adrenocortical disease. Ann Intern Med 1999;131:585-91.
665. Powell AC, Stratakis CA, Patronas NJ, et al. Operative management of Cushing syndrome secondary to micronodular adrenal hyperplasia. Surgery 2008;143:750-8.
666. Bourdeau I, Lacroix A, Schürch W, et al. Primary pigmented nodular adrenocortical disease: paradoxical responses of cortisol secretion to dexamethasone occur in vitro and are associated with increased expression of the glucocorticoid receptor. J Clin Endocrinol Metab 2003;88:3931-7.
667. Aiba M, Hirayama A, Iri H, et al. Primary adrenocortical micronodular dysplasia: enzyme histochemical and ultrastructural studies of two cases with a review of the literature. Hum Pathol 1990;21:503-11.
668. Sasano H, Miyazaki S, Sawai T, et al. Primary pigmented nodular adrenocortical disease (PPNAD): immunohistochemical and in situ hybridization analysis of steroidogenic enzymes in eight cases. Mod Pathol 1992;5:23-9.
669. Travis WD, Tsokos M, Doppman JL, et al. Primary pigmented nodular adrenocortical disease. A light and electron microscopic study of eight cases. Am J Surg Pathol 1989;13:921-30.
670. Sarlis NJ, Chrousos GP, Doppman JL, Carney JA, Stratakis CA. Primary pigmented nodular adrenocortical disease: re-evaluation of a patient with Carney complex 27 years after unilateral adrenalectomy. J Clin Endocrinol Metab 1997;82:1274-8.

Aldosterone Excess Due to Adrenocortical Hyperplasia

671. Young WF Jr. Pheochromocytoma and primary aldosteronism: diagnostic approaches. Endocrinol Metab Clin North Am 1997;26:801-27.
672. Bravo EL, Tarazi RC, Dustan HP, et al. The changing clinical spectrum of primary aldosteronism. Am J Med 1983;74:641-51.
673. Conn JW, Knopf RF, Nesbit RM. Clinical characteristics of primary aldosteronism from an analysis of 145 cases. Am J Surg 1964;107:159-72.
674. Lack EE. Tumors of the adrenal gland and extra adrenal paraganglia. AFIP Atlas of Tumor Pathology, 4th Series, Fascicle 8. Washington DC: American Registry of Pathology; 2007.

675. Neville AM, O'Hare MJ. Aspects of structure, function and pathology. In: James VH, ed. The adrenal gland. New York: Raven Press; 1979:165.
676. Güllner HG, Gill JR Jr. Beta endorphin selectively stimulates aldosterone secretion in hypophysectomized nephrectomized dogs. J Clin Invest 1983;71:124-8.
677. Rabinowe SL, Taylor T, Dluhy RG, Williams GH. Beta-endorphin stimulates plasma renin and aldosterone release in normal human subjects. J Clin Endocrinol Metab 1985;60:485-9.
678. Banks WA, Kastin AJ, Biglieri EG, Ruiz AE. Primary adrenal hyperplasia: a new subset of primary hyperaldosteronism. J Clin Endocrinol Metab 1984;58:783-5.
679. Orth DN, Kovacs WJ. The adrenal cortex. In: Wilson JD, Foster DW, Kronenberg HM, Larsen PR, eds. Williams textbook of endocrinology, 9th ed. Philadelphia: WB Saunders; 1998:517-664.
680. Young WF Jr. Primary aldosteronism: a common and curable form of hypertension. Cardiol Rev 1999;7:207-14.
681. Pitcock JA, Hartroft PM. The juxtaglomerular cells in man and their relationship to the level of plasma sodium and to the zona glomerulosa of the adrenal cortex. Am J Pathol 1973;34:863-83.
682. Kurtz I. Molecular pathogenesis of Bartter's and Gitelman's syndromes. Kidney Int 1998; 54:1396-410.
683. Simon DB, Karet FE, Hamdan JM, et al. Bartter's syndrome, hypokalaemic alkalosis with hypercalciuria, is caused by mutations in the Na-K-2Cl cotransporter NKCC2. Nat Genet 1996;13:183-8.
684. Simon DB, Karet FE, Rodriguez-Soriano J, et al. Genetic heterogeneity of Bartter's syndrome revealed by mutations in the K+ channel, ROMK. Nat Genet 1996;14:152-6.
685. Simon DB, Nelson-Williams C, Bia MJ, et al. Gitelman's variant of Bartter's syndrome, inherited hypokalaemic alkalosis, is caused by mutations in the thiazide-sensitive Na-Cl cotransporter. Nat Genet 1996;12:24-30.
686. Stein JH. The pathogenetic spectrum of Bartter's syndrome. Kidney Int 1985;28:85-93.
687. Bartter FC, Pronove P, Gill JR Jr, MacCardle RC. Hyperplasia of the juxtaglomerular complex with hyperaldosteronism and hypokalemic alkalosis. A new syndrome. Am J Med 1962;33:811-28.
688. Bartter FC. Bartter's syndrome. Urol Clin North Am 1977;4:253-64.
689. Gleason PE, Weinberger MH, Pratt JH, et al. Evaluation of diagnostic tests in the differential diagnosis of primary aldosteronism: unilateral adenoma versus bilateral micronodular hyperplasia. J Urol 1993;150:1365-8.
690. Johnson CM, Sheedy PF, Welch TJ, et al. CT of the adrenal cortex. Semin Ultrastruct CT MR 1985;6:241-60.
691. Young WF Jr. Management approaches to adrenal incidentalomas. A view from Rochester, Minnestota. Endocrinol Metabol Clin North Am 2000;29:159-85.
692. Harper R, Ferrett CG, McKnight JA, et al. Accuracy of CT scanning and adrenal vein sampling in the pre-operative localization of aldosterone-secreting adrenal adenomas. QJM 1999;92:643-50.
693. Kempers MJ, Lenders JW, van Outheusden L, et al. Systematic review: diagnostic procedures to differentiate unilateral from bilateral adrenal abnormality in primary aldosteronism. Ann Intern Med 2009;151:329-37.
694. Magill SB, Raff H, Shaker JL, et al. Comparison of adrenal vein sampling and computed tomography in the differentiation of primary aldosteronism. J Clin Endocrinol Metab 2001;86:1066-71.
695. Radin DR, Manoogian C, Nadler JL. Diagnosis of primary hyperaldosteronism: importance of correlating CT findings with endocrinologic studies. AJR Am J Roentgenol 1992;158:553-7.
696. Young WF, Stanson AW, Thompson GB, Grant CS, Farley DR, van Heerden JA. Role for adrenal venous sampling in primary aldosteronism. Surgery 2004;136:1227-35.
697. Page DL, DeLellis RA, Hough AJ. Tumors of the adrenal. Atlas of Tumor Pathology, 2nd series, Fascicle 23. Washington DC: Armed Forces Institute of Pathology; 1985.
698. Neville AM. The nodular adrenal. Invest Cell Pathol 1978;1:99-111.
699. Neville AM, O'Hare MJ. Histopathology of the human adrenal cortex. Clin Endocrinol Metab 1985;14:791-820.
700. Geller DS, Zhang J, Wisgerhof MV, Shackleton C, Kashgarian M, Lifton RP. A novel form of human Mendelian hypertension featuring nonglucocorticoid-remediable aldosteronism. J Clin Endocr Metab 2008;93:3117-23.
701. Milsom SR, Espiner EA, Nicholls MG, Gwynne J, Perry EG. The blood pressure response to unilateral adrenalectomy in primary aldosteronism. Q J Med 1986;61:1141-51.
702. Sawka AM, Young WF, Thompson GB, et al. Primary aldosteronism: factors associated with normalization of blood pressure after surgery. Ann Intern Med 2001;135:258-61.
703. Young WF Jr. Primary aldosteronism: renaissance of a syndrome. Clin Endocrinol (Oxf) 2007;66:607-18.
704. Celen O, O'Brien MJ, Melby JC, Beazley RM. Factors influencing outcome of surgery for primary aldosteronism. Arch Surg 1996;131:646-50.

705. TAIPAI Study Group, Wu VC, Chueh SC, et al. Association of kidney function with residual hypertension after treatment of aldosterone-producing adenoma. Am J Kidney Dis 2009;54:665-73.
706. Williams TA, Lenders JW, Mulatero P, et al. Outcomes after adrenalectomy for unilateral primary aldosteronism: an international consensus on outcome measures and analysis of remission rates in an international cohort. Lancet Diabetes Endocrinol 2017;5:689-99.
707. Young WF Jr. Laparoscopic adrenalectomy. An endocrinologists' perspective. Curr Opin Endocrinol Diabetes 1999;6:199-203.
708. Griffing GT, Cole AG, Aurecchia SA, Sindler BH, Komanicky P, Melby JC. Amiloride in primary hyperaldosteronism. Clin Pharmacol Ther 1982;31:56-61.
709. Ganguly A, Weinberger MH. Triamterene-thiazide combination: alternative therapy for primary aldosteronism. Clin Pharmacol Ther 1981;30:246-50.
710. Bernini G, Galetta F, Franzoni F, et al. Arterial stiffness, intima-media thickness and carotid artery fibrosis in patients with primary aldosteronism. J Hypertens. 2008;26:2399-405.
711. Born-Frontsberg E, Reincke M, Rump LC, et al. Cardiovascular and cerebrovascular comorbidities of hypokalemic and normokalemic primary aldosteronism: results of the German Conn's Registry. J Clin Endocrinol Metab 2009; 94:1125-30.

Unilateral Adrenal Cortical Hyperplasia

712. Catania A, Reschini E, Orsatti A, Motta P, Airaghi L, Cantalamessa L. Cushing's syndrome due to unilateral adrenal nodular hyperplasia with incomplete inhibition of the contralateral gland. Hormone Res 1986;23:9-15.
713. Ganguly A, Zager PG, Luetscher JA. Primary aldosteronism due to unilateral adrenal hyperplasia. J Clin Endocrinol Metab 1980;51:1190-4.
714. Josse RG, Bear R, Kovacs K, Higgins HP. Cushing's syndrome due to unilateral nodular adrenal hyperplasia. A new pathophysiological entity? Acta Endocrinol 1980;93:495-504.
715. Neville AM, Symington T. The pathology of the adrenal gland in Cushing's syndrome. J Pathol Bacteriol 1967;93:19-35.
716. Oberfield SE, Levine LS, Firpo A, et al. Primary hyperaldosteronism in childhood due to unilateral macro nodular hyperplasia. Case report. Hypertension 1984;6:75-84.
717. Sigman LM, Wallach L. Unilateral adrenal hypertrophy in ectopic ACTH syndrome. Arch Intern Med 1984;144:1869-70.

Congenital Adrenal Hyperplasia

718. White PC, New MI, Dupont B. Congenital adrenal hyperplasia (1). N Engl J Med 1987;316: 1519-24.
719. de Crecchio L. Sopra un caso di apparenzi virili in una donna. Morgagni 1865;7:154-88.
720. Biglieri EG, Kater CE. 17-alpha-hydroxylation deficiency. Endocrinol Metab Clin N Am 1991; 20:257-68.
721. Bongiovanni AM, Root AW. The adrenogenital syndrome. N Engl J Med 1963;268:1391-9.
722. Merke DP. Diagnosis of classic congenital adrenal hyperplasia due to 21-hydroxylase deficiency in infants and children. In: Geffner ME, ed.UptoDate. LK Nieman, ME Geffner, eds. Available from https://www.uptodate.com.acs.hcn.com.au/contents/diagnosis-of-classic-congenital-adrenal-hyperplasia-due-to-21-hydroxylase-deficiency-in-infants-and-children?search=congenital%20adrenal%20hyperplasia&source=search_result&selectedTitle=1~126&usage_type=default&display_rank=1#H1.
723. Speiser PW, Agdere L, Ueshiba H, White PC, New MI. Aldosterone synthesis in salt-wasting congenital adrenal hyperplasia with complete absence of adrenal 21-hydroxylase. N Engl J Med 1991;324:145-9.
724. New MI. Steroid 21-hydroxylase deficiency (congenital adrenal hyperplasia). Am J Med 1995;98:2S-8.
725. Speiser PW, Arlt W, Auchus RJ, et al. Congenital adrenal hyperplasia due to steroid 21-hydroxylase deficiency: an Endocrine Society Clinical Practice Guideline. J Clin Endocrinol Metab 2018; 103:4043-88.
726. Knudsen JL, Savage A, Mobb GE. The testicular 'tumour' of adrenogenital syndrome—a persistent diagnostic pitfall. Histopathology 1991;19:468-70.
727. Mangray S, DeLellis RA. Adrenal glands. In: Mills SE, Greenson JK, Hornick JL, Longacre TA, Reuter VE, eds. Sternberg's diagnostic pathology, 6th ed. Philadelphia: Wolters Kluwer Health; 2015:595-646.
728. New MI, White PC. Genetic disorders of steroid hormone synthesis and metabolism. Baillieres Clin Endocrinol Metab 1995;9:525-54.
729. Auchus RJ. The genetics, pathophysiology and management of deficiencies of P450c17. Endocrinol Metab Clin North Am 2001;30:101-19.
730. Hughes I. Congenital adrenal hyperplasia: phenotype and genotypes. J Pediatri Endocrinol Metab 2002;15(Suppl 15):1529-40.
731. Peter M. Congenital adrenal hyperplasia: 11 beta-hydrocylase deficiency. Semin Reprod Med 2002;20:249-54.
732. Merke DP, Bornstein SR. Congenital adrenal hyperplasia. Lancet 2005;365:2125-36.

733. New MI. Male pseudohermaphroditism due to 17 alpha-hydroxylase deficiency. J Clin Invest 1970;49:1930-41.
734. Azziz R, Kenney PJ. Magnetic resonance imaging of the adrenal gland in women with late-onset adrenal hyperplasia. Fertil Steril 1991;56:142-4.
735. Menon PS, Virmani A, Sethi AK, et al. Congenital adrenal hyperplasia: experience at intersex clinic, AIIMS. Indian J Pediatr 1992;59:531-5.
736. Al-Alwan I, Navarro O, Daneman D, Daneman A. Clinical utility of adrenal ultrasonography in the diagnosis of congenital adrenal hyperplasia. J Pediatr 1999;135:71-5.
737. Orth DN, Kovacs WJ. The adrenal cortex. In: Wilson JD, Foster DW, Kronenberg HM, Larsen PR, eds. Williams textbook of endocrinology, 9th ed. Philadelphia: WB Saunders; 1998:517-664.
738. Neville AM, O'Hare J. The human adrenal cortex: pathology and biology—an integrated approach. New York: Springer-Verlag; 1982:161-4.
739. Rutgers JL, Young RH, Scully RE. The testicular "tumor" of the adrenogenital syndrome: a report of six cases and review of the literature on testicular masses in patients with adrenocortical disorders. Am J Surg Pathol 1988;12:503-13.
740. Kim I, Young RH, Scully RE. Leydig cell tumors of the testis. A clinicopathological analysis of 40 cases and review of the literature. Am J Surg Pathol 1985;9:177-92.
741. Clayton PE, Miller WL, Oberfield SE, et al. Consensus statement on 21-hydroxylase deficiency from the European Society for Paediatric Endocrinology and the Lawson Wilkins Pediatric Endocrine Society. Horm Res 2002;58:188-95.
742. Hirvikoski T, Nordenström A, Wedell A, et al. Prenatal dexamethasone treatment of children at risk for congenital adrenal hyperplasia: the Swedish experience and standpoint. J Clin Endocrinol Metab 2012;97:1881-3.
743. Joint LWPES/ESPE CAH Working Group. Consensus statement on 21-hydroxylase deficiency from the Lawson Wilkins Pediatric Endocrine Society and the European Society for Paediatric Endocrinology. J Clin Endocrinol Metab 2002;87:4048-53.
744. Swerdlow AJ, Higgins CD, Brook CG, et al. Mortality in patients with congenital adrenal hyperplasia: a cohort study. J Pediatr 1998;133:516-20.
745. Hauffa BP, Miller WL, Grumbach MM, Conte FA, Kaplan SL. Congenital adrenal hyperplasia due to deficient cholesterol side-chain cleavage activity (20, 22-desmolase) in a patient treated for 18 years. Clin Endocrinol (Oxf) 1985;23:481-93.
746. Bose HS, Pescovitz OH, Miller WL. Spontaneous feminization in a 46,XX female patient with congenital lipoid adrenal hyperplasia due to a homozygous frameshift mutation in the steroidogenic acute regulatory protein. J Clin Endocrinol Metab 1997;82:1511-5.
747. Nakae J, Tajima T, Sugawara T, et al. Analysis of the steroidogenic acute regulatory protein (StAR) gene in Japanese patients with congenital lipoid adrenal hyperplasia. Hum Mol Genet 1997;6:571-6.

Beckwith-Wiedemann Syndrome

748. Weksberg R, Shuman C, Beckwith JB. Beckwith-Wiedemann syndrome. Eur J Hum Genet 2010;18:8.
749. Beckwith JB. Extreme cytomegaly of the adrenal fetal cortex, omphalocele, hyperplasia of kidneys and pancreas, and Leydig-cell hyperplasia: another syndrome? Los Angeles: Western Society for Pediatic Research; 1963.
750. Wiedemann HR. Familial malformation complex with umbilical hernia and macroglossia-a "new syndrome"?. J Genet Hum 1964;13:223.
751. Mussa A, Russo S, De Crescenzo A, et al. Prevalence of Beckwith-Wiedemann syndrome in North West of Italy. Am J Med Genet A 2013; 161A:2481.
752. Pettanati MJ, Haines JL, Higgins RR, Wappner RS, Palmer CG, Weaver DD. Wiedemann-Beckwith syndrome: presentation of clinical and cytogenetic data on 22 new cases and review of the literature. Hum Genet 1986;74:143-54.
753. Weksberg R, Shuman C, Caluseriu O, et al. Discordant KCNQ1OT1 imprinting in sets of monozygotic twins discordant for Beckwith-Wiedemann syndrome. Hum Mol Genet 2002;11:1317-25.
754. Cortessis VK, Azadian M, Buxbaum J, et al. Comprehensive meta-analysis reveals association between multiple imprinting disorders and conception by assisted reproductive technology. J Assist Reprod Genet 2018;35:943-52.
755. Mussa A, Molinatto C, Cerrato F, et al. Assisted reproductive techniques and risk of Beckwith-Wiedemann syndrome. Pediatrics 2017; 140:e20164311.
756. Cooper WN, Luharia A, Evans GA, et al. Molecular subtypes and phenotypic expression of Beckwith-Wiedemann syndrome. Eur J Hum Genet 2005;13:1025-32.
757. Enklaar T, Zabel BU, Prawitt D. Beckwith-Wiedemann syndrome: multiple molecular mechanisms. Expert Rev Mol Med 2006;8:1-19.
758. Weksberg R, Shuman C, Smith AC. Beckwith-Wiedemann syndrome. Am J Med Genet C Semin Med Genet 2005;137C:12-23.
759. Bliek J, Gicquel C, Maas S, et al. Epigenotyping as a tool for the prediction of tumor risk and tumor type in patients with Beckwith-Wiedemann syndrome (BWS). J Pediatr 2004;145:796-9.

760. Bliek J, Maas S, Alders M, Mannens M. Epigenotype, phenotype, and tumors in patients with isolated hemihyperplasia. J Pediatr 2008;153:95-100.
761. Bliek J, Maas SM, Ruijter JM, et al. Increased tumour risk for BWS patients correlates with aberrant H19 and not KCNQ1OT1 methylation: occurrence of KCNQ1OT1 hypomethylation in familial cases of BWS. Hum Mol Genet 2001; 10:467-76.
762. Cohen MM Jr. Beckwith-Wiedemann syndrome: historical, clinicopathological, and etiopathogenetic perspectives. Pediatr Dev Pathol 2005;8:287-304.
763. DeBaun MR, Niemitz EL, McNeil DE, et al. Epigenetic alterations of H19 and LIT1 distinguish patients with Beckwith-Wiedemann syndrome with cancer and birth defects. Am J Hum Genet 2002;70:604-11.
764. Maas SM, Vansenne F, Kadouch DJ, et al. Phenotype, cancer risk, and surveillance in Beckwith-Wiedemann syndrome depending on molecular genetic subgroups. Am J Med Genet A 2016;170:2248-60.
765. Rump P, Zeegers MP, van Essen AJ. Tumor risk in Beckwith-Wiedemann syndrome: a review and meta-analysis. Am J Med Genet A 2005;136:95-104.
766. Scott RH, Douglas J, Baskcomb L, et al. Constitutional 11p15 abnormalities, including heritable imprinting center mutations, cause nonsyndromic Wilms tumor. Nat Genet 2008;40:1329-34.
767. Weksberg R, Nishikawa J, Caluseriu O, et al. Tumor development in the Beckwith-Wiedemann syndrome is associated with a variety of constitutional molecular 11p15 alterations including imprinting defects of KCNQ1OT1. Hum Mol Genet 2001;10:2989-3000.
768. Zhang P, Liegeois NJ, Wong C, et al. Altered cell differentiation and proliferation in mice lacking p57kip2 indicates a role in Beckwith-Wiedemann syndrome. Nature 1997;387:151-8.
769. Kalish JM, Doros L, Helman LJ, et al. Surveillance recommendations for children with overgrowth syndromes and predisposition to Wilms tumors and hepatoblastoma. Clin Cancer Res 2017;23:e115.
770. Lack EE. Tumors of the adrenal gland and extra-adrenal paraganglia. AFIP Atlas of Tumor Pathology, 3rd Series, Fascicle 19. Washington, DC: Armed Forces Institute of Pathology; 1997.
771. McCauley RG, Beckwith JB, Elias ER, Faerber EN, Prewitt LH Jr, Berdon WE. Benign hemorrhagic adrenocortical macrocysts in Beckwith-Wiedemann syndrome. Am J Roentgenol 1991;157:549-52.
772. Beckwith JB. Macroglossia, omphalocele, adrenal cytomegaly, gigantism and hyperplastic visceromegaly. Birth Defects 1969;5:188-96.

Adrenal Medullary Hyperplasia

773. Carney JA, Sizemore GW, Sheps SG. Adrenal medullary disease in multiple endocrine neoplasia, type 2: pheochromocytoma and its precursors. Am J Clin Pathol 1976;66:279-90.
774. DeLellis RA, Wolfe HJ, Gagel RF, et al. Adrenal medullary hyperplasia. A morphometric analysis in patients with familial medullary thyroid carcinoma. Am J Pathol 1976;83:177-96.
775. Visser JW, Axt R. Bilateral adrenal medullary hyperplasia: a clinicopathological entity. J Clin Pathol 1975;28:298-304.
776. Bongiovanni AM, Yakovac WC, Steiker DD. Study of adrenal glands in childhood: hormonal content correlated with morphologic characteristics. Lab Invest 1961;10:956-67.
777. Naeye RL. Brain-stem and adrenal abnormalities in the sudden-infant-death syndrome. Am J Clin Pathol 1976;66:526-30.
778. Bialestock D. Hyperplasia of the adrenal medulla in hyper tension of children. Arch Dis Child 1961;36:465-73.
779. Beckwith JB. Extreme cytomegaly of the adrenal fetal cortex, omphalocele, hyperplasia of kidneys and pancreas, and Leydig-cell hyperplasia—another syndrome? West Soc Pediatr Res; 1963.
780. Beckwith JB. Macroglossia, omphalocele, adrenal cytomegaly, gigantism, and hyperplastic visceromegaly. Birth defects 1969;5:188-96.
781. Chen SX, Zhou ZQ, Zhao JS, Wang SZ, Sun ND. Catecholamine acute abdomen. A case of adrenal medulla hyperplasia accompanied by acute abdomen. Chin Med J 1989;102:811-3.
782. Wu CP. Adrenal medullary hyperplasia. Natl Med J Chin 1977;57:331-3.
783. Wu JP, Xu FJ, Zeng ZP. Adrenal medullary hyperplasia. Long-term follow-up of 15 patients. Chin Med J (Engl) 1984;97:653-6.
784. Yoshida A, Hatanaka S, Ohi Y, Umekito Y, Yoshida H. von Recklinghausen's disease associated with somatostatin-rich duodenal carcinoid (somatostatinoma), medullary thyroid carcinoma and diffuse adrenal medullary hyperplasia. Acta Pathol Jap 1991;41:847-56.
785. Wells SA Jr, Asa SL, Dralle H, et al. Revised American Thyroid Association guidelines for the management of medullary thyroid carcinoma. Thyroid 2015;25:567-610.
786. Gagel RF, Tashjian AH Jr, Cummings T, et al. The clinical outcome of prospective screening for multiple endocrine neoplasia type 2a. An 18-year experience. N Engl J Med 1988;318:478-84.

787. Keiser HR, Beaven MA, Doppman J, Wells S Jr, Buja LM. Sipple's syndrome: medullary thyroid carcinoma, pheochromocytoma, and parathyroid disease. Studies in a large family. NIH conference. Ann Int Med 1973;78:561-79.
788. Schimke RN. Multiple endocrine adenomatosis syndromes. Adv Intern Med 1976;21:249-65.
789. Sipple JH. The association of pheochromocytoma with carcinoma of the thyroid gland. Am J Med 1961;31:163-5.
790. Steiner AL, Goodman AD, Powers SR. Study of a kindred with pheochromocytoma, medullary thyroid carcinoma, hyperparathyroidism and Cushing's disease: multiple endocrine neoplasia, type 2. Medicine 1968;47:371-409.
791. Eng C, Smith DP, Mulligan LM, et al. Point mutation within the tyrosine kinase domain of the RET proto-oncogene in multiple endocrine neoplasia type 2b and related sporadic tumours. Hum Mol Genet 1994;3:237-41.
792. Lloyd RV. RET proto-oncogene mutations and rearrangements in endocrine diseases. Am J Pathol 1995;147:1539-44.
793. Mulligan LM, Marsh DJ, Robinson BG, et al. Genotype-phenotype correlation in multiple endocrine neoplasia type 2: report of the Intentional RET Mutation Consortium. J Int Med 1995;238:343-6.
794. Verdy M, Weber AM, Roy CC, Morin CL, Cadotte M, Brochu P. Hirschsprung's disease in a family with multiple endocrine neoplasia type 2. J Pediatr Gastroenterol Nutr 1982;1:603-7.
795. Williams ED, Pollock DJ. Multiple mucosal neuromata with endocrine tumours: a syndrome allied to von Recklinghausen's disease. J Pathol Bacteriol 1966;91:71-80.
796. Gagel RF. Multiple endocrine neoplasia. In: Wilson JD, Faster DW, Kronenberg HM, Larsen PR, eds. Williams textbook of endocrinology, 9th ed. Philadelphia: WB Saunders; 1998;16:27-49.
797. Korpershoek E, Petri BJ, Post E, et al. Adrenal medullary hyperplasia is a precursor lesion for pheochromocytoma in MEN2 syndrome. Neoplasia 2014;16:868-73.
798. Grogan RH, Pacak K, Pasche L, Huynh TT, Greco RS. Bilateral adrenal medullary hyperplasia associated with an SDHB mutation. J Clin Oncol 2011;29:e200-2.
799. Mete O, Asa SL. Precursor lesions of endocrine system neoplasms. Pathology 2013;45:316-30.
800. Mete O, Tischler AS, de Krijger R, et al. Protocol for the examination of specimens from patients with pheochromocytomas and extra-adrenal paragangliomas. Arch Pathol Lab Med 2014;138:182-8.
801. Diaz-Cano SJ. Clonality studies in the analysis of adrenal medullary proliferations: application principles and limitations. Endocr Pathol 1998;9:301-16.
802. Diaz-Cano SJ, de Miguel M, Blanes A, et al. Clonal patterns in pheochromocytomas and MEN2 adrenal medullary hyperplasia: histologic and kinetic correlates. J Pathol 2000;192:221-8.
803. Thomas JE, Rooke ED, Kvale WF. The neurologist's experience with pheochromocytoma. A review of 100 cases. JAMA 1966;197:754-8.
804. Carney JA, Sizemore GW, Tyce GM. Bilateral adrenal medullary hyperplasia in multiple endocrine neoplasia, type 2: the precursor of bilateral pheochromocytoma. Mayo Clin Proc 1975;50:3-10.
805. Valk TW, Frager MS, Gross MD, et al. Spectrum of pheochromocytoma in multiple endocrine neoplasia. A scintigraphic portrayal using 131I-metaio dobenzylguanidine. Ann Int Med 1981;94:762-7.
806. Maurea S, Cuocolo A, Reynolds JC, et al. Iodine-131-metaiodobenzylguanidine scintigraphy in preoperative and postoperative evaluation of paragangliomas: comparison with CT and MRI. J Nucl Med 1993;34:173-9.
807. Yobbagy JJ, Levatter R, Sisson JC, Shulkin BL, Polley T. Scintigraphic portrayal of the syndrome of multiple endocrine neoplasia type 2b. Clin Nucl Med 1988;13:433-7.
808. Padberg BC, Garbe E, Achilles E, Dralle H, Bressel M, Schröder S. Adrenomedullary hyperplasia and phaeochromocytoma. DNA cytophotometric findings in 47 cases. Virchows Arch Apathol Anat Histopathol 1990;416:443-6.
809. Carney JA, Sizemore GW, Hayles AB. Multiple endocrine neoplasia, type 2b. Pathol Annu 1978;8:105-53.
810. Kurihara K, Mizuseki K, Kondo T, Ohoka H, Mannami M, Kawai K. Adrenal medullary hyperplasia. Hyperplasia-pheochromocytoma sequence. Acta Pathol Jap 1990;40:683-6.
811. Maki Y, Irie S, Ohashi T, Ohmori H. A case of unilateral adrenal medullary hyperplasia. Acta Medica Okayama 1989;43:311-5.
812. Zhang ZX, Yu ST. Analysis of 17 cases of adrenal medullary hyperplasia. Natl Med J Chin 1979;59:95-7.
813. Asari R, Scheuba C, Kaczirek K, Niederle B. Estimated risk of pheochromocytoma recurrence after adrenal-sparing surgery in patients with multiple endocrine neoplasia type 2A. Arch Surg 2006;141:1199-205.
814. Norton JA, Froome LC, Farrell RE, Wells SA Jr. Multiple endocrine neoplasia type IIb: the most aggressive form of medullary thyroid carcinoma. Surg Clin N Am 1979;59:109-18.

5 DIFFUSE NEUROENDOCRINE SYSTEM

The diffuse (dispersed) neuroendocrine system (DNES) is composed of cells at various sites throughout the body. Some endocrine organs, such as the parathyroid and pituitary glands, are composed predominantly of DNES cells. Other endocrine organs, such as the thyroid gland, are composed predominantly of non-neuroendocrine follicular cells, but also have a component of neuroendocrine C cells. The cortical tissue of the adrenal gland is composed of non-neuroendocrine cells, while the adrenal medulla is composed of neuroendocrine cells. Other non-neuroendocrine cells in the endocrine system include steroid-producing cells of the ovary and testis. Neuroendocrine cells are dispersed in many other organs and tissues such as the breast, kidney, lung, prostate gland, larynx, skin, and gastrointestinal tract. The cells of the DNES have cytoplasmic secretory granules, and express neuroendocrine markers and specific hormones and amines (1).

The DNES concept has evolved over time and integrates the ideas of Feyrter, Pearse, Fujita, and other investigators (2–8). In 1938, Feyrter described clear or pale neuroendocrine cells in a complex distribution, particularly throughout the gastrointestinal tract and pancreas (2,9). Pearse noted that adrenomedullary chromaffin, enterochromaffin, corticotroph, islet beta, and thyroid C cells had common cytochemical properties, amine and/or peptide production, amino acid decarboxylase production, and ultrastructural secretory granules with neurons (10–14). Common cytochemical and ultrastructural characteristics of cells producing polypeptide hormones were described with the acronym APUD (amine precursor uptake and decarboxylation) in 1968 by Pearse (3,11). Pearse proposed that some of these cells (adrenal medullary cells, paraganglia, thyroid C cells, and sympathetic ganglia) were derived from the neural crest (1), although subsequent investigators have shown this not to be true for many of the cell types (3,15–18). The identification of thyroid C cell progenitors arising in the endoderm germ layer has changed the long held concept of a neural crest origin of thyroid C cells (19–21).

Pearse later revised the DNES concept to propose that 40 cell types were divided into the central hypothalamic-pituitary axis and pineal gland, as well as the peripheral gastropancreatic axis, lungs, parathyroid gland, adrenal medulla, sympathetic ganglia, skin melanocytes, thyroid C cells, and urogenital tract (1). It is now known that melanocytes are derived from the neural crest, but do not have the properties of neuroendocrine cells. Interestingly, Merkel cells have the properties described as neuroendocrine, but were not included in Pearse's system.

The paraneuron concept, as proposed by Fujita (6), described paraneurons as "receptosecretory cells producing substances of neurons." They have membrane-bound granules and vesicles containing peptides, amines, and chromogranins. Thus, immunohistochemical detection of chromogranins is a marker of paraneurons (6). Fujita noted that chromogranin immunopositivity is particularly helpful in identifying these cells, because other non-neuroendocrine cells can ectopically produce bioactive peptides or amines.

The DNES is composed of cells and tumors with secretory granules that commonly stain for chromogranin and synaptophysin, and may express a variety of peptide hormones (8). Immunohistochemistry, in situ hybridization, and polymerase chain reaction (PCR) can be used to identify gene products expressed by neuroendocrine cells (8). The updated DNES theory combines these observations and provides a unifying approach for these cells with a unique genotype and phenotype that helps explain neuroendocrine syndromes and takes into account ectopic hormone production by other cell types (1).

Table 5-1

GRANINS: CHROMOGRANINS AND SECRETOGRANINS

Granin	Gene	Location
Chromogranin A	*CHGA*	14q32.12
Chromogranin B	*CHGB*	20pter-p12
Secretogranin II	*SCG2*	2q35-2q36
Secretogranin III	*SCG3*	15q21
7B2 (secretogranin V)	*SCG5*	15q13-q14
NESP55 (secretogranin VI)	Part of *GNAS* gene locus	20q13.2
VGF (secretogranin VII)	*VGF*	7q22.1
ProSAAS (secretogranin VIII)	*PROSAAS/PCSK1N*	Xp11.23

IMMUNOHISTOCHEMISTRY OF DNES CELLS

Chromogranin/Secretogranin

The major components of the granins are chromogranin A, chromogranin B, and secretogranin II. These acidic proteins are located within the secretory granules of neuroendocrine cells. Other members of the granin family are secretogranin III, 7B2 (secretogranin V), NESP55 (secretogranin VI), VGF (secretogranin VII), and proSAAS (secretogranin VIII) (Table 5-1) (22).

The distribution of chromogranin A, chromogranin B, and secretogranin has been studied in neuroendocrine cells and tumors. Chromogranin A is one of the most extensively studied and widely used neuroendocrine markers. It has been studied in tissues and tumors throughout the body, including the breast, gastrointestinal tract, genitourinary tract, skin, lung, head and neck, and endocrine organs (23–77).

Chromogranin A is widely distributed in the secretory granules of most peptide-producing endocrine tissues. It is immunopositive in neoplastic endocrine tissues and tumors with abundant secretory granules, but may not show strong staining in neuroendocrine tumors with few secretory granules (72). Chromogranin may not always show significant immunoreactivity in Merkel cell carcinoma, neuroblastoma, small cell lung carcinoma, prolactinoma, or hindgut neuroendocrine tumors (72,78). Staining may vary in intensity among cells: chromogranin A stains parathyroid chief cells more strongly than parathyroid oxyphil cells; normal rims of parathyroid tissue may stain more intensely than parathyroid adenomas (79). The distribution and functions of granin-derived peptides is of increasing interest in physiology and medicine, and these peptides have been studied as disease biomarkers (22,80).

Synaptophysin

Synaptophysin is an integral membrane protein of the presynaptic vesicles of neurons. Unlike the larger (80 to 400 nm) secretory granules that contain chromogranin/secretogranin, smaller (40 to 80 nm) transparent-appearing vesicles contain synaptophysin. Wiedemann and Franke (81) proposed the name synaptophysin because it was initially identified as an integral membrane component of presynaptic small vesicles in synapses, and suggested it may be involved in synaptic vesicle formation and exocytosis (81,82).

Prohormone Convertases (Proconvertases)

Prohormone convertases (proconvertases) are proteolytic enzymes involved in proteolytic cleavage of prohormones to hormones (83). Nine proconvertases have been identified: PC1/3 (PCSK1), PC2 (PCSK2), PC4 (PCSK4), PACE4 (PCSK6), PC5/6 (PCSK5), PC7/8/LPC (PCSK7), furin/Pace/PC1 (PCSK3), Site1 protease/S1P/SKI (PCSK8), and NARC-1 (PCSK9) (84,85). The prohormone convertases 1/3 and 2 (PC1/PC3 and PC2) are localized to neuroendocrine tissues (86). PC1/PC3 and PC2 are useful markers of neuroendocrine tissues and tumors. The other proconvertases have a variable tissue distribution (85). For example, PC5/6 is more commonly expressed in the adrenal gland and the gastrointestinal tract (1), while PC4 is expressed in testicular germ cells (85,87). Within the central nervous system, PC5 is detected in neurons while PACE4 is found in neural and non-neural cells (88).

Leu-7 (HNK-1)

Leu-7 monoclonal antibody (HNK-1) reacts with natural killer lymphocytes and myelin-associated glycoprotein (89). Adrenal medullary cells and pheochromocytomas are immunoreactive with Leu-7 (89). In adrenal medullary cells, Leu-7 reacts with a protein in the matrices of the chromaffin granules (89). Other neuroendocrine tumors positive for Leu-7 are small cell carcinomas of the urinary bladder, colon and rectum, and lung, although Leu-7 may have lower sensitivity in identifying these tumors than other neuroendocrine markers (49,68,90–93). Leu-7 is positive in neuroblastomas, primitive neuroectodermal tumors, nephroblastoma, melanoma, small cell mesothelioma (90,94), and medullary thyroid carcinoma, but it is also immunoreactive with folliculogenic (papillary and follicular) thyroid tumors (95).

Protein Gene Product 9.5 (PGP9.5)

Protein gene product 9.5/ubiquitin c-terminal hydrolase-L1 (PGP9.5/UCHL1) is a marker of neuroendocrine and neural tissues. PGP9.5 has been studied in a variety of tumors, including medulloblastoma; carcinomas of the lung, pancreas, and gastrointestinal tract; salivary gland tumors; cellular neurothekeoma; and melanoma (96–104). PGP9.5 is a good marker of neuroendocrine differentiation, but is not entirely specific since non-neuroendocrine tumors, such as melanoma, may stain for PGP9.5. In addition to its role as a marker of neuroendocrine differentiation, PGP9.5 has been evaluated as a diagnostic marker in parathyroid carcinoma (105,106).

Gastrin-Releasing Peptide

Gastrin-releasing peptide is a mammalian bombesin-like peptide that is present in neuroendocrine cells and tumors (107–110). It is present in lung tissues and tumors, intestinal and pancreatic neuroendocrine tumors, thyroid C cells and medullary thyroid carcinomas, and some paragangliomas (107–113). Although most well-differentiated neuroendocrine tumors of the lung, pancreas, and intestine are positive for gastrin-releasing peptide, immunoreactivity is less common in poorly-differentiated neuroendocrine carcinomas, pituitary neuroendocrine tumors (PitNETs), neuroblastomas, and pheochromocytomas (110). Clinically, bombesin-related peptides/receptors are being evaluated for therapeutic roles in cancer imaging and treatment (114).

Neuron-Specific Enolase

Neuron specific enolase (NSE) is a broad-spectrum neuroendocrine marker with low specificity. NSE was identified in brain tissue extracts and found in neurons and neuroendocrine cells and tumors (115). The lack of specificity greatly limits the use of NSE as a neuroendocrine marker.

Neural Cell Adhesion Molecule (NCAM/CD56)

Neural cell adhesion molecule (NCAM/CD56) is a membrane-bound glycoprotein expressed in neoplastic and normal neuroendocrine cells (116), as well as many non-neuroendocrine tissues and tumors. In a study of gastrointestinal stromal and mesenchymal tumors, NCAM/CD56 expression was seen in 86 percent of leiomyomas, 80 percent of schwannomas, 76 percent of leiomyosarcomas, and 50 percent of epithelioid and 7 percent of spindled gastrointestinal stromal tumors (117). NCAM/CD56 immunoreactivity is variable in liposarcoma, PEComa, myofibroblastic sarcoma, and other sarcomas (117). Phosphaturic mesenchymal tumors are also positive for CD56/NCAM (118). In a study of 206 ovarian carcinomas, immunoreactivity for NCAM/CD56 was identified in 65 percent of carcinomas of all histologic types and did not correlate with chromogranin and synaptophysin immunoreactivity (119). NCAM/CD56 may help in the differential diagnosis of cholangiocarcinoma, and benign cholangiocellular lesions, and in classifying some non-neuroendocrine thyroid tumors (120–122).

CD56 is immunoreactive with natural killer cells. Thus, CD56 is of limited use as a neuroendocrine marker.

Insulinoma-Associated Protein 1 (INSM1)

Insulinoma-associated protein 1 (INSM1) is a transcription factor involved in neuroendocrine cell differentiation (123–127). In a recent study INSM1 immunopositivity was reported in 88.3 percent of 129 neuroendocrine tumors and only 1 of 27 tumors without a neuroepithelial or neuroendocrine component (128). In this study, INSM1 expression was restricted to the nuclei of neuroendocrine cells and tissues, and there was no staining in any of the adult non-neoplastic non-neuroendocrine tissues. In addition, among 113 gastrointestinal neuroendocrine neoplasms, higher expression of INSM1 was seen in midgut tumors with known metastases than those without known metastases. In a study of pancreatic tumors, INSM1 was positive in neuroendocrine tumors, but pancreatic ductal adenocarcinomas were negative (129).

Another recent study evaluated 97 neuroendocrine tumors of various types (middle ear adenoma, paragangliomas, PitNET, medullary thyroid carcinoma, olfactory neuroblastoma, small cell carcinoma, large cell neuroendocrine carcinoma, and sinonasal teratocarcinoma) and 626 non-neuroendocrine tumors of various histologic grades and sites in the head and neck (130). INSM1 showed 99 percent sensitivity and 97.6 percent specificity for neuroendocrine tumors from the head and neck. Of the non-neuroendocrine tumors in this study, alveolar rhabdomyosarcoma and SMARCB1-deficient sinonasal carcinomas were immunopositive for INSM1. In another recent study, 28 of 31 extraskeletal myxoid chondrosarcomas were also found to be immunoreactive for INSM1 (131).

PANCREAS

Histology

Neuroendocrine cells comprise 1 to 2 percent of the volume of the pancreas in adults (1,83). They are located predominantly in the islets of Langerhans, but single neuroendocrine cells and small clusters may also be associated with the pancreatic ducts and paraductal acinar cells (1). Islets are present throughout the pancreas, but random variation in the number of islets occurs among lobules (132). Additionally, the concentration of islets does not appear to be related to age, sex, body weight, or pancreatic weight, although the concentration of islets is greater in the tail of the pancreas than the head or body (133). The pattern of islet concentration is similar in patients with diabetes mellitus and pancreatitis, however, their volume may appear increased (islet cell aggregation) due to a relative decrease in exocrine pancreatic tissue and atrophy (fig. 5-1) (132,133).

Pancreatic islets can be classified as either compact or diffuse. Compact islets predominate (90 percent), measure 75 to 225 µm, are sharply circumscribed, and are composed of cells with round nuclei and pale amphophilic cytoplasm (132). Although pancreatic neuroendocrine cells are relatively uniform in size and shape, occasional cells with enlarged nuclei are seen (134). Diffuse islets measure up to 450 µm, have ill-defined borders, and are often composed of trabeculae of columnar cells with hyperchromatic nuclei and basophilic cytoplasm (132).

The volume density of neuroendocrine cells in the pancreas of neonates is up to 15 percent and in infants is 6 to 7 percent (135). Extrainsular endocrine cells are more prominent in neonates and infants than in adults. Up to 15 percent of pancreatic neuroendocrine cells in neonates are located in ducts and acini (135). Infants have fewer clusters of neuroendocrine cells than neonates (135). During development, islet-like structures appear at week 12 of gestation and initially include aggregates of insulin cells or glucagon cells, while the architecture of islets in adults is mixed (136).

Four main peptides are produced by the islet cells: insulin (beta cells), glucagon (alpha cells), somatostatin (delta cells), and pancreatic polypeptide (PP cells) (137). Each cell produces one major peptide. Insulin-producing beta cells constitute 60 to 70 percent of the pancreatic neuroendocrine cells and are most prominent in the central area of compact islets (132,137). Glucagon-producing alpha cells comprise 15 to 20 percent and are most prominent at the periphery of the islets. A few somatostatin-producing delta cells and pancreatic polypeptide-producing PP cells are scattered throughout the compact islets. The cellular constituency of diffuse islets is predominantly (70 percent) PP cells. Beta cells comprise approximately 20 percent of diffuse islets, with alpha and delta cells each comprising 5 percent of the remaining cells.

The proportions of the different types of neuroendocrine cells vary anatomically within the pancreas. For example, PP cells are more common in the posterior portion of the pancreatic head (135,138). In addition, proportions of the different cell types vary with age. For example, the proportion of insulin-secreting cells (beta cells) increases from 50 percent of neuroendocrine cells in neonates to 70 percent in adults, while the proportion of somatostatin-containing cells (delta cells) decreases from 30 percent in neonates to 10 percent in adults (135,139). In contrast, the proportion of alpha cells remains stable throughout life (135).

Diabetes Mellitus

Definition. *Diabetes mellitus* is caused by relative or absolute insulin deficiency resulting in persistent hyperglycemia.

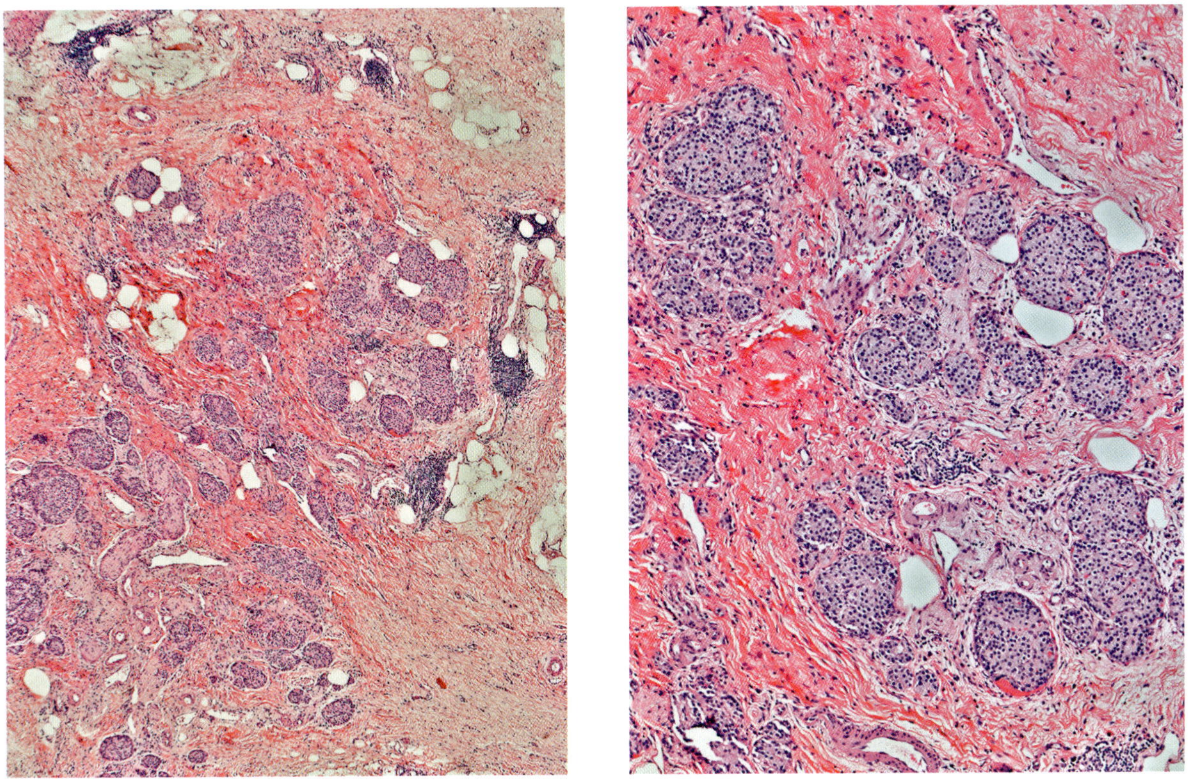

Figure 5-1

PANCREATIC ISLET CELL AGGREGATION

Left: Low-power view of pancreatic islet cell aggregation in which the pancreatic islet volume may appear increased in diabetes and pancreatitis, but the volume is similar to that of normal pancreas.

Right: The appearance of increased islet tissue is due to the relative decrease in exocrine pancreatic tissue and atrophy.

Clinical Features. Type I diabetes mellitus is an autoimmune disorder characterized by lymphocyte-mediated destruction of the beta cells of the pancreas (140). Worldwide, type I diabetes mellitus affects approximately 30 million individuals (140). Significant complications of diabetes mellitus include kidney disease, cardiovascular disease, retinopathy, and neuropathy. Diabetes mellitus is the most common cause of end-stage renal disease in the United States. Diabetic kidney disease is the main cause of premature mortality in individuals with type I diabetes mellitus (141). Cardiovascular disease occurs in 40 percent of individuals with type I diabetes mellitus by 40 years of age (142). The development of complications may have both genetic and environmental influences (141).

Type II diabetes mellitus is more common than type I, affecting approximately 400 million individuals worldwide (143). Type II diabetes mellitus is characterized by β-cell dysfunction and insulin resistance. The increasing incidence of obesity is associated with the increasing incidence of type II diabetes mellitus (144); however, both genetic and environmental influences are also involved.

Laboratory Findings. Individuals with diabetes mellitus have hyperglycemia and/or glycosuria. Other laboratory findings associated with diabetic ketoacidosis or hyperosmolar nonketotic coma may also be present. Glycosylated hemoglobin (HbA1c) levels are elevated.

Gross Findings. The pancreas may not show significant gross abnormalities in diabetes mellitus. In late stage disease, it may be decreased in weight and size (145,146).

Microscopic Findings. The pancreas in type I diabetes mellitus often shows insulitis early in disease, particularly in very young children (147). After a year or more, the islets contain

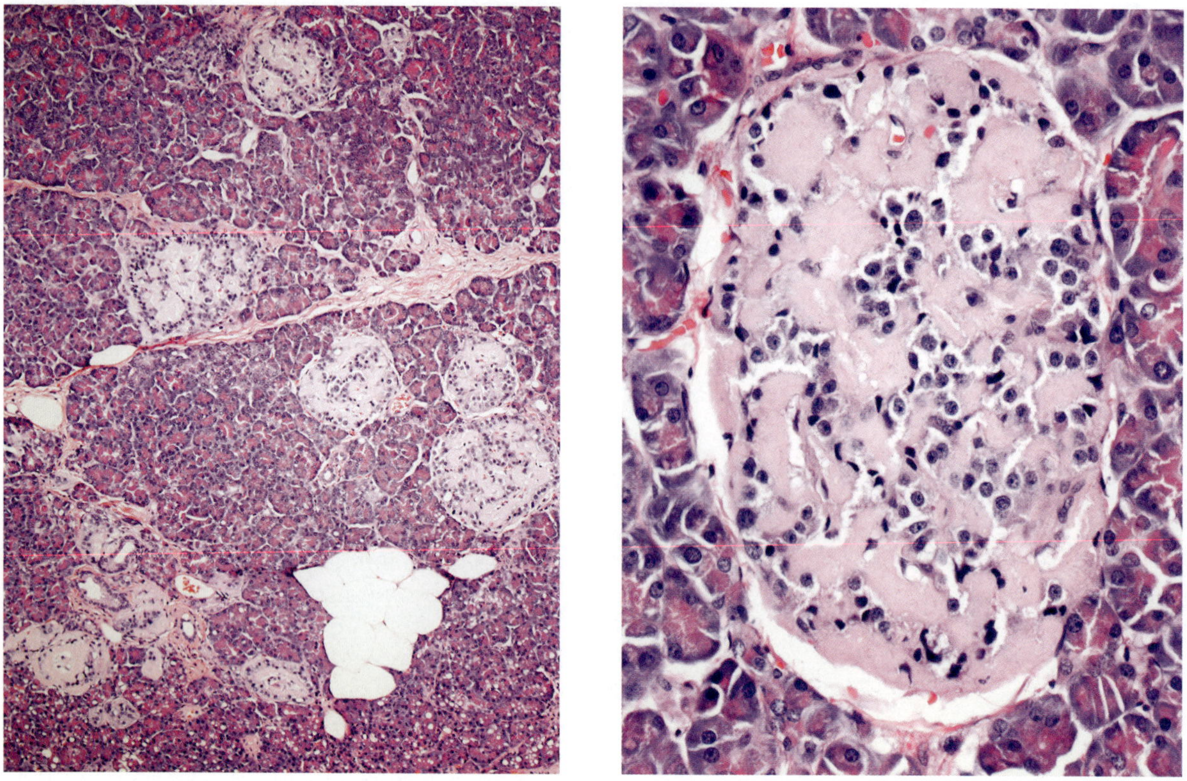

Figure 5-2

PANCREATIC ISLET CELLS IN TYPE II DIABETES MELLITUS

Left: Although generally not seen in type I diabetes mellitus, most pancreases involved by type II diabetes show involvement by amyloid.

Right: These islet cells show prominent involvement by amyloid.

glucagon, somatostatin, and pancreatic polypeptide cells, but contain few insulin-producing beta cells (148). The acinar cells are also atrophied, since insulin has a trophic effect on acinar cells (1). Unlike type II diabetes mellitus, the islets in type I disease generally do not contain amyloid.

In type II diabetes mellitus, lymphocytic inflammation (predominantly CD8 suppressor T cells and a few B cells) of the pancreas may be seen early in disease (149). Islet amyloidosis is seen in 80 to 90 percent of pancreases involved by type II diabetes mellitus, and only 2 to 3 percent of pancreases in individuals without diabetes (fig. 5-2) (150,151). The amyloid is produced by the beta cells (150).

Treatment. The treatment of type I diabetes mellitus focuses on exogenous insulin replacement with insulin analogs and mechanical technologies (152). Future possible therapies include those targeting B cells, among other autoimmune processes (140). Engineering bioartificial pancreas tissue has also been suggested as a potential future therapy (153).

The treatment of type II diabetes includes addressing underlying issues such as obesity with lifestyle modification. Oral and injectable drugs are also used to reduce and maintain glucose concentrations as well as drugs that mediate glucose absorption in the gastrointestinal tract, inhibitors of sodium glucose co-transporter 2, drugs that affect the central nervous system circadian rhythm, and insulin therapies, among others (144).

Hyperinsulinism in Infants

General and Clinical Features. Infants may have *transient* or *persistent hypoglycemia* and *hyperinsulinemia*. There is even a transient hyperinsulinemia in normal newborns. This hyperinsulinemia is usually mild and subsides in the first day or two after birth. Infants of mothers with poorly controlled diabetes mellitus, low birth-weight infants, and infants small for gestational age may have hypoglycemia and hyperinsulinemia (154).

Persistent congenital hyperinsulinemia encompasses a heterogeneous group of disorders with different molecular abnormalities (154–163). Eleven genes have been identified in association with monogenic disorders of hyperinsulinemia in neonates (154). The most common and most severe form of monogenic hyperinsulinemia in infants is due to mutation in one of the two subunits, *SUR1* or *Kir6.2*, two adjacent genes on chromosome 11p15.1 that encode the sulfonylurea receptor 1 and Kir6.2 proteins of the KATP plasma membrane channel of pancreatic beta cells (154,164–167). Approximately half of infants with the KATP form of hyperinsulinemia have a focal and potentially resectable pancreatic lesion.

Congenital hyperinsulinemia can be classified as either diffuse or focal. In a study by Lord (168) of 223 children with congenital hyperinsulinism (97 diffuse, 114 focal, and 12 other), children with diffuse hyperinsulinism had shorter gestational age, higher birth weight, younger age at presentation, higher insulin levels during hypoglycemia, and required higher glucose infusion rates than those with focal hyperinsulinism. The children with focal hyperinsulinemia presented later than those with diffuse disease and had higher rates of seizures. Children with diffuse hyperinsulinism had a median 98 percent pancreatectomy rate, and 41 percent required continued treatment for hypoglycemia. Those with focal hyperinsulinemia had median 27 percent pancreatectomy rate, and 6 percent required additional treatment after surgery.

In a series of 417 children with congenital hyperinsulinism, 298 were diazoxide unresponsive and 118 were diazoxide responsive (163). Of the 298 children with diazoxide unresponsive hyperinsulinism, 272 (91 percent) had identifiable mutations (*ABCC8, KCNJ11*, and *GCK*). Of the 118 children with diazoxide responsive hyperinsulinism, 56 (47 percent) had identifiable mutations (*ABCC8, KCNJ11, GLUD1, HADH, UCP2, HNF4A*, and *HNF1A*). Of the 298 children with diazoxide-unresponsive hyperinsulinism, 149 had focal hyperinsulinism. Monoallelic recessive *KATP* mutations were identified in 145 of 149 (97 percent) cases. In all cases, maternal transmission was excluded. Of the 122 diazoxide-unresponsive cases with diffuse hyperinsulinism, 109 (89 percent) had *KATP* mutations and 2 had *GCK* mutations (163). Of the 56 (47 percent) of children with mutation-positive diazoxide-responsive congenital hyperinsulinism, 24 (42 percent) had *GLUD1* mutations, 23 (41 percent) had dominant *KATP* (19 *ABCC8, 4 KCNJ11*) mutations, and the rest had mutations of *HADH, UCP2, HNF4A*, or *HNF1a*.

Beckwith-Wiedemann, Turner, and Kabuki syndromes are associated with hyperinsulinemia (154). Hyperinsulinemia occurs in half of neonates with Beckwith-Wiedemann syndrome. Beckwith-Wiedemann syndrome is caused by abnormalities involving a region on chromosome 11p15.4, containing the *H19, CDKN1C, KCNQ1*, and *IGF2* genes (154). When Beckwith-Wiedemann syndrome is associated with paternal 11p uniparental isodisomy, the hyperinsulinemia can be severe and associated with extensive islet adenomatosis (154).

Gross Findings. Gross histologic findings are not identified in pancreas specimens in congenital hyperinsulinemia.

Microscopic Findings. *Focal hyperinsulinism (focal adenomatous hyperplasia)* is characterized by an increase in beta cells in a focal region of the pancreas, with the rest of the pancreas appearing normal. The endocrine cell overgrowth involves greater than 40 percent of a given area (169).

In *diffuse hyperinsulinism*, enlarged islet cell nuclei (three times larger than the surrounding nuclei) are present throughout the pancreas (fig. 5-3) (169); although, in another study (170), nucleomegaly was only identified in 60 percent of cases of diffuse hyperplasia. Ductulo-insular complexes (beta cells budding off from duct epithelium) are often present in diffuse hyperplasia (170). Other less common findings in diffuse hyperplasia are septal islets, centroacinar small beta cell clusters, nucleomegaly, abundant cytoplasm, and spreading isolated beta cells (170). However, many of these features, except for nucleomegaly and abundant cytoplasm, are also seen in normal pancreas specimens (170).

One study found that frozen section examination of multiple biopsies may be able to classify many cases of hyperinsulinism intraoperatively,

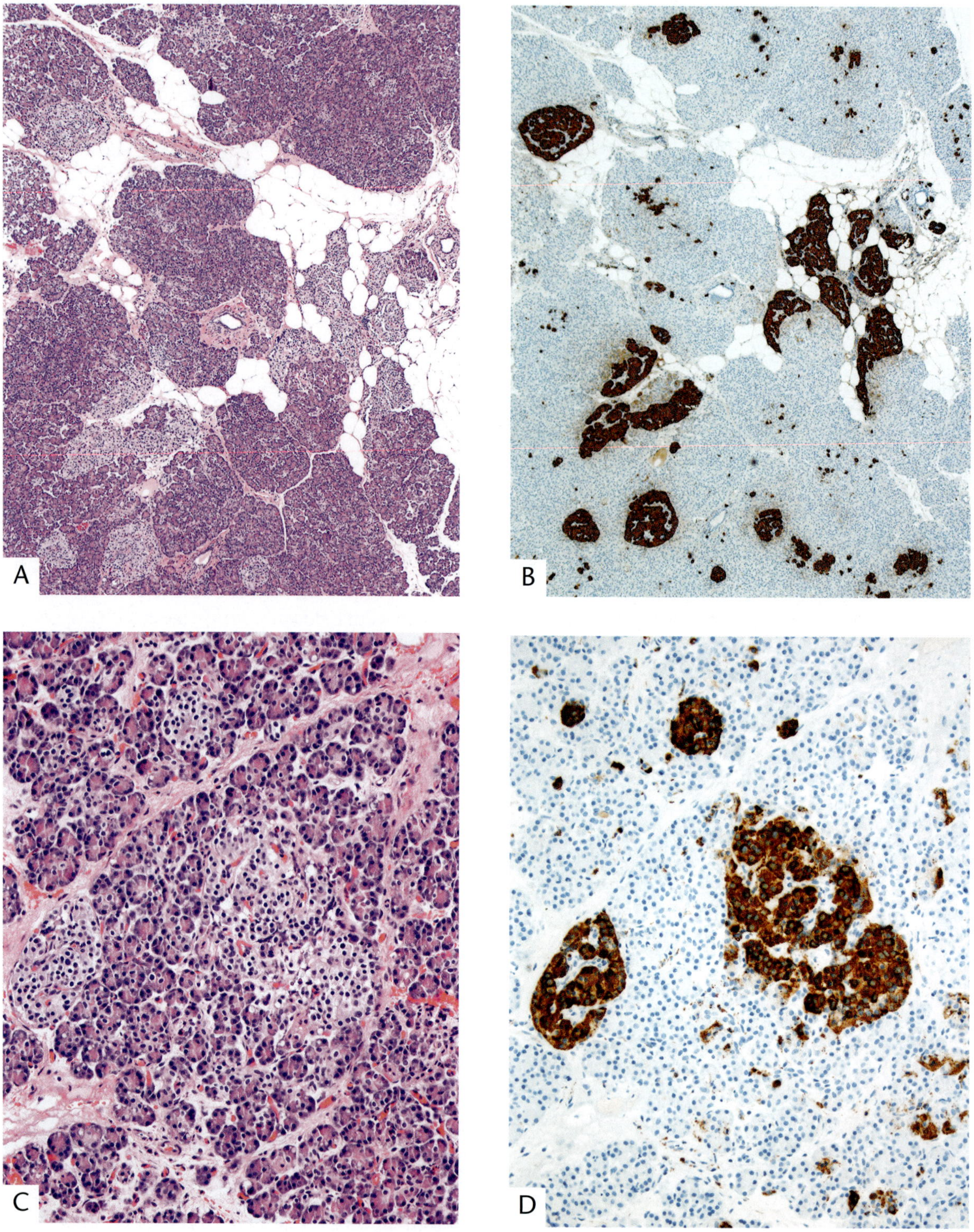

Figure 5-3

CONGENITAL HYPERINSULINEMIA

In diffuse hyperinsulinism, the pancreas shows an increase in beta cells, enlarged islet cell nuclei, and ductulo-insular complexes (A,C). The hyperplastic islets are highlighted with chromogranin immunostains (B,D).

but there are diagnostic difficulties in this approach, such as equivocally large islet cell nuclei, rare truly large islet cell nuclei in areas away from a focal lesion, large or ill-defined areas of adenomatous hyperplasia, and cases for which the histologic features in the pancreas, even on permanent sections, did not fit well for either focal or diffuse hyperinsulinism (169). Ki-67 labeling indices have been studied in focal and diffuse hyperinsulinism but the results have been controversial (170–172).

Differential Diagnosis. Hyperinsulinemia in infants can be due to a great variety of causes that need to be evaluated clinically. After imaging, it is still important to exclude pancreatic neuroendocrine tumors grossly and histologically. In a different setting, such as with pancreatitis, the volume of islets may appear increased ("islet cell aggregation") due to a relative decrease in exocrine pancreatic tissue and atrophy (133).

Treatment. Genetic testing and 18F-DOPA positron emission tomography (PET) scans are useful in identifying the cause of hyperinsulinemia in infants (154,164). Hyperinsulinemia is difficult to recognize in neonates, but its identification is important since it is associated with seizures and brain deficits. Congenital hyperinsulinism encompasses a great variety of disorders, some transient and some persistent, with different etiologies and genetic abnormalities. Thus, genetic testing and imaging modalities are helpful to determine the type of hyperinsulinism and the most appropriate management.

Noninsulinoma Pancreatogenous Hyperinsulinemic Hypoglycemia in Adults (Adult Nesidioblastosis)

Clinical Features. *Hyperinsulinemic hypoglycemia* is a rare condition, most often caused by an insulin-producing pancreatic neuroendocrine tumor (insulinoma). Pancreatic beta-cell hyperplasia *(nesidioblastosis)* accounts for approximately 4 percent of cases (173). Autoimmune hypoglycemia may also be considered in hyperinsulinemic hypoglycemia. *Noninsulinoma pancreatogenous hyperinsulinemic hypoglycemia,* known to most pathologists as nesidioblastosis, can occur as an idiopathic condition or in individuals who have had upper digestive tract bariatric surgery (174). Unlike congenital hyperinsulinemic hypoglycemia, *SUR1* and *Kir6.2* mutations are usually not identified in adult cases (174).

The possibility of noninsulinoma pancreatogenous hyperinsulinemic hypoglycemia syndrome (NIPHS) in adults is considered after an insulinoma and autoimmune hypoglycemia are excluded. Hyperinsulinemic hypoglycemia is often associated with neuroglycopenic symptoms of dizziness, confusion, seizures, and in extreme cases, coma (174). Autonomic symptoms include tremor, anxiety, palpitations, hunger, sweating, and paresthesias (174).

Gross Findings. The pancreas in NIPHS usually does not show macroscopic abnormalities or well-defined lesions. It appears grossly normal (175).

Microscopic Findings. NIPHS was originally described as a functional disorder of pancreatic beta cells. The histologic features are variable, but are mostly restricted to the pancreatic islets (175). The pancreatic ducts can show hyperplastic changes, but are often normal. The exocrine pancreatic tissue appears normal.

The islets may be enlarged to greater than 300 µm each, and the number of islets may be increased (islet hyperplasia), with an increase in beta cells (fig. 5-4) (175). The islets may also appear irregular and appear to have a lobulated composition due to the increase in beta cells. The beta cells show nuclear enlargement (fig. 5-5) and abundant clear cytoplasm, and are up to twice as large as normal islet cells (175). Ductulo-insular complexes may be seen, but are not a specific finding. Peliosis may also be seen (fig. 5-6) (176). Post-gastric bypass and idiopathic NIPHS are histologically indistinguishable (176,177). The beta cells do not show increased proliferative activity (175). An increase in insulin-like growth factor 2, insulin-like growth factor receptor 1-alpha, and transforming growth factor-beta receptor 3 expression is seen in the islets in these cases compared to the normal pancreas (176).

Differential Diagnosis. Hyperinsulinemic hypoglycemia is most often caused by an insulin-producing pancreatic neuroendocrine tumor (insulinoma). Insulinomas are often single lesions, although they can be multifocal in some cases. In contrast, pancreatic beta cell hyperplasia, on its own, is generally not tumefactive. Autoimmune hypoglycemia is also not tumefactive and should also be considered

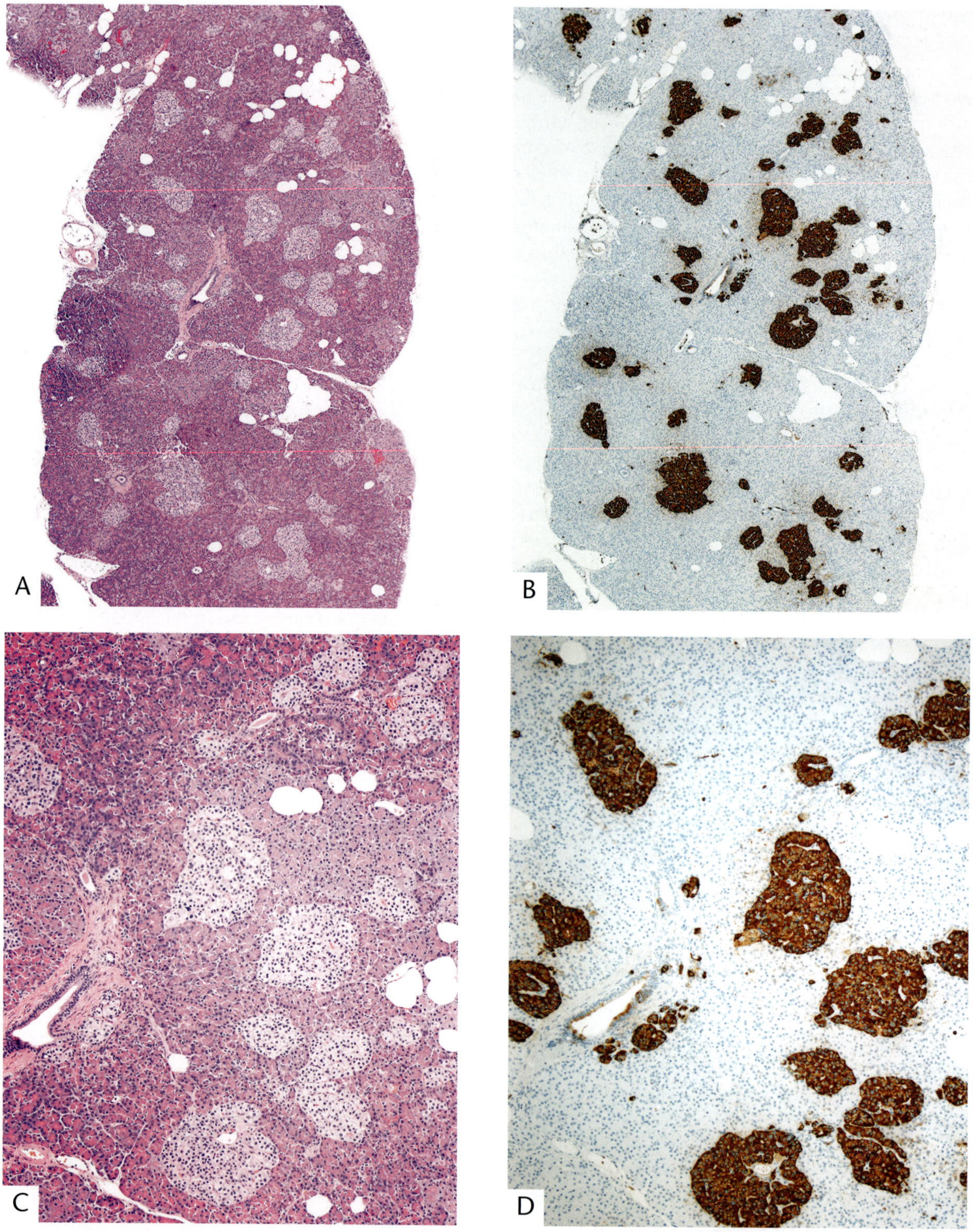

Figure 5-4

ADULT NONINSULINOMA PANCREATOGENOUS HYPERINSULINEMIC HYPOGLYCEMIA (NESIDIOBLASTOSIS)

The number of islets may be increased and the islets may be enlarged and appear lobulated, with an increase in beta cells. The hyperplastic islets (A,C) are highlighted with chromogranin immunohistochemistry (B,D).

in the differential diagnosis of hyperinsulinemic hypoglycemia. Antibodies against insulin (Hirata syndrome) or against insulin receptors (autoimmune hypoglycemia type B) must be excluded. Although already evaluated by imaging, it is still important to exclude pancreatic neuroendocrine tumors grossly and histologically. In a setting such as pancreatitis, the volume of islets may appear increased (termed "islet cell aggregation" or "pseudohyperplasia of the islets") due to a relative decrease in exocrine pancreatic tissue and atrophy of the pancreas (133).

Treatment. The treatment of hyperinsulinemic hypoglycemia can include medical and/or surgical approaches and depends on the underlying etiology.

Other Pancreatic Neuroendocrine Cell Hyperplasias

In addition to congenital and adult persistent hyperinsulinemic hypoglycemia associated with hyperplasia of insulin-producing beta cells of the pancreas, other types of pancreatic neuroendocrine cell hyperplasia can occur. *Neuroendocrine*

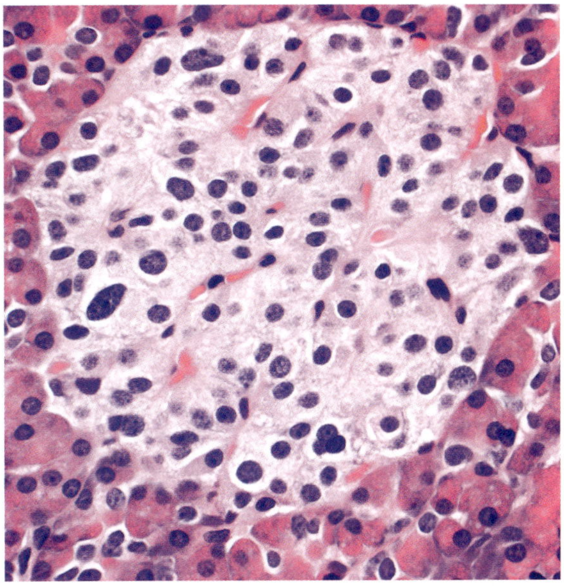

Figure 5-5

ADULT NONINSULINOMA PANCREATOGENOUS HYPERINSULINEMIC HYPOGLYCEMIA: PLEOMORPHISM

The islets are composed of beta cells with nuclear enlargement, pleomorphism, and abundant clear cytoplasm.

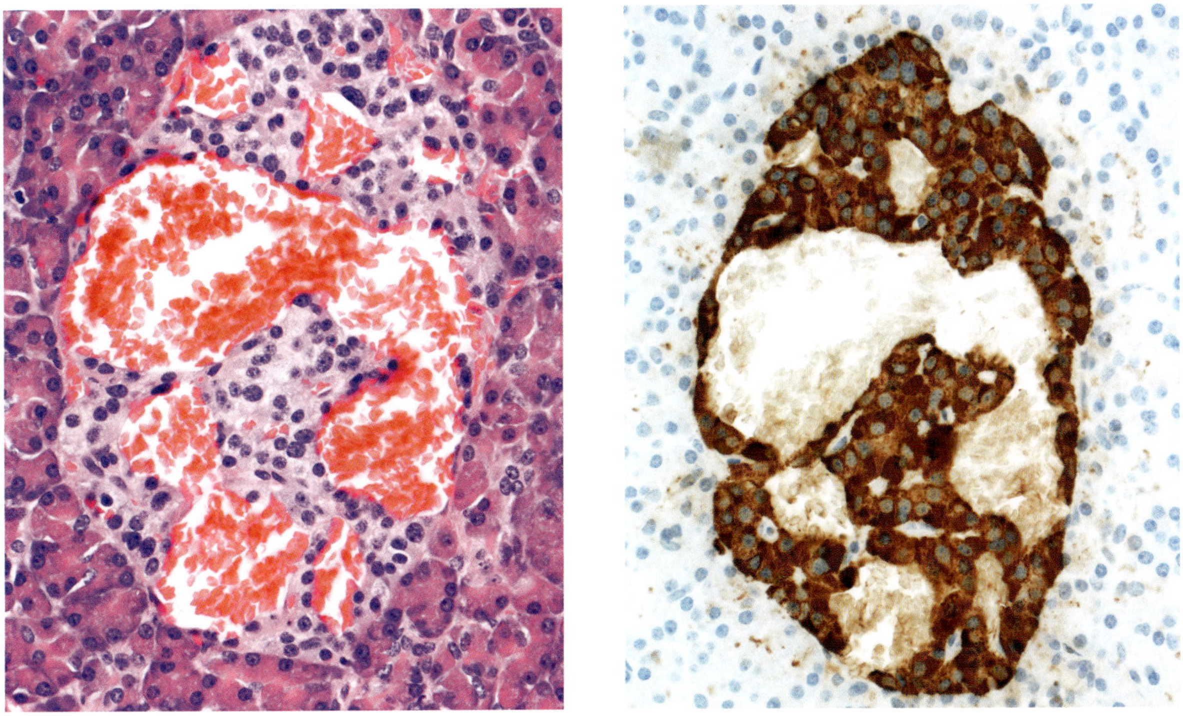

Figure 5-6

ADULT NONINSULINOMA PANCREATOGENOUS HYPERINSULINEMIC HYPOGLYCEMIA: PELIOSIS

Peliosis is not a specific finding, but is seen. (Left: H&E; right: chromogranin immunohistochemistry.)

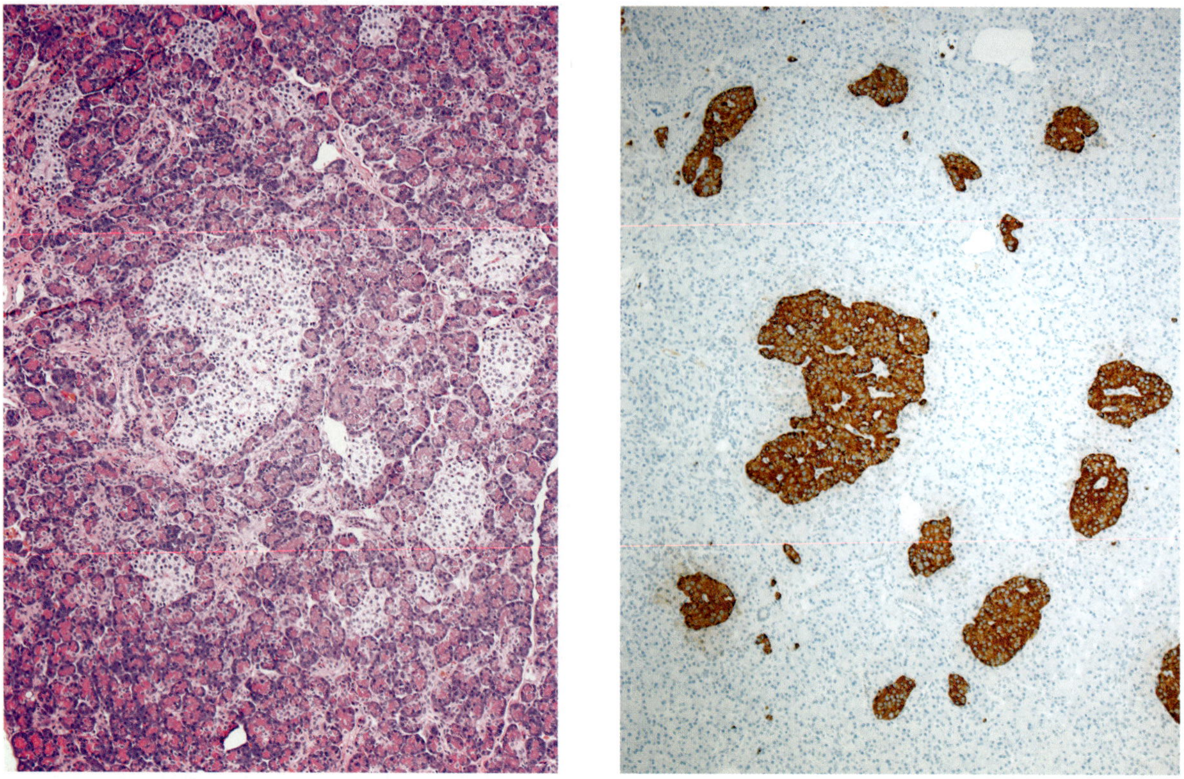

Figure 5-7

ISLET CELL HYPERPLASIA IN MULTIPLE ENDOCRINE NEOPLASIA TYPE 1 (MEN1)

Islet cell hyperplasia is seen in MEN1 and von Hippel-Lindau disease. (Left: H&E; right: chromogranin immunohistochemistry.)

cell hyperplasia involving ducts (ductulo-insular complexes) and *islet cell hyperplasia* are seen in the setting of multiple endocrine neoplasia type 1 (MEN1) (fig. 5-7) and von Hippel-Lindau disease (178,179). Islet cell hyperplasia can be seen adjacent to neuroendocrine tumors (fig. 5-8). Islet dysplasia and hyperplasia are seen in familial pancreatic neuroendocrine tumors (178).

In *islet cell dysplasia*, the islets are slightly enlarged (less than 0.5 mm), and show mild atypia and loss of normal numbers and arrangement of the neuroendocrine cell types (178). Dysplastic islets larger than 0.5 mm are classified as microadenomas (178). In *glucagon cell adenomatosis*, enlarged islets with hyperplasia of glucagon-containing cells are present in the background of multiple microadenomas and tumors (180,181). These cases occur in individuals without MEN1 or von Hippel-Lindau disease, but may be associated with glucagon receptor mutations in some cases (180,181).

GASTROINTESTINAL TRACT

Histology

DNES cells are scattered throughout the gastrointestinal tract. In the esophagus, neuroendocrine cells are identified in approximately 25 percent of individuals, with most seen in the distal esophagus. In the stomach, endocrine cells are identified among epithelial cells. In the antrum, gastrin-producing G cells are most common, while in the body of the stomach, histamine-producing enterochromaffin (EC)-like cells predominate. Serotonin-producing EC cells and somatostatin-containing D cells are also present in the antrum, and along with endothelin-containing X cells, are also present in the fundic oxyntic glands. Endocrine cells are also scattered among the epithelial cells of the small intestinal villi and crypts, with lesser numbers in the large intestinal crypts.

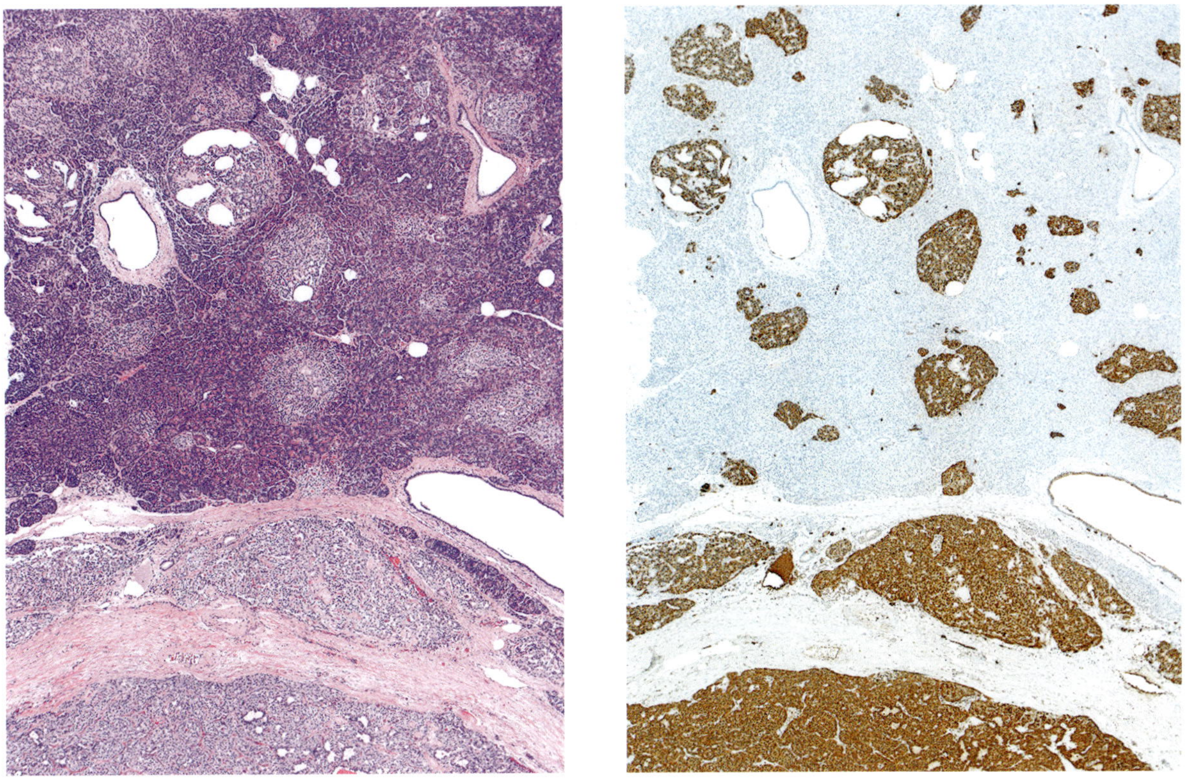

Figure 5-8

ISLET CELL HYPERPLASIA

Islet cell tumor and adjacent islet cell hyperplasia. (Left: H&E; right: chromogranin immunohistochemistry.)

The most common endocrine cell in the small and large intestine contains 5-hydroxytryptamine. Glicentin cells are the next most common endocrine cell and are most prominent in the ileum and large intestine. Many other neuroendocrine cells are identified in lesser numbers in the small and large intestines.

The neuroendocrine cells of the gastrointestinal tract produce various peptide hormones and bioamines (Table 5-2). These cells may have paracrine or neuroendocrine effects. The functions of the neuroendocrine cells of the gastrointestinal tract include growth and development; stimulating the secretion of pancreatic enzymes (cholecystokinin), insulin (gastrin-inhibiting peptide), gastric acid (gastrin), gastrin (gastrin-releasing peptide), and bicarbonate (secretin); regulation of gastric emptying and gut motility (motilin); and possibly, neurotransmitter function.

Gastric Neuroendocrine Cell Hyperplasia

Clinical Features. *Hyperplasia of gastric neuroendocrine cells* occurs in atrophic gastritis-related hypochlorhydria-associated hypergastrinemia. The findings are most prominent in autoimmune atrophic gastritis where hypochlorhydria or achlorhydria causes hyperplasia of antral G cells and hypergastrinemia, resulting in the proliferation of enterochromaffin-like (ECL) cells that produce histamine. Long-term usage of proton pump inhibitors is also associated with *ECL cell hyperplasia*.

Gross Findings. Unless gastric ECL cell hyperplasia is associated with a gastric neuroendocrine tumor, the hyperplasia may not be grossly apparent. However, when ECL cell hyperplasia occurs in the setting of chronic atrophic gastritis, the gastric mucosa may show changes of atrophic gastritis.

Table 5-2

NEUROENDOCRINE CELLS OF THE GASTROINTESTINAL TRACT[a]

Cell Type	Peptide Hormones or Bioamines
ECL (enterochromaffin-like) cells	Histamine (stomach body/fundus)
EC (enterochromaffin) cells	Serotonin, melatonin, substance P
G and IG cells	Gastrin (stomach antrum and small intestine/proximal duodenum)
D (somatostatin) cells	Somatostatin (small and large intestine)
X cells	Endothelin (stomach: body/fundus)
Glicentin cells	Glicentin (small and large intestine)
M cells	Motilin (small intestine)
PP cells	Pancreatic polypeptide (small intestine)
N cells	Neurotensin (small bowel: ileum)
S cells	Secretin (small intestine)
L cells	Glucagon-like peptides, peptide YY, and pancreatic polypeptide (small and large intestine)
P/D1 cells (Gr cells, A-like cells)	Ghrelin
I cells	Cholecystokinin (small intestine: duodenum)
K cells	Gastrin inhibiting peptide (small intestine: duodenum jejunum)
EC cells	Serotonin (stomach, small and large intestine)
VIP cells	Vasoactive intestinal polypeptide (nerves)
Others	Gastrin-releasing peptide (nerves), Pro-gamma-melanocyte-stimulating hormone, beta-endorphin, beta-lipotropin

[a]Data from references 1 and 147.

Microscopic Findings. Solcia et al. (182) classified ECL cell hyperplasia as *simple/diffuse* when there was a proliferation with greater than twice the normal number of neuroendocrine cells (compared to age- and gender-matched controls); this is most often seen in Zollinger-Ellison syndrome (ZES) and primary gastrin cell hyperplasia (182). *Linear hyperplasia* is defined as linear groups of five or more cells along the glandular basement membrane, with at least two chains of cells per millimeter of mucosa. Clusters of five or more cells in the epithelium measuring less than 150 µm in diameter are classified as *micronodular hyperplasia.* At least one micronodule is present per millimeter of mucosa. *Adenomatous hyperplasia* involves clusters of more than five adjacent micronodules in the lamina propria.

Micronodular hyperplasia and adenomatous hyperplasia are both seen in the setting of autoimmune atrophic gastritis. Adenomatous hyperplasia is also often seen in the setting of MEN/ZES. Dysplasia of ECL cell proliferations occurs in the progression to neuroendocrine neoplasms (fig. 5-9) and include: enlarging nodules with nests of cells greater than 150 µm; fusing micronodules without intervening basement membranes; microinvasive lesions infiltrating the lamina propria; and nodules with new stroma formation and microlobular or trabecular architecture (182).

Prognosis. Gastric ECL cell hyperplasia may progress to dysplasia and is a precursor of type I and type II gastric neuroendocrine tumors (NETs) involving the corpus-fundic region of the stomach. Most gastric NETs are type I and occur in the setting of autoimmune chronic atrophic gastritis (183–187). Women in the sixth decade of life are most frequently affected (183–188). The autoimmune chronic gastritis causes loss of parietal cells and achlorhydria, which stimulates gastrin-producing G cells in the antrum and duodenum, resulting in hypergastrinemia; hypergastrinemia, in addition to growth factors, promotes histamine- producing ECL cell proliferation in the corpus-fundic region of the stomach

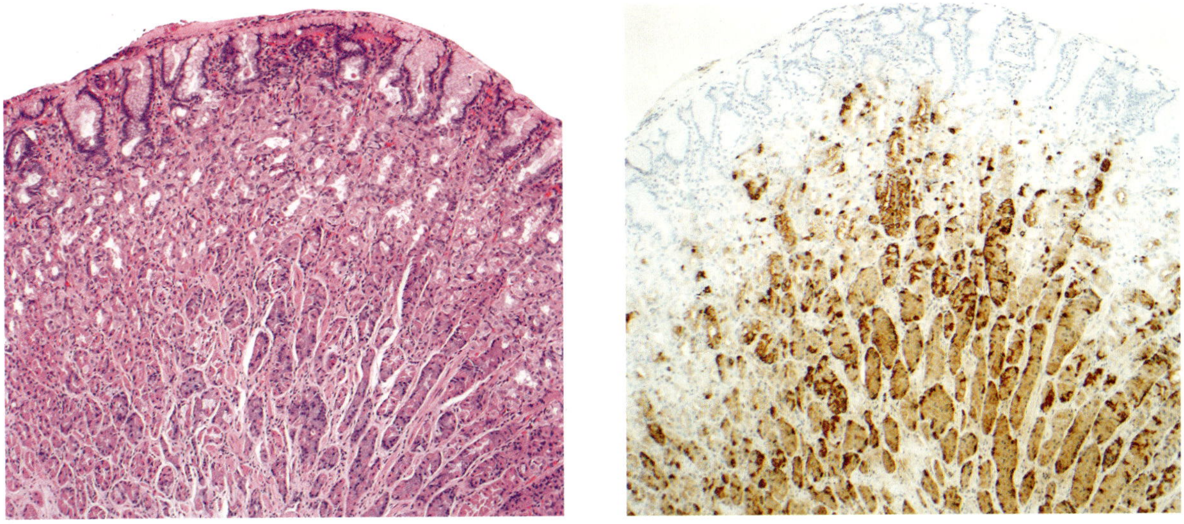

Figure 5-9

ZOLLINGER-ELLISON SYNDROME/ECL CELL HYPERPLASIA

ECL cell hyperplasia can be seen in Zollinger-Ellison syndrome (ZES) and primary gastrin cell hyperplasia (Left: H&E stain; right: chromogranin immunohistochemistry).

(183–186). The ECL cell hyperplasia is followed by dysplasia and multiple small ECL cell neoplasms (type I gastric NETs) (183,184,186,188).

Type II gastric NETs occur in the setting of MEN1/ZES, with hypergastrinemia resulting from duodenal gastrin-producing NETs usually in the setting of MEN1 (179,186, 187,189). Type II gastric NETs are thought to be caused by loss of heterozygosity of the *MEN1* gene (11q13), since individuals with ZES but without MEN1 do not develop these tumors (186,189). Type II gastric NETs are not associated with atrophic chronic autoimmune gastritis or achlorhydria, and occur equally in females and males.

Type III gastric NETs are sporadic tumors that usually occur in men around age 55 and are not associated with atrophic gastritis or MEN1 (186,187,190). Type IV gastric NETs are poorly differentiated tumors, not related to chronic atrophic gastritis, and only rarely associated with MEN1 (186,190,191).

Small and Large Intestine Neuroendocrine Cell Hyperplasia

The neuroendocrine cells in the small intestine may be associated with hyperplasia and have the potential to progress to neuroendocrine neoplasms (179). In the duodenum, hyperplasia of G (gastrin-producing) cells and D (somatostatin-producing) cells can occur in the setting of MEN1 and may be associated with gastrinomas (179,192). The mucosal crypts and Brunner glands may show an increase in neuroendocrine cells (178). The hyperplasia may be diffuse, linear, or micronodular, including enlarged nodules with solid growth, and microinvasive lesions in the lamina propria. Sporadic gastrin-producing NETs are often associated with ZES or unspecified symptoms (192).

The most common location for MEN1/ZES-associated gastrinomas is the duodenum (fig. 5-10) (179,193,194). Gastrinomas are the most common NET in the duodenum. Other neuroendocrine neoplasms of the duodenum include somatostatinomas, gangliocytic paragangliomas, pancreatic polypeptide-producing tumors, nonfunctional NETs, and poorly differentiated neuroendocrine carcinomas (195).

In a recent study of 203 neuroendocrine neoplasms of the duodenum and ampullary/periampullary region, gastrinomas, nonfunctional NETs, and poorly differentiated neuroendocrine carcinomas were identified predominantly in the proximal duodenum, while somatostatin-producing NETs and gangliocytic paragangliomas were most commonly identified in the ampullary/periampullary region (195). Neuroendocrine tumors also occur in the jejunum

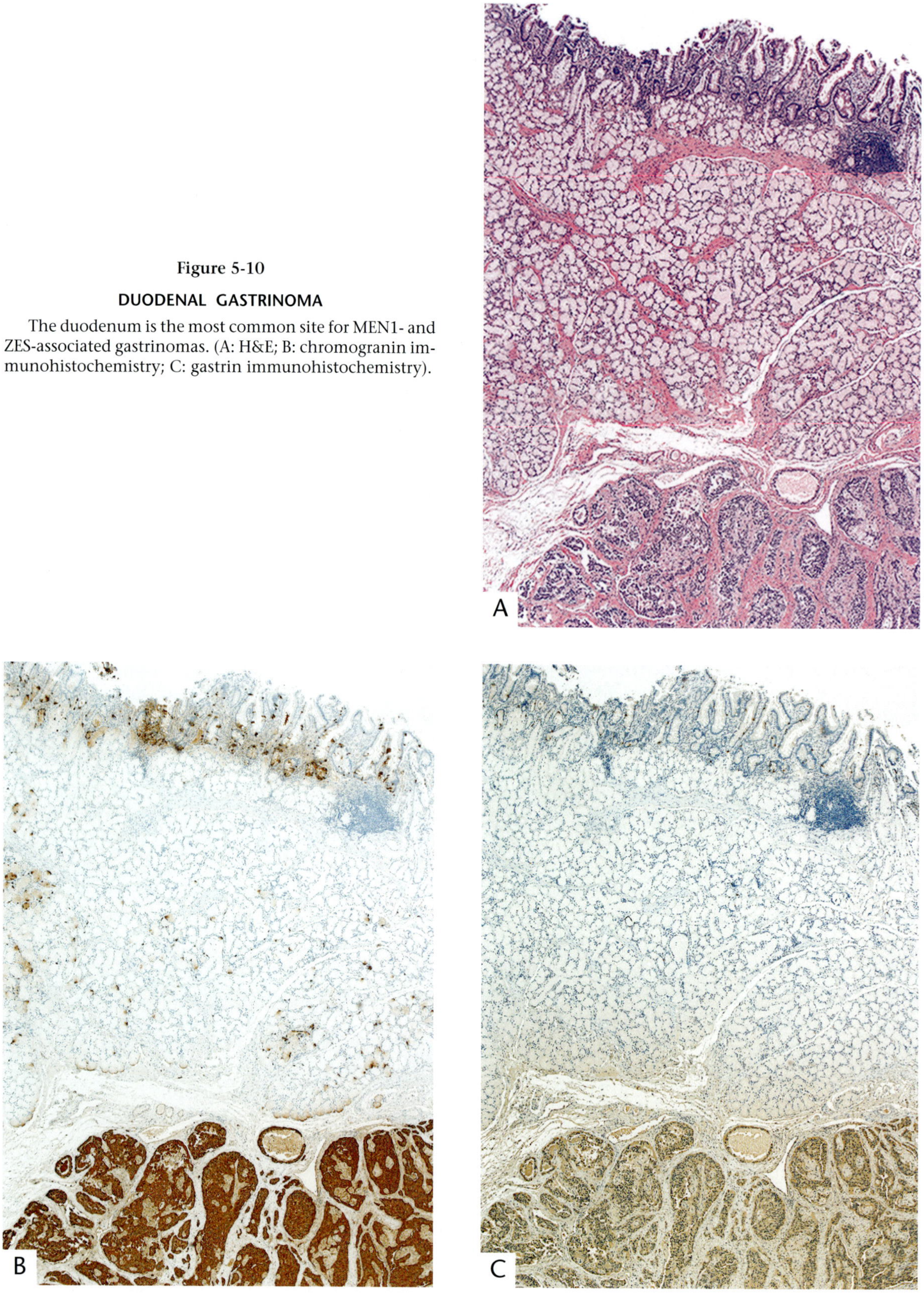

Figure 5-10

DUODENAL GASTRINOMA

The duodenum is the most common site for MEN1- and ZES-associated gastrinomas. (A: H&E; B: chromogranin immunohistochemistry; C: gastrin immunohistochemistry).

and ileum, and these are usually EC (serotonin-producing) cell tumors or possibly tumors arising from L cells. It has been suggested that neuroendocrine cell hyperplasia and neoplasia in the large intestine may be associated with inflammatory bowel disease (196–199).

EXTRA-ADRENAL PARAGANGLIA

Histology

The extra-adrenal paraganglia are located along the distribution of the para-aortic and paravertebral axis. This distribution follows that of the sympathetic nervous system. Many of the extra-adrenal paraganglia are not visible grossly.

Histologically, the extra-adrenal paraganglia may or may not be circumscribed or may appear partially encapsulated. They are composed of ganglion cells and neural tissue. Intermixed adipocytes and connective tissue can be seen. Foci of chromaffin tissue may also persist in extra-adrenal paraganglia in adults. Extra-adrenal paraganglia tissue is positive for chromogranin A and synaptophysin. S-100 protein marks the dendritic processes of sustentacular cells.

Hyperplasia of Extra-Adrenal Paraganglia

Hyperplasia of the paraganglia results from an increase in chief cells and sustentacular cells.

Clinical Features. The carotid body is the best known extra-adrenal paraganglionic tissue to be involved by hyperplasia, but other paraganglia may also show hyperplasia. *Carotid body* and *vagal paraganglia hyperplasias* result from chronic hypoxemia (200). High altitude-associated hypoxemia is associated with chief cell hyperplasia of the carotid body (201). The magnitude of enlargement of carotid bodies at high altitudes increases with age, and may be associated with progressive chemoreceptor insensitivity (201). Carotid body hyperplasia is also associated with cystic fibrosis and cyanotic heart disease (202). Hyperplasia of extra-adrenal paraganglia may occur in the aorticopulmonary paraganglia of infants dying from sudden infant death syndrome (SIDS) (203).

Gross Findings. In an autopsy study, Lack (200) showed the combined mean weight of carotid bodies in six individuals with chronic hypoxemia was 47.6 mg, which was much greater than the control group (25.3 mg). In an autopsy study of 213 children and young adults, Lack (202) showed the average combined weights of carotid bodies in those with cystic fibrosis and cyanotic heart disease were significantly greater than those of the control group (202).

Microscopic Findings. Hyperplastic paraganglia show an increase in chief cells and sustentacular cells. The lobule dimensions of carotid body paraganglia and vagal body paraganglia from individuals with chronic hypoxemia are larger than those from the normal control group (200). An increase in sustentacular cells and carotid body volume has been noted in individuals with enlarged carotid bodies after longstanding cardiopulmonary disease, such as cardiac hypertrophy (204). The area occupied by parenchymal cells is significantly greater in carotid bodies of individuals born and living at high altitudes compared to those at sea level; the hyperplastic cells appear to be chief cells and may be vacuolated (201). When the histology of hyperplastic carotid bodies from individuals with hypoxemia and right ventricular hypertrophy due to emphysema was compared with the hyperplastic carotid bodies of individuals with left ventricular hypertrophy due to hypertension, the histologic features were similar (205). A proliferation of type II sustentacular cells and compression of central cores of type I chief cells were noted in both groups. An increase in the number, mean lobule diameter, and total glomic tissue volume is reported in aorticopulmonary paraganglia from 23.8 percent infants who died of SIDS compared to age-matched controls (203).

Lack (202) found a decrease in dense core neurosecretory granules in carotid bodies from individuals with cystic fibrosis as well as a decrease in chief cell argyrophilia (202). However, no differences in morphometry, ultrastructure, or catecholamine content was noted in carotid bodies from victims of SIDS compared to controls (202).

Differential Diagnosis. Unlike paraganglioma, hyperplasia of extra-adrenal paraganglia is not neoplastic. Hyperplasia of extra-adrenal paraganglia is generally a response to a stimulus such as hypoxia.

Treatment. Treatment of extra-adrenal paraganglia hyperplasia is generally based on treating the underlying disorder.

PULMONARY NEUROENDOCRINE CELLS

Histology, Anatomy, and Physiology

Pulmonary neuroendocrine cells are present in the lungs as single cells, usually associated with the small airways, and as clusters of cells referred to as neuroendocrine (neuroepithelial) bodies (9,206). Neuroendocrine bodies at bifurcation points of branching airways or bronchioalveolar junctions are referred to as nodal neuroendocrine bodies (207). Neuroendocrine bodies in the airway between branch points are internodal (207,208). Single neuroendocrine cells have been shown by imaging to have directed migration distally (even when a closer bifurcation point is present in another direction) and cluster predominantly at bifurcation points forming neuroendocrine bodies, although some neuroendocrine cells cluster at interbifurcation areas (208).

Pulmonary neuroendocrine cells are involved in lung development, and are more prominent in fetal and neonatal lung than adult lung. Pulmonary neuroendocrine cells also function as chemoreceptors. These cells produce serotonin, (5-hydroxytriptamine [5-HT]), and peptides such as bombesin and calcitonin that have a role in lung development (209).

Pulmonary neuroendocrine cells are elongated and flask shaped, with apical cytoplasm reaching the airway lumen, or are elongated cells with lateral dendritic-like cytoplasmic processes extending along the basement membrane of the airway (210,211). These cells seem to react to stimuli from the airway (such as hypoxia) via microvilli on the apical membrane, which contacts the airway lumen, or by pores between Clara cells (212). The elongated pulmonary neuroendocrine cells without apparent luminal contact, but with dendritic-like cytoplasmic processes, are referred to as closed-type stimulus, mediated by mechanical stretch channels leading to 5-HT release (209).

Pulmonary neuroendocrine cells and bodies are involved in oxygen sensing and chemotransmission of hypoxia stimulus (211). Hyperplasia of pulmonary endocrine cells results from the hypoxia of high altitude (213), chronic lung disease states, and congenital lung disorders. Patients with bronchopulmonary dysplasia, lung hypoplasia due to diaphragmatic hernia, SIDS, congenital central hypoventilation syndrome, childhood emphysema, and neuroendocrine cell hyperplasia of infancy may have pulmonary neuroendocrine cell hyperplasia (211,214–219). The role of pulmonary neuroendocrine cells is also being studied in the pathogenesis or pathophysiology of bronchial asthma, cystic fibrosis, and pulmonary hypertension (211,215,219).

Diffuse Idiopathic Pulmonary Neuroendocrine Cell Hyperplasia

Definition. According to the fifth edition World Health Organization (WHO), *diffuse idiopathic pulmonary neuroendocrine cell hyperplasia* (DIPNECH) is defined as a multifocal hyperplasia of pulmonary neuroendocrine cells associated with tumorlets. It is considered a preinvasive condition that may develop into carcinoid tumors. DIPNECH is often accompanied by mild chronic lymphocytic inflammation and bronchiolar fibrosis with constrictive features (constrictive bronchiolitis) (220).

Clinical Features. Individuals over a wide age range are affected, but DIPNECH most commonly occurs in the fifth to sixth decades and affects women more often than men (220–222). In a study of 30 cases and a review of the literature of 169 additional cases of DIPNECH, three clinical presentations were noted (222). DIPNECH was a somewhat unexpected finding in individuals who presented with carcinoid tumors and multiple small pulmonary nodules on imaging. A second group of individuals presented with multiple small pulmonary nodules that were concerning for metastatic disease on imaging. The third group presented with respiratory symptoms (222). Most individuals present with symptoms of chronic cough and shortness of breath, while others present with radiographic findings, often noted incidentally, when undergoing imaging for another reason (220). Radiographically, 81 percent of individuals with DIPNECH have multiple pulmonary nodules, with or without a lung mass (221).

Gross Findings. DIPNECH typically does not result in any macroscopic abnormalities, but may occur in association with a carcinoid tumor.

Microscopic Findings. Microscopically, DIPNECH is characterized by small groups or a single layer of pulmonary neuroendocrine cells that may form aggregates and protrude

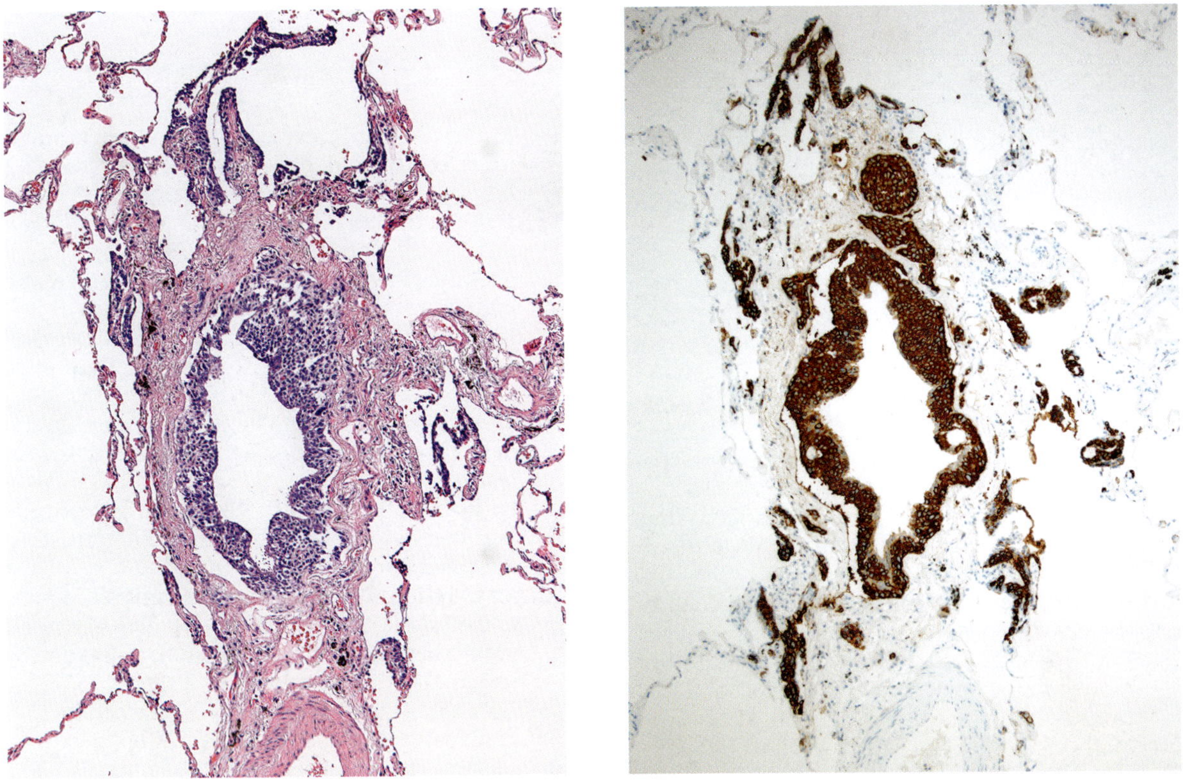

Figure 5-11

DIFFUSE IDIOPATHIC PULMONARY NEUROENDOCRINE CELL HYPERPLASIA (DIPNECH)

Pulmonary neuroendocrine cells forming a single layer or small aggregates protrude into the lumen. (Left: H&E; right: chromogranin immunohistochemistry.)

into the bronchial lumen (fig. 5-11) (220). The neuroendocrine cells may also invade across the basal lamina as tumorlets (220).

Specific criteria for the diagnosis of DIPNECH and carcinoid tumorlet have not been clearly defined. One group has suggested a minimum requirement of multifocal neuroendocrine cell hyperplasia in addition to three or more carcinoid tumorlets for a diagnosis of DIPNECH (223). Others have studied the histologic features in relation to the clinical presentation (224). Compared to asymptomatic/incidental cases, individuals with symptomatic disease were younger, more likely to have mosaic attenuation and multiple small nodules on imaging, and more likely to have a higher number of foci of linear neuroendocrine proliferations and tumorlets (224).

Differential Diagnosis. DIPNECH is distinguished from reactive pulmonary neuroendocrine cell proliferations because they have a known underlying etiology and are not associated with carcinoid tumors.

Prognosis. DIPNECH usually has a good prognosis. Although the neuroendocrine tumors may be slowly growing, progression to high grade neuroendocrine neoplasms such as large cell neuroendocrine carcinoma and small cell carcinoma has not been reported. The constrictive bronchiolitis may progress to respiratory failure requiring transplantation, but the majority of patients remain clinically stable (220,222).

REFERENCES

Immunohistochemistry of DNES Cells

1. Lloyd RV, Douglas BR, Young WF Jr. Endocrine diseases. Atlas of Nontumor Pathology, Series 1, Fascicle 1. Washington DC: American Registry of Pathology; 2002:315.
2. Feyrter F. Die Peripheren endokrinen (parakriken) Drusen. In: Kauffman E, Staemler M. Lehrbuch der speziellen pathologischen anatomie, vol. 11-12. Berlin: De Guyrter; 1969.
3. Pearse AG. The APUD cell concept and its implications in pathology. Pathol Annu 1974;9:27-41.
4. Pearse AG. The APUD concept and hormone production. Clin Endocrinol Metab 1980;9:211-22.
5. Pearse AG. The diffuse neuroendocrine system: peptides, amines, placodes and the APUD theory. Prog Brain Res 1986;68:25-31.
6. Fujita T. Present status of paraneuron concept. Arch Histol Cytol 1989;52(Suppl):1-8.
7. DeLellis RA, Wolfe HJ. The polypeptide hormone-producing neuroendocrine cells and their tumors: an immunohistochemical analysis. Methods Achiev Exp Pathol 1981;10:190-220.
8. Lloyd RV. Overview of neuroendocrine cells and tumors. Endocr Pathol 1996;7:323-8.
9. Feyrter F. Uber diffuse endokrine epitheliale organe. Leipzig: Barth;1938:1-62.
10. Carvalheira AF, Welsch U, Pearse AG. Cytochemical and ultrastructural observations on the argentaffin and argyrophil cells of the gastro-intestinal tract in mammals, and their place in the APUD series of polypeptide-secreting cells. Histochemie 1968;14:33-46.
11. Pearse AG. Common cytochemical and ultrastructural characteristics of cells producing polypeptide hormones (the APUD series) and their relevance to thyroid and ultimobranchial C cells and calcitonin. Proc R Soc Lond B Biol Sci 1968;170:71-80.
12. Pearse AG. The calcitonin secreting C cells and their relationship to the APUD cell series. J Endocrinol 1969;45(Suppl):13-4.
13. Pearse AG. The cytochemistry and ultrastructure of polypeptide hormone-producing cells of the APUD series and the embryologic, physiologic and pathologic implications of the concept. J Histochem Cytochem 1969;17:303-13.
14. Rost FW, Polak JM, Pearse AG. The melanocyte: its cytochemical and immunological relationship to cells of the endocrine polypeptide (APUD) series. Virchows Arch B Cell Pathol 1969;4:93-101.
15. Pearse AG, Polak JM. Neural crest origin of the endocrine polypeptide (APUD) cells of the gastrointestinal tract and pancreas. Gut 1971;12:783-8.
16. Andrew A. Further evidence that enterochromaffin cells are not derived from the neural crest. J Embryol Exp Morphol 1974;31:589-98.
17. Le Douarin NM. The neural crest. Cambridge: Cambridge University Press; 1982.
18. Le Douarin NM. On the origin of pancreatic endocrine cells. Cell 1988;53:169-71.
19. Johansson E, Anderson L, Ornros J, et al. Revising the embryonic origin of thyroid C cells in mice and humans. Development 2015;142:3519-28.
20. Kameda Y. Cellular and molecular events on the development of mammalian thyroid C cells. Dev Dyn 2016;245:323-41.
21. Nilsson M, Williams D. On the origin of cells and derivation of thyroid cancer: C cell story revisited. Eur Thyroid J 2016;5:79-93.
22. Bartolomucci A, Possenti R, Mahata SK, Fischer-Colbrie R, Loh YP, Salton SR. The extended granin family: structure, function, and biomedical implications. Endocr Rev 2011;32:755-97.
23. Eriksson B, Oberg K, Skogseid B. Neuroendocrine pancreatic tumors. Clinical findings in a prospective study of 84 patients. Acta Oncol 1989; 28:373-7.
24. Hartschuh W, Weihe E , Yanaihara N. Immunohistochemical analysis of chromogranin A and multiple peptides in the mammalian Merkel cell: further evidence for its paraneuronal function? Arch Histol Cytol 1989;52(Suppl):423-31.
25. Martin EM, Gould VE, Hoog A, Rosen ST, Radosevich JA, Deftos LJ. Parathyroid hormone-related protein, chromogranin A, and calcitonin gene products in the neuroendocrine skin carcinoma cell lines MKL1 and MKL2. Bone Miner 1991; 14:113-20.
26. Weiler R, Fischer-Colbrie R, Schmid KW, et al. Immunological studies on the occurrence and properties of chromogranin A and B and secretogranin II in endocrine tumors. Am J Surg Pathol 1988;12:877-84.
27. Bussolati G, Gugliotta P, Sapino A, Eusebi V, Lloyd RV. Chromogranin-reactive endocrine cells in argyrophilic carcinomas ("carcinoids") and normal tissue of the breast. Am J Pathol 1985; 120:186-92.
28. Bussolati G, Papotti M, Sapino A, Gugliotta P, Ghiringhello B, Azzopardi JG. Endocrine markers in argyrophilic carcinomas of the breast. Am J Surg Pathol 1987;11:248-56.
29. Capella C, Usellini L, Papotti M, et al. Ultrastructural features of neuroendocrine differentiated carcinomas of the breast. Ultrastruct Pathol 1990; 14:321-34.

30. Locurto P, Antona AD, Grillo A, et al. Primary neuroendocrine carcinoma of the breast a single center experience and review of the literature. Ann Ital Chir 2016;87:S2239253X.
31. Pagani A, Papotti M, Hofler H, Weiler R, Winkler H, Bussolati G. Chromogranin A and B gene expression in carcinomas of the breast. Correlation of immunocytochemical, immunoblot, and hybridization analyses. Am J Pathol 1990;136:319-27.
32. Papotti M, Macri L, Finzi G, Capella C, Eusebi V, Bussolati G. Neuroendocrine differentiation in carcinomas of the breast: a study of 51 cases. Semin Diagn Pathol 1989;6:174-88.
33. Rosen LE, Gattuso P. Neuroendocrine tumors of the breast. Arch Pathol Lab Med 2017;141:1577-81.
34. Sapino A, Papotti M, Righi L, Cassoni P, Chiusa L, Bussolati G. Clinical significance of neuroendocrine carcinoma of the breast. Ann Oncol 2001;12(Suppl 2):S115-7.
35. Scopsi L, Andreola S, Pilotti S, et al. Argyrophilia and granin (chromogranin/secretogranin) expression in female breast carcinomas. Their relationship to survival and other disease parameters. Am J Surg Pathol 1992;16:561-76.
36. Shin SJ, DeLellis RA, Ying L, Rosen PP. Small cell carcinoma of the breast: a clinicopathologic and immunohistochemical study of nine patients. Am J Surg Pathol 2000;24:1231-8.
37. Uccini S, Monardo F, Paradiso P, et al. Synaptophysin in human breast carcinomas. Histopathology 1991;18:271-3.
38. Yang X, Cao Y, Chen C, Liu L, Wang C, Liu S. Primary neuroendocrine breast carcinomas: a retrospective analysis and review of literature. Onco Targets Ther 2017;10:397-407.
39. Buffa R, Mare P, Gini A, Salvadore M. Chromogranins A and B and secretogranin II in hormonally identified endocrine cells of the gut and the pancreas. Basic Appl Histochem 1988;32:471-84.
40. Lloyd RV, Lin L, Kulig E, Fields K. Molecular approaches for the analysis of chromogranins and secretogranins. Diagn Mol Pathol 1992;1:2-15.
41. Pagani A, Papotti M, Abbona GC, Bussolati G. Chromogranin gene expressions in colorectal adenocarcinomas. Mod Pathol 1995;8:626-32.
42. Atasoy P, Ensari A, Demirci S, Kursun N. Neuroendocrine differentiation in colorectal carcinomas: assessing its prognostic significance. Tumori 2003;89:49-53.
43. Lloyd RV, Schroeder G, Bauman MD, et al. Prevalence and prognostic significance of neuroendocrine differentiation in colorectal carcinomas. Endocr Pathol 1998;9:35-42.
44. Mori M, Mimori K, Kamakura T, Adachi Y, Ikeda Y, Sugimachi K. Chromogranin positive cells in colorectal carcinoma and transitional mucosa. J Clin Pathol 1995;48:754-8.
45. Park JG, Choe GY, Helman LJ, et al. Chromogranin-A expression in gastric and colon cancer tissues. Int J Cancer 1992;51:189-94.
46. Swatek J, Chibowski D. Endocrine cells in colorectal carcinomas. Immunohistochemical study. Pol J Pathol 2000;51:127-36.
47. Syversen U, Halvorsen T, Marvik R, Waldum HL. Neuroendocrine differentiation in colorectal carcinomas. Eur J Gastroenterol Hepatol 1995; 7:667-74.
48. Wang DC, Wang LD, Jia YY, et al. Immunohistochemical study on endocrine-like tumor cells in colorectal carcinomas. World J Gastroenterol 1997;3:176.
49. Wick MR, Weatherby RP, Weiland LH. Small cell neuroendocrine carcinoma of the colon and rectum: clinical, histologic, and ultrastructural study and immunohistochemical comparison with cloacogenic carcinoma. Hum Pathol 1987;18:9-21.
50. Amorino GP, Parsons SJ. Neuroendocrine cells in prostate cancer. Crit Rev Eukaryot Gene Expr 2004;14:287-300.
51. Conlon JM. Granin-derived peptides as diagnostic and prognostic markers for endocrine tumors. Regul Pept 2010;165:5-11.
52. Cox ME, Deeble PD, Lakhani S, Parsons SJ. Acquisition of neuroendocrine characteristics by prostate tumor cells is reversible: implications for prostate cancer progression. Cancer Res 1999;59:3821-30.
53. di Sant'Agnese PA, de Mesy Jensen KL. Neuroendocrine differentiation in prostatic carcinoma. Hum Pathol 1987;18:849-56.
54. Palmer J, Venkateswaren V, Fleshner NE, Klotz LH, Cox ME. The impact of diet and micronutrient supplements on the expression of neuroendocrine markers in murine lady transgenic prostate. Prostate 2008;68:345-53.
55. Schmid KW, Helpap B, Totsch M, et al. Immunohistochemical localization of chromogranins A and B and secretogranin II in normal, hyperplastic and neoplastic prostate. Histopathology 1994;24:233-9.
56. Sommerfeld HJ, Partin AW, Pannek J. Incidence of neuroendocrine cells in the seminal vesicles and the prostate—an immunohistochemical study. Int Urol Nephrol 2002;34:357-60.
57. Turbat-Herrera EA, Herrera GA, Gore I, Lott RL, Grizzle WE, Bonin JM. Neuroendocrine differentiation in prostatic carcinomas. A retrospective autopsy study. Arch Pathol Lab Med 1988;112:1100-5.
58. Yuan TC, Veeramani S, Lin FF, et al. Androgen deprivation induces human prostate epithelial neuroendocrine differentiation of androgen-sensitive LNCaP cells. Endocr Relat Cancer 2006; 13:151-67.

59. Leone BE, Taccagni GL, Dell'Antonio G, Cantaboni A. Chromogranin A as a marker of neuroendocrine histogenesis of tumours: an immunoelectron microscopic study with considerations about the influence of fixation and embedding media on immunolabelling. Basic Appl Histochem 1990;34:143-53.
60. O'Connor DT, Burton D, Deftos LJ. Immunoreactive human chromogranin A in diverse polypeptide hormone producing human tumors and normal endocrine tissues. J Clin Endocrinol Metab 1983;57:1084-6.
61. Schmid KW, Fischer-Colbrie R, Hagn C, Jasani B, Williams E, Winkler H. Chromogranin A and B and secretogranin II in medullary carcinomas of the thyroid. Am J Surg Pathol 1987;11:551-6.
62. Portel-Gomes GM, Grimelius L, Johansson H, Wilander E, Stridsberg M. Chromogranin A in human neuroendocrine tumors: an immunohistochemical study with region-specific antibodies. Am J Surg Pathol 2001;25:1261-7.
63. O'Connor DT, Deftos LJ. Secretion of chromogranin A by peptide-producing endocrine neoplasms. N Engl J Med 1986;314:1145-51.
64. Oba H, Nishida K, Takeuchi S, et al. Diffuse idiopathic pulmonary neuroendocrine cell hyperplasia with a central and peripheral carcinoid and multiple tumorlets: a case report emphasizing the role of neuropeptide hormones and human gonadotropin-alpha. Endocr Pathol 2013;24:220-8.
65. Deftos LJ, Linnoila RI, Carney DN, et al. Demonstration of Chromogranin A in human neuroendocrine cell lines by immunohistology and immunoassay. Cancer 1988;62:92-7.
66. Eriksson B, Arnberg H, Oberg K, et al. Chromogranins—new sensitive markers for neuroendocrine tumors. Acta Oncol 1989;28:325-9.
67. Gazdar AF, Helman LJ, Israel MA, et al. Expression of neuroendocrine cell markers L-dopa decarboxylase, chromogranin A, and dense core granules in human tumors of endocrine and nonendocrine origin. Cancer Res 1988;48:4078-82.
68. Linnoila RI, Mulshine JL, Steinberg SM. Neuroendocrine differentiation in endocrine and nonendocrine lung carcinomas. Am J Clin Pathol 1988;90:641-52.
69. Takegahara K, Sato A, Ibi T, Inoue T, Usuda J. Atypical carcinoid localized at the bronchus accompanied by diffuse idiopathic pulmonary neuroendocrine cell hyperplasia in the distal lung: a rare case report. J Thorac Dis 2017;9:E774-8.
70. Walts AE, Said JW, Shintaku IP, Lloyd RV. Chromogranin as a marker of neuroendocrine cells in cytologic material—an immunocytochemical study. Am J Clin Pathol 1985;84:273-7.
71. Weiler R, Feichtinger H, Schmid KW, et al. Chromogranin A and B and secretogranin II in bronchial and intestinal carcinoids. Virchows Arch A Pathol Anat Histopathol 1987;412:103-9.
72. Wilson BS, Lloyd RV. Detection of chromogranin in neuroendocrine cells with a monoclonal antibody. Am J Pathol 1984;115:458-68.
73. O'Connor DT. Chromogranin: widespread immunoreactivity in polypeptide hormone producing tissues and in serum. Regul Pept 1983;6:263-80.
74. O'Connor DT, Burton D, Deftos LJ. Chromogranin A: immunohistology reveals its universal occurrence in normal polypeptide hormone producing endocrine glands. Life Sci 1983;33:1657-63.
75. Cohn DV, Elting JJ, Frick M, Elde R. Selective localization of the parathyroid secretory protein-I/adrenal medulla chromogranin A protein family in a wide variety of endocrine cells of the rat. Endocrinology 1984;114:1963-74.
76. Rindi G, Buffa R, Sessa F, Tortora O, Solcia E. Chromogranin A, B and C immunoreactivities of mammalian endocrine cells. Distribution, distinction from costored hormones/prohormones and relationship with the argyrophil component of secretory granules. Histochemistry 1986;85:19-28.
77. Deftos LJ, Woloszczuk W, Krisch I, et al. Medullary thyroid carcinomas express chromogranin A and a novel neuroendocrine protein recognized by monoclonal antibody HISL-19. Am J Med 1988;85:780-4.
78. Al-Khafaji B, Noffsinger AE, Miller A, DeVoe G, Stemmermann GN, Fenoglio-Preiseret C. Immunohistologic analysis of gastrointestinal and pulmonary carcinoid tumors. Hum Pathol 1998;29:992-9.
79. Tomita T. Immunocytochemical staining patterns for parathyroid hormone and chromogranin in parathyroid hyperplasia, adenoma, and carcinoma. Endocr Pathol 1999;10:145-56.
80. Troger J, Theurl M, Kirchmair R, et al. Granin-derived peptides. Prog Neurobiol 2017;154:37-61.
81. Wiedenmann B, Franke WW. Identification and localization of synaptophysin, an integral membrane glycoprotein of Mr 38,000 characteristic of presynaptic vesicles. Cell 1985;41:1017-28.
82. Wiedenmann B. Synaptophysin. A widespread constituent of small neuroendocrine vesicles and a new tool in tumor diagnosis. Acta Oncol 1991;30:435-40.
83. Muller L, Lindberg I. The cell biology of the prohormone convertases PC1 and PC2. Prog Nucleic Acid Res Mol Biol 1999;63:69-108.
84. Oliva AA Jr, Chan SJ, Steiner DF. Evolution of the prohormone convertases: identification of a homologue of PC6 in the protochordate amphioxus. Biochim Biophys Acta 2000;1477:338-48.
85. Seidah NG, Benjannet S, Pareek S, et al. Cellular processing of the nerve growth factor precursor by the mammalian pro-protein convertases. Biochem J 1996;314(Pt 3):951-60.

86. Portela-Gomes GM, Grimelius L, Stridsberg M. Prohormone convertases 1/3, 2, furin and protein 7B2 (Secretogranin V) in endocrine cells of the human pancreas. Regul Pept 2008;146:117-24.
87. Mbikay M, Raffin-Sanson ML, Tadros H, Sirois F, Seidah NG, Chretien M. Structure of the gene for the testis-specific proprotein convertase 4 and of its alternate messenger RNA isoforms. Genomics 1994;20:231-7.
88. Seidah NG, Chretien M. Proprotein and prohormone convertases: a family of subtilases generating diverse bioactive polypeptides. Brain Res 1999;848:45-62.
89. Tischler AS, Mobtaker H, Mann K, et al. Anti-lymphocyte antibody Leu-7 (HNK-1) recognizes a constituent of neuroendocrine granule matrix. J Histochem Cytochem 1986;34:1213-6.
90. Michels S, Swanson PE, Robb JA, Wick MR. Leu-7 in small cell neoplasms. An immunohistochemical study with ultrastructural correlations. Cancer 1987;60:2958-64.
91. Mills SE, Wolfe JT 3rd, Weiss MA, et al. Small cell undifferentiated carcinoma of the urinary bladder. A light-microscopic, immunocytochemical, and ultrastructural study of 12 cases. Am J Surg Pathol 1987;11:606-17.
92. Blomjous CE, Thunnissen FB, Vos W, de Voogt HJ, Meijer CG. Small cell neuroendocrine carcinoma of the urinary bladder. An immunohistochemical and ultrastructural evaluation of 3 cases with a review of the literature. Virchows Arch A Pathol Anat Histopathol 1988;413:505-12.
93. Gaffey MJ, Mills SE, Lack EE. Neuroendocrine carcinoma of the colon and rectum. A clinicopathologic, ultrastructural, and immunohistochemical study of 24 cases. Am J Surg Pathol 1990;14:1010-23.
94. Mayall FG, Gibbs AR. The histology and immunohistochemistry of small cell mesothelioma. Histopathology 1992;20:47-51.
95. Fucich LF, Freeman SM, Marrogi AJ. An immunohistochemical study of leu 7 and PCNA expression in thyroid neoplasms. Biotech Histochem 1996;71:298-303.
96. Cruz-Sanchez FF, Rossi ML, Hughes JT, Esiri MM, Coakham HB. Medulloblastoma. An immunohistological study of 50 cases. Acta Neuropathol 1989;79:205-10.
97. Mori M, Yamada K, Takagi H, Shrestha P, Lee S. Protein gene product 9.5 (PGP9.5) immunoreactivity in salivary gland tumors. Oncol Rep 1996; 3:249-54.
98. Banerjee SS, Agbamu DA, Eyden BP, Harris M. Clinicopathological characteristics of peripheral primitive neuroectodermal tumour of skin and subcutaneous tissue. Histopathology 1997; 31:355-66.
99. Hibi K, Westra WH, Borges M, Goodman S, Sidransky D, Jen J. PGP9.5 as a candidate tumor marker for non-small-cell lung cancer. Am J Pathol 1999;155:711-5.
100. Wang AR, May D, Bourne P, Scott G. PGP9.5: a marker for cellular neurothekeoma. Am J Surg Pathol 1999;23:1401-7.
101. Tezel E, Hibi K, Nagasaka T, Nakao A. PGP9.5 as a prognostic factor in pancreatic cancer. Clin Cancer Res 2000;6:4764-7.
102. Yamazaki T, Hibi K, Takase T, et al. PGP9.5 as a marker for invasive colorectal cancer. Clin Cancer Res 2002;8:192-5.
103. Takase T, Hibi K, Yamazaki T, et al. PGP9.5 overexpression in esophageal squamous cell carcinoma. Hepatogastroenterology 2003;50:1278-80.
104. Bouwens L. Islet morphogenesis and stem cell markers. Cell Biochem Biophys 2004;40(Suppl):81-8.
105. Kruijff S, Sidhu SB, Sywak MS, Gill AJ, Delbridge LW. Negative parafibromin staining predicts malignant behavior in atypical parathyroid adenomas. Ann Surg Oncol 2014;21:426-33.
106. Truran PP, Johnson SJ, Bliss RD, Lennard TW, Aspinall SR. Parafibromin, galectin-3, PGP9.5, Ki67, and cyclin D1: using an immunohistochemical panel to aid in the diagnosis of parathyroid cancer. World J Surg 2014;38:2845-54.
107. Yamaguchi K, Abe K, Kameya T, et al. Production and molecular size heterogeneity of immunoreactive gastrin-releasing peptide in fetal and adult lungs and primary lung tumors. Cancer Res 1983;43:3932-9.
108. Bostwick DG, Roth KA, Barchas JD, Bensch KG. Gastrin-releasing peptide immunoreactivity in intestinal carcinoids. Am J Clin Pathol 1984;82:428-31.
109. Bostwick DG, Roth KA, Evans CJ, Barchas JD, Bensch KG. Gastrin-releasing peptide, a mammalian analog of bombesin, is present in human neuroendocrine lung tumors. Am J Pathol 1984;117:195-200.
110. Bostwick DG, Bensch KG. Gastrin releasing peptide in human neuroendocrine tumours. J Pathol 1985;147:237-44.
111. Nomori H, Shimosato Y, Kodama T, Morinaga S, Nakajima T, Watanabe S. Subtypes of small cell carcinoma of the lung: morphometric, ultrastructural, and immunohistochemical analyses. Hum Pathol 1986;17:604-13.
112. Sunday ME, Wolfe HJ, Roos BA, Chin WW, Spindel ER. Gastrin-releasing peptide gene expression in developing, hyperplastic, and neoplastic human thyroid C-cells. Endocrinology 1988;122:1551-8.

113. Sunday ME, Wolfe HJ, Roos BA, Chin WW, Spindel ER. Gastrin-releasing peptide gene expression in small cell and large cell undifferentiated lung carcinomas. Hum Pathol 1991;22:1030-9.
114. Moreno P, Ramos-Alvarez I, Moody TW, Jensen RT. Bombesin related peptides/receptors and their promising therapeutic roles in cancer imaging, targeting and treatment. Expert Opin Ther Targets 2016;20:1055-73.
115. Tapia FJ, Polak JM, Barbosa AJ, et al. Neuron-specific enolase is produced by neuroendocrine tumours. Lancet 1981;1:808-11.
116. Jin L, Hemperly JJ, Lloyd RV. Expression of neural cell adhesion molecule in normal and neoplastic human neuroendocrine tissues. Am J Pathol 1991;138:961-9.
117. Agaimy A, Wunsch PH. Distribution of neural cell adhesion molecule (NCAM/CD56) in gastrointestinal stromal tumours and their intra-abdominal mesenchymal mimics. J Clin Pathol 2008;61:499-503.
118. Tajima S, Fukayama M. CD56 may be a more useful immunohistochemical marker than somatostatin receptor 2A for the diagnosis of phosphaturic mesenchymal tumors. Int J Clin Exp Pathol 2015;8:8159-64.
119. Bosmuller HC, Wagner P, Pham DL, et al. CD56 (neural cell adhesion molecule) expression in ovarian carcinomas: association with high-grade and advanced stage but not with neuroendocrine differentiation. Int J Gynecol Cancer 2017;27:239-45.
120. Gutgemann I, Haas S, Berg JP, Zhou H, Buttner R, Fischer HP. CD56 expression aids in the differential diagnosis of cholangiocarcinomas and benign cholangiocellular lesions. Virchows Arch 2006;448:407-11.
121. Abd El Atti RM, Shash LS. Potential diagnostic utility of CD56 and claudin-1 in papillary thyroid carcinoma and solitary follicular thyroid nodules. J Egypt Natl Canc Inst 2012;24:175-84.
122. Ceyran AB, Senol S, Simsek BÇ, Sagiroglu J, Aydin A. Role of cd56 and e-cadherin expression in the differential diagnosis of papillary thyroid carcinoma and suspected follicular-patterned lesions of the thyroid: the prognostic importance of e-cadherin. Int J Clin Exp Pathol 2015;8:3670-80.
123. Zhu J, Wang H, Ramelot TA, et al. Solution NMR structure of zinc finger 4 and 5 from human INSM1, an essential regulator of neuroendocrine differentiation. Proteins 2017;85:957-62.
124. Jia S, Wildner H, Birchmeier C. Insm1 controls the differentiation of pulmonary neuroendocrine cells by repressing Hes1. Dev Biol 2015;408:90-8.
125. Osipovich AB, Long Q, Manduchi E, et al. Insm1 promotes endocrine cell differentiation by modulating the expression of a network of genes that includes Neurog3 and Ripply3. Development 2014;141:2939-49.
126. Gierl MS, Karoulias N, Wende H, Strehle M, Birchmeier C. The zinc-finger factor Insm1 (IA-1) is essential for the development of pancreatic beta cells and intestinal endocrine cells. Genes Dev 2006;20:2465-78.
127. Xie J, Cai T, Zhang H, Lan MS, Notkins AL. The zinc-finger transcription factor INSM1 is expressed during embryo development and interacts with the Cbl-associated protein. Genomics 2002;80:54-61.
128. Rosenbaum JN, Guo Z, Baus R, Werner H, Rehrauer WM, Lloyd RV. INSM1: a novel immunohistochemical and molecular marker for neuroendocrine and neuroepithelial neoplasms. Am J Clin Pathol 2015;144:579-91.
129. Tanigawa M, Nakayama M, Taira T, et al. Insulinoma-associated protein 1 (INSM1) is a useful marker for pancreatic neuroendocrine tumor. Med Mol Morphol 2018;51:32-40.
130. Rooper LM, Bishop JA, Westra WH. INSM1 is a sensitive and specific marker of neuroendocrine differentiation in head and neck tumors. Am J Surg Pathol 2018;42:665-71.
131. Yoshida A, Makise N, Wakai S, Kawai A, Hiraoka N. INSM1 expression and its diagnostic significance in extraskeletal myxoid chondrosarcoma. Mod Pathol 2018;31:744-52.

Pancreas

132. Klimstra DS, Hruban RH, Pitman MB. Pancreas. In: Mills SE, ed. Histology for pathologists. Philadelphia: Lippincott Williams & Wilkins; 2012:777-816.
133. Wittingen J, Frey CF. Islet concentration in the head, body, tail and uncinate process of the pancreas. Ann Surg 1974;179:412-4.
134. Lecompte PM, Merriam JC Jr. Mitotic figures and enlarged nuclei in the islands of Langerhans in man. Diabetes 1962;11:35-9.
135. Rahier J, Wallon J, Henquin JC. Cell populations in the endocrine pancreas of human neonates and infants. Diabetologia 1981;20:540-6.
136. Jeon J, Correa-Medina M, Ricordi C, Edlund H, Diez JA. Endocrine cell clustering during human pancreas development. J Histochem Cytochem 2009;57:811-24.
137. Pelletier G. Identification of four cell types in the human endocrine pancreas by immunoelectron microscopy. Diabetes 1977;26:749-56.
138. Orci L, Malaisse-Lagae F, Baetens D, Perrelet A. Pancreatic-polypeptide-rich regions in human pancreas. Lancet 1978;2:1200-1.
139. Orci L, Stefan Y, Malaisse-Lagae F, Perrelet A. Instability of pancreatic endocrine cell populations throughout life. Lancet 1979;1:615-6.

140. Smith MJ, Simmons KM, Cambier JC. B cells in type 1 diabetes mellitus and diabetic kidney disease. Nat Rev Nephrol 2017;13:712-20.
141. Sandholm N, Groop PH. Genetic basis of diabetic kidney disease and other diabetic complications. Curr Opin Genet Dev 2018;50:17-24.
142. Tuomilehto J, Borch-Johnsen K, Molarius A, et al. Incidence of cardiovascular disease in type 1 (insulin-dependent) diabetic subjects with and without diabetic nephropathy in Finland. Diabetologia 1998;41:784-90.
143. Morris AP. Progress in defining the genetic contribution to type 2 diabetes susceptibility. Curr Opin Genet Dev 2018;50:41-51.
144. Kahn SE, Cooper ME, Del Prato S. Pathophysiology and treatment of type 2 diabetes: perspectives on the past, present, and future. Lancet 2014;383:1068-83.
145. Maclean N, Ogilvie RF. Observations on the pancreatic islet tissue of young diabetic subjects. Diabetes 1959;8:83-91.
146. Lohr M, Kloppel G. Residual insulin positivity and pancreatic atrophy in relation to duration of chronic type 1 (insulin-dependent) diabetes mellitus and microangiopathy. Diabetologia 1987;30:757-62.
147. Kloppel G, In't Veld PA, Komminoth P, Heitz PU. The endocrine pancreas. In: Kovacs K, Asa SL, eds. Functional endocrine pathology. Oxford: Blackwell Sciences; 1998:415-87.
148. Gepts W, De Mey J. Islet cell survival determined by morphology. An immunocytochemical study of the islets of Langerhans in juvenile diabetes mellitus. Diabetes 1978;27(Suppl 1):251-61.
149. Bottazzo GF, Dean BM, McNally JM, Mackay EH, Swift PG, Gamble DR. In situ characterization of autoimmune phenomena and expression of HLA molecules in the pancreas in diabetic insulitis. N Engl J Med 1985;313:353-60.
150. Westermark P, Wilander E, Westermark GT, Johnson KH. Islet amyloid polypeptide-like immunoreactivity in the islet B cells of type 2 (non-insulin-dependent) diabetic and non-diabetic individuals. Diabetologia 1987;30:887-92.
151. de Koning EJ, Fleming KA, Gray DW, Clark A. High prevalence of pancreatic islet amyloid in patients with end-stage renal failure on dialysis treatment. J Pathol 1995;175:253-8.
152. Subramanian S, Hirsch IB. Intensive diabetes treatment and cardiovascular outcomes in type 1 diabetes mellitus: implications of the diabetes control and complications trial/epidemiology of diabetes interventions and complications study 30-year follow-up. Endocrinol Metab Clin North Am 2018;47:65-79.
153. Orive G, Emerich D, Khademhosseini A, et al. Engineering a clinically translatable bioartificial pancreas to treat type I diabetes. Trends Biotechnol 2018;36:445-56.
154. Stanley CA. Perspective on the genetics and diagnosis of congenital hyperinsulinism disorders. J Clin Endocrinol Metab 2016;101:815-26.
155. James C, Kapoor RR, Ismail D, Hussain K. The genetic basis of congenital hyperinsulinism. J Med Genet 2009;46:289-99.
156. Kalish JM, Boodhansingh KE, Bhatti TR, et al. Congenital hyperinsulinism in children with paternal 11p uniparental isodisomy and Beckwith-Wiedemann syndrome. J Med Genet 2016;53:53-61.
157. Kapoor RR, Flanagan SE, Arya VB, Shield JP, Ellard S, Hussain K. Clinical and molecular characterisation of 300 patients with congenital hyperinsulinism. Eur J Endocrinol 2013;168:557-64.
158. Kowalewski AM, Szylberg L, Kasperska A, Marszalek A. The diagnosis and management of congenital and adult-onset hyperinsulinism (nesidioblastosis)—literature review. Pol J Pathol 2017;68:97-101.
159. Martinez R, Fernandez-Ramos C, Vela A, et al. Clinical and genetic characterization of congenital hyperinsulinism in Spain. Eur J Endocrinol 2016;174:717-26.
160. Meissner T, Beinbrech B, Mayatepek E. Congenital hyperinsulinism: molecular basis of a heterogeneous disease. Hum Mutat 1999;13:351-61.
161. Minakova E, Chu A. Congenital hyperinsulinism. Pediatr Ann 2017;46:e409-14.
162. Rahman SA, Nessa A, Hussain K. Molecular mechanisms of congenital hyperinsulinism. J Mol Endocrinol 2015;54:R119-29.
163. Snider KE, Becker S, Boyajian L, et al. Genotype and phenotype correlations in 417 children with congenital hyperinsulinism. J Clin Endocrinol Metab 2013;98:E355-63.
164. Senniappan S, Arya VB, Hussain K. The molecular mechanisms, diagnosis and management of congenital hyperinsulinism. Indian J Endocrinol Metab 2013;17:19-30.
165. Nestorowicz A, Inagaki N, Gonoi T, et al. A nonsense mutation in the inward rectifier potassium channel gene, Kir6.2, is associated with familial hyperinsulinism. Diabetes 1997;46:1743-8.
166. Aguilar-Bryan L, Nichols CG, Wechsler SW, et al. Cloning of the beta cell high-affinity sulfonylurea receptor: a regulator of insulin secretion. Science 1995;268:423-6.
167. Thomas PM, Cote GJ, Wohlik N, et al. Mutations in the sulfonylurea receptor gene in familial persistent hyperinsulinemic hypoglycemia of infancy. Science 1995;268:426-9.

168. Lord K, Dzata E, Snider KE, Gallagher PR, De Leon DD. Clinical presentation and management of children with diffuse and focal hyperinsulinism: a review of 223 cases. J Clin Endocrinol Metab 2013;98:E1786-9.
169. Suchi M, Thornton PS, Adzick NS, et al. Congenital hyperinsulinism: intraoperative biopsy interpretation can direct the extent of pancreatectomy. Am J Surg Pathol 2004;28:1326-35.
170. Lovisolo SM, Mendonca BB, Pinto EM, Manna TD, Saldiva PH, Zerbini MC. Congenital hyperinsulinism in Brazilian neonates: a study of histology, KATP channel genes, and proliferation of beta cells. Pediatr Dev Pathol 2010;13:375-84.
171. Kassem SA, Ariel I, Thornton PS, Scheimberg I, Glaseret B. Beta-cell proliferation and apoptosis in the developing normal human pancreas and in hyperinsulinism of infancy. Diabetes 2000;49:1325-33.
172. Sempoux C, Guiot Y, Dubois D, et al. Pancreatic B-cell proliferation in persistent hyperinsulinemic hypoglycemia of infancy: an immunohistochemical study of 18 cases. Mod Pathol 1998;11:444-9.
173. Anlauf M, Wieben D, Perren A, et al. Persistent hyperinsulinemic hypoglycemia in 15 adults with diffuse nesidioblastosis: diagnostic criteria, incidence, and characterization of beta-cell changes. Am J Surg Pathol 2005;29:524-33.
174. Davi MV, Pia A, Guarnotta V, et al. The treatment of hyperinsulinemic hypoglycaemia in adults: an update. J Endocrinol Invest 2017;40:9-20.
175. Kloppel G, Anlauf M, Raffel A, Perren A, Knoefel WT. Adult diffuse nesidioblastosis: genetically or environmentally induced? Hum Pathol 2008;39:3-8.
176. Rumilla KM, Erickson LA, Service FJ, et al. Hyperinsulinemic hypoglycemia with nesidioblastosis: histologic features and growth factor expression. Mod Pathol 2009;22:239-45.
177. Service GJ, Thompson GB, Service FJ, Andrews JC, Collazo-Clavell ML, Lloyd RV. Hyperinsulinemic hypoglycemia with nesidioblastosis after gastric-bypass surgery. N Engl J Med 2005;353:249-54.
178. Mete O, Asa SA. Precursor lesions of endocrine system neoplasms. Pathology 2013;45:316-30.
179. Kloppel G, Anlauf M, Perren A, Sipos B. Hyperplasia to neoplasia sequence of duodenal and pancreatic neuroendocrine diseases and pseudohyperplasia of the PP-cells in the pancreas. Endocr Pathol 2014;25:181-5.
180. Henopp T, Anlauf M, Schmitt A, et al. Glucagon cell adenomatosis: a newly recognized disease of the endocrine pancreas. J Clin Endocrinol Metab 2009;94:213-7.
181. Al-Sarireh B, Haidermota M, Verbeke C, Rees DA, Yu R, Griffiths AP. Glucagon cell adenomatosis without glucagon receptor mutation. Pancreas 2013;42:360-2.

Gastrointestinal Tract

182. Solcia E, Fiocca R, Villani L, Luinetti O, Capella C. Hyperplastic, dysplastic, and neoplastic enterochromaffin-like-cell proliferations of the gastric mucosa. Classification and histogenesis. Am J Surg Pathol 1995;19(Suppl 1):S1-7.
183. Bordi C, D'Adda T, Azzoni C, Ferraro G. Pathogenesis of ECL cell tumors in humans. Yale J Biol Med 1998;71:273-84.
184. Kloppel G, Anlauf M, Perren A. Endocrine precursor lesions of gastroenteropancreatic neuroendocrine tumors. Endocr Pathol 2007;18:150-5.
185. Kloppel G, Heitz PU. Classification of normal and neoplastic neuroendocrine cells. Ann N Y Acad Sci 1994;733:19-23.
186. Kloppel G. Classification and pathology of gastroenteropancreatic neuroendocrine neoplasms. Endocr Relat Cancer 2011;18(Suppl 1):S1-16.
187. Rindi G, Bordi C, Rappel S, La Rosa S, Stolte M, Solcia E. Gastric carcinoids and neuroendocrine carcinomas: pathogenesis, pathology, and behavior. World J Surg 1996;20:168-72.
188. Bordi C, D'Adda T, Azzoni C, Pilato FP, Caruana P. Hypergastrinemia and gastric enterochromaffin-like cells. Am J Surg Pathol 1995;19(Suppl 1):S8-19.
189. Debelenko LV, Emmert-Buck MR, Zhuang Z, et al. The multiple endocrine neoplasia type I gene locus is involved in the pathogenesis of type II gastric carcinoids. Gastroenterology 1997;113:773-81.
190. Scherubl H, Cadiot G, Jensen RT, Rosch T, Stolzel U, Kloppel G. Neuroendocrine tumors of the stomach (gastric carcinoids) are on the rise: small tumors, small problems? Endoscopy 2010;42:664-71.
191. Bordi C, Falchetti A, Azzoni C, et al. Aggressive forms of gastric neuroendocrine tumors in multiple endocrine neoplasia type I. Am J Surg Pathol 1997;21:1075-82.
192. Rosentraeger MJ, Garbrecht N, Anlauf M, et al. Syndromic versus non-syndromic sporadic gastrin-producing neuroendocrine tumors of the duodenum: comparison of pathological features and biological behavior. Virchows Arch 2016;468:277-87.
193. Anlauf M, Garbrecht N, Bauersfeld J, et al. Hereditary neuroendocrine tumors of the gastroenteropancreatic system. Virchows Arch 2007;451(Suppl 1):S29-38.
194. Anlauf M, Perren A, Kloppel G. Endocrine precursor lesions and microadenomas of the duodenum and pancreas with and without MEN1: criteria, molecular concepts and clinical significance. Pathobiology 2007;74:279-84.

195. Vanoli A, La Rosa S, Klersy C, et al. Four neuroendocrine tumor types and neuroendocrine carcinoma of the duodenum: analysis of 203 cases. Neuroendocrinology 2017;104:112-25.
196. Greenstein AJ, Balasubramanian S, Harpaz N, Rizwan M, Sachar DB. Carcinoid tumor and inflammatory bowel disease: a study of eleven cases and review of the literature. Am J Gastroenterol 1997;92:682-5.
197. Gledhill A, Hall PA, Cruse JP, Pollock DJ. Enteroendocrine cell hyperplasia, carcinoid tumours and adenocarcinoma in long-standing ulcerative colitis. Histopathology 1986;10:501-8.
198. Kanada S, Sugita A, Mikami T, Ohashi K, Hayashi H. Microcarcinoid arising in patients with long-standing ulcerative colitis: histological analysis. Hum Pathol 2017;64:28-36.
199. McNeely B, Owen DA, Pezim M. Multiple microcarcinoids arising in chronic ulcerative colitis. Am J Clin Pathol 1992;98:112-6.

Extra-Adrenal Paraganglia

200. Lack EE. Hyperplasia of vagal and carotid body paraganglia in patients with chronic hypoxemia. Am J Pathol 1978;91:497-516.
201. Arias-Stella J, Valcarcel J. Chief cell hyperplasia in the human carotid body at high altitudes; physiologic and pathologic significance. Hum Pathol 1976;7:361-73.
202. Lack EE, Perez-Atayde AR, Young JB. Carotid body hyperplasia in cystic fibrosis and cyanotic heart disease. A combined morphometric, ultrastructural, and biochemical study. Am J Pathol 1985;119:301-14.
203. Ramos SG, Matturri L, Biondo B, Ottaviani G, Rossi L. Hyperplasia of the aorticopulmonary paraganglia: a new insight into the pathogenesis of sudden infant death syndrome? Cardiologia 1998;43:953-8.
204. Sivridis E, Pavlidis P, Fiska A, Pitsiava D, Giatromanolaki A. Myocardial hypertrophy induces carotid body hyperplasia. J Forensic Sci 2011;56(Suppl 1):S90-4.
205. Heath D, Smith P, Jago R. Hyperplasia of the carotid body. J Pathol 1982;138:115-27.

Pulmonary Neuroendocrine Cells

206. Froelich F. Die Helle zelle der bronchialschleimhaut und ihre beziehungen zum problem der chemoreceptoren. Franz Z Pathol 1949;60:517-59.
207. Hoyt RF Jr, McNelly NA, Sorokin SP. Dynamics of neuroepithelial body (NEB) formation in developing hamster lung: light microscopic autoradiography after 3H-thymidine labeling in vivo. Anat Rec 1990;227:340-50.
208. Noguchi M, Sumiyama K, Morimoto M. Directed migration of pulmonary neuroendocrine cells toward airway branches organizes the stereotypic location of neuroepithelial bodies. Cell Rep 2015;13:2679-86.
209. Pan J, Copland I, Post M, Yeger H, Cutz E. Mechanical stretch-induced serotonin release from pulmonary neuroendocrine cells: implications for lung development. Am J Physiol Lung Cell Mol Physiol 2006;290:L185-93.
210. Van Lommel A, Bollé T, Fannes W, Lauweryns JM. The pulmonary neuroendocrine system: the past decade. Arch Histol Cytol 1999;62:1-16.
211. Cutz E, Yeger H, Pan J. Pulmonary neuroendocrine cell system in pediatric lung disease-recent advances. Pediatr Dev Pathol 2007;10:419-35.
212. Van Lommel A. Pulmonary neuroendocrine cells (PNEC) and neuroepithelial bodies (NEB): chemoreceptors and regulators of lung development. Paediatr Respir Rev 2001;2:171-6.
213. Gosney JR. Pulmonary neuroendocrine cells in species at high altitude. Anat Rec 1993;236:105-12.
214. Alshehri M, Cutz E, Banzhoff A, Canny G. Hyperplasia of pulmonary neuroendocrine cells in a case of childhood pulmonary emphysema. Chest 1997;112:553-6.
215. Cutz E. Hyperplasia of pulmonary neuroendocrine cells in infancy and childhood. Semin Diagn Pathol 2015;32:420-37.
216. Cutz E, Perrin DG, Hackman R, Czegledy-Nagy EN. Maternal smoking and pulmonary neuroendocrine cells in sudden infant death syndrome. Pediatrics 1996;98(Pt 1):668-72.
217. Carr LL, Kern JA, Deutsch GH. Diffuse idiopathic pulmonary neuroendocrine cell hyperplasia and neuroendocrine hyperplasia of infancy. Clin Chest Med 2016;37:579-87.
218. Cutz E, Perrin DG, Pan J, Haas EA, Krous HF. Pulmonary neuroendocrine cells and neuroepithelial bodies in sudden infant death syndrome: potential markers of airway chemoreceptor dysfunction. Pediatr Dev Pathol 2007;10:106-16.
219. Dovey M, Wisseman CL, Roggli VL, Roomans GM, Shelburne JD, Spock A. Ultrastructural morphology of the lung in cystic fibrosis. J Submicrosc Cytol Pathol 1989;21:521-34.
220. Rossi G, MacMahon H, Marchevsky AM, Nicholson AG, Snead DR. Diffuse idiopathic pulmonary neuroendocrine cell hyperplasia. In: WHO Classification of Tumors Editorial Board. WHO classification of tumours: thoracic tumors, 5th ed. Lyon: IARC; 2021.
221. Marchevsky AM, Walts AE. Diffuse idiopathic pulmonary neuroendocrine cell hyperplasia (DIPNECH). Semin Diagn Pathol 2015;32:438-44.

222. Wirtschafter E, Walts AE, Liu ST, Marchevsky AM. Diffuse Idiopathic pulmonary neuroendocrine cell hyperplasia of the lung (DIPNECH): Current Best Evidence. Lung 2015;193:659-67.
223. Marchevsky AM, Wirtschafter E, Walts AE. The spectrum of changes in adults with multifocal pulmonary neuroendocrine proliferations: what is the minimum set of pathologic criteria to diagnose DIPNECH? Hum Pathol 2015;46:176-81.
224. Trisolini R, Valentini I, Tinelli C, et al. DIPNECH: association between histopathology and clinical presentation. Lung 2016;194:243-7.

Index*

Pituitary Gland

*In a series of numbers, those in boldface indicate the main discussion of the entity.

Parathyroid Gland

Thyroid Gland

Adrenal Gland

Diffuse Neuroendocrine System